A Practical Guide to

NATURE STUDY

Third Edition

C.J. Fitzwilliams-Heck, Ph.D.

A Practical Guide to Nature Study
is dedicated to my family of naturalists . . .

Cover image © Shutterstock.com

Kendall Hunt
publishing company

www.kendallhunt.com
Send all inquiries to:
4050 Westmark Drive
Dubuque, IA 52004-1840

PAK ISBN: 978-1-7924-6600-7
Text alone ISBN: 978-1-7924-6601-4

Published in the United States of America

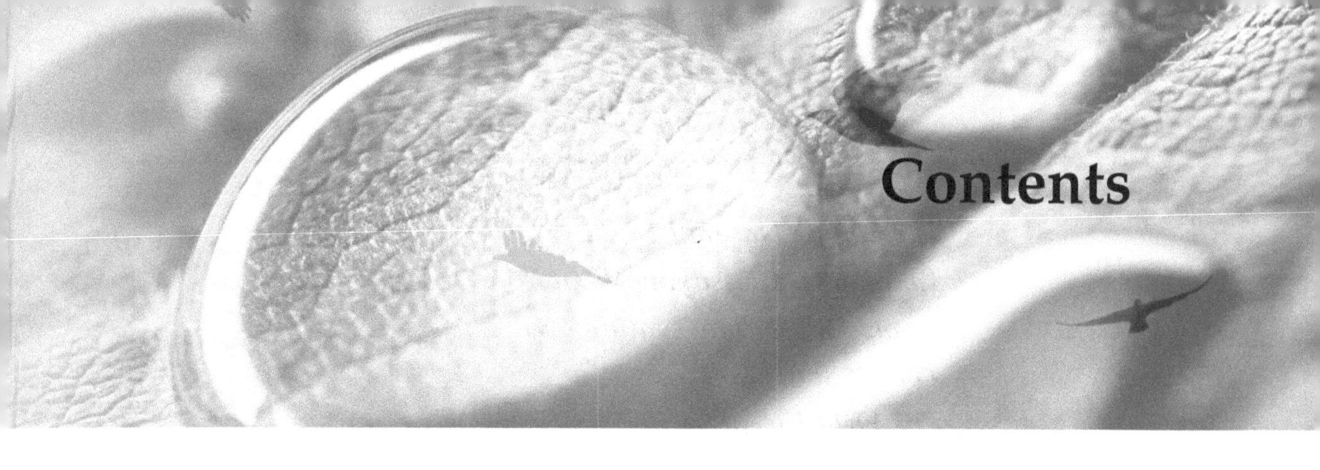

Contents

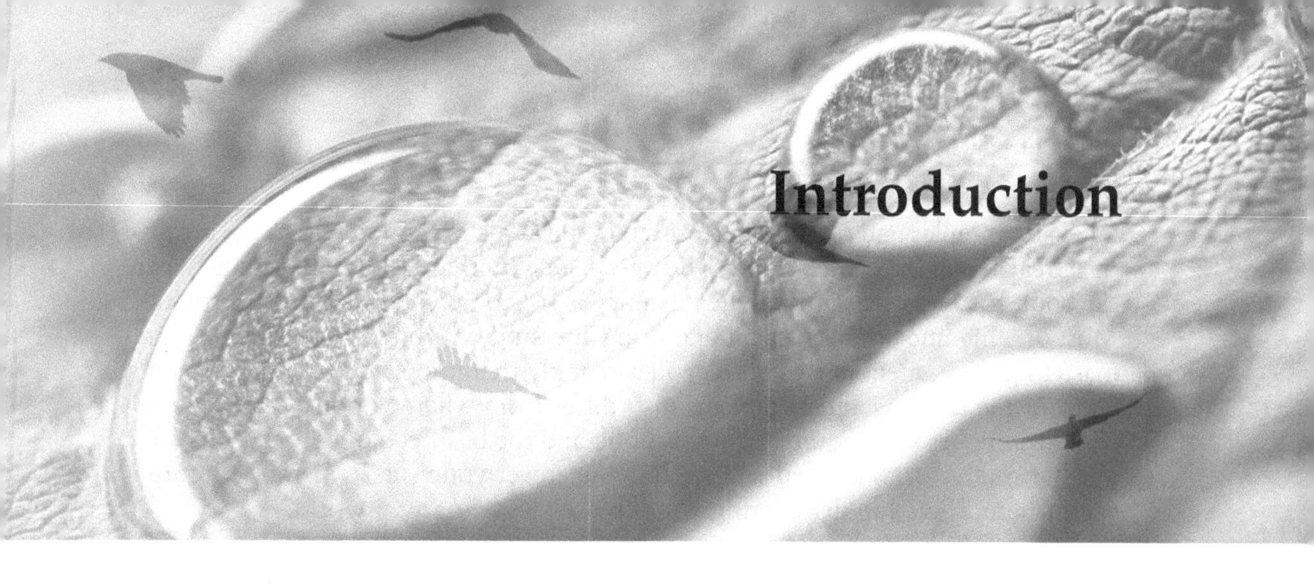

Introduction

". . . nature study is a science, and is more than a science; it is not merely a study of life, but an experience of life."

—Anna Botsford Comstock

A Practical Guide to Nature Study fills a gap in the literature dedicated to nature-based education. Whether you consider yourself a novice or seasoned naturalist, your nature study will benefit from this unique and holistic approach. As a student of nature, you will build naturalist skills, become more ecologically aware, learn basic natural histories of common flora and fauna, and explore regional field guides. Through the nature journal exercises presented, you will make connections necessary for interpreting your surrounding landscape. The result can improve your awareness, understanding, and appreciation of nature. Developing your naturalist skills may create a sense of stewardship for the natural world and help you make a positive difference in protecting it.

After becoming familiar with the layout of the book, start with "Chapter 1: Building Your Naturalist Skills" to hone techniques to develop the habits of a naturalist. Move through this chapter at the pace reasonable for your skill level while delving into other areas of the book to build your personal nature study. The material in the text was based on personal experiences and over 20 years of extensive interpretive naturalizing and teaching nature-based education in higher education and within the community. To help build awareness and understanding of the natural world, the philosophy used in this book has an emphasis on experiential, place-based approaches, and social-ecological concepts to teach nature study. The learning starts with wherever you are in nature—looking out your window, walking on a sidewalk, or spending time on your porch, in a park, forest, or along a river. There is always something to learn from the world around you.

The examples and organisms featured in this book are found within the areas of the Great Lakes basin and Upper Midwest. The map that follows represents where the book focuses its attention when examples of habitats and species are provided. Most of a lifetime has been spent exploring and studying much of this region. After mastering the content in this book, you can apply that knowledge to other systems. Some of the habitats and organisms may differ from where you live, but the naturalist techniques you acquire from this book and the basic ecology, botany, zoology, and natural histories learned will apply to wherever you visit. Remember, nature is everywhere. Have fun discovering your "place" in the natural world.

Happy Naturalizing!
C. J. Fitzwilliams-Heck, Ph.D.
Big Rapids, MI
2020

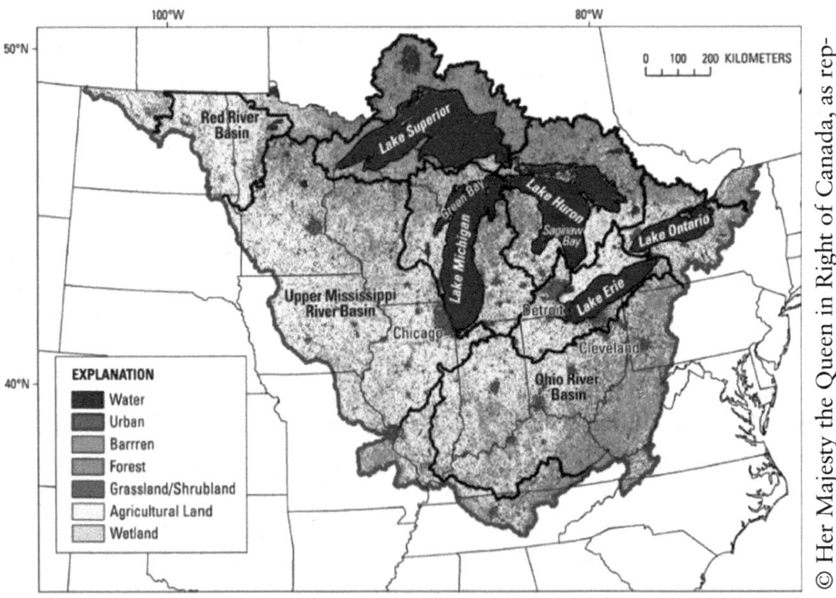

The Great Lakes basin and Upper Midwest. Use the map as a reference for where this book focuses its examples. The dark lines indicate drainage basins or watershed boundaries. The states include Michigan, Ohio, Indiana, Wisconsin, Minnesota, Kentucky, and parts of Tennessee, West Virginia, Pennsylvania, New York, Iowa, Missouri, North Dakota, and the southern Canadian province of Ontario. Note the color-coded legend of land use for the region. Based on the map, how could you interpret the landscape for where you live or would like to visit?

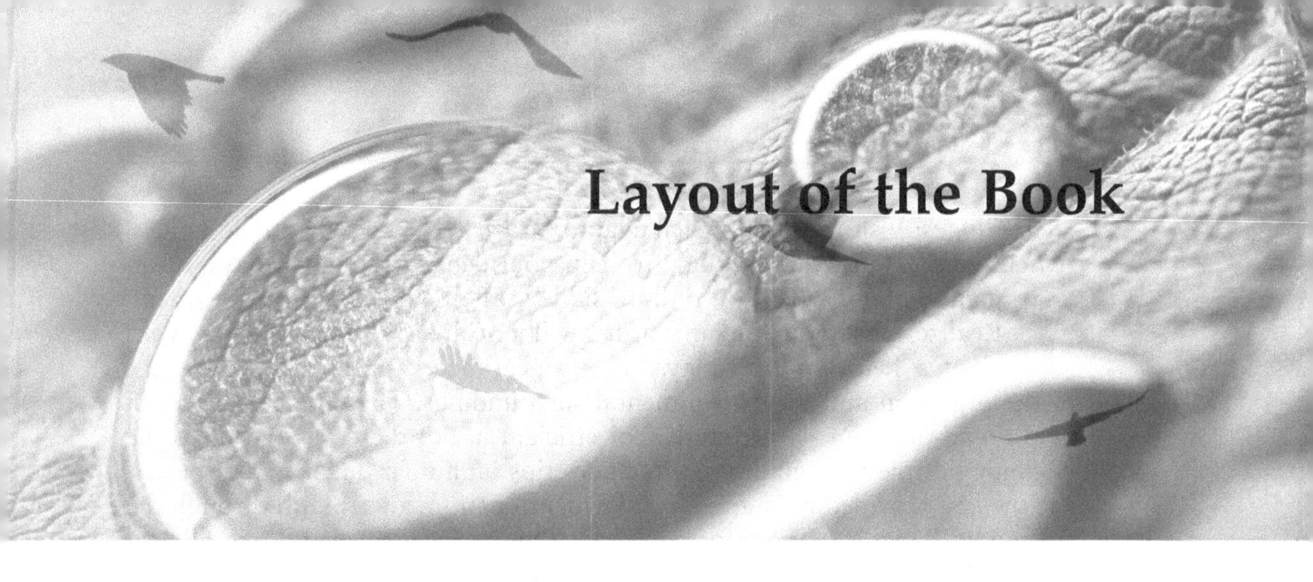

Layout of the Book

By picking up this book, you have taken the first step in a lifelong journey of nature study. First, realize the contents presents a useful approach to gaining an awareness and basic understanding of the environment, and simple methods for applying nature study principles. Although not a comprehensive or technical book, it is truly *A Practical Guide to Nature Study*. Now, how will you use the book efficiently? Learn the layout first so that you can navigate the material quickly and effectively.

Follow this sequence for getting acquainted with your new nature study book:

1. *Table of Contents:* Read through the list of contents to discover topics the book includes. Refer to it often to discover aspects of nature you would like to learn more about and where to find the information.

   You will find that the text is broken into parts, chapters, sections, and subsections.

   Each of the three parts focuses on a theme. In *Part I*, you will discover approaches for building your naturalist skills, examining a landscape, and identifying types and features of habitats. *Part II* shares more of the science in nature study with essential ecological principles to apply to your place in nature. For *Part III*, find information on the basic biology of organisms and field guides to identify species.

2. *Color-coded tabbed pages:* Notice that the corners of each page have a color. *Part I* has green-colored tabs, *Part II* has blue, and *Part III* has different colored tabs divided by organism type. On some tabs, you will notice a word present which indicates the topic of the page, or you will see an icon on the colored tab to represent that a Natural History Journal Page is found there or that a certain species type is pictured on that page in the field guide section.

3. *Field Guides:* Visit the end of each chapter in *Part III* to discover field guides. A field guide is just that, a guide to help you identify and learn basic information about what you find outside "in the field" or any environment. Realize what you find in nature will not always match exactly with what a field guide displays. The appearance can depend on the age of the organism and its environmental surroundings. These field guides will help round-out your nature study. For efficient use of these sections, read the beginning of the field guides that includes tips on how to identify the species, use the key provided for more effective identification, and thumb through the collection of pictures and information for the species displayed. It would also be useful to refer to the corresponding section in the book for further background information about the organism.

4. *Natural History Journal Pages:* Go to the end of any field guide. Here you will find prompts to complete a report on a local species of your choice. These Natural History Journal Pages (NHJP) provide a template for discovering details of an organism that interests the naturalist while at the same time applying concepts from that section in the book.

5. *Nature Journal:* Throughout the text, there are numerous green shaded boxes for Nature Journal (NJ) activities. Students love the idea and practice of having a journal. The NJ is introduced in Chapter 1.1 and is carried throughout the remainder of the book. Choose the activities that make the most sense for your study. A lifetime can be spent completing and building on these NJ prompts. Remember, the story of nature is always unfolding!

6. *Review Questions:* Presented at the end of a chapter or section, you can visit the blue boxes of review questions to test your understanding of the concepts. It is a great way to check whether you grasped what was presented.

7. *Index:* As a quick reference, look up a word, concept, or species in the index at the back of the book. Some topics will have multiple pages listed. To get the best understanding of your subject, visit all occurrences where your subject is found.

8. *References:* If you want to learn more about a topic, refer to the section at the end of the book for resources that inspired or informed each chapter.

Now that you have "met" your book, start reading Chapter 1 and other topics of interest to begin your nature study. Enjoy!

Nature Appreciation Survey

Before going further, take the following short survey. The survey is a way to assess your engagement, appreciation, and basic knowledge of nature. Select the best answer choice for each question. Be completely honest with yourself for each response. **Take this survey at the beginning of your nature studies and after a set amount of time spent completing the activities in this book.**

1. **What do you enjoy doing the most outside? Check all that apply.**

__ Backpacking/Camping
__ Canoeing/Kayaking
__ Swimming
__ Bicycling
__ Running
__ Inline Skating
__ Walking
__ Gardening/Yardwork
__ Sitting
__ Horseback Riding
__ Hunting
__ Fishing
__ Trapping
__ Sport Activity (golf, tennis, skiing, soccer, etc.)
__ ATV/Snowmobile
__ I do not enjoy being outdoors.
__ Other activities? Please specify ________________.

2. **What is my main motivation for engaging in outdoor activities? To _____.**

__ have fun
__ exercise
__ relax
__ observe and enjoy nature
__ be with friends/family
__ escape busy schedules or personal issues
__ discover new places
__ educate myself
__ challenge myself
__ compete with others
__ improve the beauty of my home
__ I tend to avoid outdoor activities

3. On average, I spend ___ each day outdoors.

__ > 3 hours __ 1.5 hours __ 45 minutes __ 15 minutes
__ 2 hours __ 1 hour __ 30 minutes __ < 15 minutes

4. Does cold weather prevent me from spending time outdoors?

__ Always __ Usually __ Sometimes __ Never

5. When I'm outside, I am _____aware of the plants, animals, landscape, and weather conditions around me.

__ always __ usually __ sometimes __ not

6. [The term, "nature," in the following statement refers to an area that does not directly involve humans]. If I see something in nature that catches my attention, I will _____.

__ spend time thinking about what I observed
__ tell a friend or family member about my experience
__ usually dismiss my observations almost immediately
__ be surprised that I witnessed something taking place in nature

7. After spending time outdoors, I feel _____ .

__ more relaxed and refreshed
__ the same as I always do
__ glad to be indoors

8. How confident am I interpreting the ecology of a given landscape?

__ Very confident __ Somewhat confident __ Not confident

9. How often do I refer to field guides or internet to identify or increase my knowledge about an organism?

__ Always __ Usually __ Sometimes __ Never

10. How confident am I in identifying trees near my home?

__ Very confident __ Somewhat confident __ Not confident

11. How confident am I in identifying insects outside near my home?

__ Very confident __ Somewhat confident __ Not confident

12. How confident am I in identifying fish in my state/province?

__ Very confident __ Somewhat confident __ Not confident

13. How confident am I in identifying amphibians (frogs, toads, salamanders) near my home?

__ Very confident __ Somewhat confident __ Not confident

14. How confident am I in identifying reptiles (snakes, turtles, lizards) near my home?
__ Very confident __ Somewhat confident __ Not confident

15. How confident am I in identifying birds near my home?
__ Very confident __ Somewhat confident __ Not confident

16. How confident am I in identifying mammals near my home?
__ Very confident __ Somewhat confident __ Not confident

Nature Journal: Reflect on your answers for the Nature Appreciation Survey in your Nature Journal. How do you feel about your results? How could you improve your engagement, appreciation, or basic knowledge of nature? At what level would you classify your naturalist intelligence? If you spend at least 20 minutes outdoors on most days (or at least look or think about the outdoors that long each day), share or reflect on what you discovered in nature during those times, and know the flora (plants) and fauna (animals) where you live and their roles and interactions, then you have a high naturalist intelligence. If your time spent in nature is about 10 minutes a day or knowledge of nature around you is not comprehensive or do not have an interest in various parts of nature, then you could consider working at improving your naturalist intelligence. Read on for further insight and ideas, and consider focusing on areas where you could grow as a naturalist. There is always something new to explore and learn about in nature!

PART I:

Discovering Nature

Those who dwell, as scientists or laymen,
among the beauties and mysteries
of the earth, are never alone or weary of life.

—Rachel Carson

Part I, *Discovering Nature* is full of tips and tricks to hone your naturalist skills. You will also find information and insights on features of the landscape to help with interpretations of your surroundings. Using Part I in conjunction with other parts of the book will likely make the most sense for your nature study. Start with Chapter 1 and build your skills as you refer to other parts of the book to go deeper in understanding the world around you.

1 Building Your Naturalist Skills

Think about how incredible it would be to actually know and understand the local flora and fauna (plants and animals). Honing observation skills can lead you to a better understanding of what is happening around you on many levels. In Chapter 1, you will build and refine your abilities for observing nature. Discover what it means to be a naturalist and start to develop these lifelong skills. By doing so, you may start to appreciate and care about your natural surroundings further and act toward the preservation and restoration of the place you live.

1.1 Nature Awareness

To begin your nature study, raise your awareness and collect tangible information of your surroundings. To learn more about that place, wherever you are, simply use your senses of sight, hearing, smell, and touch (being wary, of course, of the unfamiliar and potentially harmful). Practice stillness and moving slowly and quietly to aid in your observations. Let's learn more about defining nature and exploring ways to pursue your nature study.

Defining Nature

Although we use the word "nature" freely and can usually recognize it, how do you define it? **Nature**, or the natural environment includes living things, such as: plants, animals, fungus, protists, bacteria; and nonliving things, such as: rocks, soil, weather conditions, and nutrients available for organisms. The **built environment** includes areas constructed by humans. In nature, we find that the natural environment is often times influenced by the built environment and vice versa. Learning about these connections can deepen our understanding of the world we live in. This is how we will think of nature.

What Is a Naturalist?

As we become more aware of nature and find ourselves making interpretations or identifications within the natural world, we might think of ourselves as a naturalist. A **naturalist** is a person who strives to understand the species they encounter and how those organisms potentially interact in their surroundings. It is somebody who has, or works toward developing the knowledge and skills to study the living and nonliving parts of nature. They have an interest that develops into a habit to focus on what is happening in the out-of-doors anyplace and anytime.

Your Naturalist Intelligence

You are not born as a naturalist. Based on your experiences over time, your interests and knowledge develop for a multitude of things. People learn in a variety of ways, and will find different interests based on what they excel at or what seems to come naturally to them. When the combination is right, a person tends to remember details and may attach deeper meaning to their experiences.

Consider researching the types of learning styles and intelligences to discover what fits with your tendencies. These following questions will help you think about how you learn best. Are you a visual or auditory learner? Do you need to touch or experience something before truly understanding it? What are your interests? Do you enjoy or need music in your life? Do you thrive on social connections? Are pictures necessary to communicate effectively? Maybe you need physical activity to feel at ease. Are you nature smart, and find comfort with being in the outdoors? Aiming to develop your naturalist intelligence can encompass all learners and intelligences. Think about how nature can have something of interest for everyone.

There are many ways you could view a **naturalist intelligence**. For our purposes, we will consider it as the ability to discriminate and have sensitivity toward living and nonliving things in nature. Having "nature smarts" indicates you have a keen awareness of your surroundings and can detect any subtle changes or patterns that may emerge in the environment. A naturalist intelligence means you enjoy and appreciate the outdoors. You find yourself reading and watching what you can about nature-related themes. The time spent outdoors and self-directed nature learning drives you to identify and interpret happenings in the natural world. You will want to do what you can to protect the environment in which you live.

Becoming a Naturalist

Perhaps by now you are thinking you are already a naturalist, or you want to develop your nature smarts further. Simply put, to become a naturalist, develop an awareness of nature. Read and watch stories with nature-related themes.

Spend time in nature. The best practice is to find a place outdoors where you can sit on a regular basis. This outside place to sit can be anywhere. Have this "sit spot" convenient to your home so you can access it at any hour and feel safe and comfortable. Tune into your natural surroundings every time you are outside—even if it is just for a minute. Think about that for a moment... When was the last time you were outside, relaxed, and focused on what was happening in nature all around you?

When you are outdoors (and anytime you look through a window), make observations about the world around you using as many of your senses as possible. Try to focus on the natural world. Learn to describe those observations in detail. Discuss them with somebody or write them down. From these experiences, generate questions to help you focus your nature studies in the future and to develop a deeper awareness around you. Try to identify the living and nonliving things around you, and do research on what you observed.

Based on your observations and research, you can start to make interpretations of the ecological interactions in your landscape and how people may have impacted the area. To **interpret** means you can demonstrate your understanding by explaining or reframing what you observed. You are attaching meaning and revealing relationships of your observations. You can share your interpretations with others or reflect on a personal level. To guide you through this approach for learning about nature, you will find ideas throughout this book to incorporate into a **Nature Journal**. Simply designate a notebook or a three-ring binder for your detailed nature observations, questions you develop based on your observations, field investigations conducted, research on things of interest, and personal reflections to help process or make connections of anything that materialized from being outdoors. You will create an impressive record of the natural world around you with insight that nobody else will have experienced.

Relating Nature to Your Life

Let us make some connections between spending time in nature and everyday living or lifelong skills. Nature-based education could potentially lead to one or more of the following results on a personal or interpersonal level. As you read through the possibilities, think about if and how nature can impact you, and work toward developing the effects or practices of interest. These connections can help you identify reasons to spend time in nature.

1. *Inner Peace.* Find a sense of calm in nature. Stress management research shows that the natural world can lead to a feeling of inner peace and happiness not directly associated to personal circumstances. Striving for 20 minutes outside each day or 120 minutes outdoors in a week's time is ideal for feeling

the health benefits of nature. If stress begins to wear you down or if you feel on edge for no particular reason, then try spending time outside going for a walk or sitting in a comfortable place outdoors. Nature can offer at least some relief to overwhelming feelings that seem to impair thinking or judgment. In some places, medical professionals have prescribed time outdoors to help people improve their mental well-being and to lower blood pressure.

2. *Mental Focus and Creativity.* Nature may help you feel "unstuck". People may have difficulty starting a project or thinking clearly after spending lengthy hours writing or at the computer. When you find yourself mentally blocked, try time outdoors walking around with no agenda except to be present in the moment. Studies show exposure to the stimulations of nature can help the mind focus. You may be surprised at your mental clarity after a dose of nature. Artistic moments can also emerge after time spent with nature. The sights, sounds, smells, and textures can translate into many forms of artistic interpretations. Experiences outdoors can inspire you in ways that may be unexpected.

3. *Environmental Stewardship.* By developing your naturalist intelligence, you build your knowledge and skills through intentional time spent outdoors and studying the natural, geological, and cultural histories of an area. Through these experiences, an awareness of the surroundings develops, an interest in nature may arise, and an appreciation for that place and its inhabitants could start to build. Your newfound awareness can turn into an environmental sensitivity where you can notice a shift in your surroundings toward the better or worse. You will also start to understand and relate to environmental issues in the media. This can lead to making a positive difference locally by volunteering in initiatives to preserve, protect, or restore habitats in various ways (e.g., sorting recycling, litter cleanups, stream monitoring, bird inventories, etc.). This process describes **stewardship**, and the betterment of the environment and people's well-being in the community can result.

   Becoming a lifelong steward of our **natural resources** can be the likely result of someone who considers themselves a naturalist. Natural resources are things found in nature that can be used or valued by humans. **Renewable resources** replenish naturally (i.e., sun, wind, water, air), whereas **nonrenewable resources** form slowly or do not replenish as fast humans use them (i.e., coal, petroleum). We need nature to survive – clean and sustainable forests, waters, soil, and air, as well as the plants and animals that depend upon them. However, becoming a steward of nature does not occur overnight. It may take various influences, exposures, and experiences related to nature to motivate or inspire environmental stewardship. Presumably, the foundation begins with an interest in an aspect of nature, from there one begins to acquire further knowledge and skills associated with the natural world, and will eventually want to act on providing a service to benefit nature like changing behaviors or participating in activities to advance the conservation of our natural resources. If these habits persist through time, then one can consider themself an environmental steward.

4. *Strengthening Leadership Skills.* As you master your naturalist skills, your acute awareness enables you to observe multiple things simultaneously. You can recognize the conditions and resources present and interpret how they might influence the life in the area. Having the ability to sit and move quietly, and to observe and track animals will help you empathize and develop a sensitivity toward predicting how they behave. Applying these skills to interacting with society can help you learn to view the world from another human's perspective. You can start to develop a sensitivity about how others may be feeling, what they may be thinking, and how they might react to a given situation. This ability to make keen observations is important in many professions such as in business, criminal justice, teaching, media productions, social work, politics, and parenting. These skills can improve your chances at understanding people in a way that could lead to a mutually benefiting outcome.

5. *Self-Reflection.* While discovering the patterns of nature, there is a tendency to reflect inward. What starts out as a drive to learn about plants and animals can ultimately turn into what they can teach us. The lessons learned include being alert, to take only what we need, be prepared, stay focused, and use resources wisely. In many cultures, natural elements are highly respected and held sacred, or serve as guides in life's choices. From the perspective of time spent alone in nature, sitting quietly without distractions of people, pets, or technology will often direct a person to become more self-aware and reflective. This may help with personal growth and a variety of interpersonal relationships.

6. *Critical Thinking.* As a naturalist, you have the capacity to see patterns on many different levels within any given ecosystem. You will have trained yourself to observe the complexities in any environment. This way of viewing the world carries over into your personal life, society, and in politics. By using concrete data from observations, applying your in-depth knowledge of an area and its inhabitants, and maintaining an open and inquisitive mind you may start applying these skills in everyday scenarios. Honing your naturalist intelligence opens a world of opportunity.

Nature Journal: Consider spending time researching the benefits of nature.

- Reflect in your Nature Journal about your mood, stress level, creativity, and motivation before, during, *and* after time spent outdoors. Include how long you spent outside, describe the type of outdoor activity, and discuss highlights of what captured your attention throughout the duration of your time outdoors.
- Read current scientific literature related to the connection to nature, effects of nature, ecotherapy, etc. Record information learned and include the citations for future reference (research APA formatting or provide author, year, title; or title of website, date, web address).

Where to Focus Your Attention

At this point, you are probably eager to get outside and start naturalizing. Some people might feel at a loss of how to begin or perhaps you need to freshen your approach to observing nature. The secret is to simply spend time outside and turning your attention to all things natural – the wind, sky, plants, bugs, birds, etc. To develop your naturalist intelligence further on your path to becoming a naturalist, the key is to have a curiosity about anything you observe. This will lead you to wanting to learn more about it.

Asking questions based on your observations helps you go deeper with your understanding of nature. Read each of these questions slowly and think about your answer. How you respond will help you identify when and where to spend time outside, and to recognize your true interests when you are in nature. How much time in the day could I spend outdoors? Where can I go outside that is convenient to where I live or work? Am I prepared for the length of time I want to spend and the conditions outside? What usually catches my attention when I am outside? What do I want to learn about nature? What were some of my questions from the last time I was outdoors? Was I curious about what and where birds are? Where do the squirrels live? What is the general health and characteristics of the trees? Having a question before heading outside can help you keep your focus on nature while you are out there. Try starting with these questions: *What are the things in nature that catch my attention outside my home? How are the plants and trees different than the last time I observed them? What are the birds doing right now?*

While answering these questions outside, you may find more questions coming to mind that lead you to observing nature more closely. We all have our own interests, so each person's nature study journey will be unique. In nature, there is always something new to realize. There is a great sense of satisfaction when you can start making connections in nature. Reading the landscape helps the naturalist discover how everything is connected.

Think of any given area as having multiple dimensions. People usually watch where they are walking, so why not start there? Consider what exists underfoot. From there, work your way up and all around where you stand. What is happening with the plants, trees, birds, insects, other animals, or any fungus around you? I hear the birds, but do not see them and wonder where they nest. I see the same number of squirrels each day and wonder their range throughout the day and where they sleep. By knowing the characteristics and qualities of plants and animals around your home, you can make plans to focus your attention on them and start to get answers to your questions. You may even begin to understand some of their behaviors, recognize patterns, and predict relationships in nature.

Observing Animals in the Wild

All of this is exciting, but what if you do not notice any animals near where you live or you never observe any when you are outside. Think about how much time you truly spent trying to find animals, and whether you were out during the best time or in an area that would attract animals. What can a naturalist learn by observing animals? You can learn something new about that animal every time you observe it in its natural habitat. They may reveal their movement patterns, methods of communication, diet, or survival strategies. Animals can also teach you things about the landscape by watching interactions with other animals, plants, trees, the land or water it traverses, and how it responds to other conditions it encounters. The bird alarm call you hear every time you step into the yard or new habitat is likely a blue jay signaling your arrival, or letting you know of a new intruder in the area like a cat or other human.

Actually seeing an animal in the wild is thrilling. Stillness will help you observe more deeply because the animals may not sense your presence or do not feel threatened. Practice being still right outside your home in the morning and late evening. Sit in a comfortable place and just let nature reveal itself. Learn the Sense Meditation as described in the next section, *The Simple Naturalist*. This will help you hone your senses and focus entirely on the moment. It will help improve your chances of actually witnessing animals in nature. Another idea to increase wildlife viewing is to spend time in an area with a variety of plants and preferably a water source that animals might frequent. Use this book to start gaining further insight on how to find and view more wildlife when outdoors.

To expand your observational abilities and to compare your observations between different settings, explore the types of habitats convenient for you to visit nearby. Read Chapter 3 to help you learn to interpret the habitats you encounter. What are the types of animals that you observe living there? Who else might live in the area?

Research the basic information for each of the animal types you identified within the field guide sections of this book for fish, amphibians, reptiles, birds, and mammals. Discover the animals' behavior and preferred habitat to increase the chances of seeing it. Knowing about a particular animal's habits and requirements will lead you to where they might be found. A **preferred habitat** is where the animal spends most of its time. Its basic needs are met there for most of the year – food, shelter, and potential mate. Try to think like your animal of interest based on the knowledge you have about it. To really increase your chances of observing wildlife, regularly practice the tips and skills you learn in this book and from other resources.

Read the following ideas for improving your chances of observing animals.

Sight extensions. Here are some ideas for increasing your power of seeing wildlife.

1. Binoculars, spotting scope (mounted), and zoom lenses on cameras will bring the surroundings into an up-close, sharp focus. The trick to using these devices is to find your target (e.g., bird) with the unaided eye ... do not take your eyes off it. While still watching the animal, bring the binoculars or scope to your eyes. Use the focusing dials to adjust your view. Even if you know what something is in the distance, practice getting the object in view. You will be thankful you had this practice when your skills are tested when an exciting, unknown bird appears.
2. A magnifying glass or hand lens helps you examine nearby objects even more closely. Avoid shadows and find where your object comes into focus by moving it or your head for the best view. Keep this magnifier in your jacket pocket or in the nature bag you grab when heading outside. Also, an inexpensive field or pocket microscope can offer a convenient closer view of your specimen. These often have a built-in light.
3. To see identifying characteristics underneath something hard to reach, like the underside of a fungus, use a small mirror or the blade of a knife in position to see the reflected image.
4. Use a mirror tilted on a telescopic pole to see inside nests above you or things otherwise out of reach. Use caution to not disturb nestlings, and do not do this when the adult is in the nest. If the nest is close enough, sometimes you can reach your camera up to it and photograph inside the nest. Using a digital camera or smartphone camera can help you view the scene first before taking a picture. Consider using binoculars first from different vantage points to see your target of interest.
5. Cellular phones can be used for any of the ideas above with a quality good enough for getting a better look at something. It can also be used as a recording device and resource for finding more information about what you observed through its lens.
6. At night, cover your flashlight lens with a red filter (secured tissue paper or plastic wrap works well). Red goes undetected by many animals (often seen as black) and they will not notice your light. The filtered light will still allow you to see where you need to step safely and will not scare away nocturnal animals.

Get close. It takes practice and patience to get close to animals without frightening them. To help prevent unwanted situations, keep something between you and the animals like a tree, hill, or building. Sometimes animals may feel threatened and could be aggressive. Animals may be territorial during mating season and protective of their young. Take extra caution during these times. Use binoculars to get extra close yet keep a safe distance. Tread lightly across the landscape by leaving no trace or evidence of your disturbance. Stay on property where you have the right or permission to explore. Consider these ideas for getting closer to wildlife in their natural settings:

1. Follow tracks and trails. Look off the main trail for bent grasses or pathways.
2. Walk upwind by keeping the wind blowing in your face. Many animals have a keen sense of smell and could detect you coming if the wind were at your back.
3. Have the sun at your back so that the light makes it difficult for animals to see in your direction. This is also good practice for photography.
4. Move when they move. Advance when they move their head away to eat, groom themselves, or when they become focused on something else. If an ear twitches, body tenses, or head raises, then you should freeze.
5. Take to the water. Walk an existing boardwalk through a nearby wetland or skirt or sit along its margin. A paddle craft such as a kayak or canoe can move silently to places where animals may be drinking, hunting, etc. in a lake or river. They may not expect danger from the water, so be careful not to startle animals.
6. Find where there could be high visitation by animals. Consider trail intersections, water borders, den sites, natural springs, or a hilltop with a good view. Sit quietly and camouflaged (blend in and become part of the landscape) off to the side of the site.

The following three practices should be done at a minimum to reduce any stress on wildlife. Only conduct these actions where you think the animal may be, ensure other nature watchers know of your activities, and only carry these out for 30 second intervals and not extending beyond 5 minutes. Be subtle with low volume.

7. Create a bird's alarm call. This is called "pishing." Do this where you already hear birds making a chirping call. This is effective during breeding season. Put your teeth together and exhale the word "pish" a few times. The birds may fly in to investigate.
8. Squeak. Kiss the back of your hand to produce squeaks that sound like a rabbit or mouse in its death throes. Predators like foxes, coyotes, bobcats, hawks, or owls may come to investigate. When this happens, it could produce a mobbing of the predator. This is when birds flock to the predator to make it leave the territory.

9. Imitate calls. Imitating a call or a recording of it can bring in the animal. This works well with the black-capped chickadee (*Poecile atricapillus*).

Wildlife Through the Seasons

If you want to view animals, it is also helpful to know when they are active. Consider the season. Animals, especially birds, are less elusive (very active) in *spring* during the mating season while trying to attract a mate. Prior to trees leafing out in the spring is a great time to view wildlife. Wetlands will attract birds during migration while vernal pools (temporary ponds devoid of fish) are where amphibians often breed. Many discoveries can be found in and around these natural features, and in some regions there may be **citizen**, or **community science** opportunities to monitor them during their short duration in the habitat. Citizen science, or now more widely known as community science, is an initiative offered by a variety of environmental organizations, natural resource agencies, and universities for nonprofessional scientists to volunteer to monitor habitats or collect wildlife data. The results will help advance scientific research, increase the public's understanding of science, and can lead to the improvement of local environments.

There are other wildlife viewing considerations and opportunities throughout the year. *Spring* in the region brings new plant growth, hibernating animals awaken, birds who had migrated in the fall return, and these animals and more are active finding food and breeding. This is an exciting time for watching wildlife. In *summer*, wildlife young are often born and parents may be busy and protective of the offspring. Because of this and the fact that leaves are full on trees, animals may be more difficult to find. *Fall* offers plentiful food in the form of nuts and berries produced by shrubs and trees. Animals busy themselves in these areas by storing or gorging in preparation for winter. In *winter*, the landscape is often void of leaves and can have a covering of snow. Both instances offer clear views or glimpses of animals active during this potentially harsh season. Finding animal tracks in the snow is especially fun.

Another timing opportunity to observe wildlife is to take advantage of rainy days. If it is warm and raining, you will increase your odds of seeing amphibians—toads, frogs, and salamanders travel, and seeing birds as they bathe in puddles or forage in grass. Flooding in an area can also displace small animals from their homes such as worms, insects, amphibians, reptiles, and rodents. Opportunistic predators will take advantage of these situations. To go deeper in your nature study, think about the fundamental cause of different conditions.

Why do we have seasons?

The Earth's tilt is responsible for seasons. As the Earth revolves, or orbits counterclockwise around the sun, the Earth's tilt of 23.5° relative to its orbit remains the same throughout the year. However, the orientation of

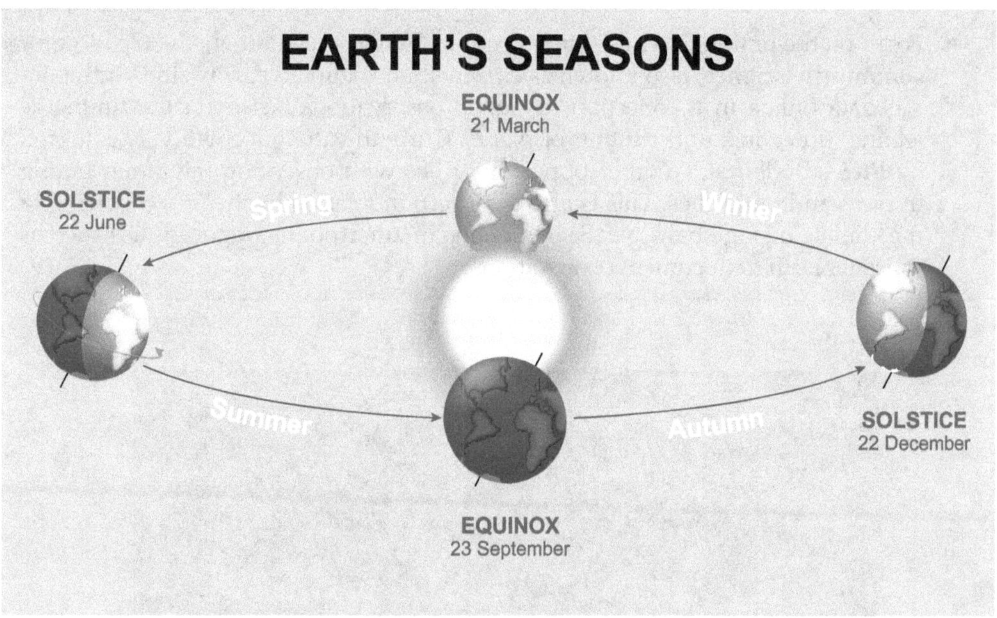

Illumination of the Earth at the beginning of each of the four seasons. The Northern and Southern Hemispheres exhibit opposite seasons as the planet revolves around the sun. The Northern Hemisphere experience the beginning of summer in June whereas the Southern Hemisphere begins winter at that time.

***Nature Journal:* Phenological Studies.** Tracking and studying seasonal changes over time is called **phenology**. Flowering times of plants, animal migrations, and human food supplies depend on the timing of naturally occurring events. To scientists, changes in these timings can indicate a climate shift. Some ideas for phenology study could include monitoring the following things through the years in your Nature Journal:

- Photograph a tree (preferably the one you "adopt" in Chapter 2) the first day of each month.
- Note the first appearance of butterflies, in particular monarchs and migratory bird species returning in the spring. Record their suspected last appearance before departing in the early fall (or when noticeably gone).
- Create a table with the headings of "firsts." Start with the year as your left column, then create headings based on the "first" time you notice something like when it frosts for the first time in the fall or when it snows for the first time, or perhaps in the spring it is when you notice the first migratory bird species appear or buds bursting, and then summer days over 90 degrees, etc.—it can be anything you think is noteworthy. Provide the date of each one of those "firsts" (or "lasts"). Later, graph the trends you observed over the years for each weather phenomenon and species occurrence.

- Focus on the timing of when things occur and contribute data that you collect for **community science,** also known as citizen science initiatives. Provide bud-opening data online in a collection of public observations based on the timing of leafing, flowering, and fruiting of plants. Currently, the information is collected by ProjectBudburst. Another opportunity lies with observing migration timing of birds and butterflies. This is also a concern in relation to changes occurring to the climate on a global scale. Search online or smartphone opportunities such as JourneyNorth to document observations.

How does nature change through the year? Document it. Choose a tree or shrub that grows near your home and photograph it each season or monthly.

the Earth with respect to the sun changes as we orbit. In the Northern Hemisphere, the Earth is tilted away from the Sun at the winter solstice (around December 21) and will experience cool to cold temperatures depending on where you are located. At the summer solstice (about June 21), the northern hemisphere is tilted toward the Sun and will experience warm to hot temperatures. The spring equinox (March 21) and fall equinox (September 21) will have milder conditions with equal daylight to night hours. These seasonal variations will affect plants and animals in terms of their growth, behavior, and overall life cycle.

Naturalizing Day and Night

Time of day will also influence animal activity. Dawn and dusk will have peak animal movement. The more you can spend time outdoors at these times, the greater the odds of seeing the most variety of animals. Experiment with that statement. Spend time outside counting different species observed at daybreak for at least 20 minutes. The next day, hopefully with the same weather conditions, do the same thing but two hours later. Repeat throughout the day. See more ideas and types of field investigations in Chapter 2. Some animals, like white-tailed deer will be most active during these times (**crepuscular**) and this is when daytime (**diurnal**) and nighttime (**nocturnal**) animals might cross paths. This can make for an exciting time in nature.

Why do we have day and night?

The Earth's spin, or rotation plays a significant role in day and night hours. Location and season will also serve as factors in the number of daylight hours. In general, as we rotate toward the sun, daylight and typically warmer temperatures are observed compared with when we rotate away from the sun at night when we observe darkness and relatively cooler temperatures. The number of daylight hours will affect the behavior of plants and animals. This is called photoperiodism. The timing of leaf drop and emergence of leaves, and animal migration times, daily activity, and mating are major events cued by the amount of light present. Other things to add to your nature study are the phases of the moon and star constellations.

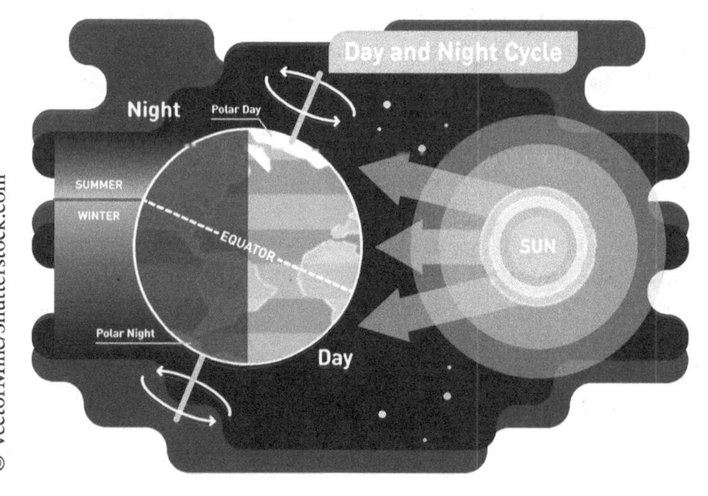

The Earth experiences day and night as it rotates on its axis. ***Activity:*** To determine how much time you have until the sun sets, outstretch your arm in front of you with your fingers together and palm facing you. Count the number of fingers between the sun and the horizon. Each finger equals about 15 minutes (i.e., four fingers is an hour until sunset).

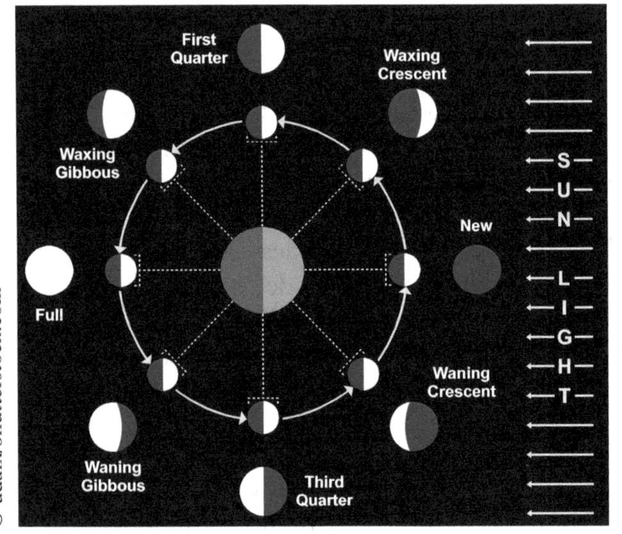

A top-down view of the Earth (at the center) with its perspectives of the moon phases. Track the moon during the month by searching online and observing it in the sky at different times of the night and day. You will find it easier to navigate under a full moon than a new moon. Research how the full moon affects nature and wildlife.

Nature Journal: Here are a few fun ideas to help you understand some of the effects the revolving or rotation of the Earth can have on nature. Record your findings and reflections in your Nature Journal. On a sunny day, trace or track shadows every 15 minutes or hour to see how the shadows change. Do this near the solstices and equinoxes at the same time of day to discover differences. Also, research how to create your own sun dial to tell time.

You can continue to use shadows as a resource to find directions! Use two rocks and a stick. Place a tall enough stick in the ground so it stands straight up 2–3 ft. (~1 m). Position the rock where you find the tip of the stick's shadow (this will be west). Allow at least 10 minutes to pass and place the second rock at the new position of the stick's shadow tip (east). Remember the saying, "never eat soggy waffles" to help recall the clockwise order of the cardinal directions: north, east, south, west. You can stand between the rocks and face due north (east to your right and west to your left) or position yourself in the direction you need to head.

For night activities, moon phase tracking and identifying groups of stars and constellations can help you navigate and understand animal activities. Constellations to learn first can include the Small and Big Dippers (Ursa Minor and Ursa Major), Orion, and Cassiopeia. The large, distinct shape of the Big Dipper will help you find true north. Draw an imaginary line straight through the stars of the dipper's outer edge and straight to the Little Dipper's handle end—the North Star (see the following image).

Find the Big Dipper in the sky (it will have different orientations to the horizon during different seasons). Can you find the North Star? Print a planisphere from the internet to help you map other stars throughout the year.

Improving Your Observation Skills

So far, you have already learned many things you can apply to help you become a naturalist. Other than spending time in nature and studying about it, honing your observational skills will serve you well to boost your naturalist intelligence. Reflecting about nature experiences in your journal will help get you to where you want to be much faster. Recording your observations, pondering and questioning them, and making sketches will create a meaningful record for you to build on. You can improve your observation skills even further through simple things like practicing stillness, doing a Sense Meditation (see the fun and practical Simple Naturalist section later in this chapter), trying to increase your peripheral vision, walking barefoot, imitating animals, and developing a sensory memory. Learn to be quiet and move without being noticed. Blend in with your surroundings to not disrupt nature.

Without moving or using any other device, you can expand your ability to see what is happening around you. Your **peripheral vision** is what you see out of the corner of your eyes. Practice blurring your vision to increase your field of view. By relaxing the eyes while looking forward, try focusing on a fixed image to the side. You can expand your peripheral vision so you can remain still and watch what may be happening next to you.

By taking off your shoes, you can heighten your sense of touch and other perceptions. Going barefoot can help you slow down and quiet your walking. It also forces you to be more present in the moment with every step. Another form of movement could involve imitating what you observed. This forces you to make interpretations. How does the bird fly, or how did the raccoon walk? Think about what you can learn just by imitating!

Lastly, developing a **sensory memory** is good practice for learning to look more closely and recall details. After time spent making observations or just being outdoors, try to describe something in detail like a tree, the arrangement of the habitat or trail, or discuss your experiences as you had encountered them and recall the details. Try recollecting details from using different senses to retell your time in nature. Taking pictures and creating a **photo journal** of your nature excursions can help document what you found most intriguing. You can use these images as prompts to tell your story and to look more closely at something you had not noticed previously. You could also link the photo journal to an electronic map of the area (currently, smartphone apps exist for this capability) or associate the photos with a hand-drawn map of where you visited. Sharing your nature stories helps recall observations and can make the experience stay in your memories for much longer than if you did not reflect personally or interpersonally.

Making better observations can result from simply asking, "What can I observe about this ___ (tree, bird, cloud, weather, track, soil, landscape, etc.)?" Exhaust your senses for each element you are observing (be wary of using the senses of touch and taste when in nature). If you are at a loss for new observations, another trick is to make more than one observation of a particular object. For example, you might notice a pine tree. What did you observe about, on, around, and within that tree? Perhaps you noticed the size, shape, and color of the tree. Did you get sticky from touching the bark? How about that small yellow and black bird on the pine tree limb? Or noticing needles getting blown off the tree by the wind, or the fact you heard a squirrel chattering repetitively within the tree, or the fresh smell of the pine needles? Different senses were used to make new observations about a particular object, in this case it was the pine tree. Also, notice how the adjectives and adverbs were used, or could be used to help create the mental picture.

To truly start interpreting observations, you first need to make thorough descriptions. For example, a large, brown squirrel (~10 in.; 25 cm) with a bushy tail sat on a high branch and made a "chip-chip" sound. It suddenly ran down the tree head first and ran up a nearby tall tree without leaves. Those were relatively thorough observations. What else is there to know? Try asking interrogative questions of who, what, where, when, why, and how in relation to what you observed. The intent is not necessarily to follow the squirrel, but to simply cause you to look more closely and derive new ideas and interpretations. This could help lead to further research and field investigations on squirrel behavior near where you live. Always support claims with detailed observations and research. Learn a more formal approach to collecting data for your nature study in Chapter 2's section, *Scientific Thinking*. Deliberate tracking of your nature study will lead to a deeper understanding that results in accurate identifications and plausible interpretations.

Making Better Identifications

Making good identifications begins with paying close attention to the details of your observations. Venture outside and find something of interest – a fungus, plant, leaf, insect, or other animal. Study it. Now, turn around or walk away. What types of things do you remember from your species of interest? You should be able to recollect important details to help you make an identification or interpretation of what you saw. Draw it from memory and label what you think is essential for identification. Show it to somebody to see if they can identify it. If you described what you observed to somebody, could they draw it accurately and identify it? If you or another person cannot make an accurate identification from this exercise, the results will at least reveal where you need to improve your observation skills. The goal is to provide enough details in your recollections or drawings for anyone to make an accurate identification. Size, color, identifying features, and location of your observation will often be the foundation of what you need to start making an identification.

A quick and easy way to make size estimates in the field is to compare what you see with something familiar – like your hand, fist, forearm, the length of your arm, the span of both arms, and your height. Measure these body parts and you will have a good size approximation of what you observed. When referring to the color of something, first report the most dominant color that covers most of the organism. Next, note the other colors you observe and where they are located on the organism. The underside of plants and animals is often a different shade than its topside. You will also want to focus on any markings that you see such as stripes, patches, or spots. Where are these features and what color are they? Lastly, take note of the surroundings where you observed your species. Habitat and region will be important in narrowing your search for the species in question. All these observations can lead to an informed identification.

Learning the basic anatomy, physiology, behavior, and taxonomy of what you are trying to identify can help you recognize differences between species. In Part III of this book you will find the basic biology of fungus, plants, and animals. In the field guide sections of the book, you will also be introduced to a practical approach for identifying each of these groups, and a collection of common, unique, or invasive species in the region. Review these sections before and after spending time outside and anytime you are curious about a species. This will help you build your identification skills.

1.2 Naturalist Activities

If you practice the tips and complete the activities in this book, you are on your way toward a lifelong study of nature. In this section, you will find effortless activities you can easily commit to so you can become more connected to nature. Try practicing one activity each week then move on to the next activity, or do a couple activities every day or try them all at once every day or every other day. Find what works best for you.

The activities are listed in order from the simplest to the most complex in terms of what can be considered the most attainable naturalist goals. Use the previous section, "Nature Awareness", to help get the most out of the experiences. By practicing the following activities, you will develop a routine and naturalist habits for everytime you are in nature. Record and organize everything in a designated notebook or binder—your **Nature Journal.** This will be a useful resource for you as you journey through places in nature.

Preparing for the Outdoors

Prior to going outside, consider the weather and potential conditions you may encounter during your outing. Dress accordingly and bring additional gear to help protect you from the elements and stave off hunger and dehydration. If you are just heading out your door and staying close to home, the extent of preparedness would be less than if you ventured further or if you wanted to ensure a long, comfortable visit in nature.

When selecting clothing, cover as much skin as possible—head to toe. Regardless of whether it is warm or cold outside, these are good guidelines to follow. Wear layers to allow for proper insulation or removal of garments. A base layer T-shirt, long-sleeved layer, insulating jacket, and a windbreaker offer the best combination of clothing to wear. Varying the types of layers based on the season or potential weather conditions will help with your comfort level. Polyester in summer and wool in winter tends to function best. Cotton will hold moisture and is slow to dry, which can be dangerous in terms of potential hypothermia if you spend extended periods of time outdoors in cold weather, especially when backpacking.

Consider bringing a small backpack or hip pack (fanny pack) to carry any of the gear mentioned above and other nature adventure gear. Other items to consider for excursions that take you away from home (or beyond shouting distance) include: water, a tasty energy bar, sunscreen, bug spray, whistle, bandana, alcohol wipes, bandages, a pocket knife, a small tarp or heavy trash bag for a makeshift shelter or poncho, a bundle of small diameter rope, compass, magnifying glass or hand lens, small notebook, pencil, and binoculars. Have these things ready to go so that it takes no effort to spend an extended length of time in nature. On the other hand, many nature study activities do not require any of those things if you are right outside your home! Additional things to consider while on your outdoor excursions: remember to let somebody know when and where you will be, do not trespass, leave no trace of where you were, bring a map, and have a charged phone. When you are ready to spend time outside, this means that your needs are met and you are prepared for the conditions and responsibility. You will ensure your safety, and respect others around you and the environment. These simple things set you up for absolute enjoyment in the outdoors.

The Simple Naturalist

Perhaps you are thinking you need more motivation to go outside or to have guidance in your nature study. With anything new or when trying to create a habit, it is important to start with easily attained goals so you experience success and build on those successes and skills. How can this idea be applied to creating a nature study habit? Work toward completing each of the following activities in this *Simple Naturalist* section. Here you will find basic, straightforward naturalist practices to incorporate on a regular basis – these ideas can easily be applied daily!

The activities start very basic to build your awareness and focus toward nature in the most convenient way possible. From there, a little more time and effort is required to complete an activity – but nothing is difficult to do. These *Simple Naturalist* activities are for the novice or the expert. Consider starting your nature study with *The Simple Naturalist* if you are just beginning your naturalist journey, feel uncomfortable or bored with nature, or want to observe nature more closely, or perhaps you are a seasoned naturalist and want direction for focusing on new things.

The Simple Naturalist

Before reading the activities below, **on a scale from 1 to 5, how would you rate your naturalist skills?** Give yourself a 1 if you do not think you are very observant and not in tune with your natural surroundings. A 5-rating would indicate you are extremely observant and consider yourself having a high naturalist intelligence. A 3-rating would indicate average abilities. _____

1. **1-Minute Window.** Open your curtains or shades of a window in your home (if feasible). At the least, find **1 minute** every day to be still and simply **look outside a window.** Increase this to many 1-Minute Window opportunities throughout the day. Eventually, increase the duration. This is a great activity for when you cannot get outside. Create a comfortable and inviting space so that you are apt to do this activity. Take advantage of

your spare time to learn more about what is happening around your home, work, or school by observing nature through all your windows. This is a quick and easy way to get some naturalizing in!

Choose a place out your window where you want to go the next time you can be outdoors. Also realize, while having the curtains or blinds open, we get more natural daylight! This may do wonders for your overall mood. From an environmental conservation perspective, consider passive heating in the winter and cooling in the summer—sometimes it is best to keep the blinds open or closed, respectively during peak sunlight hours. Otherwise, have the windows visible to view and study nature. Do this every day. If possible, open the window to get fresh air and hear the sounds of nature. What did you observe? Write it down in your Nature Journal or tell somebody about it.

2. **Pre-Pause.** Picture yourself doing this, then actually try it out. Every time prior to stepping outdoors, simply take a deep, conscious breath and move silently to the outdoors. Once outside, focus your attention on what is happening in nature around you. Remain aware of your surroundings as you move toward your destination. Use the following exercises to help tune in to your natural surroundings. At the very least while outdoors, practice the 1-minute outdoor vacation daily (see below).

3. **1-Minute Outdoor Vacation.** Same as number 1 above but spend your 1-minute observing nature in the out-of-doors. Think of the time you are outside as your personal time - a vacation or holiday. Be present and focused on what is happening around you during that minute. Think about those observations. Share your experience with somebody or record it in your Nature Journal. Increase this to a 5-minute Outdoor Vacation (or longer) when you feel like 1 minute does not suffice.

4. **Sense Meditation.** Have you ever felt overwhelmed or things are happening too quickly around you? Sense meditation is a great sensory reset that will help bring you into the present moment and hone each sense. While outside, remain still and close your eyes. You may need to sit down or support yourself for this activity. With your eyes closed, focus on the sounds around you. In what direction is the loudest sound? The quietest sound? Can you identify those sounds—natural or humanmade? How does the weather feel on your skin? What direction is the wind coming? How intense is it? What are the smells around you? When you open your eyes, your sense of sight (usually a person's dominant sense) is heightened and you will feel more alert to your surroundings.

5. **Nature Stroll.** Take at least **5 minutes to slowly walk around outside** near where you live or where you are visiting (the length of time will likely increase as you go deeper into your nature study). Do the Nature Stroll alone (that means no other people, pets, or electronic devices will join you!). Move at half the speed you walk normally. Try to move quietly. Pause every 50 ft. (15 m) or so, and especially stop to make observations of things that interest you. You will instantly observe more by following this advice of slowing down and stopping regularly. Be open and aware during this experience. If you become caught up in thinking or worrying, that's fine for awhile – and you may find a way to work through some tough times. Eventually, gently bring your attention back to your senses. What do you see? Hear? Can you feel a breeze? How does the ground feel beneath your feet? Focus on the plants, trees, birds, chipmunks, clouds, rocks, etc. around you. For our purposes, the intent of the Nature Stroll is to focus on anything you find that is part of the natural world. Often times during these walks, you may find a clear mind results. Possibilities are endless on a Nature Stroll. Share the accounts of your walk with somebody or write about it in your Nature Journal.
6. **Where is North?** Find north, then find landmarks around you in that direction. From there, look to the east, south, and west. Throughout the day, ask yourself to locate north. Before heading to a new location, determine which direction you are headed. This will help you stay oriented to your place in the landscape and help with predicting time of day.
7. **Sit Spot.** Commit to **at least 5 minutes a day alone at a designated "sit spot"** near where you live (increase the time as you become more comfortable outside and advance with your nature study). The Sit Spot is a place that is quick and easy to get to and comfortable to sit for observing your surroundings—your front porch, side stoop, back deck, or maybe a lawn chair or blanket placed in a peaceful location in your yard or somewhere on your property, at a park, or at work. The Sit Spot extends all around you and relates to all things tangible – encompassing approximately 3 feet, or 1 meter in all directions for what you can observe. It also includes what you can sense beyond your reachable space. Quietly sit and open your senses while in your new special place.

   What did you experience with your five senses? What did you observe about the weather? Birds? Trees? Insects? Mammals and other wildlife? Seasonal observations?

   After time at your Sit Spot, share your experiences with someone or write them in your Nature Journal. Describe the time you spent outside.

What caught your attention? What were some questions you had about what you observed? How did you feel before, during, and after your time outdoors? What sense did you use most often (see, hear, smell, taste, touch)? How could you improve your observations in the future?

Apply what you learn in this book to what you experience at your Sit Spot. You may need to create another Sit Spot later where you can make deeper nature investigations. For your nature study, you will want to designate a Sit Spot at the center of your larger Nature Area where you can explore and inventory your observations. More ideas and details are provided in the next activity and toward the end of Chapters 1 and 2. Look for other Sit Spots nearby and wherever you go. Having multiple Sit Spots are great for making nature observations, interpretations, and comparisons between locations. As you work through this book, apply all the concepts you discover to your Sit Spot.

8. **Nature Area.** The area that surrounds your Sit Spot is the Nature Area. Conversely, you could find a place to explore then claim a Sit Spot at the center. The Nature Area should be a nearby location, preferably right outside your home. It needs to be accessible and where you feel safe and comfortable walking. Try to choose a place that has plant life. A variety of plant life will attract more wildlife and potentially provide more interest to you as the observer and nature investigator. Perform a Nature Stroll around the area where you will focus your attention for future nature study activities. You will create a more defined Nature Area at the end of this chapter.

9. **I *Can* Draw.** Drawing is another way to sharpen observation skills and pique curiosity. You do not have to be an artist to make drawings of what you see around you. Creating something quick and somewhat legible is all that is necessary to give another dimension to your nature study. Choose a natural artifact of interest, observe it up close, and draw it.

   Using a pencil is best so that you can erase markings. Make a simple line drawing of what you observe. First, get the overall shape or outline of the object on paper. Think basic shapes like circles, squares, triangles. You can erase unnecessary lines later. Next, provide labels for colors, markings, and sizes of features. These are important for interpreting the drawing and as reference for further understanding and identification of species. The goal is about 10 minutes for a decent sketch while in the field. Feel free to stay longer if you have the time. Details of your drawings can be filled in later with your field guide. Taking a photograph can also be a useful reference for providing details to sketches. Consider searching the internet for video tutorials on sketching certain aspects of nature.

Blind Contours. A detailed sketch or photo is not always possible if time is limited in the field, or when trying to get details of an unknown animal that could likely move at any moment. In these cases, you may need to draw "blind contours." Give it a try.

Go to a place where you are likely to see an animal. This may be while looking out the window, at your Sit Spot or anywhere else. Have a pencil and your Nature Journal or separate sheet of paper ready for drawing. Do not stop looking at the animal. As you are looking at the animal, draw it—*quickly.* Keep looking at the animal. This should take no more than 2 minutes. *Without lifting your pencil and not looking at your paper*, draw the animal's outline and mark details you think would be relevant for identification (e.g., stripes, spots, etc.)! This is where the term "blind" comes into the activity. Remember to look at and consider the outline of the animal's body in terms of simple shapes (e.g., small oval head with large triangular ears, larger oval body, and a blocky or rectangular tail). Once you get the main features, look at your paper to note other observations such as size and color.

© addillum/Shutterstock.com

The picture you see on this page is an example of a blind contour drawing of a bird in flight. Getting the general shape of your subject will help make an identification later on.

Next, do a better job. Without taking too much time, redraw the animal so details are clear. As you spend time drawing, you will hopefully develop a curiosity about the animal (or feel free to do use this technique for a plant!). Think about the animal. Ideas for deeper investigation may eventually develop. Place the animal in a wider context. Where did it come from before you saw it? How is it connected to other animals, plants, and people in the area? Where do you think it is going?

10. **Nature Mysteries.** To help you stay interested in spending time in nature and its study, keep track of Nature Mysteries. What has caught your attention outdoors? Did you stop what you were doing because something piqued your interest? What puzzles you about this phenomenon,

object, or species? This can lead you to looking more closely at the surroundings, asking more questions, learning more, making connections to other parts of nature, thinking about cause and effect, solving problems, or simply growing a deeper appreciation for what you observed. Keep track of these curiosities in your Nature Journal.

11. **Observation Record.** Keep record of the animals and plants you encounter around your Sit Spot and throughout your daily activities. No need to know the name of what you observe. Simply make detailed observations to help you look up the organism later. Use the sample chart in this chapter to get you started or copy it in your Nature Journal. Start to organize your Nature Journal for what makes sense for your study.
12. **The Common Bird.** Based on your observations over time, determine the **most common bird species you encountered.** If you do not know the type of bird, use the field guides in this book to help you identify and learn basic information about it. What is its full common name as seen in this book (e.g., "black-capped chickadee"). What can you learn about and from the bird?
13. **The Common Mammal.** Now, do the same for the **most common mammal species you discovered** around your Sit Spot or near your home. Remember, you can identify whether an animal is a mammal if it has the presence of fur. Try to find it in this field guide. What is the full common name of the mammal? For example, instead of just "chipmunk," it would be the "Eastern chipmunk" or maybe it is a "thirteen-lined ground squirrel". What can you learn about and from this animal?

After completing the 13 activities within The Simple Naturalist section above, reevaluate your naturalist skills. On a scale from 1 to 5, what is your personal rating of your observation skills and naturalist intelligence. A 1 indicates not being very observant and not in tune with your natural surroundings. A 5-rating would indicate you are extremely observant and consider yourself having a high naturalist intelligence. _____

Revisit each Simple Naturalist activity and give yourself ratings on how well you completed each task after about a week. Continue to practice these exercises over time.

Taxonomy: Classifying What You Observe

To help communicate about the species we observe, a system for naming all nature's organisms exists. Thanks to the Swedish botanist in the 1700s, Carl Linnaeus developed a way to make sense of and organize a taxonomic system. Taxonomy is the naming, identifying, and classifying of organisms. This system helps describe relationships between species. The taxonomy of species can reveal information about the structure, appearance, and ecological role of species.

To understand relationships among and between species, learning the hierarchical taxonomic levels is key. Going from the broadest level to the most specific is as follows: domain, kingdom, phylum, class order, family, genus, and species. Humans are in the same domain as plants and fungus! Organisms in the same domain have some common characteristics but may be distantly related. Those organisms found in the same species are the most closely related.

Remembering the taxonomic levels in their proper arrangement can be daunting, but very helpful in understanding connections among species. This is when memory tricks can help. Try using the first letter of each taxonomic level in descending order to create a sentence with words that start with the same letter. For example, you could use the nonsensical sentence: Do kings play chess on finely grained sand? Use the following diagram as a quick reminder of the arrangement of the levels. Try committing it to memory.

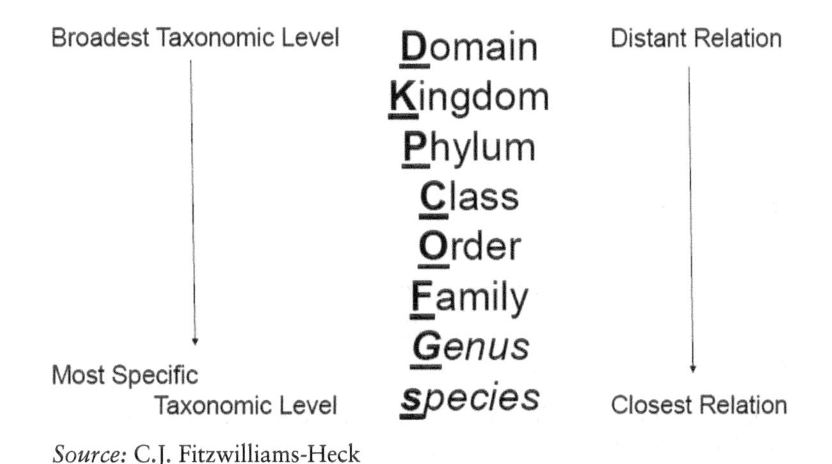

Source: C.J. Fitzwilliams-Heck

What are the Three Domains and Where Do the Kingdoms Come In? You may have heard of the levels, domain and kingdom – but what do you know about them? All life can be classified into one of three domains: Archaea, Bacteria, or Eukarya. Archaeans appeared first in the evolutionary time scale, then bacteria and finally eukaryans. Domains Archaea and Bacteria are grouped as **prokaryotes,** or single-celled organisms with no distinct nucleus nor specialized organelles, whereas **eukaryotes** of Domain Eukarya, do have those features.

Archaea: These organisms are microscopic, unicellular, anaerobic (lives without oxygen) prokaryotes (no nucleus or membrane-bound organelles) that live in extreme environments (i.e., high temperatures, low or no oxygen, hypersaline conditions, etc.) and still exist today.
Bacteria: Microscopic organisms that are unicellular and **aerobic** (requires oxygen) or **anaerobic** (does not require oxygen) prokaryotes (which includes cyanobacteria, or those photosynthesizing species formerly known as blue-green algae).
Eukarya: Species that are either unicellular or multicellular, aerobic, eukaryotes (have a true nucleus and membrane-bound organelles). Naturalists spend most of their time studying this domain. There are four **kingdoms** within Eukarya: Protista, Fungi, Plantae, Animalia.

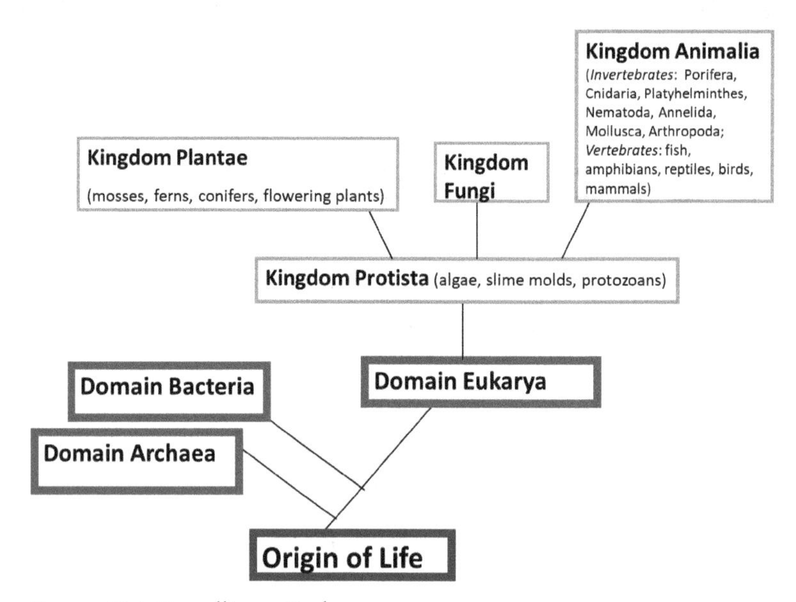

Source: C.J. Fitzwilliams-Heck

From some origin of life, you see the three domains arise and the four kingdoms from Domain Eukarya.

Of the kingdoms, Protista appeared first in the fossil record. Protists evolved into the other three kingdoms over time – Plantae, Fungi, and Animalia. Protists consist of plant-like, fungus-like, and animal-like organisms such as algae, slime molds, and protozoans, respectively.

Scientific Names. Latin words are used to communicate names of species within the global scientific community and among naturalists. Speaking in one universal language helps prevent misunderstandings. Often, one species has multiple common names. For example, the Latin or scientific name, *Populus tremuloides* is a tree with the common names quaking aspen, trembling aspen, popple, poplar, and likely others. This can get confusing when trying to have a productive conversation or when trying to understand somebody's written account of this tree. The scientific name is always binomial (has two words for the name) and tells a story. The first name, or genus, is a noun. When writing a scientific or an organism's Latin name, always capitalize the first letter and underline it or type it in italics. The second name is the species and is an adjective to describe a characteristic of the organism. This name will never be capitalized but will also appear underlined or italicized. In the tree example, *Populus tremuloides*, the name indicates it is in the poplar genus and the species name indicates the tree has a trembling nature. This tree has a leaf and stem structure that catches the slightest wind to send its leaves in a unique trembling or quaking pattern. See the field guide of trees in Chapter 6 to learn more about this species.

The more you pay attention to and use the Latin names, and consider the taxonomy of organisms, the more you will start to understand and appreciate characteristics and connections in nature. Use the chart template in the following section to organize your observations.

Animal Observation Chart

You are striving to open your senses to the environment around you. You are trying to become more familiar with the animals you encounter. **Create a list of the animals you witness in nature (start with observing only 5)—limit observations of domestic animals to one (e.g., pet/farm/zoo animals). Do NOT include HUMANS in your observations.** Use the field guide sections in this book to help you properly identify organisms. Discovering their Latin or scientific name will help you recognize relationships among the animals. Make it a game—it will be fun—you are on your way to becoming a naturalist. **Use different animals for each entry and future investigations.** This will broaden your knowledge base. If you encounter the same species exhibiting unusual behavior or a different morphology, then feel free to record it—as long as the distinction is clear. **Fill in or complete the following chart in your Nature Journal or in a digital file. The Animal Observation Chart can also be used as a foundation for a field investigation** (see Chapter 2).

Animal Observation Chart

Date	Time	Weather Conditions	Animals	Latin/Scientific Name	Species Description OR Sign/Evidence of Animals	Activity/ Behavior Observed/ Suspected	Habitat/ Ecosystem
	a.m./ p.m.	*temperature, cloud cover, precipitation, wind, etc.*	*Common Name; start with what you know, e.g., "deer," then research the full common name, "white-tailed deer"*	*Genus species name—write it as it appears in the field guide—*Odocoileus *virginianus. Order or family name acceptable for organisms difficult to identify—e.g., some insects: Diptera/* Muscidae	*size, colors, and identifying/ notable characteristics—spots, stripes, tail length, etc. describe type of clue observed of its presence—tracks, scat, nest, etc.; give size and features*	*What was it doing? or What do you think it was doing? Try to include the observation and potential interpretations of its behavior*	*Describe where you found it. What type of habitat was it in? Do not just say "yard" or "campus"! Provide the types of vegetation nearby to describe the habitat*
1.							
2.							
3.							
4.							
5.							

Source: C.J. Fitzwilliams-Heck

Answer questions on the next page about your animal observations.

Consider your animal observations and answer the following questions #6-9 in your Nature Journal.

6. How do you think the weather influenced the behavior of one of the animals you observed? Describe your claim and provide support of your observation. In other words: state the animal observed, what it was doing, and describe why you think the weather affected its behavior.
7. Describe how one of your animal's surroundings (habitat) affected the behavior of the animal. Include a description of the habitat and how that animal used and was affected by those surroundings.
8. How could you improve the chances of viewing more animals in the future? List at least two ways to improve the opportunity to see wildlife where you visited.
9. Create five questions about what you observed. What were your Nature Mysteries about the animals observed? I wonder ... who, what, where, when, how, and why ...

Charting Plant Observations

You can create a similar chart for your records of plant and tree observations. Start with the plants just outside your door. Trees are plants, so you can include them in this list or keep a separate list depending on what makes sense for your situation and purposes. Treat the following prompts as headings for columns in your chart or section entries in your Nature Journal. Describe what the plants look like. Do a quick *sketch* of the leaves, flowers, seeds, bark, etc. What is the *soil* like where it grows—sandy, mixed grains, dark? How much *sun* does the plant receive—full, partial, shaded? Use Chapter 6 to start trying to identify the plants and trees. Keep in mind that often times in landscaping, plants are hybridized or grown in a special way for certain conditions. These types of plants are not typical in natural landscapes and may not appear in this field guide. In that situation, refer to the internet or free smartphone apps to help with identification. Based on your observations and research, what do you think is its *taxonomy*? Provide the plants' full common name (e.g., "common plantain" or "eastern white pine"). What is its Latin or scientific name (e.g., "*Plantago major*" or "*Pinus strobus,*" respectively)? What is the plant's family name? Making accurate identifications requires keen observations and can open a new understanding and appreciation for "place".

Going Deeper into Nature

In this section, the activities will take more time but will offer you in-depth opportunities to discover your surroundings. Read these insightful practices to determine where you need to develop further. To truly have a master naturalist's mentality and capabilities, everything in this section should be completed and practiced. By practicing and completing what is presented in this section, you will have made interpretations of your landscape, have achieved a naturalist's foundation, and discovered your "place" on a social-ecological scale.

Interpreting the Landscape. The landscape refers to the characteristics of the surrounding ground and waters. This section will help you start looking at a scene in nature with a thorough approach. With practice, you might start making these types of observations automatically in any given scenario. Read the questions to guide you through initial interpretations of an area. Notice that you begin with a large scope and get more specific to help with thinking about interconnections.

Interpreting the Landscape

1. What are the most obvious things you notice about the scene?
2. What could be said about the plant life that you see?
3. What animals or signs of animals are observed?
4. Is the land hilly or flat?
5. What is noticeable about the soil?
6. How does the water runoff across the land? Where is the nearest water body?
7. What are the potential interactions between the plants and animals you observed and the surroundings? In other words, how could the area be used for food, water, shelter, or breeding?
8. How do humans interact and affect the landscape?

Consider the following photo you see here. It was taken on the author's property where most of the inspiration came for writing the book. Let's use the questions to help guide our interpretations. What major observations can you make in the picture? The obvious things you may have noticed could have included the dirt pathway, the log with worn away bark and broken root exposure, and the scene is mostly green. For the plant life, there is green moss in the foreground, small ferns in the center, deciduous trees of various sizes in the distance, and a clearing with herbaceous plants to the left. Animal evidence is difficult to see in the picture, but insects likely exist in the log (some insect galleries may be visible) and other animals could use the path. The soil is dark. The land is fairly flat with a slight incline to the right draining runoff to the area to the left where it looks like it could be wet.

What else could you say about the scene? Interpretations based on your observations could be that this is a growing season in a mesic forest with moist, rich soil. Humans and animals likely frequent the area due to the obvious pathway. This could be a transition zone or edge habitat where it is the coming together of two or more habitats. Sunlight can penetrate the forest floor at least along the back, left

Source: C.J. Fitzwillimas-Heck

side of the area allowing for the growth of new plants needing full sun. A variety of vegetation exists to provide potential food and shelter for the animals. The log could house a variety of invertebrates and decomposers that could also provide another source of food for animals. The more you observe, know, and research will lead to a deeper understanding of place. Chapters 2 and 3 offers more insight on how to read the landscape. Start with the questions presented in this section first, and build on your interpretations as you progress through this book.

The Naturalist's Foundation: A Basic Checklist of Achievements. These goals will encourage you to look more closely at the natural world. The checklist format may prove a motivation toward accomplishing each task or activity. You will notice that some tasks could be completed during one trip outdoors. Review the list in its entirety prior to embarking on your outdoor adventures. By completing the list, you will know your "place" more intimately with an awareness of its flora, fauna, conditions, and resources. You may also find yourself tuned into nature and having the desire to build on your understanding of the natural world. Consider recording your progress here in the textbook, in your Nature Journal, and utilizing the web or phone application for iNaturalist to record and organize your observations (other online resources also exist that can help you store and share information).

The Naturalist's Foundation

1. 7 days of 1-minute outdoor vacations __ __ __ __ __ __ __
2. 7 days of Sense Meditations __ __ __ __ __ __ __
3. 7 days of 5-minute Nature Strolls (at least that long)
   __ __ __ __ __ __ __
4. 7 days of 5-minutes at Sit Spot (at least; same/different places)
   __ __ __ __ __
5. Identify 5 signs of the season or seasonal change:
   - ______________________, ______________________,
     ______________________, ______________________,
     ____
6. Learn 3 different cloud types for your region:
   - ______________________, ______________________,
     ____
7. Observe 3 different wind directions: __ __ __
8. Find 5 local Nature Mysteries:
   - ______________________, ______________________,
     ______________________, ______________________,
     ____
9. Map the natural and built environment around your Sit Spot (where is north?) __ __ __
10. Learn 5 birds that live near you:
    - ______________________, ______________________,
      ______________________, ______________________,
      ____
11. Learn 5 mammals that live near you:
    - ______________________, ______________________,
      ______________________, ______________________,
      ____

12. Watch a crow/raven for at least 5 minutes.
13. 7 days spending at least 5 minutes observing a bird/birds
    - _ _ _ _ _ _ _
14. 7 days spending at least 5 minutes observing a squirrel
    - _ _ _ _ _ _ _
15. Learn 5 local trees:
    - ________________, ________________, ________________, ________________, ________________
16. Learn 5 local plants:
    - ________________, ________________, ________________, ________________, ________________
17. Learn 2 local rock types:
    - ________________, ________________

Discovering Place. To truly understand and care for a landscape, we need to learn about the potential present-day interactions, possible past relationships, and have the ability to make conceivable predictions. We need to ponder connections between animals, plants and animals, plants and animals and their surroundings, and how humans fit into the picture. When we think about people's interactions within a landscape, we call this **social-ecology** and this helps identify your place in the landscape.

As a naturalist and conscientious citizen of the place you call home, it is imperative to be knowledgeable about how you and society fit into the ecosystem. **"Place"** can be your location at the moment, where you live, the place you enjoy spending time, the city, watershed, state, region, or the planet. The following questions are meant to help you start to identify what you know and what you need to research on a local level. By understanding the answers to these basic questions about the region in which you live, work, or go to school you will start to make deeper and more meaningful connections to your place. This will also serve as a foundation for helping you understand where you live and what you can do to help protect it on a social-ecological scale.

Discovering Place

Read through the following questions. Mark the ones you do not know. Note where you might find the answers, and then do the research.

1. In what major watershed do you live?
2. Describe the watershed's social-ecology.
3. What is the local watershed organization? Website?
4. What is the size of the human population in the area where you live? Is the population growing, stable or declining and why?
5. What is the major bedrock and common soil types for where you live?
6. What is the most dominant and climax vegetation where you live?
7. What vegetation was there circa 1800?
8. Who were the indigenous people in the area? Discuss their lifestyle and history.
9. What is the source of water for drinking, bathing, and washing?
10. Where does wastewater go?
11. Where does the local trash go?
12. What is the local recycling system(s)?
13. Where does local recycling go once it is picked up and leaves the recycling center?
14. Where does most of your food come from (i.e., geographic area, processed versus unprocessed foods, animal versus plant foods)?
15. Name one of your state representatives.
16. Name one of your state senators.
17. What body(s) oversee(s) or make decisions regarding public land in your city or county?
18. Who is your drain commissioner? What is their job?
19. Who is one of your city council members?
20. How large is your home (approximately) or living space?

21. From what source(s) is most of your energy for the electricity in your home?
22. From what source(s) is most of your energy for heating in your home?
23. From what source(s) is most of your energy for cooking in your home?
24. How many kilowatt hours of electricity does your household use per week?
25. How many miles (kilometers) do you drive in a typical week? What is your average miles (km) per gallon you achieve with your vehicle?
26. How often do you bike to get where you need or want to go? Use public transportation (bus, train)?
27. Name your State's Soil, Bird, and Tree. Where can you find these species in the region?
28. Name three endangered species in your state. Describe where they live (habitat type), and what issues have led to their classification as endangered.
29. Name three non-native species that are present in your state and who are also posing resource management challenges. Describe (for each species) the challenges posed.
30. Name five of the most common bird species seen in the winter near your home.
31. List any venomous reptiles that can be found as native wildlife in your state. How can you identify the animal(s)?
32. List poisonous or harmful plants found nearby and describe and draw each.
33. Name common animals found in the region for the categories of: fish, amphibian, reptile, bird, and mammal.
34. Name three major constellations that can be seen in your night sky. Where are they positioned in the sky, and during what season?
35. What is your ecological footprint. You can research this as an online activity. What were the results? Where do you need to improve? Record your thoughts.

In your Nature Journal, reflect on how well you know your "place" based on your reactions or responses from the questions. What do you need to look up? Spend time researching and elaborating on your answers from "Discovering Place". What were the surprising discoveries from your research? Visit Chapter 2 to help you elaborate on the concept of place. You are on your way to becoming a responsible, well-informed naturalist! With this knowledge, you can make better-informed decisions regarding the environment and may appreciate nature more than when you started your research of "place".

The Sit Spot

Naturalists have kept journals of their field observations through the centuries. The nature journal gets you involved in watching processes, patterns, shapes, cycles, and changes in your own immediate landscape. E. O. Wilson once noted, "The creative process is at the heart of natural history observation ... it involves the illustrator directly in what he [she] observes." Learning to create a visual record of your observations will prove useful for remembering and understanding your experience in nature.

Designating a Sit Spot. Now it's time to get more serious about designating a Sit Spot and developing a routine when you visit it. If you have started to implement the activities described in *The Simple Naturalist*, then you may have already identified a Sit Spot location. Go back and refresh your memory on what was presented there. In this section, you will formalize this special place where you can go to decompress and apply all your nature study activities (it's also fun to find and use many Sit Spots).

- Prepare to spend at least 30 minutes outside discovering your Sit Spot and making observations. Wear weather appropriate clothes and footwear for traversing the terrain. Anticipate various conditions, bring water, sunscreen, bug spray, rain gear, hat, gloves, or an extra clothing layer. Also, consider bringing a whistle if you are in a secluded area. Bring something to sit on. Have a pencil, eraser, and paper. **Review the rest of the activities in this section *before* heading out, so that you know what to be thinking about and to make the most of your time.** Let somebody know if you will be away from home, where you will be, and what time you expect to be back.
- Plan where you want to go. Consider a nearby natural area. Leave your dog at home so that you can fully experience *your* time in nature. Your area to visit could simply be right outside your door and has been designated as your Nature Area. It does not have to be "in the woods" or far from civilization.

If possible, venture away from human activity, or go outdoors during hours when it is not busy. The ideal times to avoid people *and* to increase chances of seeing animals are early morning or late evening. Always be **smart and safe!** Be respectful of private property. Be respectful of nature—tearing or pulling plants or branches is not your purpose, and be wary around potential animal homes (adult animals, especially parents can be very territorial or aggressive). Lastly, but importantly, leave no trace within the natural landscape you spent time in.

- From the times when you ventured outside to sit and enjoy nature, you hopefully discovered one place you can adopt as your Sit Spot. The ideal **Sit Spot** should be appealing to your senses, is convenient for you to get to on a regular basis, has diverse vegetation (a variety of plants and trees), and has the potential for wildlife visits. The presence of water will also increase your chances of seeing wildlife. **Choose a Sit Spot where you can walk 100 paces in four directions – this is your Nature Area.** If this ideal "sit"uation does not exist nearby, simply get outside and find a comfortable place to sit. Keep the tips previously mentioned in mind.
- You will become totally immersed in your surroundings each time you are in this area—either sitting or walking. Make yourself comfortable. If you visit your Sit Spot with a friend, clarify that this is a quiet, reflective time for truly understanding this place in time. They will hopefully appreciate the experience as well! Ideally, you would make this *your* time to get the most out of it.
- **Take pictures.** This will enhance your experience outdoors and will help you see detail for species identification or to interpret the landscape. When taking pictures of objects up-close, use a commonly known object or a finger in the photograph to help the viewer determine the relative size of the item. Also, consider the best lighting on your subject and avoid shadows. Early morning and evening offer good back lighting for photographs.

Sit Spot Activities. By completing the activities described in this section each time you visit your Sit Spot, you will develop the habit of looking at the landscape more closely. Writing the information in your Nature Journal will allow you to reflect on and notice trends in nature.

1. **The Basics.** Always walk slowly and enjoy your Nature Stroll to your Sit Spot. Once settled, do a Sense Meditation to help hone your senses (see the notes in "The Simple Naturalist"). At your Sit Spot, silently observe the happenings around you (at least 15 minutes).

As described in the following sections of the Sit Spot Routine, while seated, you can also record your streams of consciousness, current conditions, observations, generate questions, make a sketch of something of interest, and then reflect on the experience. Alternatively, try recollecting all of these details *after* your time spent at your site. Try both approaches. You might be amazed at how much you recall. This can also help you realize what was truly salient about your experience. On the other hand, you might find it enjoyable to complete the entire exercise while relaxed at your Sit Spot.

Details of what you can include in the **Sit Spot Routine** are as follows:

2. **Stream of Consciousness.** Do this practice each time you go to your Sit Spot or just prior to heading outside. Sit. Get comfortable. Relax. Take a deep breath. Sit here for at least 10 minutes before doing anything else. Now, just write. Let your streams of consciousness flow out of you. Sometimes you discover being in nature brings more clarity to your mind. The first stage to making good initial nature observations is to not think. Do not pause to think about anything you are writing, just allow your thoughts to come out—whatever they may be. *Fill at least half of a page* in your Nature Journal or elsewhere. This can be a very cathartic experience. More than likely your thoughts, feelings, or opinions about whatever is on your mind will surface or what is happening around you may be the topic of your entry. This exercise will clear your thoughts that are currently on your mind or are deeply rooted. After writing your stream of consciousness thoughts, you can focus more fully on the moments you discover in nature.

3. **Current Conditions.** Each time spent at your Sit Spot or another area outdoors, focus on the current conditions and write them in your Nature Journal. The conditions include the date, time, location, atmospheric occurrences, description of the land, and water sources. Thinking about these nonliving, or **abiotic** factors, helps us think about what could affect the living things (**biotic** factors) in the area. Provide enough detail to help you support any claims and develop conclusions. If it is your first time to your Sit Spot, complete the entire Current Conditions Chart in this chapter.

   Cloud Types. To help you describe and interpret atmospheric conditions, consider the cloud types you observe and their location in the sky. Spend time cloud watching and try to classify them using this diagram.

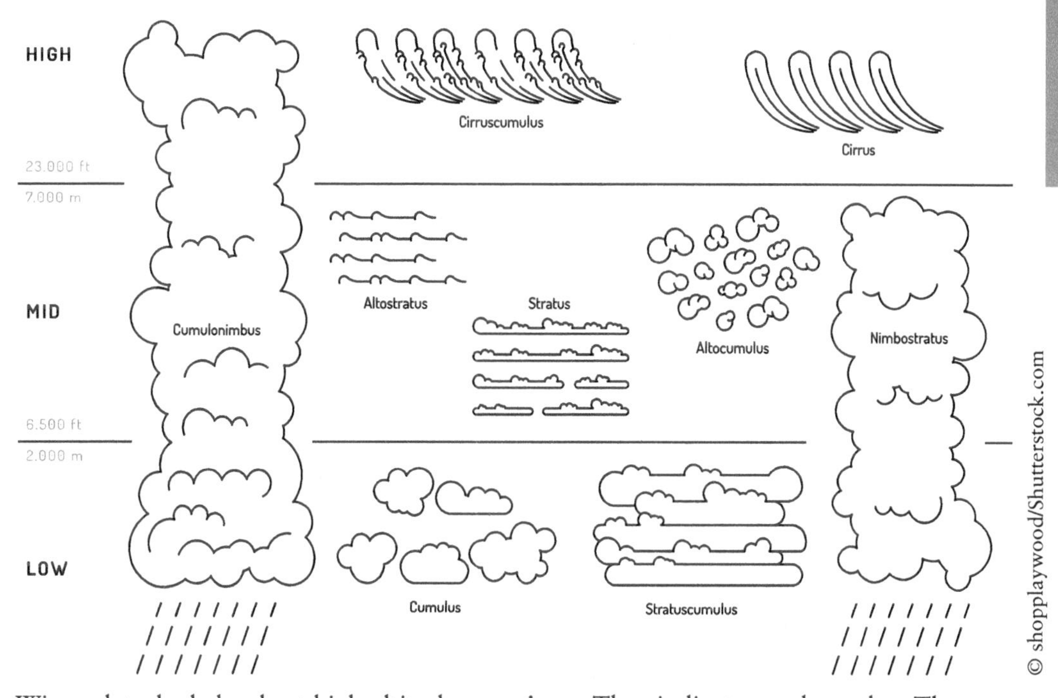

Wispy, detached clouds at high altitudes are **cirrus**. They indicate good weather. The **cumulus** clouds are large and fluffy with a flattened base and often indicate fair weather. **Alto** clouds give a rippling appearance to the sky when many of these rounded types are found mid altitude. If you note altocumulus in the morning, it may storm in the afternoon. **Stratus** clouds are flat, hazy, and found at low-mid altitudes and can indicate light precipitation. Stratocumulus are denser and light rain or snow may ensue. **Nimbus**, or darkened clouds indicate potential precipitation. A towering flat-topped **cumulonimbus** cloud or "thunderhead" means strong winds and possible heavy rain and hail possible. Start tracking what clouds you observe associated with the weather. What percentage of the sky above you is covered by clouds? Do animals behave differently during different cloud cover? If you want to figure the time until an approaching storm reaches you, count the number of seconds between seeing a flash of lightening and hearing the thunder. The storm is 1 mile away for every 5 seconds counted. In other words, take your number of seconds and divide by 5.

Wind Speed. What about the wind? From which direction is it blowing? Do you think it has an effect on the current animal behavior? Based on your observational skills, use the Beaufort Wind Scale chart pictured below to make predictions about wind speed, wave height, and the water and land conditions.

BEAUFORT WIND SCALE

Beaufort Number	Description	Wind speed	Wave height	Sea conditions	Land conditions
0	Calm	< 1 knot < 1 mph < 2 km/h	0 ft 0 m	Sea like a mirror	Smoke rises vertically
1	Light air	1–3 knots 1–3 mph 2–5 km/h	0–1 ft 0–0.3 m	Ripples	Direction shown by smoke drift
2	Light breeze	4–6 knots 4–7 mph 6–11 km/h	1–2 ft 0.3–0.6 m	Small wavelets	Wind felt on face
3	Gentle breeze	7–10 knots 8–12 mph 12–19 km/h	2–4 ft 0.6–1.2 m	Large wavelets	Leaves and small twigs in constant motion
4	Moderate breeze	11–16 knots 13–18 mph 20–28 km/h	3.5–6 ft 1–2 m	Small waves	Raises dust and loose paper
5	Fresh breeze	17–21 knots 19–24 mph 29–38 km/h	6–10 ft 2–3 m	Moderate waves	Small trees and leafs begin to sway
6	Strong breeze	22–27 knots 25–31 mph 39–49 km/h	9–13 ft 3–4 m	Large waves	Large branches in motion
7	High wind, moderate gale, near gale	28–33 knots 32–38 mph 50–61 km/h	13–19 ft 4–5.5 m	Sea heaps up	Whole trees in motion
8	Gale, fresh gale	34–40 knots 39–46 mph 62–74 km/h	18–25 ft 5.5–7.5 m	Moderately high waves	Twigs break off trees
9	Strong/severe gale	41–47 knots 47–54 mph 75–88 km/h	23–32 ft 7–10 m	High waves	Slight structural damage
10	Storm, whole gale	48–55 knots 55–63 mph 89–102 km/h	29–41 ft 9–12.5 m	Very high waves	Trees uprooted, considerable structural damage
11	Violent storm	56–63 knots 64–72 mph 103–117 km/h	37–52 ft 11.5–16 m	Exceptionally high waves	Widespread damage
12	Hurricane force	≥ 64 knots ≥ 73 mph ≥ 118 km/h	≥ 46 ft ≥ 14 m	Exceptionally high waves, sea is completely white	Devastation

Current Conditions Chart. Complete this template for the current conditions each time you visit your Sit Spot or Nature Area. After repeating this chart a few times, you will start to automatically tune in to these abiotic factors around you when spending time outside. If you are revisiting a place, you only need to record conditions in A-F that have changed since your last visit.

CURRENT CONDITIONS

A. **Today's Full Date:** ____________________

B. **Boundaries of Location** (what defines the corners or edges of the Nature Area?):

____

- City, State: ____________________
  Directions or landmarks for finding location: ____________________
  ____
- What are the natural and built environments within the Nature Area? ______
  ____
- How do people use the area? Think in terms of an individual's use and how natural resources are used: ____________________
  ____

C. **Duration of Visit:**

- Arrival Time: _______a.m./p.m.
- Sun Position (direction and relative height in sky—low, medium, high):
  _______ _______
- Departure Time: _____a.m./p.m.
- Sun Position (direction and relative height in sky—low, medium, high):
  _______ _______

D. **Atmospheric Conditions:**

- Cloud Cover (% of sky covered) ____________________
- Canopy Coverage (if applicable; % of sky above covered by trees): ________
- Cloud Description: ____________________
- Temperature (°F/°C): ____________________
- Relative Humidity (% or low, medium, high): ____________________
- Wind Direction (from north, east, south, west?): ____________________

CURRENT CONDITIONS (continued)

- Wind Speed (~mph; kph): ____________________
- Last Rain/Snow Event: ____________________

E. **Geosphere:**

- Topography (e.g., hilly, flat, varied, etc.): ____________________
- Soil Type (e.g., sandy, clay, loam, organic rich): ____________________
- Ground Surface Description: ____________________

F. **Hydrosphere (water sources):**

- List nearest sources (dew, groundwater, within plants, in air, puddles, stream, river, lake): ____________________
- Location of water: ____________________
- Snow depth and description (if applicable): ____________________
- Temperature (if applicable): _____ °F/°C
- Description of water (if applicable; level of turbidity, pH, etc.): ____________________

____

4. **Observations.** Focus on the natural surroundings as much as possible. Limit observations about the built environment or humans, unless you think they are having an obvious impact on the natural environment (positive or negative effects). After some time, provide descriptions of three words or more in your Nature Journal of what you observed at your Sit Spot (words like a, the, and, or it do not count as part of the three words). Use adjectives and adverbs to create accurate accounts of the observations. Provide your narrative for as many things as you can with the ultimate goal of at least 10 observations on this and future visits.

5. **Questions.** In your Nature Journal, create questions based on what you observed. Simply write queries about what you find intriguing or curious. What do you wonder? At this time, do not focus on questions requiring answers based on evidence. See tips on how to improve making observations in the previous section, "Discovering Nature." Try choosing one of your observations of a natural object, plant, or animal. Now, use the questioning prompts of who, what, where, when, how, and why to think about your item

more critically and from a naturalist's perspective. These questions can also form the foundation for an actual field study.

6. **Sketch.** From your observations, choose something in nature that interests you to draw in your Nature Journal. Get up close to the natural object to see and record details. If sketching an animal, do a blind contour initially then redraw it so that its characteristics are obvious. Further instructions on making drawings in the field can be found in the "The Simple Naturalist" section.

7. **Reflection.** Processing the experience can help you go deeper into your nature study. Use the prompts to help you reflect on your visit. How did you feel before, during, and after your time outdoors? What caught your attention, or what did you find most interesting during your time outdoors? What sense did you use most often (seeing, hearing, smell, or sense of touch)? How could you improve your observations in the future? What do you want to focus on next time you visit your Sit Spot?

8. **The Nature Area and Mapping.** By designating the Nature Area around your Sit Spot, you will hone your naturalist knowledge and skills by focusing your attention on one specific area. The best way to define this area is to create a map. The data you collect in this area will help you track the changes through the seasons and over the years. This is a study in phenology. Chart where and when certain observations of species or bloom times exist. Plot where you find different types of vegetation or habitats to help with interpreting areas where you could see animals. Note places where you have seen animal activity. Regardless of how you use the map, you need to start by actually drawing it! A "simple map" is usually done first and displays only the obvious features of the area. The "detailed map" will involve more naturalizing to complete a visual story of a given place. With either map, draw it from a bird's-eye view or as if looking at the area as if you were viewing it from above.

   - *Simple Map.* Create a quick diagram of where you spent your time. Provide significant landmarks (natural and built) so you or another person could recognize the area when visited again. Draw simple shapes and symbols to communicate what exists in your surroundings. Include where you made salient observations so they may be potentially observed again. For clarity, label items you draw or provide a legend. Where is North?

   - *Detailed Map.* To designate a *formal* Nature Area where you would make focused observations, allow at least 30 minutes. Read further to learn how to define the boundaries and what details to include. After defining the area, you can conduct a Field Inventory for observations of interest (see the next section).

The detailed map of your Nature Area will give you the opportunity to focus your attention and really become intimate with one place.

- *Finding the Boundaries.* From your Sit Spot, walk **100 steps heading north.** If possible, you will **do this in all four directions**, north, east, south, west—or whatever distance or direction makes sense for your Nature Area. Before setting out in each direction, **pick a landmark to walk toward (e.g., prominent tree, boulder, stump, building, etc.)**. **If you are restricted in one direction** due to an impassable body of water or a building or barricade, do your best to estimate out the 100 steps (250 ft.; 76 m) to make distant observations, make the barricade the endpoint, or make more observations in another direction. **In your Nature Journal, start creating a map of the area by drawing the landmarks at the far points of the Nature Area in each direction.** Note the distance you traveled to each landmark. The boundaries should be easy for anyone to identify at a glance (especially you!).
- *Adding Details.* Here you will complete the base map for your Nature Area to be used in all subsequent visits. Keep it simple yet include all relevant features. You want it easy to remember and recreate. Taking a picture of the end product or photocopying it could also offer efficiency for future use. However, by drawing it each time can help you focus on any changes that may occur over time.

  After your boundaries and landmarks are determined: (a) **add prominent landscape features and obvious and permanent structures to your map** (use symbols to designate habitat types, waterways, hills, boulders, footpaths, roads, picnic table, buildings, etc.); (b) provide **labels for what you draw and use a legend** to help key out the symbols used; (c) provide **a scale** for the map so the approximate size of the area is recorded; **and** (d) always provide where **north** is oriented (typically, maps are drawn so that north is positioned toward the top of the page).

The Field Inventory

This is a chance to really get to know your Nature Area. If you have the time after creating your detailed map as described above, continue with an inventory of your observations, otherwise create a separate Current Conditions Chart for another time you visit your Sit Spot. The Field Inventory will approximately take at least 30 minutes to complete. Photographs from the experience are encouraged. Regardless, the following description is how you could go about the survey of the area.

- *Notations.* From your Sit Spot, start walking slowly toward one of your designated landmarks. On a fresh copy of your Simple or Detailed Map, you will plot where you make noteworthy nature observations (use all your senses when appropriate). First, mark *where* you made the observations on the map by using the number 1. On a different sheet of paper in your Nature Journal, record that number 1 and note what caught your attention (e.g., 1—wind; this notation could also be used on the map as long as it does not clutter the map making it hard to interpret). Later, you will elaborate on that observation to include a narrative of at least three words (e.g., 1-wind: strong westerly wind began). See previous tips regarding improving your observation skills. Continue your Nature Stroll toward that first landmark and record at least four more observations—mark the map with the appropriate numbers and write a corresponding note on what was observed (e.g., 2—muck, 3—red bird, etc.). Spend time elaborating on descriptions that may help you with an interpretation or identification later on. All observations should have a description that includes at least an object, adjective, or verb and adverb when relevant.

- *20 Observations.* You will walk in the other directions toward each landmark or walk toward your Sit Spot—depending on which way you are travelling. Continue your numbering so that you have five observations in each direction. At the end of your visit, you will have numbered at least 1–20 observations on your base map. Each of those numbers on the map will correspond to 20 observations written in your Nature Journal.
- *Reflect.* After completing your inventory observations, write about how you felt about the experience. What were the most interesting observations you made? Based on the observations made today, what types of things are you the most curious about or wonder?
- *Name It.* After your first field inventory, name your Nature Area. Think about what your special nature retreat looks like—your Sit Spot, its surrounding area, and the types of organisms that live there (or might visit the area). Come up with a unique, relevant, descriptive name to help personalize the area (e.g., Willow Creek—named for the large willow trees growing along a wetland creek). Write it down. Provide an explanation why you chose this name.
- *Revisit.* Conduct a field inventory the same date each month or better yet, twice per month. Or even better, aim for 1 or more times per week! Use a new Detailed Map each visit. If possible, print or make copies of your [unnumbered] map ahead of time so your journaling time is more efficient.

1.3 Reflecting on Chapter 1—Building Your Naturalist Skills

The content in Chapter 1 will prove useful throughout your nature study development.

In your Nature Journal, reflect on how you feel after reading through the chapter. Did you practice everything presented? What did you practice? What do you want to try in the future? Do you feel a sense of accomplishment? How well do you think you developed your naturalist skills? After completing the activities in the chapter and after studying the rest of the book, retake the "Nature Appreciation Survey" to see how your score may have changed since working on developing the fundamental skills of a naturalist. Discuss your results. In what areas do you still need improvement? What areas of nature are you most interested in learning right now?

What's Next? The following chapters delve into thinking more deeply about nature interpretations, discovering habitats of the region, and cover the fundamental ecology, basic botany and zoology, and natural histories of common flora and fauna. Additional naturalist activities will be included in each section to add to your personal Nature Journal. Spend the time building your knowledge and skills to further develop your naturalist intelligence. You may find yourself utilizing the entire book in one sitting as you discover new ways of observing and interpreting nature.

Review Questions

1. What is nature? Explain what a naturalist does.
2. Discuss how nature can benefit a person.
3. What types of questions will you try using the next time outdoors to help you improve your observational or identification skills?
4. Why do we have seasons?
5. What is a phenological study?
6. Define citizen, or community science, and give an example.
7. How does day and night occur?
8. Reflect on your results of the Nature Appreciation Survey.
9. Using the tips and practices from the chapter, create a detailed plan for how you will build your naturalist skills.
10. What is taxonomy? Name the 3 domains. From where do the 4 kingdoms arise? Name them. What is a scientific name? How is it written?
11. What will you focus on for going deeper into nature?
12. How many answers did you know in the Discovering Place section?
13. Describe your Sit Spot and Nature Area. Complete a Current Conditions chart. Use the cloud chart and Beaufort wind scale to interpret the day. Map your setting according to the directions.

2 Nature as Your "Place"

In Chapter 1, you learned how to go deeper with your observations of nature and ways to discover a sense of "place". Anywhere you go, there is a story of that "place". The more you know about a place's history, culture, and environment the more you might relate to, sympathize, and care for that particular place. To help you understand the places in nature you explore, a practical approach to viewing a landscape is provided throughout this book. The scaffolded activities and essential principles for nature study that are covered helps you critically examine any "place" you visit to make a myriad of connections.

Activity: *Reread the section in Chapter 1, "Interpreting the Landscape". Now, stand outside and look around you to answer those questions. You are reading the landscape of your "place". Your answers will reveal the shape of the land, its relationship to water, the plant and animal life around you, how it all might be connected, and humans' place in that landscape. How do you feel after becoming more acquainted with your "place"?*

By the end of Chapter 2, you will start to build your knowledge of nature fundamentals and strengthen your awareness of the surrounding landscape. The last section, "Scientific Thinking" will provide insight to the inquiry process and a collection of nature study applications. The landscape perspective you develop can lead to taking more responsibility toward protecting the areas in which you live, work, and play.

2.1 Making Connections—An Ecology Primer

As a naturalist, you spend a considerable amount of time thinking about the potential interactions and relationships that may exist in nature. With your naturalist intelligence, you also have developed ecological thinking. Wait. What is ecology? Read through the brief introduction that follows and visit Chapter 4 for further details.

Ecology Defined

While learning to build your naturalist skills, did you realize you had begun to think about ecological relationships? **Ecology** is the study of interactions between living things, and the interactions the living things may have within their environment. The environment includes the living (biotic) and nonliving (abiotic) factors. The biotic factors around you could include plants, animals, fungus, and even bacteria. Abiotic factors potentially influencing the living things might include the weather, soil type, and water. When you start to think about how the biotic and abiotic things might interact with one another, then you are thinking about ecological relationships. Having this ability to think about connections can also help you think about potential consequences of our actions on the environment. Once we understand ecology, we may be able to start living more sustainably with the use of the Earth's natural resources.

An Ecological Perspective

Viewing nature through the lens of the four laws, or more widely referred to as the rules of ecology may help naturalists and environmentally conscious people move forward in their goals. These **rules of ecology** can help us think about the functioning and interconnections that exist in an ecosystem. The perspectives are relevant and bring meaning to your nature study. The rules of nature on which this book focuses are: (1) everything is connected; (2) there is no away in nature; (3) nature always wins in the end; and (4) there is no such thing as a free lunch.

1. In the first rule, ***everything is connected to everything else*** refers to the direct or indirect effects organisms have on one another and their surroundings. This rule often pertains to food web dynamics, or who eats who in nature. The populations of living things fluctuate over time because of this. Another common connection is when something happens to a habitat—a tree falls over, the presence of new vegetation growth, a flood, or human disturbance can impact all who live in the area. Human interactions with nature can have a positive or negative effect in the area. The positive effect may help stabilize or balance

population sizes when considering feeding relationships. Along with natural animal predation, hunting laws and limits, and land management will help stabilize animal and plant populations so that none of them overruns the habitat in which they live. However, if a part of a natural ecosystem is stressed, it can lead to other problems. For example, the burning of fossil fuels (i.e., cars, factories, etc.) is overloading the global carbon cycle and is exacerbated by the elimination of trees, which is a major player in the uptake of carbon. This issue triggers changes to climate, weather, ice cover, ocean acidification, sea levels, and crop yields to name a few. All of these issues create changes in the dynamics and balance of plants, animals, and society.

2. The second rule—***there is no away in nature***, or that everything must go somewhere—refers to the realization that the byproducts or disposal of things no longer wanted will need to go someplace in the environment. Harmful materials or amounts of material that accumulate in nature can cause issues where it is or once it is extracted or consumed. Examples are endless but include nonbiodegradable products, hazardous waste, and factory discharges. These things take different forms and may accumulate in detrimental quantities within, on, or around the land, water, and air in which we depend.
3. In the third rule of ecology—***nature always wins in the end***, or nature knows best—is a reminder that no matter what humans try to do to change the natural landscape, nature will find a way to prevail. Consider the homes built on a floodplain. The rising river after a heavy rain or snowmelt may eventually flood the home. This also occurs if you build along a bluff or cliff's edge. Erosion will find a way to destroy the foundation of the building. Learning to work with nature and understand its forces can prevent disastrous situations.
4. The fourth rule focuses on there is ***no such thing as a free lunch***—or for every gain, there is a loss. This ecological rule exemplifies the previous three rules. The global ecosystem functions as a connected whole. Anything altered within or extracted from nature will eventually have an effect on the system, and nothing gained or lost is not subject to an expense on the functioning of something within the surroundings. The ecosystem will respond or change for every action taken within it. In areas untouched or unaffected by humans, nature functions in a healthy and balanced way (see Chapter 4 for more details on biogeochemical cycles). Animals expend energy when acquiring and consuming food. Plant and animal populations shift in response to predation. Nature functions differently when humans alter the landscape for personal gain—eliminating trees, damming rivers, mining, etc. A system not operating efficiently and effectively

signifies a warning that humans have likely altered nature too much. Think about the possible consequences of any action taking place in an ecosystem. The best approach is to reduce personal impact on the environment and to also help others in your community minimize their impact.

Ecological Knowledge

Understanding ecological relationships and the rules of ecology may help *improve our environment*. If we can identify what could cause pollution or introduce non-native species, we could prevent it from happening. We could also recognize when nature is not balanced or functioning optimally. Lakes overgrown with algae or a landscape covered with one plant species could be indicators. These concerns may affect water we depend on for drinking, recreation, or the overgrown area could now harbor a vector for disease. Concerns about the environment leads to potential issues of *public health*. Nature provides natural services with examples like flood control by wetlands, plant attributes that contribute to medicines, and predicting a Lyme disease increase due to high acorn yield. Understanding what nature can do for people could help prevent the destruction of these areas, which ultimately helps protect us. This understanding also serves as the foundation for decisions made in *natural resource management*. Scientific and ecological principles are used to inform sustainable practices and solutions in forestry, agriculture, fishing, and managing wildlife.

Ecology's Scale of Life

When considering a landscape and the relationships in nature that may form, also consider the scales of life within that area to help you understand how the rules of nature may develop. The scale's simplest level is the individual and expands to the individual's population, community, ecosystem, and biosphere. See Chapter 4 for further details of these concepts, and in the meantime, begin immediately developing your observational skills for the first level of this scale.

Start with the simplest level of life, an individual. This one organism—a bird, frog, ant, tree, etc.—has certain features, or **adaptations** that help it get along in its environment. It has behaviors like mating rituals or reproductive habits, or sleep patterns that aid in its survival. Physiological functions like its respiration may also tell us something about the organism. Taking the time to observe one organism in nature could lead to a deeper understanding of the place you are spending time. Chapter 1 discussed many activities revolving around this concept of making good, meaningful observations.

Nature Journal: While visiting your Sit Spot, Nature Area, or just going for a Nature Stroll, which animals or plants were the easiest to observe and study? Make as many observations as you can about that *individual* animal or plant—the single organism. What questions come to mind about this animal or plant? Search this book and other resources for information about this organism. Use the Natural History Journal Pages template found in the corresponding section for your organism. Revisit the exercises in Chapter 1 to develop a stronger connection to the place in nature where you discovered this animal or plant.

Widen your lens on the Scales of Life (see Chapter 4 for more details). Now, consider how your chosen individual organism potentially interacts with the rest of its species in the area. Try to describe these interactions with just its own kind—or within its population. If you are studying the green frog, then how does it behave around other green frogs? How does the green frog fit into its community? Meaning, how does it potentially interact with other species? Consider what the green frog eats and what could eat it. On the ecosystem level of the scale, you start to put your organism in a larger context and think about how the elements or abiotic factors may also affect the organism's lifestyle. The weather, soil type, water quality, etc. will influence how that individual behaves. On the biosphere level, you are considering the presence of, impacts on, and roles of all life on a global scale.

Learning to consider the broader perspective in terms of influences on organisms can seem overwhelming. Once you start training your observational skills, thought process, and interpretations to make connections, then focusing in and out on different scales of life can become fun and exciting. A whole new world will reveal itself. You may start to perceive a landscape like never before. How do you think your Nature Area became what it looks like today?

2.2 Landscape Shapers—Glaciers, Climate, Winter, Fire, Humans

To think like a naturalist and to get to know your place in nature, you will also need to think about what has shaped the landscape in your Nature Area. The five landscape shapers focused on here include a brief overview of glaciers, climate, winter, fire, and humans. Regardless of where you are, it is likely that at least three of these factors have altered the area. In the Great Lakes region, these five things have physically transformed nature through time. As you learn about each landscape shaper, determine if and how it has impacted your Nature Area.

Glaciers

The Great Lakes basin and Upper Midwest was glaciated about 20,000 to 12,000 years ago and became ice-free about 9,000 years ago. Geologists compare modern or recently glaciated areas to those whose landscape was glaciated thousands of years ago.

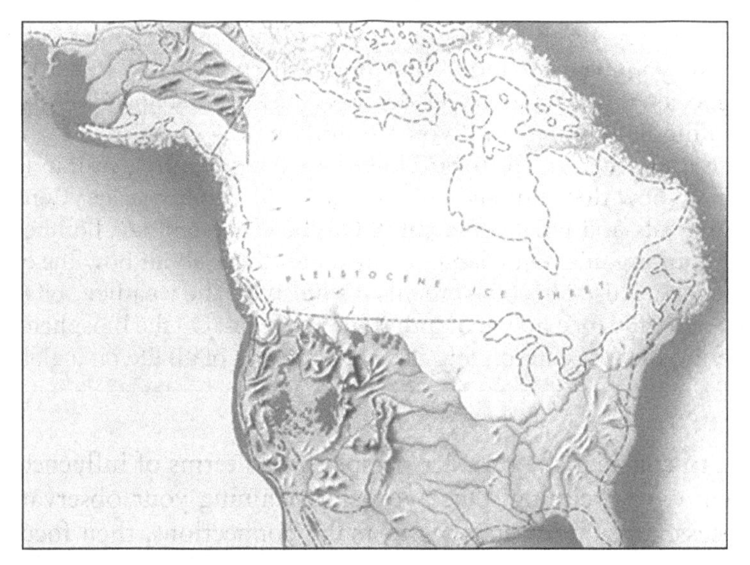

Source: Schlee, J. (USGS).

During the last glaciation in the Pleistocene Epoch, the Laurentide ice sheet covered the northern part of North America.

The enormous weight of the mile-high glaciers once covering the Great Lakes region put stress on the landscape. The glaciers bulldozed mountains and gouged the Earth to cut steep slopes and u-shaped valleys as it moved. River valleys and its systems are a result of this action. As the glaciers melted, or retreated northward, it left other evidence of its existence. Torrents of its meltwater held mixed sediment and rock that would drop out of its stream. The resulting hills or **moraines** consist of glacial **till** or a mix of gravel. For example, most areas of Michigan's Lower Peninsula have its bedrock or foundational parent rock of limestone covered with mile-deep till. Ice had also settled and melted over the entire landscape to create a lake-splattered scene. The presence of random boulders or glacial **erratics** also indicates glaciation. These large rocks will be the first to settle out of the meltwaters. In some areas of Michigan, especially in the western Upper Peninsula, at Porcupine Mountains State Park, or in Ohio on Kelley's Island we see evidence of glacial grooves on solid, bare rock that show the general direction the glaciers retreated. Eventually, the ebb and flow of the glacier allowed life to take hold. As the last ice mass retreated north, plants and animals from unglaciated areas were free to migrate into newly exposed areas.

Climate

Life is a product of the climate and depends upon the climate. This is not to be confused with the weather. The **weather** is what is happening right here, right now. The weather includes temperature, atmospheric pressure, humidity, precipitation, sunshine, cloud cover, wind direction, wind speed, etc. and are short-term properties at a particular place and time. **Climate** is a region's general pattern of atmospheric or weather conditions over a long period. It has been determined by the average precipitation and average temperature of the area, which is greatly influenced by the latitude, altitude, and ocean currents. Knowing about a region's climate will help the naturalist understand about its effects on where organisms live, how they live, and what they eat. More discussion on climate can be found in Chapter 4.

Sun Angle. The sun's angle is also significant for understanding an area's climate. Depending on where on the globe you are will determine the intensity of the sun's rays. At the equator there will be direct sunlight making it warmer than other areas on the planet. As you move north and south of the equator you have less direct, or oblique sun rays hitting the planet for less intense sunlight at those points making regions progressively cooler moving north and south from the equator. The average relative temperatures over many years in these regions stay somewhat consistent over the years but may fluctuate seasonally.

Biomes. Major communities of flora and fauna that occupy a large, given area is considered a **biome.** Biomes change as you go from the equator to the poles in a somewhat predictable pattern. This **latitudinal zonation** would have a tropical rainforest biome at the equator and as you move northward you will then encounter predominately broadleaf, deciduous forests, then needleleaf or coniferous forests, a tundra biome, and finally a barren ice and snow landscape at the poles. This pattern is often also observed as you climb in altitude. The **altitude (vertical) zonation** in a mountainous region is where you could have a tropical rainforest at the base and the biomes progress similarly to that of latitudinal zonation as you climb.

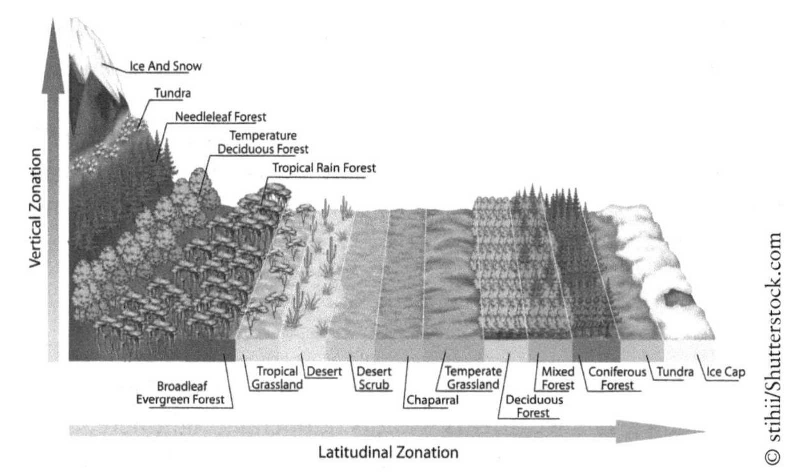

Biomes change with latitude and altitude due to physical properties of the air. Have you ever experienced vertical (altitude) zonation as depicted in the figure? What biomes have you visited and where was it in terms of its latitude?

Microclimate. Found anywhere you go, microclimates are fun places for naturalists to discover and ponder. **Microclimates** are conditions at specific localized areas. You can find diverse life forms in unexpected places that have different conditions than the surrounding environment such as with varying temperatures, soil types, moisture, or sun and wind exposure than the surrounding area. Often this is due to a varied landscape or topography. Microclimates also exist in the built environment, such as around buildings and other hardscapes. In natural hilly landscapes, you will find microclimates around and on that hill. South-facing sides of a building or slopes of a hill (also called **aspects**) will receive the most solar energy. The naturalist can predict warm, dry conditions in those areas compared with its opposite side or the northern aspect where the area does not receive as much sunlight. Considering the microclimates in an area will help the naturalist discover species or experiences sought. If you know a plant grows in cool, shaded areas you might search on the northern aspect before the other side of the hill. In the colder months, you may want to consider opening up your south-facing windows to let the sun warm your home. Better yet, in the winter, find a microclimate outdoors in the warm sunshine—remember your sunscreen!

Winter

In the Great Lakes basin and Upper Midwest region, winter can be harsh and considered the ultimate test for all who live in the area. Some places during the winter can experience temperatures below zero degrees Fahrenheit (or −18 degrees Celsius), sometimes for extended periods. Inland lakes often freeze this region, allowing for great ice fishing, while very small lakes can freeze to the extent of sandwiching some of its animals. The snowiest areas can have snow accumulation beyond 25 ft. (7.6 m) deep.

A car covered with snow. When have you been affected by the snow?

Snow cover can make it difficult for animals to find food in the snow, but it can also give some animals an advantage. On a snowpack, high branches become reachable food sources for snowshoe hare (*Lepus americanus*) or white-tailed deer (*Odocoileus virginianus*). Snow is also considered a safe place for small mammals. The insulating property of snow and the ground allows the snow to melt slightly where they join. At the bottom of a snowpack, a haven can exist for mice, voles, moles, and shrews. They can tunnel throughout this area where it is relatively warm, rarely reaching below 20 °F (6.7°C) and they will have reliable sources of water. The small mammals have room to roam with no wind, good food supply, and very few predators (except on the occasion where a weasel or an owl uses their keen senses to prey upon them). The snow tunneling creatures sometimes surface for more food when weather permits but become susceptible to more predation on the surface.

Another hazard in terms of winter and wildlife is when the snow is deep, we get a thaw and then it refreezes. This creates a hardened crust at the top of the snowpack and becomes hazardous for long-legged wildlife, like those in the deer family. It takes more energy to travel through these conditions and if they fall through, the animal could injure a leg. When the weather is this adverse, animals do not travel far. They will find places where trails are maintained by humans or trails where they created a well-worn pathway. Typically, coniferous forests (among the evergreens) are ideal because the snow is not deep. There are minimal winds due to the windbreak that the conifers create, and the canopy of these trees will trap some heat. This example of a microclimate can offer a refuge from the cold temperatures and wind.

A moose (*Alces alces*) in a snow-covered forest.

Fire

Another landscape shaper is fire. Fire is a naturally occurring force in nature. Humans often suppress fire due to its proximity to where we live. Fire can be started by people, but usually begins with lightning. If an area is dry, then it will burn fast. During dry seasons, fire becomes difficult to combat. When considering fire in nature, we should recognize the incredible job it serves. Fire releases nutrients from the organic debris on the ground almost immediately. Decomposition takes several years to break down materials in the soil to release the essential nutrients for life. **Nutrients** are the essential food or chemicals needed by organisms for life and growth. Examples of nutrients are carbohydrates, fat, protein, and minerals (i.e., nitrogen, phosphorus, magnesium, calcium, potassium, sodium, chloride, etc.).

Green vegetation is burned off from a fire—the understory first and then the canopy if the fire intensifies. After the fire, rootlets can easily take hold and penetrate the mineral-rich soil to reach water and newly released nutrients.

This allows for a relatively quick reset or recovery time. The fire scar usually pulsates with life again a year later.

Certain plants and animals depend on fire. For example, the jack pine (*Pinus banksiana*) requires fire for the release of its seeds held within its tightly closed cones. The Kirtland's warbler (*Dendroica kirtlalndii*), a threatened migratory bird will only nest in jack pines of a certain age in Michigan. Jack pine forests are managed in Michigan with controlled burns and harvesting to allow for this tree and bird to survive. The natural propagation of these trees could occur if we did not suppress fires or eliminate large stands of trees.

In the absence of fire, a certain type of forest will persist. Examples could include a forest with shade-loving species like white spruce (*Picea glauca*), balsam fir (*Abies balsamea*), and sugar maple (*Acer saccharum*) that grow beneath pioneer species like aspen, birch, and cherry. This arrangement can persist until something causes a tall tree to topple over and those shade lovers can flourish in the sunlight and nurture their seedlings. Professional management of prairies and forests through controlled burns can help with the maturing of the area and the elimination of unwanted nonnative and invasive species.

Humans

When nature cannot function properly, humans are typically to blame. People have the tendency to have a negative impact on our environment. Some major examples include that from 500 years ago to the present, a slash-and-burn agriculture mentality exists. Land is cleared to farm or to build upon. This was also observed with the reaping of white pines from Michigan in the 1800s during the big lumbering boom.

In recent times, wetland draining occurs for development of built environments. Regulations apply in some areas, but this varies based on place or political jurisdiction. This does not bode well for the ecological services that wetlands provide (see Chapter 3).

The most obvious, and perhaps the one thing we have control over is pollution. We pollute the air with our gas-powered vehicles and the factories that manufacture products we buy. We pollute our waterways with the runoff around our homes and from farming areas. Minimizing these impacts can not only improve air and water quality, but can also slow global warming and climate change (see Chapter 4 on biogeochemical cycles). Nature is having difficulty keeping up with these taxing overloads of unnatural occurrences (see Chapter 4's section on biodiversity and conservation). We are making changes to nature at an accelerated rate and some ecosystems cannot recover quickly enough. Ecosystems may be lost entirely if we do not think before acting.

Car exhaust is a form of pollution.

Nature Journal: Reflect how each of these landscape shapers (glaciers, climate, winter, fire, and humans) have impacted your Nature Area and to what extent. Make interpretations from observations, support your ideas and discover more about local land use through researching online or at your local library.

2.3 The Watershed—A Broad, Yet Local Perspective

A watershed perspective makes sense. Unlike political boundaries on a map, sharp boundaries in nature are often blurred. However, boundaries in nature become apparent when viewing the landscape through a watershed lens. The land contours, flow of surface water, and habitat types often follow the natural watershed divides. Connections and relationships become more obvious. It might be too late to use watershed maps to designate political boundaries for counties, states, and countries but using them to designate property lines could promote unity.

What Is a Watershed?

A **watershed** is an area that surrounds and "sheds" water into a larger body of water. Other names may include "catchment area" or "drainage basin." We can make the interpretation of a landscape that the river or lake also extends onto land. Anything that falls onto the land within the boundaries marked by the highest ridgelines will drain into the respective water body—the lowest point in the surrounding landscape. These **watershed divides** serve as a marker for what constitutes each watershed. Things happening to the land and water within a watershed will affect what happens downstream. Take some time to study the major features of drainage basins (watersheds) below. Refer to the illustration of the regional map in the book's introduction or in Chapter 3.1 to see the watersheds you are in. Considering what you know about watersheds and making connections in nature, reflect on the following questions.

DRAINAGE BASINS

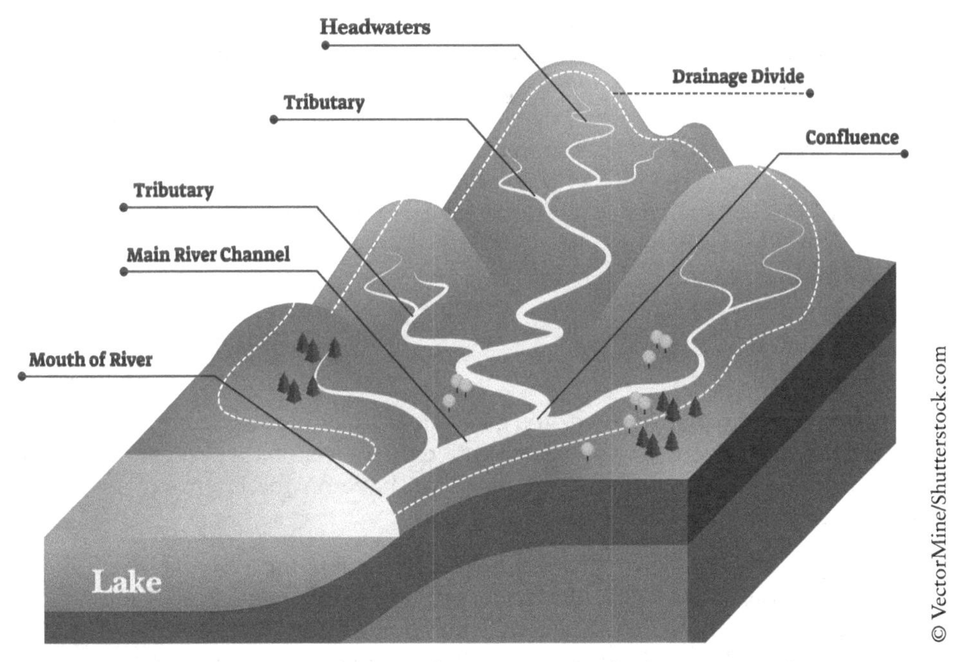

Anywhere you go, you are in a drainage basin, or watershed. You are either upstream or downstream from somewhere or someone. Learn more about terrestrial and aquatic ecosystems in Chapter 3.

> ***Nature Journal:*** What is the importance and relevance of watersheds? How can this knowledge be applied to our lives? How can having a watershed perspective be important as a citizen of the watershed? How can having a watershed perspective be useful in natural resource management?

Scales of Place

In Chapter 1, you began to think about your "place." You answered questions about what you knew of the social, environmental, and political aspects of where you live. In this chapter, you will begin to think about your "scales of place." As in the scales of ecology, the scales of place help you read and interpret a landscape while focusing in and out of the different levels. Later in the chapter, you will also think about how to investigate some of these places of interest.

Nature Journal: Where are you right now? Identify your current scales of place to help you interpret your location and the potential connections that could exist.

- Location (where you are right now)
- Subwatershed (think about the nearest small stream/tributary)
- City name
- County name
- Major river watershed (this would be the largest river near you in which the small stream you mentioned previously would eventually drain into)
- Great Lake watershed basin (Huron, Ontario, Michigan, Erie, Superior)
- State/province where you live
- Which ocean watershed are you a part?
- Country
- Hemisphere
- Planet Earth

By knowing these scales of place, you can start to view the landscape and the world through these various lenses and perspectives. Connections will seem more apparent and the positive or negative impacts you have on the environment becomes more obvious.

For further discovery, you will want to review a map and research the following information online or at the library to provide a thorough description of your watershed. Additionally, see Chapter 3 for more information on river characteristics. How would you describe the watershed in which you live? Discuss how the land is used throughout the watershed, from its origin to where it empties. Describe the watershed's social-ecological characteristics. Meaning, how are people and types of people and their job positions affected, and how do they respond or alter the landscape? See Chapter 1 for more details on the social–ecological concept. How are the plants and animals affected by the happenings in their watershed? Any information you can provide about your watershed will help you understand your place more deeply.

In this exercise, locate a **topographical map** online or from your library—sometimes referred to as a topo map for short. A topographical map shows **contour lines** indicating the characteristics and elevation of the relief across the landscape. Frequent reference to this type of map can help you navigate anywhere. The features illustrated help you gain your bearings. Lines close together indicate steeper terrain, closed circles are the high points, circles with dashes pointing inward are depressions, and waterways and roads are marked accordingly. Watching an instructional video on reading topographical maps can help solidify concepts. Also, watch a video on learning to use a compass. These are essential tools and skills of a naturalist for discovering a "place".

Contour lines show points that are on the same level.

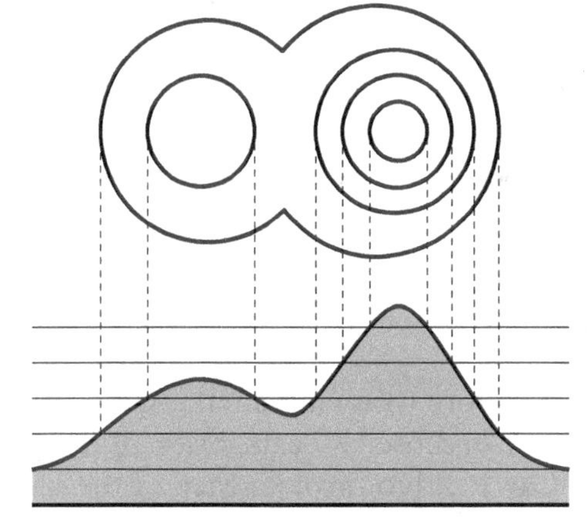

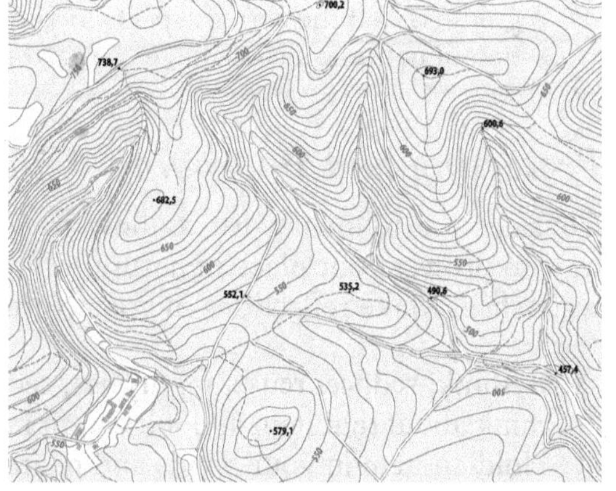

The first image shows the top-down and profile views of two hills of different heights. The second image is a detailed topographical map. Choose a location on this map. Describe the topography around it. In which direction would water travel across the landscape? Where is the high ground? Figure out the elevation.

Making It Real

Do what you can to make the topics in this book relevant, or "real" in your life. You started to make nature more tangible in Chapter 1. Starting and implementing naturalist practices daily will make the point of why we should care about the environment a little more obvious. Create meaningful experiences when outdoors and try to apply it to different aspects of what you do each day and the decisions you make.

Nature Journal: To help you glean the greatest benefits from your nature study, reflect on each of the prompts provided to reveal how you can apply the material you learn. Some of your answers may be easier for you to discuss than others on the list, yet you should strive to find connections to each element of your life.

- What do you want to learn from this book or your nature study?
- Apply the nature study material to your:
  - Schooling/learning objectives
  - Naturalist goals
  - Career/workplace
  - Family
  - Social life
  - Recreational time
  - Political views
- After completing some or all of your nature study activities from this practical guide, revisit the above aspects in your life. Describe if and how your application of nature concepts have changed or broadened.

Protecting and Improving Watershed Quality

Consider the following list as how to minimize your ecological footprint and how to protect your place in nature. All things listed here will help improve your watershed. Make reflections in your Nature Journal.

1. **Rethink, reduce, reuse, repurpose, recycle.** Before throwing anything "away," think about that item and if it can fit into one of those "R" categories. ***Activity:*** Hold an item in your hands and recite, "Can I … [go through each of the five R's listed above] this item?" More than likely, the answer will be "yes" to one of the R's. Follow through with the alternative to sending it to the landfill.

2. **Refuse! Minimize or eliminate plastic.** Avoid single-use plastic. Find alternatives. Before making purchases, ask yourself what the item is made from. Although plastic is versatile and extremely useful, as in medical supplies, it has been shown to cause detrimental impact to the environment. Most plastics will not decompose, but will break down into small particulate plastics, or **microplastics**. Microplastics have entered aquatic food webs (in the ocean, lakes, and rivers). Microplastics get ingested by invertebrate animals, fish, and birds causing the starvation of certain species. ***Activity:*** Research microplastics and its impact on the area nearest you. Inventory what you use on a daily basis made from plastic. Challenge yourself to make purchases that use less or no packaging and look for alternatives to what you typically would purchase or use that is plastic. What types of changes did you make? How did you feel about these alternatives?

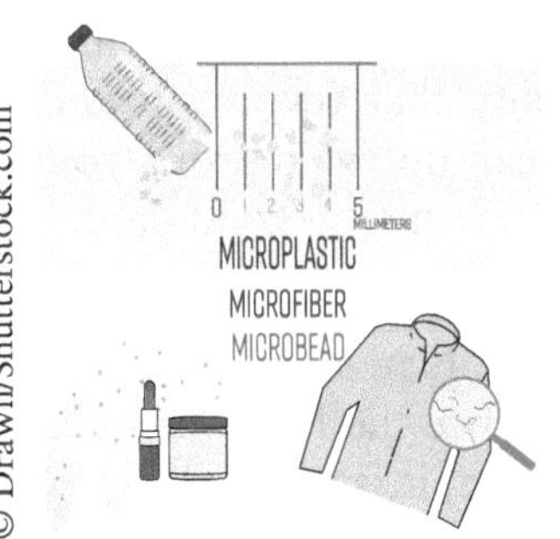

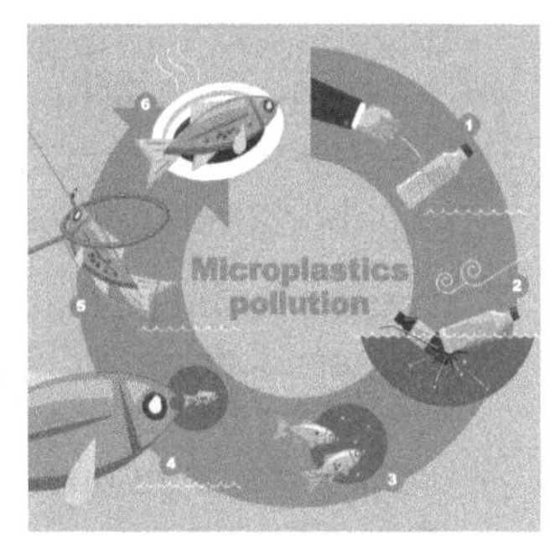

Common household items will only degrade to an extent. Microplastic, microfibers from clothing, and microbeads from cosmetics enter our waterways. This **micropollution** is found globally at varying levels but end up in our food chain and are potentially consumed by humans.

3. **Do I need it?** Overbuying, or not truly needing an item yet purchasing it anyway can perpetuate the production of nonbiodegradable items. ***Activity:*** Before buying something, consider these questions, "Do I actually need this item? Do I love it?" If the answer is no, move on—you can save your money and potentially a piece of nature. How did you feel about this practice after trying to implement it?

4. **Buy local.** Supporting local farms and small businesses will not only help your local economy, it will also help protect the environment. Once you begin to think about where food and manufactured goods originate, how it is grown or how it is made, and what it takes to bring it to you will illustrate the resources used and the potential impact it may have on the environment. ***Activity:*** Choose any item in your household that you buy frequently. What resources were used and affected in its production? Where did it originate? Who worked to make the product a reality for you? How far has it traveled to reach you? Research how much carbon dioxide was burned for it to get to you. Where could you buy the same or a similar product near where you live?

5. **Do not litter.** There is no away in nature. Trash that you dispose of improperly can choke or starve wildlife, potentially cause release of toxins into the environment, and looks terrible in any place. ***Activity:*** Simply, do not litter. Refer to number 1 above for alternative ideas to throwing it away in the trash (putting it in a trash can is still better than littering). Start taking inventory of trash and recycling receptacles in places you frequent. What would it take to get more receptacles in these locations?

6. **Pick up litter.** This will help with aesthetics and prevent any of the issues discussed previously. ***Activity:*** Pick up trash wherever you go to help clean and protect nature. Practice careful collecting. Dispose of the litter in trash or recycle bins.

Who knows, you might find some treasures along the way! Look for community science or stewardship opportunities in your community that revolve around litter clean-ups. You could also initiate a city-wide clean-up to improve your neighborhoods and natural places, and to work toward a common good deed.

7. **Native plantings.** Planting diverse trees, shrubs, wildflowers, and grasses native to your location will promote habitat health and balance. Native plantings will help clean our air and water, stabilize the soil, and enhance the biodiversity of ecosystems. If nonnative (invasive) species exist in an area, it is likely they will spread uncontrollably causing the displacement of the natives in which other organisms depend. You will read more about the importance of native plants in Chapters 2–4. ***Activity:*** Plant native plants! Plant natives around your home and participate in native plantings in your community. Buy your native plants in the spring from your local conservation district or search online for a nearby nursery.

Native prairie wildflowers will help stabilize the soil with their deep, perennial roots and will attract pollinators and native wildlife.

8. **Educate.** Share your knowledge and skills with family members and friends. Help them interpret the landscape and identify how humans impact nature. Discuss what each of us can do to help support nature's cycles and minimize practices that may harm the quality or sustainability of our natural resources.

9. **Become a member.** Research the types of environmental groups in your watershed. At the very least, start a membership with your local watershed council. These nonprofit organizations have the welfare of the people and natural resources at the forefront of their mission. You will learn ways you can help support initiatives near you that protect the land and water from degradation.

10. **Volunteer.** Through your watershed council membership, other nonprofit organization membership, or state/federal/provincial natural resource agency you can find opportunities to offer your time toward making a positive difference in nature near you. Similarly, you may have a local conservation district established under state law or a university extension that provides natural resource management programs through technical assistance and tools to manage and protect your resources locally. Investigate to learn what you can do to help.

2.4 Scientific Thinking

Naturalists understand their "place" and think scientifically. To think like a scientist, try acting like a kid! Children are born scientists. They are curious about their surroundings. They wonder what things are and how they work. Kids want to find out answers by eagerly observing or poking at it for a response. Children learn about the environment through exposure and experimentation. Strive toward getting [re]acquainted with this behavior. The goal of learning about the subject of science is not to make every person a scientist, but to ensure people are scientifically literate. To accomplish this, individuals need to maintain their understanding of the natural world. The best way to start (or continue!) this mindset is to remain curious and know where and how to search for answers.

Scientific thinking simply asks and answers scientific questions. These are based on observations and doing experiments. A scientific mind searches for cause and effect relationships in nature. Any judgment is unbiased. Claims will also not include feelings or beliefs. Three basic practices can help ensure good science takes place.

1. Pay attention to **empirical evidence**. This includes all the observable facts based on your senses—what you can see, hear, touch, and smell. This makes any evidence concrete and others can potentially experience it, therefore making the evidence repeatable.
2. Strive to accomplish **logical reasoning skills**. This does not come easily to people. You are not born with it. Logic and reasoning are studied in philosophy, math, and science courses and books. Logic is based on empirical evidence.
3. Learn to have a **skeptical attitude**. Constantly question conclusions and beliefs you or somebody else derives. Examine all evidence, arguments, and reasons. This also means you are open to new evidence and rational beliefs.

Inquiry Process

Many of us are familiar with the "scientific method." This has mostly been taught in a very linear process of making an observation, asking a question, doing background research, constructing a hypothesis, testing it through an experiment, analyzing the data, drawing conclusions, and accepting or rejecting your hypothesis. These steps offer a good guideline for doing science. However, we must acknowledge the dynamic process of scientific inquiry.

Scientific inquiry includes that traditional science process, but also combines the process with extensive scientific knowledge, critical thinking and scientific reasoning to develop new scientific knowledge. Tangible or empirical observations from repeated investigations are used to create logical explanations and to answer your research questions. Many factors that interconnect and follow a cyclical format exist for getting to this point. Scientists and naturalists will carry out this process throughout a lifetime.

The experimental design and data collection process for a scientist, or in our context, a naturalist tends to develop throughout the course of a study. To answer the research question, extensive and continuous planning and investigation need to occur. Throughout the observation and exploration phase, thorough details of the experiences need recording. Patterns and relationships will likely emerge from the investigations, and the researcher needs to reflect on and organize these occurrences. New explorations and approaches can result throughout the study, and the person should have an open mind to the possibilities. Once data is collected, an analysis of the results and discussion of conclusions can follow. Through sharing the information with others, new ideas, questions and methods of exploration can also emerge. This demonstrates a scientific method rooted in inquiry.

Realize that we use scientific inquiry every day without needing to think about the underlying science of it. You make observations all the time and that often leads to the process explained above. Think about times you thought: "why is it like that?," "how does this happen?," "why doesn't this work?," or "what is the best ...?"

By practicing the three scientific thinking skills and an inquiry process, you will be ready to present your results and conclusions. With this information, you want to at least provide your claim, evidence, and reasoning. State the "answer" or claim to your question. Discuss what you discovered from your repeated observations and provide exhaustive explanations for your claim.

Nature Journal: Practice developing a claim, providing evidence, and supplying rational reasoning. What was a question you pondered recently? It does not necessarily have to relate to nature study. What were the steps you used to find the answer? What were your observations? What was your conclusion? What were the possible reasons for your results (the why and how the evidence supports the claim)? In the next section, you will learn various ways to apply scientific inquiry to your nature study. You will really start to go deeper into your understanding of the world around you.

Nature Studies

A naturalist studies the natural world. Typically, large-scale organisms like fungus, plants, and animals will be observed and pondered. To go deeper into nature, working like a scientist and using the inquiry process will help the naturalist understand their surroundings. Examples of a nature study's focus may include taxonomy (the naming of or classifying organisms), ecology, and behavior. Read the following sections to become acquainted with nature studies you will want to try.

Natural Histories

To learn more about the local flora and fauna, you will need to spend time in nature observing your surroundings and researching the organisms you encountered. Chapters 5–8 focus on possible species you may encounter. Make it a goal to become familiar with what exists at your Sit Spot and then focus on the species within your Nature Area. In the corresponding chapters, you will find Natural History Journal Pages. These are informal report templates for researching non-native invasives, fungus, plants, trees, arthropods, fish, amphibians, reptiles, birds, and mammals. Find and follow the most common of those species near you. Describe what it looks like, where you find it, what its relationship is with the surroundings, and identify what species it is. The natural history templates will help you focus on all these characteristics, behaviors, and more to become an expert of your Nature Area and move you closer to your naturalist goals.

Tree Adoption

Adopting a tree is another great nature study idea! Trees serve a significant ecological service in an ecosystem. They help improve the quality of our air, water, and soil while also providing food and shelter for animals. Wouldn't it be great to really get to know the trees near you? Think about the trees within your Nature Area, at your Sit Spot, or another nearby location where you have noticed a tree. To clarify, a tree has a woody trunk (typically single) and branches and will stand 13 ft. (4 m) or taller.

Choose a living tree that has captured your attention (realize that deciduous trees are still alive in the winter months even though they have shed their leaves; look for buds or ensure good quality bark to help you know they are still living). Did the tree's height or width get your attention? Its shape or how it moves in the wind? Was it the leaves or bark that caught your eye? If no tree comes to mind, take a stroll through your Nature Area or around your home with the intention of finding a tree to "adopt."

This will be a long-term study of your special tree as it ages and changes. Tree adoption is a great phenology study because it helps the naturalist focus their attention throughout the year, and it can provide clues about seasonal changes over time. The intent is to not only help you deepen and expand your knowledge about trees, but to also develop or strengthen your connection and appreciation for them.

1. For this activity, select a nearby living tree of interest. The tree would preferably be found in your Nature Area. The tree must be taller than you, and at least one branch needs to ideally be within reach. Describe where the tree is located so that you could find it again. Include its position on your Nature Area Map.
2. Why did you choose this tree? What caught your attention about it?
3. What is the month, date, and year? Record this information each time you make observations about your tree. Ideally, you would make observations of your tree each month to record how it changes each season and through the years. Consider monitoring the timing of bud bursts, leaf emergence, color changes, leaf drop, seed or cone development, etc.
4. Take photographs of the tree. Create a special file for this tree on your smartphone, camera, or computer and take a picture of it each month. If you are typing a report for this tree, copy the photo into your document each time you report on your observations.
5. Create bark and leaf rubs. Use a piece of paper on top of the trunk's bark and use the entire length of crayon (with its paper peeled off) or charcoal to rub over the tree's bark. This will emphasize the texture of the bark. If the leaves or needles are available, remove one from a branch (or if you are confident of any of the leaves on the ground belonging to it, then use one of those) and apply the same technique to acquire a nice copy of the leaf.
6. Describe your tree and its overall health. Include identifying characteristics that would help you identify the type of tree you are studying (see Chapter 6). Also, classify your tree's health (excellent, good, fair, poor) and why you gave the tree that rating.
7. Determine the tree's circumference and diameter. This information can help you determine the tree's ecological services or benefits. Use a flexible measuring tape or a string and wrap it around the tree at your chest height. Record its size in inches or centimeters. If you used a string, lay it out to measure using a ruler or yard/meter stick. To calculate the diameter, divide the circumference by π (3.14). Next, visit an online tree benefits calculator to learn the tree's value for helping reduce stormwater, its property value, energy savings benefits, carbon dioxide storage, and air quality contributions. Summarize the results. There are many free online tree and forest assessment tools available.
8. What is the tree's height? You can use a variety approaches to answer this question. Examples include an application for your smartphone, using a

stick the length of your arm (an early logger's method), use a pencil as a measuring device, or making your own clinometer. These methods work best if the tree is significantly taller than you, and the ground is relatively level.

To use the aforementioned "stick method" for measuring a tree's height, follow these steps:

- Find a stick (or similar object) the same length as your arm.
- At arm's length, hold the stick at a 90-degree angle straight out in front of you.
- With the tree in view, walk backwards until the tip of the stick lines up with the top of the tree.
- Once your stick is aligned as describe above, the approximate height of the tree will be the distance your feet are from the tree.

For a more accurate method for determining a tree's height, you can research how to make your own **clinometer** or by following the trigonometry steps, $h = (\tan A \times d) + \text{eye height}$; h is height, A is the angle to the top of the tree, and d is the distance between you and the tree.

9. In Chapter 6, complete the corresponding Natural History Journal Pages for your tree species. Use the information in that chapter and other resources to deepen your understanding of the tree and to help you identify it. Complete the prompts in the NHJP template for a typical specimen of your species.
10. After you have taken a series of photographs through or between seasons, reflect on the observable changes and ecological interactions you witnessed. What did you enjoy the most about the tree study? Add key observations of the tree over time and how the tree had changed in each photo (differences). What were any interactions you observed between other trees, plants, animals, and humans? What were, if any, surprises with the results?
11. Find another tree of a different variety. If you selected a conifer for your adopted tree, then choose a deciduous tree to learn more about now. Complete the same questions above for the deciduous tree.

Field Investigation

Focusing on Nature Mysteries you want to solve makes for intriguing nature studies. A field investigation in science is the systematic process of collecting data anywhere outdoors for the purpose of understanding the surrounding environment further. Do you have any burning questions or things you are curious about that you observed at your Sit Spot or Nature Area? Spend at least 15 minutes looking outside, at one of your Sit Spots, Nature Area, or go on a Nature Stroll.

You can keep things very simple for the field investigation, yet still meaningful, by focusing all of your attention on your Sit Spot. To choose a more focused approach at your Sit Spot, outstretch your arms (or designate it at 3 sq. ft.; 1 sq. m), from soil to sky, and as far as you can see and hear from that single point – this can be your study site. For details on examining a broader study area or specific approaches for your Sit Spot site, keep reading further. Wherever you decide to explore and conduct a study, immerse yourself in your observations, move slowly, and be curious about your surroundings.

- *After* your time outside, recall your experiences from the very moment you set out on your Nature Stroll, all the way to the time you were done with your outdoor experience.
- Follow the "Sit Spot Routine" as described in Chapter 1 for recording current conditions, observations, and Nature Area maps.
- To help you start thinking deeper about what you observed, write questions associated with each of your observations. You will hopefully discover the types of things you are truly interested in or a puzzle you want to solve in nature. Refer to Chapter 1 for help developing the skills of making good observations and asking questions. What are you curious about regarding what you observed in nature?

  QUESTIONS – Create questions based on what you observed. Simply share what you find intriguing about one or more of your observations. What do you wonder? At this time, do not focus on questions requiring answers based on evidence.

  1.

  2.

  3.

  4.

Question Types

What types of questions did you ask? Realize that not all questions can serve as a testable field investigation. Label the types of questions you developed above as either:

Book/Internet Research

Why Questions, Life Pondering, or Always Wondered

Descriptive, Comparative, OR a Correlative Field Investigation

If you are not sure if you have a potential field investigation, read the next sections for tips on what makes a question testable.

NOT Field Investigation Questions

With your question about nature in mind, first determine if it can be researched and if there is already an actual answer for it. Ask yourself how you would collect the data to answer the question.

a. If any of your questions could be answered by conducting *research in a book or on the internet*, then you know what you posed is better suited for background research on something of interest (e.g., What does a baby deer look like?). This can lead to amazing insights, discoveries, and a lifelong interest in something. However, it may not be a question you can investigate in the field or does not contribute to the body of knowledge that already exists. Can you look up the answer to your question? If the answer is yes, then try to think about other questions you could ask about that species or occurrence.

b. If any of your questions were *"why" questions*, then these would be difficult if not impossible to determine as a field study or may be inconsequential to nature studies (e.g. Why is the fawn (baby deer) by itself?). Although this type of question can get you to investigate more closely, you cannot always know the true reason for something.

c. *Life pondering questions* or things that you *always wondered* about may not have attainable answers in a nature study project (e.g. What would the world look like if there were no deer?). These types of questions tend to surface when you are in nature, and can lead to more questions about what you observe. Sometimes these types of questions can also help with environmental sustainability models.

All questions have value and can lead to a deeper awareness and understanding about the world around you. Keep reading to find out the questions you can answer with a nature study.

Field Investigation Questions

Any questions you can answer through an observational nature study will be considered a field investigation. Any field investigation question will fit into one of these three categories to help you understand nature. Be sure your field investigation questions are practical and attainable. Specify details in your questions to identify organisms or conditions you are focusing on, and the place and time of your study.

a. *Descriptive.* A descriptive study will focus on describing observations of a setting in nature. Answering what, when, or how questions would likely indicate this type of study. Questions need to include observable or measurable variables in a given area that can be mapped, written, or quantified (e.g., How many x are in a given area? Where does x nest in a given area? What are the behaviors of x in the Nature Area?).

   Consider your study site and read these sentence stem ideas for a descriptive study (create one testable question): What type of [trees/birds] are found within the study site between 8-8:30 a.m. in mid-April? What is the relative abundance of each tree/bird/squirrel species in the designated area (relative abundance is an estimate of the number or percent of species based on data collected)? What is the species diversity found in [a certain area] of the study site during the evening in May? Animal signs can also be included in your data collection if you can identify that it belongs to a specific species or animal type.

   What are the behavior types observed for an animal type during a certain time and season? If possible, choose a species you know exists in numbers greater than 4, and relatively dependable for viewing. Consider these animals as possible subjects: common songbird of your choice, crows, gulls, geese, ducks, deer, chipmunks, or squirrels.

   How can you extend the study further? Create potential comparative and correlative questions from your descriptive field investigation that you could carry out.

b. *Comparative.* This type of study builds on formal or informal observations of a descriptive study. It demonstrates a comparison made between two or more plant/animal groups, times of activity, or locations of where things occur. Observations, questions, and reflections from your study site will help guide your inquiry. How can you build on your descriptive study? Here are ideas to help get you started – Are there more x than y at the study site (e.g., American goldfinches than black-capped chickadees)? Are there more bird species at 7 a.m. than 7 p.m.? Does the [animal species] show more alert type of behavior than other types of behavior? Will the [animal species] exhibit more behavior types in the evening than in the morning? Are more alarm-type behaviors exhibited by the white-tailed deer in the field compared to when they are in the forest? The question possibilities are endless.

c. *Correlative.* A cause and effect relationship are sought in this type of field investigation. What types of abiotic factors do you think affected what you observed in your comparative study? For correlative studies, two or more variables are observed and tested to determine a pattern. Think about relationships between animals, animals and plants, plants and other plants, animals/plants and abiotic factors like the temperature, wind, or precipitation. For example – What is the relationship between x and y (e.g., temperature and number of birds)? How does x change when y changes? What is the relationship between [animal species] behavior and an increase in relative humidity before precipitation?

For additional guidance, refer to the free, online resource for educators listed at the end of the book that is from the Association of Fish and Wildlife Agencies and Pacific Education Institute (Otto et. al, 2015).

Conducting a Field Investigation

You have thought about and analyzed the different types of questions you could have about nature. Now, it is time to refine your inquiries to those that are testable in the field. Start with the most basic field investigation to become acquainted with your Nature Area.

Based on your observations and interests, create a descriptive question for a field investigation that you will actually carry through at your study site. Use one of the questions you already developed or construct a new question for your study. It is difficult to create a meaningful field investigation and experience if you have not spent time in the study site. However, a descriptive study can help you explore the area with intent, but the question you have in mind should at least be based on research of what you could discover in the habitat.

Be sure you can carry out all aspects of the question. Is the "answer" something you can simply look up on the internet or in a field guide? If so, you need to make it specific to your site and something that you are curious about but can only answer by making observations at your site. A good field investigation question tells exactly what the investigator/naturalist is trying to determine—nothing more and nothing less. You may need to wordsmith or reword your question as you make more observations and refinements in your methods for finding answers in nature. At this time, also start thinking about potential comparative and correlative questions related to your descriptive field investigation question.

The more time you spend in nature, the more you will want to learn about the occurrences around you. Keep track of ideas in your Nature Journal.

Use the following table to help you think about potential studies.

FIELD INVESTIGATION QUESTIONS – *Develop testable field investigation questions related to wildlife (any plants, animals, fungus, or other organisms existing in the wild). The study should be possible to do within a period reasonable for a given time frame. Build your comparative study upon the descriptive study and build your correlative study upon those studies.*	**Considerations or Concerns** *regarding the setting or the investigators* *(human interference, lack of experience, time, access, etc.)*	**Materials** *needed to complete the studies* *(list everything you would need)*	**Extensions** *such as technology, tips, or additional activities* *(computer programs, apps, reminders, other ideas for help with completing the study)*
Descriptive:			
Comparative:			
Correlative:			

METHODS - How would you carry out the field investigations? *Provide the materials and steps needed to gather the data to answer your questions. See the next section to help you identify the best approach for collecting data. Number each step. Use enough detail so the study can be replicated.*
Which essential questions does your field investigation help answer? Explain. *Does your study help answer any of the following: What defines the environment of your study area? What is the quality or health of the environment? What are humans' relationship with the environment in the study area or nearby? How can we sustain this environment? What is your personal role in the preservation of the area?*

Potential Methods

The first part of this section introduces the basic things to think about for your field investigation setup. Next, you will read about the simple field methods of finding organisms; using quadrats, study plots, and transect lines; and then you can learn about conducting simplified studies in animal behavior and species diversity. Once you determine your field investigation questions, list the materials you need to conduct each of your scientific inquiries (don't forget the Current Conditions template from Chapter 1 for each visit to your site). Read through the following sections to help you outline the steps of how you will perform the study. These are your methods.

Consider collecting enough data to carry out your comparative and correlative studies related to the initial descriptive investigation. Depending on the questions, this can be a relatively easy addition to your studies. Choose or modify the materials and procedures that best fit the type of data you need to collect to answer your question.

Review the details of what to include in a culminating field investigation report found later in this section. Knowing what you need to report on will help you be more efficient with your time in the field and analyzing the data. It will give you more perspective of what to think about and do for your study. Carry out the descriptive field investigation and make notes or adjustments in your outline to ensure enough details with the procedure. Spend adequate time in the field collecting your data. The first time to the site tends to take the longest. Depending on your research question, visit the site 5–10 times, and try to incorporate 5–10 replications. Regardless of the study's focus, you will want to collect relevant and ample data to increase the credibility of your claims. Keep good notes in your Nature Journal. Details of the experimental design will help with replication by others. Be thorough and record everything.

Many different field techniques exist. The methods applied will depend on your target species, level of personal expertise, and the type of data and results required or desired. For our purposes, we focus on practical field inventory methods any naturalist should be able to perform. First, let's think about where to find living things in nature.

Where to Look for Organisms. If unfamiliar with where you might find certain species, review this quick summary of possible locations. Also, peruse the corresponding sections in this book for further information to help you make the most of your experience.

Fungus: at the base of trees or stumps, in woodchips, lawns, in dead hardwood trees; after frequent rains in warm or mild weather

Plants: in gardens, along paths, trails, railroad tracks, fields

Trees and Shrubs: in natural spaces – forests, woodlots, edges, parks, yards; look for groupings and variety of sizes

Snails and Slugs: gardens, under rotting wood, woodchips, near or in rocks, in leaf litter

Insects and Arachnids: near water bodies, shaded paths, shaded logs, moist soils, near rotting carrion (dead animals), compost heaps, grassy areas, along roadsides, sidewalks, forest edges, woodlands, near flowers, in gardens

Amphibians and Reptiles: moist soils, under fallen trees, leaf litter, under logs, near or in water; reptiles found basking in the sun on logs or rocks on warm days

Birds: trees, shrubs, tall grasses, among a transition between habitats, near water sources, open air, bird feeders

Mammals: near sheltered areas like shrubs, holes and burrows, trees, gardens, near water

Quadrats. For plant studies (or studies involving microorganisms or insects), randomly put quadrats out into your habitat of focus to estimate the number of organisms or species that are in that area. Quadrats can be any size with 1 square meter or 50 cm by 50 cm or 25 cm by 25 cm (and similar sizes in feet). You can construct a quadrat from plastic pipes, use a hula hoop, or cut a string to the required size. Consider your study area and choose a quadrat size that makes sense to encompass a good representation of the study's test subjects within it. However, your quadrats should be tossed or placed randomly to avoid bias in data collection and to find a good sampling of the area. Many samples need to be taken in the study site so that at least half of the area is inventoried.

Study Plots. Use study plots for inventorying relatively large portions of land to determine the biotic and abiotic factors that exist in your study site. This field method would also lend itself to calculating the relative abundance of species and species diversity. Examine your Nature Area map to help you randomly place these plots in a relevant area (e.g., an entire quarter of your site or 30-sq. ft.; 10 m).

Within the plot, you would slowly walk (half your normal pace) in a serpentine or s-shaped pattern. Repeat in the other designated sections of your site for at least a total of four plots. Record your observations related to your field investigation questions on your map of the study site and on a data table. Remember to examine all layers of the study site if applicable, from soil to sky. You may be repeating this procedure depending on your questions.

Transect Lines. Transects are used for determining relative abundance and species diversity along a gradient or linear pattern. They provide a good sample of how communities may change through the habitat. Consider what type of data you are interested in collecting to determine where to designate your 5–10 lines. You can use your Sit Spot location outward in straight, marked lines to your Nature Area boundary (as you designated on your field inventory map) or 10-m transects (30-ft. line) placed in relevant locations. Make observations along that line as it relates to your field investigation questions. Examine what crosses the line from soil to sky. Your inventory might include vegetation type(s) or presence of animals and animal signs if you can determine who it belongs to in relation to your study (e.g., tracks, chewed food, scat, exoskeleton, bones, fur, nest, etc.). Record your findings on your inventory map of the study site and on a data table. You may need to repeat this procedure depending on your questions.

Species Observations Data Table Examples

Using these tables or something similar would help the field investigator or naturalist inventory unknown species encountered to identify later. Adjust it accordingly.

This table is often used in descriptive field investigations. It can help you keep track of the species encountered at your study site, especially if you did not know the names of the organisms. Assigning a number to the species description would help maintain consistency as you move through your study. Identification can happen later. Plants, trees, and insects are common examples.

Date:

Location:

Species No. or Code	Description	Sketch/Other Notes

Use this type of data recording chart when inventorying animals that move in and out of your field of view. Typically used for bird counts, but can be used for other organisms and especially in species diversity studies. Make a bar graph of your results. Create a comparative field investigation by comparing your data to different times of day, the seasons, or habitats.

Date:

Location:

Species Name or Description	Record of the Number of Species Observed at any Given Time During a 15-Minute Observation Period (or other time frame)	Highest Number of Each Species Observed During Inventory Period

Types of Field Studies - Animal Behavior and Species Diversity

The three field investigations discussed previously have seemingly endless study type possibilities. All the questions are rooted in observation, especially descriptive studies. Comparative and correlative studies take things further and you will incorporate mathematical differences, averages, or other statistics to answer their corresponding research questions. From the information in the Current Conditions Chart you collect with each observation period, and the other relevant data compiled in the tables or Nature Journal, you can easily complete your study and answer all three field investigation questions. For additional ideas, read this section on conducting animal behavior studies, and for calculating and analyzing data for a species diversity study.

Animal Behavior Studies. What animals have you noticed lately in your Nature Area? Have you ever wondered what they do in your habitat? Scientists and naturalists study animal behaviors to not only learn more about a given animal, but it also sheds light on humans and the human connection to the environment.

Behavioral studies are essential to conservation biology. They are crucial to understanding how to preserve species during the ongoing negative impact humans have on the biosphere. Scientists and naturalists study animals for various reasons. They observe feeding behaviors, habitat selection, mating behaviors, and their social relationships. The more you know about the animals in a habitat, the more you will know about the social-ecological connections of its ecosystem.

If you are curious about how certain wild animals behave, or you want to make comparisons of behaviors at different times of the day or during different weather conditions, or if you want to attempt to correlate behaviors to other events, then you should consider an animal behavior study. Ideally, you would have more than four of the wild animals you want to focus on in their natural habitat to make it a meaningful study. The more individuals you observe, the better the data will be to analyze and draw valid conclusions compared with watching one or a few animals for a long period of time. However, just observing one animal can provide data and can pique interest in learning more about the animal, its entire species, and how it fits into its community and ecosystem. If observing wild animals is not feasible, or if you want to get started right away on a behavior study, start by watching your pet cat or dog. Other ideas are to observe birds at feeders, mammals at a zoo, or watch an online webcam of animals in their habitat.

Choose a location that makes sense to observe your study subjects. Getting better acquainted with the animals in your Nature Area would be ideal. Visit in the morning and evening when there is likely the most animal activity. Other ideas could include a nearby habitat with varied vegetation types and a water source like a lake, pond, river, creek, wetland, or even a bird bath. Ensure your presence does not alter the species' behavior but try to get to the closest and safest range possible. Try to sit downwind from where you expect to observe the animals. This will help you go undetected to the animals and can increase your odds of witnessing more of their natural behaviors. For some studies, you may also need binoculars. Dress for the weather and bring a chair, cushion, or blanket to get comfortable.

Observe your animals of focus (e.g., songbirds, crows, gulls, geese, ducks, deer, chipmunks, squirrels) for at least 15 minutes to describe behaviors they exhibit. Describe the behavior of the animals observed without interpretation. Meaning, if you see a squirrel burying an acorn, do not assume he is storing (**caching**) food for the winter. He is simply burying an acorn. Make a list of the behaviors you observe of the squirrel during the 15 minutes (use the sample chart that follows for reference). Repeat this 15-minute observation period at least 5 different times or as necessary to provide reliable data for your purposes.

Another format for behavior studies can include the use of trail cameras to capture animal activity. This is called a **video trap**. Think about research questions you could generate from this data. The camera can capture hours of video footage triggered by movement (the sensitivity to motion can be adjusted on some trail cameras). Who visits the study site? At what times and weather? What are the behaviors of those animals? Experimenting with the setup and how you view the results will also pique your curiosity about what happens in nature when you're not looking. Strategically place cameras within your study site where you think it would record animals. Practice proper etiquette and do not have the camera facing toward or focused on somebody else's property.

Tally the number of times you see the animal exhibit the different behaviors you had identified in your initial visit to the study site. What you are doing is creating an ethogram. An **ethogram** is a data table where you describe initial behavior observations (see the examples provided later in this section). Use that animal behavior list you create as a foundation for subsequent observations. To make the most efficient use of your time in the field, research the animal for the behavior study before going out for initial observations. Complete the corresponding Natural History Journal Pages for the animal. With this knowledge, you can study the animals more critically when you observe them. On the other hand, observing an animal's behaviors as if you were the first person to ever witness it can sometimes be life changing. With either approach, researching the animal's natural history is important to conducting a meaningful behavior study. From your initial observations and doing background research on the animal, you will be able to generate testable questions that pertain to your study site, contribute to a larger body of knowledge about the area or species, and hopefully the questions are also of interest to you.

Discussion of Animal Behavior Study. You will describe all behaviors you observe and the context in which they occurred. Only after you have built a catalog of the behaviors in terms of types, frequency, duration, and context, will you be ready to quantify the data. When reporting, discuss behaviors in relation to the animal's actions and not the projected or perceived function. For example, if an eastern gray squirrel (*Sciurus carolinensis*) buries an acorn, it should be described in its entirety without interpretation. Specifically, an eastern gray squirrel dug in the ground below an oak tree, placed an acorn in the hole, and covered it with dirt. Do not describe it as "stored food for later." It may be a plausible hypothesis, but not a direct observation (unless you witnessed the same squirrel retrieve it at a later time and eat it). It is difficult to not project an **adaptive function**, or an effective skill necessary for survival in our environment (e.g., communication, planning, etc.).

Another perception difficult to overcome is **anthropomorphizing** or ascribing human characteristics to nonhuman beings. If not careful, this can lead to misconceptions about the plant or animal. However, anthropomorphizing can help pique the interest of your audience or help describe something initially so that it may be more relatable. Always clarify an observed behavior from anthropomorphizing by using words like seems like a ..., similar to ...

The 1897 naturalist and author, Neltje Blanchan described an interpretation of an American robin's (*Turdus migratorius*) vocalization as it tends to the young from the nest: "Love, contentment, anxiety, exultation, rage—what other bird can throw such multifarious meaning into its tone?" The author projected human emotions that she thought the bird felt. Listen to an American robin's song and you could almost attach meaning to the melody to describe the pattern you hear. However, anthropomorphizing should never be used in formal descriptions of plants or animal behavior or of inanimate objects. State what was observed. The song of the American robin has an evenly spaced warble or string of a clear assemblage often described as *cheerily, cheer up, cheer up, cheerily, cheer up*. The "cheer" in this example does not pertain to a disposition, but it describes the literal sound the bird makes.

Behavior Examples. Here are some descriptive words to use to describe behaviors of birds and mammals. Birds: flying, flocking, bathing, preening, walking, branch hopping, foraging, singing, alarm calls, feeding, drinking, territorial behavior, etc. Mammal behaviors might involve: walking, running, eating, drinking, lying down, territorial behavior, grooming, mating, etc.

What to Report. On your data collection page and final report, you will want to include the following information. In terms of context of the behaviors, include the animal's location (e.g., habitat type, proximity to major features of the habitat), other organisms present (same and different species), abiotic conditions (time, temperature, light; see Chapter 1 for the Current Conditions template), the individual's characteristics (male/female, size, nutritional state, other notables), and describe the animal's previous behavior (in other words, what the animal was doing before the observation you noted). Furthermore, what can you objectively state about what follows a certain behavior? Was there a measurable reward (food eaten, predator avoided, new position claimed or one lost)? How much time was expended in the behavior? What behavior type was observed the most? The least amount? Did there seem to be a trend or pattern for what the animal was doing? What are any personal interpretations you have of the observations? What other field investigation questions can you create?

Behavior Data Tables. As described above, you will first spend time at your study site making observations of your animal of interest. Then, you will need to have a systematic approach for recording the observations. In this section, you will find examples of data tables you can use when conducting an animal behavior study.

Ethogram. Create an ethogram to begin your study, develop field investigation questions from your observations, and use your ethogram to collect and analyze the data. The ethogram below is for gray squirrels.

You can use this chart to study the gray squirrels, or other squirrel species near you. You may find other behaviors that could be clustered with existing behaviors. Feel free to modify the chart. During your next visit to observe the gray squirrels at your study site, you may also notice something new. Take good notes. Bring your ethogram each time you set out to make observations. **To make an ethogram for an animal of your choosing, use the same chart as a template but with your own actions and descriptions created.**

An Ethogram: Behavior Descriptions in a Field Study of Eastern Gray Squirrels (*Sciurus carolinensis*)

Location (city, state): Big Rapids, MI
Habitat (type; surrounding habitat): swamp, mature deciduous forest transition
Date: August 10, 2020 **Time:** 8:30 a.m.
Weather: sunny; 75 degrees F

Behavior No.	Action	Description	Sketch/ Other Notes
1	Resting	No movement, eyes closed	
2	Chirp	Vocalization, short duration, quiet between pulses	
3	Chatter	Vocalization, long duration and continuous	
4	Freeze	No movement, eyes open, erect posture	
5	Running	Moving at quick, regular speed	
6	Saltate	Jumping locomotion, as compared with running	

Location:
Habitat:
Date: **Time:**
Weather:

Behavior No.	Action	Description	Sketch/ Other Notes
1			
2			
3			
4			
5			

Once you have your ethogram, choose a behavior record table below that best fits your field investigation question to record your data. Modify the tables to make sense for the data you collect. The best layout often emerges as the study progresses.

Behavior Record. A tally is used to record each time one of the behaviors is observed during the designated time frame. The behavior number would have been determined in a previous observation period and recorded in your ethogram (see the above example). Some naturalists find it useful to use one- or two-word reminders of what behavior corresponds to what number. The "action" column is often added to the tables (see ethogram example). In the second table provided in this section, five different observation columns are present. Tallies are recorded in a single observation column for each observed behavior type (that corresponds to the number given to it in an initial visit). Alternatively, the activity type can be described along with the numbers in the first behavior column, and the column for observation period 1 could be used as descriptions for those activities observed during your initial visit. Provide the date, time, and conditions at the top for each observation period and tally the behaviors.

Behavior No	Behavior Observation Tallies	Notes

Animal Species:

Location (city, state):

Habitat (type; surrounding habitat):

Number of Occurrences for Each Observation Period					
Behavior No.	Observation Period 1 Date: Time: Conditions:	Observation Period 2 Date: Time: Conditions:	Observation Period 3 Date: Time: Conditions:	Observation Period 4 Date: Time: Conditions:	Observation Period 5 Date: Time: Conditions:
1					
2					
3					
4					
5					
6					
Notes:					

Behavior Observation Sheet. In the previous data recording format, behaviors were based on observations made during a data gathering period and recorded on an ethogram. The following table shows how to set up another approach for tracking the animal behaviors observed. Use this data collection method if any of your field investigation questions (descriptive, comparative, or correlative) may have something to do with time or if you want more detail from your study. Provide a check mark for every behavior exhibited within each 30-second time frame. You may also want to include a description or commentary of the correlating behavior or what happened during the time interval in general. You would want to create this table in your Nature Journal where you have more space to make notes. The methods described here are intended to help you begin your behavior studies, and modifications are expected based on your question and preferences.

Animal Type:

Date:

Time:

Conditions:

	Behavior Type				
Time (every 30 seconds up to 15 minutes)	1	2	3	4	Notes
0:30					
1:00					

To reflect on and analyze your behavior study results, visit *The Field Investigation Report* at the end of the next section.

Species Diversity Studies. Another type of field investigation is one that involves quantifying the different organisms encountered in a designated area. This is a species diversity study and can be used as an indicator of a habitat's quality. Various models and indices exist for calculating species diversity. In this practical guide, we focus on what most people could do or would have interest in doing.

For the field component of your investigation, use study plots or transects for your procedure to collect data regarding species diversity, or the variety of species that exist at a given site. The procedure selected will depend on your field investigation question, the species of focus, and habitat. The **Simplified Diversity Index (SDI)** measures how diverse the vegetation, animal types, or a combination of all living things are at the study site, or the diversity between sites. **To calculate SDI, take the total number of different species observed divided by the total number**

of all species. The index reflects both species richness, relative abundance, and species evenness. The index will be a decimal number between 0 and 1. The closer the diversity index is to 1, then the more the habitat is diverse and healthy. More details on interpreting the data can be found later in this section.

When collecting the data, if you cannot make a positive identification, then give it a label with a detailed description of the species so that you can assure consistency and a future identification. Also, always include a Current Conditions Chart for each visit.

The Species Diversity Field Investigation. Use the example in the table as a guide for collecting and analyzing your own data. Create your own table in your Nature Journal and modify as needed.

Species Type	Total No. of *Different* Species (species richness)	Species Type and Total No. of Each	Total No. of Each Species Group	Relative Abundance in the Study Site (total no. of each species group/total no. of all species × 100 = %)
Mammals	2	Gray squirrel = 3 Eastern chipmunk = 2	5	Mammals = 5/27 = 0.19 × 100 = 19%
Birds	4	American goldfinch = 9 Black-capped chickadee = 5 Red-winged blackbird = 4 Blue jay = 3	21	Birds = 21/27 = 0.78 × 100 = 78%
Insects	1	Mourning cloak = 1	1	Insects = 1/27 = 0.04 × 100 = 4%
Total no. of different species	7	**Total no. of all species**	27	**SDI for study site = 0.26**

Analyzing the Data. What does it all mean? Different things can be determined from the data collected as seen in the table. When writing your final report on the field investigation, state the raw data trends and make comparisons. See the information in this chapter on what to include in a final report. To truly understand how to perform the calculations and what you can determine from the data collected, it is best to experience the process. To make it an easy exercise for practice, sit outside, look out your window, or open a bag of colorful candy or dried beans to collect data on species diversity.

You can identify the **species richness**, or how many different types of species were observed in the inventory. In the example in the table, you will see the species richness for mammals was 2, birds were 4, and insects 1. The **relative abundance** refers to the number of individuals per species represented as a percent for each type present. We would want to know how common a species was in relation to the other species in the defined area. You could also do this within each animal group. To calculate SDI, take the total number of different types of species divided by the total number of all individuals found.

To determine the relative **species diversity** in the study sites, use the SDI as you observed in the calculations on the example table and compare with the descriptions that follow.

SDI value of 1 indicates high diversity

SDI value of 0.5 indicates area is relatively diverse

SDI value of 0 indicates no diversity

SDI of a healthy forest would typically range around 0.70–0.80

SDI of an agricultural field would typically range from 0.02 or less

An SDI closer to 1, typically indicates a diverse sample. A lower value for the index may indicate many of the same species (a high **species evenness**). An index with a higher value may be due to many different species, with only a few of each of the species present (high species richness). The example in the table had an SDI of 0.26. This indicates relatively little diversity. How could species diversity change over time? What could promote diversity in a habitat? See Chapters 3 and 4 to learn more about the importance of diversity and what can be done to increase it in an area.

Discuss the Trends. Organizing data in a readable format helps identify trends and draw conclusions. From the example given in the completed SDI table, discuss each animal group in turn and then make comparisons among all the data. Compare the total number of each animal group (species richness). Which species had the most individuals observed? The least number of species? What were the other species richness values? What was the relative abundance for the animal groups (or plants, depending on your focus)? Compare relative abundance values. What was the SDI for the site? If part of your research question, was there a difference in animal numbers between times of day? How does the SDI compare between different animal or plant groups? How does the SDI compare between locations within the study site? If studied, what was different between habitat type? What is the relationship, if any, between the number of a species and the temperature? Have fun discovering ways to view the data. Great revelations and new ideas may emerge! Keep reading for further help with interpreting the data.

Considerations of Your Results. Think about your field investigation results in the context of the bigger picture. Most species are habitat-specific, and as you enlarge the research area, it will likely include more species. The diversity of species is dependent on pockets of habitat such as a group of trees, a creek or water feature, a home with a wild garden or even a home with a landscaped garden. Animals seek food and shelter in areas where there is adequate habitat, which includes the type, variety, and arrangement of plants in an area. Frequently, the types of plants in a site are affected not only by the types of activities that take place on the site, but also on the surrounding habitats and how the landscaping on the site may be managed. Always consider the surrounding influences on a site. Know the "place". Often you will discover a correlation between plant and animal diversity.

When the Numbers Do Not Make Sense. Sometimes, anomalies in the data can occur. For example, if the diversity index is 1 and only one species of its kind was found, then that does not indicate true diversity. While this anomaly does not change the diversity index, when making the calculation for the whole area, scientists use a variety of tools and data to adjust and accommodate for these types of occurrences. Completing a Current Conditions inventory for each visit to the study site can help you make logical conclusions. Provide any of these observations that seem relevant as support to your results or claims.

What Affects the Species Diversity Index? A scientist measuring biodiversity of plants or animals in an area would not count every plant or animal in an area. A scientist would inventory the designated area using random sampling techniques. Random sampling affects the index. Random samples do not cover all areas, you may not have collected data from enough sample areas, some people have more experience identifying differences in plants and animals, etc. Also, technique or expertise may factor into the quality of results collected. Do your best to collect data consistently, and to tally accounts and calculate correctly. If you are working as a field investigation team, spend time before the study begins to calibrate or align everyone's techniques to help ensure consistency. Include this as a step in your methods section of the report.

The Field Investigation Report

The write-up of the field investigation will include the following information. Keep notes on all of these details throughout the study to help make this process easier and more thorough:

1. *Site Description:* Provide a description of the surrounding environment and map of the study site (photographs are also a good addition).
2. *Observations:* Discuss detailed observations from within the study site leading to the field investigation.
3. *Field Investigation Question(s):* State your field investigation question(s) (descriptive, comparative, correlative) that you researched.
4. *Variables:* What were the variables in your study (anything that can change or be varied or controlled is an **independent variable** and a **dependent variable** is the condition measured or observed)?
5. *Background:* Research what you are studying (e.g., your test subject, habitat, land use, weather trends, etc.)—complete the Natural History Journal Pages as found in the corresponding section(s) of this book.
6. *Hypothesis:* Based on your observations and background research, what do you predict will happen with what you are investigating, or the type of results you expect?
7. *Current Conditions:* Record the abiotic factors or conditions during each visit to your study site. Use the table found in Chapter 1. You may want to modify the template to consolidate the data for all the visits into one table.
8. *Methods:* List all the *materials* needed to carry out the investigation (think about what you need in sequential order). Describe the *steps* for collecting data to help answer and understand your field investigation question(s). Consider using one of the methods described in this chapter. Your claims or conclusions will strengthen the more times you visit the site (more data over time). This is a good goal—be consistent with the time of day you visit, how you collect the data, the duration, and location unless your study's focus directs you otherwise. Note all these aspects of the study in the steps of your methods.
9. *Results:* Display all your data in organized tables with a description provided on top of each. Provide various ways to represent your data (averages, modes, percentages, estimates, etc.). Use correct graphics or "figures" to display the data (bar graphs, pie chart, maps, line graph). Provide a descriptive caption below your figures. State your findings without drawing conclusions. Make comparisons and point out the highlights and trends observed.

10. *Discussion:* All scientific studies tell us something and help us understand a piece of nature.

    a. What does other research show for similar studies? Cite related findings of at least two or more reputable scientific studies. Use the most recent edition of APA Style to cite your sources.

    b. What essential questions did your field investigations help answer? How does it help answer any of the following: What defines your environment—plants, animals, and abiotic conditions that exist and interact in the area? How would you describe the health or quality of the environment—excellent, good, fair, poor—how do you know? What is humans' relationship with the environment—observations that could support positive or negative impacts in the area? How can we sustain our environment—evidence of human influence that was implemented at the site that seems to help it thrive?

    c. What may have affected your study's results? Explain.

11. *Conclusion:* After reviewing your results and discussion, what can you conclude? Restate and answer your investigative question(s), provide examples of your data that supports your claim. Be sure to apply the information to only your study site, and ensure any supporting data and claims aligns with your question.

12. *Future Investigations:* After conducting the study, what other questions are you curious about that relate to the study you just completed answering? Consider creating new field investigation questions for related descriptive, comparative, and correlative studies.

13. *Personal Role:* What is your personal role in the preservation of the area? Discuss things you could do now, and what you suggest you or others could do in the future to encourage a healthy environment at your study site.

2.5 Reflecting on Chapter 2—Nature as Your "Place"

Have you worked through all the activities so far to help train you as a naturalist? By having an ecological perspective to interpret your watershed, you can identify connections across a landscape on many scales. What are the landscape shapers near you? How do humans impact your watershed? The knowledge and skills acquired in this book will help you know your "place" more intimately and raise awareness of the types of things that construct any given location. This builds your naturalist intelligence. The process can be very time consuming, but it has intrinsic rewards that can be applied to environmental stewardship initiatives. Always try to look at your place with openness and curiosity, and to practice scientific thinking and inquiry to seek answers to Nature's Mysteries. Keep good notes in your Nature Journal so you can build upon your discoveries. Just think about how well you would know "place" if you completed all the activities in this book! What a great way to learn to really appreciate the world around you. Move to the next chapter to learn more about becoming a naturalist and interpreting any habitat around you. Consider perusing the Table of Contents for what else you can explore and apply to your nature study.

Review Questions

1. Explain what is meant by having an ecological perspective. What are the rules of ecology?
2. Discuss effects of the 5 landscape shapers on the Great Lakes basin and Upper Midwest region.
3. What is a microclimate?
4. Describe what is meant by an organism's nutrients.
5. What is a watershed? Diagram the main features of a drainage basin.
6. Discuss a plan you will try to implement to protect and improve the quality of your watershed.
7. What is micropollution? What are examples?
8. What are three basic practices to help ensure good science takes place? Explain the inquiry process. Describe examples of possible nature studies.
9. Provide an example of a descriptive, comparative, and correlative field investigation questions.

3 Parts of Nature: Terrestrial Habitats and Freshwater Systems

In the last chapter, you gained a broad, landscape perspective of "place" yet also made the concepts relevant on a personal level. Chapter 3 focuses more on the features of a place. There are various things affecting nature at any given time—weather, soil conditions, animals, and people. Recognizing the possible factors and how they could influence living things helps the naturalist interpret observations and make predictions about a given site's habitat and systems.

In Chapter 3, you will learn about the regional landscape, classifying and interpreting habitats, terrestrial ecosystems, basic geology and soil systems, and freshwater ecosystems. Learn these basic principles to help prepare you with the knowledge to understand and interpret nature to reveal its functioning of multiple scales of life. During your time studying an area, you may develop a greater appreciation for that environment.

3.1 Regional Overview

Every place on the planet has certain conditions and characteristics indicative of that particular region. By understanding these regions in terms of their general climatic conditions, topography, geology, and potential microclimates, we can start to make sense of the flora and fauna that may exist in the area. Although these areas or zones have differences, as a naturalist, you should be able to recognize basic principles. Apply your naturalist skills and foundational knowledge anywhere you go to start making sense of your surroundings. Wherever you are, making thorough observations, asking questions, and seeking and sharing answers will get you far with expanding your naturalist intelligence.

Regardless of where you may live, the following sections will provide insight on how to improve your interpretations of what you may find in nature in any region. As a reference point, examine the following map for where this book specifically covers—the Great Lakes basin and Upper Midwest region.

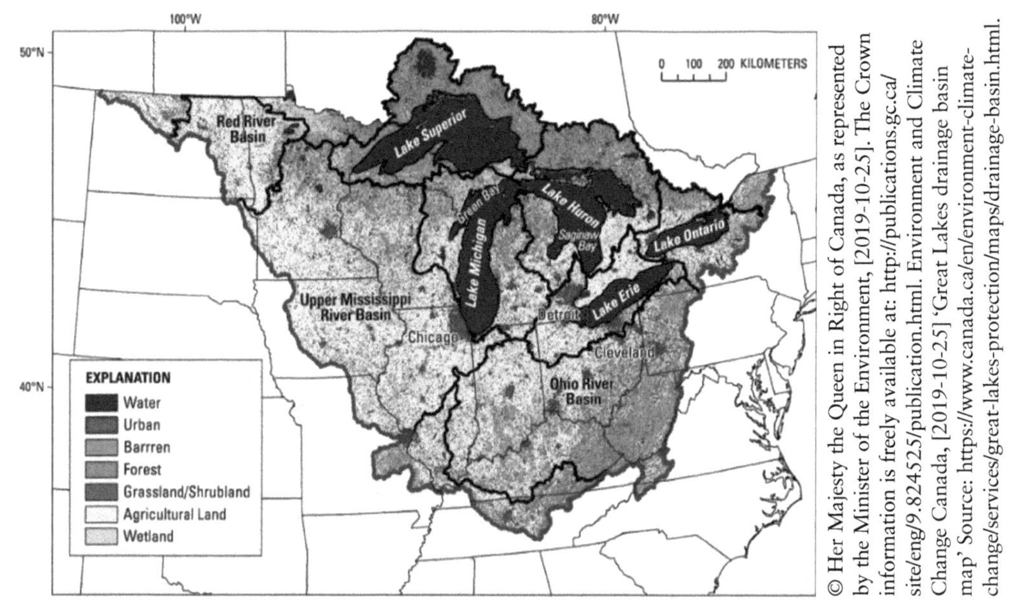

Note the northernmost area keyed as forested land. Specifically, it would have the classification of a northern forest. It has an extensive **boreal** forest consisting of long, cold winters and short, warm summers. Conifers will dominate the relatively harsh conditions, with species from pine, spruce, fir, larch, and hemlock families. A high density of lakes also is found across this portion of what is called the Canadian Shield, or the largest exposure of Precambrian rock. You will find **tundra** to the north of the northern forest and **temperate forests** south of this region on our map. Temperate forests have mild to moderately humid conditions with somewhat dense and forest cover. On our regional map, most of this region has been transformed into agricultural land.

3.2 Classifying and Interpreting Habitats

To help us understand and communicate what we discover in nature, we need to have a language that other naturalists can interpret. Remember what you learned in Chapter 1 about taxonomy – there is a science of naming all living things based on observations and those organisms are organized into a corresponding classification system. The naming of nature's habitats is similar but has much more flexibility in its classification based on your interpretations. The higher your naturalist intelligence, the more accurate the habitat name you provide.

Habitat Defined

A habitat is a natural setting where living things (**biota**) are found. Habitats either exist as a **terrestrial** (land-based) or **aquatic** (water-based) home for a species. A species' needs for survival are found in a habitat – food, water, cover, and the space it needs to live. A *good* habitat, also known as a **preferred habitat**, is where all of the species' needs are met. It provides a hospitable climate, a reliable source of food and water, and ample places for plants to grow and animals to rest, hunt, play, hide, breed, and raise young. "*Pure*" habitats are hard to come by due to disturbances and succession (more details later in the chapter). A mixture of habitat types is more likely and sometimes can be difficult to name or classify a habitat.

Classifying a Habitat

Look across the landscape of your Nature Area or any outdoor location. What features are most obvious? Do you see buildings, people, or natural spaces? If you have more business buildings, cars, and people than natural spaces with plants and trees, then you are likely in an **urban** setting. Urban areas will have greater than 1000 people per square block. If you are near a large city and are surrounded by many houses and apartment complexes yet still have green spaces, then you might say you are in a **suburban** area – this is a much smaller population density. A **rural** area would have the least human population density than the other settings described, and people typically live farther away from each other with an abundance of natural areas surrounding the living spaces or town. Knowing this level of classification for your "place" helps put your observations into context and can lead to more interpretations. Begin to tell your story about the place you are in. Create an image of the surroundings that includes the natural and built environments. From there, an actual classification of your habitat will emerge.

Classifying a Habitat

To classify or name a habitat, identify the dominant landscape features and plant life. Answer the following questions to help you create a descriptive name to classify your habitat in question. The goal is to incorporate your answers into a name that serves as a concise narrative for communicating the details of the surrounding landscape. Realize that the prompts will lead you to discover a practical description of your habitat yet may not match technical classifications in a formal scientific community. However, the classification you develop can essentially convey the same thing. To provide suitable responses, refer to the rest of this chapter to gain a fuller understanding of terrestrial and aquatic ecosystems, and definitely spend time in the habitat you want to classify.

1. Is your setting urban, suburban, rural, or natural?
2. Describe the landscape in terms of its topography. Is it more elevated than the surrounding area? This could be considered an upland.

Hilly or mountainous could also be descriptions used here if appropriate (mountainous terrain is not common in the Great Lakes basin and Upper Midwest region). If in a low-lying area, then you could simply say lowland. In the end, you might feel this description is irrelevant to the habitat's classification.

3. Are you considering a habitat that is entirely land-based (terrestrial)? If so, you could use these terms to describe it: pioneer community, field, prairie, savanna, shrubland, edge, forest. If in a forest, is it young or mature? Is it a deciduous, coniferous, or mixed forest?
4. Do you have water in your habitat? If so, what aquatic ecosystem classification would you give it? Is it a lake, pond, river, stream, or wetland? Wetlands could be classified further into marsh, swamp, bog, or fen. With assessing aquatic habitats, it is also important to consider and communicate the surrounding terrestrial landscape. You would share two habitat classifications with the most dominant habitat listed first and the addition that states something like it is "surrounded by", or "with" the other habitat type you define.

Now, read the words you chose to answer the prompts about your habitat (e.g., a natural lowland mixed forest). This could be used as your habitat classification. If the wording in your "name" seems awkward, feel free to eliminate, rearrange, or add the most relevant descriptive words to create a meaningful habitat classification name. This exercise gives you a foundation so that you can figure out the plants and animals that may live in the area and the types of ecological services they may provide the habitat (e.g., oxygen producer, soil stabilizer, food source, predator, etc.). Knowing these things can also help you determine the organisms' connection to you, your role in the ecosystem, and what could potentially happen if the habitat was disrupted. Think about how your life would be affected if an entire habitat were eliminated. What would happen to the plant and animal food sources? How would the soil, water, and air quality be affected? For further research into habitats, research your state's or province's natural areas, histories, or features inventory.

Nature Journal: Go to your Sit Spot (see Chapter 1 for details on to select and inventory it). Using the prompts in this section on classifying habitats and from other parts of this chapter that may be relevant to your site, create a descriptive classification of your habitat. Draw a picture of the habitat and label the key features you used to classify it. Remember, the habitat's main features and the plant features are what gives the habitat part of its "name". To test the quality of your habitat classification, share the name you gave it with somebody and have them describe what they think it looks like. Next, show them your sketch of the habitat. Have them describe or interpret what they see. You may want to do some editing after the interactions.

© Abelle Photography/Shutterstock.com

What habitat classification can you come up with for the photograph? You could have interpreted it as a mature, coniferous upland. With this descripton, a person can visualize or know some key features for interpreting the area even without witnessing it firsthand. You can investigate your habitat further online at your state's or region's natural features inventory site.

Once you complete the habitat classification activity in your Nature Journal, it will likely become an automatic behavior to classify habitats as you see them. To help improve your naturalist practices, spend time making observations, investigating your surroundings, asking questions, and researching your interests to help you understand the type of habitat you are in. Realize, there are many classifications that exist for any given area but learning the basics can help lead you to discovering further information.

A Habitat's Wildlife

Studying habitats in different regions can help us predict what animals could be present in areas with similar conditions. This can be a lengthy process. How about starting with what you can learn from what you observed? The presence of an animal in an area indicates the site has certain features it needs to survive. If you observe an animal outside your window or in your Nature Area, consider what it might need from its surroundings. Completing Natural History Journal Pages for the animal you observed will offer insight on what it could be doing there. If you do not witness it "doing" anything to make any obvious, direct associations to its habitat, then another way to make logical inferences is to make note of the plant communities in the area. **Plant communities** symbolize the arrangement of food, water, space, and cover that signify the type of habitat. Knowing the plant types can tell us about: soil type, amount of moisture in the ground, climate (on micro - or macroscales), topography, past use or occurrences in the area, and resident organisms. All of these aspects of the area based on the plants will determine what else grows there and what types of animals live in the area. Visit Chapter 6 to learn more about plants.

Nature Journal: Consider the dominant vegetation, or plant types at your Sit Spot, Nature Area, or where you are currently. What types of plants do you see mostly (even if it does not seem like "a lot" of plants) - grasses, flowers, shrubs, or trees? Based on that dominant plant type, what can you predict about the potential animals in the area? What evidence and reasoning do you have to support that claim? Use the remaining sections of this chapter to help you make interpretations of the habitat, Chapter 6 to learn about plants, and Chapters 7 and 8 cover animals you might think are using the plants in the area. Thinking about these types of questions helps generate ideas for a field investigation (see Chapter 2). What examples can you think of for a descriptive, comparative, and correlative studies?

Animal species and populations cross over to other habitats to satisfy all their needs. Red-tailed hawks, for example, nest in dense forests yet hunt along or over open fields. Amphibians, such as salamanders are born in the water then leave to live under logs and leaves in the forest. The habitats provided in the field guide sections of this book are the preferred habitat for each species. This would be the area where the species spends most of its time and often where they breed and raise young—typically spring, summer, and fall seasons.

Over evolutionary time, animals have developed more efficient ways to use resources available to them. Examples of these **adaptations** include varying types of claws, beaks, eye position, etc. More details on evolution are shared in Chapter 4. Organisms with behaviors and traits compatible with the habitat in which they exist will make the best competitors. These individuals will remain in the habitat where they can best survive and reproduce. What is not as important for survival in a preferred habitat is usually not well developed in comparison with those features utilized most in that environment. Forest dwellers have obstructed visibility so are not known for keen vision. They will have well developed senses of smell, sharp hearing, and means for vocalization.

The health of a habitat is often reflected in the diversity of the vegetation. A variety of plant species and ages will offer a wide array of potential resources for animals. When diversity exists along with the complementing richness of species, the more balanced the habitat. Meaning, when the population numbers of species are kept under control without any overrunning the others. Diverse habitats also tend to better tolerate stresses like storm damage, disease, invasive species, and human disturbances than areas with a low species. Chapter 4 goes into these community dynamics more in detail. You can also consider performing a field investigation to determine species diversity in or between areas.

You will find two main types of species living in a habitat, **generalists** and **specialists**. Generalists, or tolerant species tend to be ubiquitous (found almost everywhere), and adaptable to conditions with flexible means of survival. Examples include the dandelion, American robin, coyote, channel catfish, and green algae.

© LeManna/ Shutterstock.com

© Karel Bartik/ Shutterstock.com

Dandelions (*Taraxacum officinale*), a nonnative species has grown in North America for centuries. These generalists can tolerate wet, dry, sunny, or shady conditions. The rainbow trout (*Oncorhynchus mykiss*), on the other hand (also a nonnative, but was introduced to the Great Lakes) is a specialist that requires cool to cold water with high levels of oxygen. Therefore, they prefer the fast-moving water (read more about aquatic ecosystems later in the chapter) and behaviors and population numbers change with any change to those requirements.

Specialists or intolerant species compete very well in their habitat, and survival drops when certain means are not met. The species will either move away or die off in the area. The Kirtland's warbler, American beaver, trout, and watercress are examples. Indicators of a healthy habitat will have tolerant and intolerant species to create a diverse, balanced, high-quality living conditions. If just tolerant species exist in a habitat, then it may indicate a concern and further investigation of the abiotic conditions, available resources, more species sampling and monitoring is necessary to determine the extent of the issue raised.

Nature Journal: Choose an animal in the last chapter of the book. Study the picture and read its description. What is its preferred habitat? Identify at least one adaptation it has to help it survive in that environment. How is the adaptation used? Do you think it is a generalist or specialist? Explain your reasoning. Discuss your prediction of how the animal's population would respond to the habitat losing its dominant vegetation type. What do you think would happen if the animal did not exist in the habitat?

Habitat Signs and Stimuli

Features of a habitat can attract animals and improve diversity. Consider the habitat surrounding your Sit Spot. There are features around you that will attract animals. These signs or stimuli include a habitat's structure, vertical layers, horizontal zones, complexity, patchiness, edge, size, special features, and the presence of other organisms (Benyus, 1989). In the Nature Journal activity that follows this section, you will try to find and explain these examples around you.

Structure. The structure of a habitat is the general look of the environment. It will include its overall shape of the habitat in terms of the vegetation. Having a mix of grasses, shrubs, and mature trees is usually more important than the type or species of plants. Height and density would also dictate what visits or remains in the habitat. Color, diversity, and general features of the terrain also make up the structure. Visual clues may have little to do with survival, but can serve as indicators if a place is good for finding food, breeding, or raising young.

Vertical Layers. Every habitat can be looked at in terms of its variety of vertical layers. Consider what exists and occurs at every level from the lowest to highest areas of the habitat—from the soil to the sky. Each layer has its own inhabitants adapted to take advantage of that zone (*niche* or ecological roles).

Horizontal Zones. Looking across an area you will find distinct areas offering wildlife a slightly different combination of food, water, space, and cover.

The soil type, plants, and sun availability may change as you view the landscape. Vertical and horizontal zones allow many species to occupy the same habitat without competing for the same resources.

Complexity. As habitats age, its structure changes by adding more vertical layers or horizontal zones leading to a more complex system. As this occurs, resident wildlife will also change. Species complexities change with in habitats as it progresses from grassland to shrubland to mature forest.

Patchiness. Within a habitat, disturbances (windstorms, insect and disease outbreaks, floods, rockslides, logging, or fires) can disrupt the habitat and add patches of an earlier stage of forest development. In other words, a clearing often opens after a disturbance occurs and can sometimes add diversity to the area. Patchiness can accommodate animal species that need both young and mature vegetation.

Edge. The edge forms where two types of habitats or vegetative stages meet. This is also called a **transition zone** or **ecotone.** Examples include a marsh turning into a shrub swamp, or simply the area between aquatic and terrestrial habitats; or a field blending into a forested area. In these zones, you will find abundant vertical and horizontal variety. This lends itself to species from surrounding communities frequenting and joining edge species. Often times, the edge will be the most diverse area when considering what exists in a given landscape.

Size. Some species have certain size requirements for nesting or breeding (i.e., Kirtland's warbler). Specialists that need large, undisturbed habitats are called area sensitive. Humans tend to subdivide large expanses of land leaving "islands" of good habitat surrounded by roads, buildings, farms, etc. These smaller islands support fewer species and become more vulnerable to predators and competitors in the area around the island. If the island or population is jeopardized, a nearby suitable habitat may not exist, and species die thereby lowering the overall species diversity in the habitat.

Special Features. A requirement needed for certain species to visit or remain in a habitat is called a special feature. It could be the presence or absence of something in the habitat. Examples include, but not limited to snags (standing dead trees), food plants, singing posts, and water access. If a habitat lacks one essential ingredient for an animal's survival or reproduction, the species will not stay there long.

Other Organisms. An organism's success in a habitat can also depend upon its interaction with other organisms in the surrounding community. The presence or absence of a species can depend upon potential competitors, predators, parasites, or disease organisms. No plant or animal lives in isolation. It lives in an environment changed or formed by other plants and animals that live there.

The test of a good habitat would include questions like: Can a species keep itself alive there? Can it successfully reproduce there? Where human habitats overlap with natural ones, the challenge is to maintain enough prime habitat for everyone through conscientious natural resource management (see Chapter 4).

> ***Nature Journal:*** Visit your Sit Spot to complete this activity. You can identify the aforementioned signs and stimuli of habitats anywhere. Even if you are in a city environment, you are in a habitat for wildlife. Discuss details of the habitat signs and stimuli that exist in your field of view – structure, vertical layers, horizontal zones, complexity, patchiness, edge, size, special features, and the presence of other organisms. Address each feature listed. If an example of something is not apparent, you may need to visit the Nature Area at a different time of day or increase the amount of time you spend at the site to become more familiar with the nature happenings around you. Think about the animals that could be in the area. How do you think they might "perceive" the habitat? This activity will help you look more closely at a habitat and interpret the types of interactions that may occur between the animals and their habitat.

3.3 Terrestrial Ecosystems

General categories of habitats on land, or terrestrial habitats can be identified as: primary communities, field (prairie, grassland), savannas (parkland), edge (shrubland, transition zone), and forests (varieties of coniferous, deciduous, or mixed). The abiotic factors, especially levels of temperature and moisture will determine these **life zones.** A region's relatively predictable weather patterns and known soil types gives rise to certain resident species. This makes it relatively easy to make comparisons of species among similar life zone regions. The temperature can range from cold to temperate to mainly hot. On average, moisture levels would be dry, moderate, to wet. These conditions determine the plant communities and subsequent animal complexes. Realize habitats will not remain the same through time and the ecosystem will shift. This means that potential interactions and relationships will change with the changing communities.

Succession

The replacement of one community of organisms (plant or animal) by another in a somewhat orderly and predictable manner is referred to as **succession.** Two types of progression exist, primary and secondary succession. The difference between the two types along with the features of each **sere** (or stage) will follow. No matter where you are, try to find examples of these types of succession and seres around you. With this practice, you will start to recognize clues of what may have occurred in the past where you stand, and help you make predictions of what type of habitat the place could transform into in the future.

Primary Succession

Look for barren surfaces. **Primary succession** proceeds in an area not previously occupied by a community. It will begin with new soil formation. Examples of when soil begins forming in an ecosystem include newly exposed rock outcrops or cliffs, windward sides of sand dunes, newly exposed glacial till, after volcanic eruptions, or on a log or animal carcass. These are areas where it is void of other plant life. Up to five successional stages (or seres) can exist throughout the Great Lakes and Upper Midwest region.

Sere One. A pioneer, or the first community will get established on a barren substrate. Lichens will typically colonize bare rock, log, or landscape first in primary succession. Organic material from dead lichens accumulate in the cracks of rock and will absorb moisture. This forms the foundation of soil building.

Lichens growing on a boulder. Over time, plants and natural weathering of the rock will start to break it down to make minerals more attainable as nutrients.

Sere Two. Mosses and ferns are the next plants to colonize the area. These plants will shade out pioneer species and replace them. Organic matter builds up as a result of decomposition of the dead plant matter.

Mosses and ferns dominating the growth on the rocks.

Sere Three. Herbaceous (nonwoody) plants will eventually grow as the soil builds and will shade out the mosses and ferns. Organic matter from decaying vegetation contributes to the soil building.

Mixed wildflowers of goldenrod (*Solidago* sp.) and *Aster* sp. Their presence indicates favorable conditions for growth, such as adequate soil quality, moisture level, and sun intensity.

Sere Four. Perennial grasses, shrubs, pines, and immature oaks will be found in this stage that will shade out herbaceous plants. More organic matter accumulates in sere four.

© Karen Benjamin/Shutterstock.com

A prairie landscape with grasses and shrubs.

Sere Five. In this mature, or climax community you will find a self-perpetuating, "stable" assemblage of organisms (especially plants) that undergoes little change in species composition and is considered to persist as an equilibrium state. Climax speciation can vary. In parts of the Great Lakes region it can be conifers, or oak-hickory, or beech-maple communities.

Primary succession can take approximately 1,000 years to go from pioneer to climax species because of the need to create soil. A disturbance will cause a change in the system and progression. Succession will not always create the same community that existed before a disturbance. Why not? Consider the disturbances from natural and human-caused events (see the following "Secondary Succession" section). Conditions in the ecosystem become altered after a disturbance and therefore will change the predictability of "what was next." However, knowing examples of pioneer plants and animals will help identify what could emerge. Once the successional clock is reset, we say succession continues in a secondary nature because of the pre-existing soil.

Another type of primary succession, **degradative succession** uses a substrate that never served in that role previously. Dead trees, carcasses, droppings, plant galls, tree holes, etc. provide a substrate on which fungi and invertebrates feed on the dead organic matter. This material becomes suitable for plants and animals to

The vertical layering of a mature beech-maple forest with an understory will likely stay suppressed until there is an opening in the canopy. Forests in the region typically do not have trees with large diameters due to their relatively young age. Intense lumbering of the old growth forest, or forest clear cutting of the fully mature, very tall and wide trees occurred in the late 19th or early 20th centuries in the region. Large trees also may not exist because of a lack of cutting. Trees compete for resources so those growing close to each other will tend to direct energy to height not girth.

A rotting log illustrates degradative succession.
A great place to explore!

live, succeed each other, and eventually disappear. Energy and nutrients are most abundant in the early stages of succession and decline as succession progresses to the latter seres.

Secondary Succession

Secondary succession progresses similarly to primary succession except that this area was once supported by vegetation. The soil and organic debris still exist, and soil nutrients are usually high. This allows for succession to happen relatively quickly in approximately 200 years or less. This type of progression occurs following major **disturbances**. Natural disturbances include fire, wind, animals, flooding, diseases, insect or fungal infestations; and human-caused events involve fire, harvesting, pollution, development, and exotic species introductions. Often, sere three is where growth regenerates—with herbaceous plants.

Community Characteristics

When comparing natural communities in the beginning and later stages of succession, certain characteristics are somewhat predictable and help the naturalist interpret the surroundings. In a pioneer community, you can expect at least a somewhat harsh environment. The inefficient energy consumption counters with an increasing biomass. Some overall nutrient loss exists due to this early stage. Typically, low species diversity exists yet fluctuations within the community are common. The dominant species type is the one with high reproductive and mortality rates.

At the climax, or mature stage of succession you will find a favorable environment with a stable biomass. Energy consumption and nutrient cycling are efficient, and a high species diversity exists. The community experiences fewer fluctuations than earlier seres. The species type that will dominate the area is one whose populations are determined by the carrying capacity of the area. More details can be found on populations and communities in Chapter 4.

Succession and Animal Life

Animals in areas change as vegetation changes. Food and shelter may be altered for some species. If the conditions are unfavorable for a species or individual, it will either leave or die. Structural characteristics of vegetation typically influence species more than species composition. The key to having a diversity of wildlife is having a heterogeneous landscape with contiguous habitat patches of various successional stages. Forested ecosystems are dependent upon disturbance for renewal and to provide biological diversity.

Field to Forest Habitats

Read the basic information about the major types of habitats in the region to help identify and interpret the area in question.

Primary Community. You know if you are in a primary community if bedrock, gravel, or sand exposures are observed with some plant communities established that are indicative of the environment. Locations include the shoreline of the Great Lakes, and rivers that help weather and expose these surfaces.

Field/Prairie. A field will refer to an area dominated by herbaceous plants that grew after an abandoned farm field or a clearing made for human purposes was left to grow untouched. Nonnative plants often become established in these areas because of disturbances to the land. This type of popular terrestrial ecosystem has replaced native prairies in the region. Prairies have very little (less than 5%) or no woody shrubs or tree cover. Native grasses, sedges, and forbs comprise plant communities. Prairies experience 10–30 in. (up to 76 cm) of rain per year, and have thick, fertile soil with deep perennial plant roots. Noteworthy prairie animals include organisms that help to mix and aerate the soil—such as ants, moles, mice, skunks, and badgers. Due to flowering plants, pollinators also abound.

Savannas. This habitat has a relatively open landscape with some shrub and tree cover (from 5% to 60%). This **parkland** has an open, or a "park-like" look with many grasses, other herbaceous plants, and sporadic trees throughout the area. Fires clear out underbrush and stimulate native plants. More than 99% of the prairies and savannas in the Great Lakes and Upper Midwest region have been converted to agriculture, eliminated for construction of human needs, or grown into forest in the absence of fire. Prescribed burns, selective logging, and planting native plants in abandoned farms or golf courses will encourage these ecosystems to persist.

Shrublands. Also referred to as the edge or as an ecotone, these areas have mostly shrubs that typically serve as a transition zone between habitats and results in usually a very diverse ecosystem. Shrubs are woody plants with multiple branches at the ground level and reach to just above 13 ft. (up to 4 m). Shrubland species of interest include low-nesting birds such as indigo bunting (*Passerina cyanea*), gray catbird (*Dumetella carolinensis*), eastern towhee (*Pipilo erythrophthalmus*), song sparrow (*Melospiza melodia*); and **mesopredators**, medium-sized predators usually because of the lack of bigger predators in the area. Raccoons, opossums, snakes, squirrels, jays, and crows will all eat eggs and young birds within the shrubland. With the rise of the coyote (*Canis latrans*) populations in many areas, how could this help the suffering shrubland bird population?

Forests. The term forest is usually only applied where trees are more than 6 ft. (2 m) tall and the tree canopy shades more than 20% of the ground surface. Interaction and interdependence of plants and animals are common features of a forest. Plants provide food and shelter for animals, bacteria and fungi break down forest litter, and birds help with the pollination of flowering plants. Every organism depends on other living and nonliving elements of the forest system.

The type of forest in an area is mainly influenced by rainfall and temperature (climate). In the Great Lakes and Upper Midwest region, the "temperate" forest dominates the area. The average temperature changes significantly throughout the year but remains moderate with cold winters and warm summers. Forests in temperate regions can be classified into the general categories of coniferous, deciduous, or mixed. More specific naming can occur as two or more dominant tree species are identified.

Vertical Layers in a Forest. Part of some animals' preferred habitat entails the behavior of the animal. When are they active? Where do their activities take place? What are they doing in these areas? Looking at a forest, spend time making observations at each of the levels. Locate and discuss your observations from the ground or forest floor up through to the herbaceous layer, understory, and finally to the top of the tree canopy.

Tree Tolerances. The first trees to colonize an area are usually shade-intolerant species and must have full sunlight to do well. Pines, black locust, cherry, poplar, sweetgum, sassafras, and sumac will grow in these areas. Once some sun coverage occurs, they cannot persist and shade-tolerant species will begin to grow and eventually take over the canopy—such as oaks, hickories, and beech.

Animals in Forests. You can find deer, mice, squirrels, rabbits, and a variety of birds in the forest. Look and listen for nesters who occupy tree cavities, canopy-dwelling species, and those existing in the detritus-based food webs. If looking for interior forest obligates, you might find numerous neo-tropical migrants such as the scarlet tanager (*Piranga olivacea*) and the ovenbird (*Seiurus aurocapillus*). To find these birds, look for contiguous tracts of forest during summer. Can you guess which of the two bird examples is usually heard rather than seen? What else can you find?

Having a diverse representation of animals indicates a likelihood of a variety of ecological services provided. To support that high animal species diversity, an assortment of plants and varied landscape may exist. Each organism has a **niche** or job to carry out in the ecosystem. Species are adapted for the habitat in which they spend most of their time. To learn more information about ecology and evolution, review the basic concepts in Chapter 4.

Where would we be without plants? Jobs or services provided by plants include adding oxygen to the surrounding air or water, taking in carbon dioxide, stabilizing soil to prevent erosion, and providing nutrients or shelter to animals. The animals help ensure their targeted plants or prey animals do not outcompete others and dominate the landscape. An imbalance of species causes the ecosystem to shift toward low species diversity and ecosystem functioning. Other ecosystem services of animal niches include aerating soil, eating pests of other animals, pollinating flowers, carrying plant seeds, creating homes for other animals, and helping break down dead organic matter.

Humans, another form of animals—often have destructive tendencies dealing with nature, but we have a sense of control to make positive choices. People enjoy the beauty of a healthy forest. They may decide to retreat to these places for peace and solitude while simply relaxing, walking, bird watching, or camping. Other recreation practices might involve trail riding on off-road vehicles or snowmobiles, or people might enjoy hunting or trapping game species (be sure to follow your state or province's laws). The more people who get out and enjoy nature, the more people who will want to protect our natural resources.

Use the natural history and field guide sections of this book to learn more about the fungus, plants, and animals that exist in the field to forest habitats. The only thing constant about an ecosystem is that they never stop changing!

Forestry Management Goals

The ultimate goal of managing forests focuses on encouraging plant diversity that supports diverse animal communities. Related to human interest and the health of the ecosystem, management ensures sustainable monitoring of all biotic and abiotic factors in and around the forest. Forest management encourages forest regeneration to help with providing us with a continued supply of the natural resources while creating multiple age distributions through "edge" habitats and horizontal and vertical structures. This scenario adds more micro-environments and accelerates the system's metabolism and nutrient cycling. Management decisions are based on scientific, social, and political concerns.

3.4 Basic Geology and Soil Types

As a naturalist, you also pay attention to the lay of the land—the topography, rock exposures, and soil type. You think about what you are observing and why it exists in that location or the patterns you see. **Geology** is the science that deals with the earth's physical structure and composition, its history, and the processes that act on it. Most of this surface geology you see in the Great Lakes and Upper Midwest region is considered a glacial landscape as discussed in Chapter 2. Much of the substrate is glacial till with features like moraines, bedrock exposure, river valleys, and former lakebeds throughout the area.

Rock Types

Observe a rock up close with your magnifier. **Rocks** are composed of a variety of mineral types. What do you notice about the grains of the minerals?

Minerals. A rock's minerals have a unique composition of one or more chemical elements that bond in a predictable manner. A quick classification of minerals can involve determining if you have a metallic or nonmetallic mineral. This is based on the mineral's luster, or how shiny it seems. Other things to observe about minerals include: color, texture, crystal structure, if you can create a colored streak on unglazed porcelain, and how hard it is (Mohs' hardness scale (H) 1–10) in relation to scratching it with your fingernail (H = 2.5), a penny (H = 3), a piece of glass or knife (H = 5.5), or a steel file (H = 6.5).

Another thing to consider with rocks is to identify which of the three rock types it may be—igneous, sedimentary, or metamorphic. All rock types undergo continual transformation as illustrated in the rock cycle. Through time, rocks are exposed to different conditions causing them to appear as they do. Location, time, temperature, pressure, erosion, transport, and deposition contribute to the appearance and composition of the mineral grains within a rock.

Igneous Rocks. Igneous rocks will form from the molten rock or magma that pushes up (**intrusive rock**) or is ejected (**extrusive rock**) to the earth's surface. This hardened lava has an interlocking mosaic of crystals of varying sizes. Intrusive rocks, like granite and gabbro have large grains whereas extrusive rocks have smaller grains as obsidian and basalt.

Sedimentary Rock. The rock type mostly found within the Great Lakes and Upper Midwest region consists of **sedimentary** rock. Small grains of mud, sand, gravel, or fragments of plants/animals have become cemented together in layers over time.

These grains or fragments have eroded from rocks and eventually bond into plate-like layers or may have pebbles cemented in a nonuniform pattern. Common examples would include mudstone (shale), sandstone, limestone, conglomerates, breccia, and coal.

Metamorphic rock. Metamorphic rock forms when existing sedimentary or igneous rocks are put under extreme pressure to cause the minerals to change in appearance or composition. You may find coarse minerals of wavy or stretched bands, and others have fine-grained minerals. These rocks are difficult to break. Examples you may find include gneiss, schist, quartzite, mica, slate, and marble.

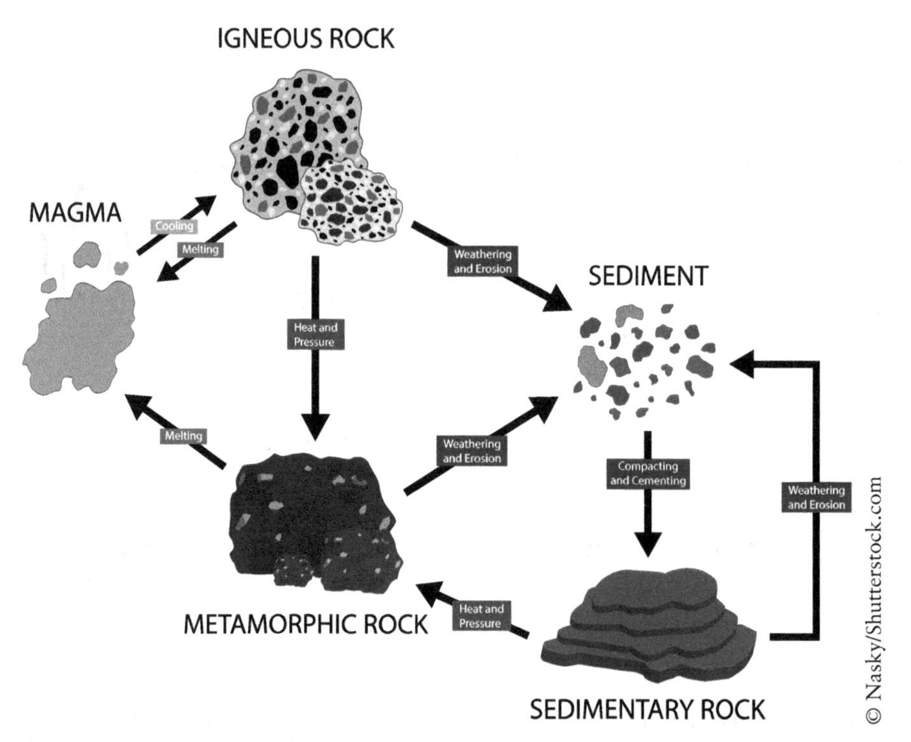

The rock cycle shows the transformations a rock can undergo.

Fossils. Another fun type of rock to find contains fossils. Fossils are the preserved remains or imprints of plants and animals that died long ago. Typically, hard components of an organism like bones, teeth, or wood will infuse with minerals to make them rock-like. The fossils can stay preserved in stone for thousands to hundreds of millions of years. Fossil hunting is best in sedimentary rock.

Fossils

A Fossil is the remains or marks of an organism that lived a long time ago, and covered with mud, and under right conditions converted into rock

Body Fossil

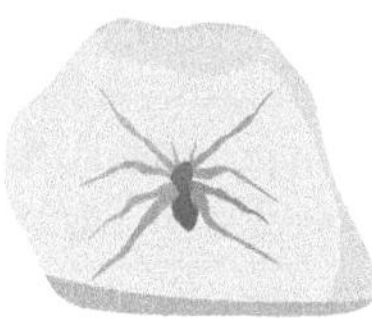

Body Fossil is formed when organism or any parts of it are kept in rock

Mold Fossil

Mold Fossil is the cavity that remained in rock after an organism died and decayed

Cast Fossil

Cast Fossil is formed when a fossil cavity is gradually filled with minerals

Trace Fossil

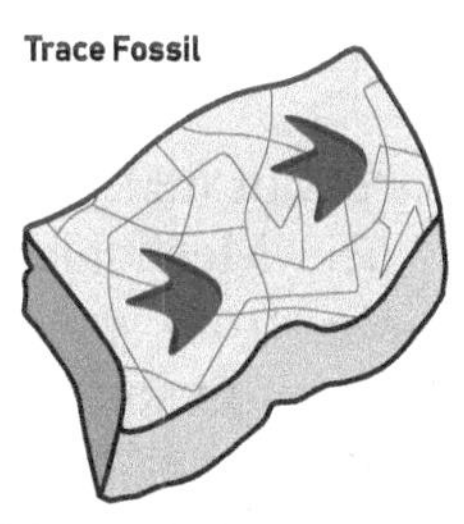

Trace Fossil is formed when organism leave traces or marks on a wet mud which converted later into rock, such as footprints, or a leaf left in a wet mud

Nature Journal: Find a rock, preferably one found in its naturally occurring place and not part of landscaping (but that would suffice!). If available, use a magnifier to make observations about the rock. Include size, colors, luster, and shapes of grains in the rock. Discuss the type of habitat where you found the rock. In what type of habitat do you think the rock originated? Make a sketch of your rock and diagram it with labeled features. Is it igneous, sedimentary, or metamorphic? Next, research the rock type more specifically. Determine its composition. Discuss how you think the rock was formed. What types of rocks are naturally found near where you live? Find a good field guide or internet source for identifying those nearby rocks and fossils.

Soil Basics

Nature as we know it would not exist without soil. Soil is the interface between the solid earth and the rest of the universe! Soil is not "dirt," it is a living ecosystem (dirt is soil where you do not want it). The study of this complex substrate is termed **pedology**. Soil is a natural product formed and synthesized by the weathering of rocks and the activities of organisms. Most plants need soil to survive, so we definitely want to encourage good soil formation.

The best soils in the world are found in the temperate zone where land has optimal temperature and precipitation. Colder areas have a shorter growing season with slow decomposition rates and poor soil. Deserts have very low precipitation with thin and porous soil. In the tropics, an area with a long growing season has high decomposition rates but minerals and nutrients are used by the many plants to create a thin upper soil layer of poor quality. In grasslands or prairies, mild climates exist with low moisture. Here, you will find a high turnover rate of nutrients thereby creating rich topsoil where you may find it 1–2 m thick in our region. Forests, contrarily, have trees that require a vast amount of nutrients from the soil. In a coniferous forest, the needlelike leaves are acidic and will create poorer soil whereas a deciduous forest will tend to have better soil quality.

Soil Formation. The natural process of making soil involves mechanical (physical) and chemical weathering of rock. Mechanical weathering includes the wearing down of rock into smaller pieces from water, wind, temperature, and roots. Chemical weathering of rock will result in soil formation from rainwater filtering through it and soil organism activities that combine with water, oxygen, acids, and dead tissues.

Five soil-forming factors should be considered when trying to classify the soil: parent material, climate, biotic factors, temperature, and time. The parent material is the starting material that affects composition of the resulting soil. It can include bedrock (solid rock making up Earth's crust), glacial deposits (**drift**), wind-blown sand and silt (**eolian**), or stream deposits (**fluvial**). Climate-related factors involve temperature, rainfall, elevation, and latitude. The climate also influences the regional plants and animals that contribute to soil formation.

Biotic factors would include input and degradation from organisms. Topography will influence weathering in terms of water runoff, drainage, and erosion. The weathering, accumulation, decomposition, and mineralization of the soil will take time.

Nature Journal: Look for signs of life in the soil. Use a hand trowel to create a hole to see what you may find. You can also move leaf litter or logs to see what exists in the decaying layer beneath it. You can also create an area made with a small piece of plywood (~1 m^2). Simply lay it flat on the ground in a natural area where you have permission to do so. Remove it regularly to record what you find beneath the wood. You will likely eventually find a variety of soil-creating organisms.

Characteristics of Soil. Consider the color and texture of soil. You can make interpretations from these characteristics. For soil color, if you find light soils, it may indicate acidic conditions or the presence of light minerals such as carbonates or quartz, dark soils are rich in organics and conducive to plant growth, and yellow to red tones indicate the presence of iron. Color also helps determine soil horizon distinctions and moisture content in the soil.

Soil has many factors contributing to its color. The parent material minerals relate to color. Older soils tend to be lighter and often more red than newer soil. The climate may encourage the leaching of minerals through the soil, remove soil coatings, or even enhance red. Topography differences can account for difference in color. Uplands are brown and red, and lowlands are grayer. Vegetation types alter color, such as conifers having less color due to acidic conditions and greater leaching compared with prairies with organic, darker colors.

Texture will help you classify the soil further based on its grain size. This will determine its water-holding capacity and the types of vegetation tolerant of these conditions. There are various methods for identifying soil type, but a simple test by feel will be a good start. Take a pinch of soil and place it in your palm with some water. Rub it to determine the relative amount of grit (sand) present. Is it half of your sample? Is it entirely smooth? Make some deductions from the information provided here and investigate further online. The United States Department of Agriculture (USDA) will have details on how to identify soils more accurately.

Nature Journal: At your Nature Area, what type of soil exists throughout? You may have differing soils if you have varied relief or topography, or if you observe changes in the habitat's vegetation. Observe a sample of soil under your magnifier. What colors do you see? Describe the grain shapes. How big are the grains? Feel the soil between your thumb and forefinger as described above for determining relative texture. Does it feel smooth, gritty, or a combination of the two? You can visit the USDA website to conduct an actual soil survey for your site, or research soil types at your local library.

To analyze soil particles further, you can separate the grains by actual size. Variation in size and shape of soil particles ranges from the largest to smallest types as follows: gravel is >2 mm; sand 0.05–2 mm; silt 0.002–0.05 mm; clay is <0.002 mm; loam is soil with a relatively even mix of pore and particle sizes. Loam with a pH close to neutral is the most desirable for agriculture.

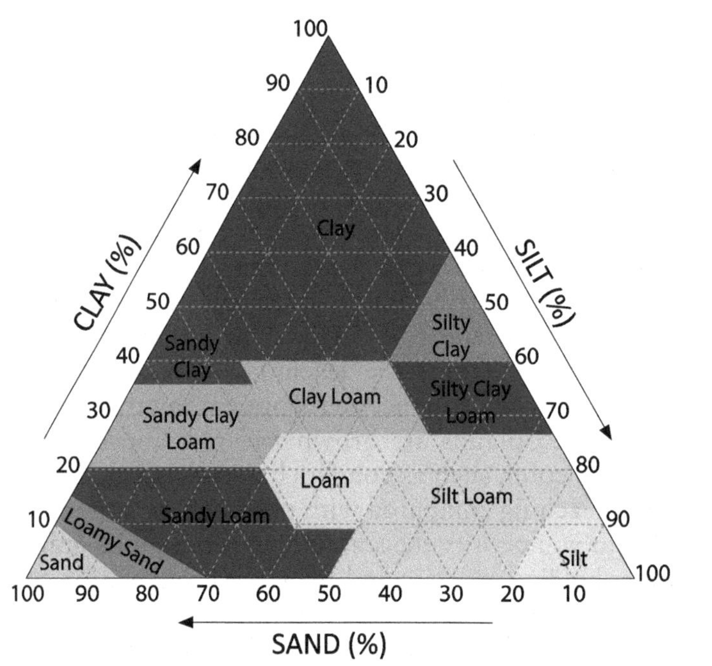

Use the soil chart and the steps described in the next Nature Journal activity to help you determine your soil sample's texture and classification.

> ***Nature Journal:*** Separate the soil textures. Pour a cup of soil into a quart or more of water with a little liquid soap. Allow for the soil particles to settle. This process may take several minutes to hours. Measure the thicknesses of the distinct layers that are formed. Sample each layer to make observations under a magnifier.

Texture also determines the porosity and permeability of the soil. **Porosity** is the amount of pore space between soil grains. **Permeability** is the rate at which water moves through the soil profile. Fine particles will have low porosity and high water retention, whereas coarse particles will have larger air spaces between the grains so that less water is held and will pass through the soil. A soil with fine particles will have small pores which makes it difficult for water to pass through, so it has low permeability. Generally, as grain size increases, so does permeability. Porosity depends more on the range of different grain sizes in the soil. If all grains are about the same size, then porosity is about the same for fine grained (many small pores) and coarse grained (fewer large pores) soils. In soils with many different grain sizes, porosity is reduced because small grains fill in the pores between large grains.

> ***Nature Journal:*** Select a location in your Nature Area where the soil is exposed. Pour a cup of water (250 ml) on the surface, and time how long it takes for it to infiltrate entirely. Now, do the same thing for an area covered with vegetation. How could these rates affect the surrounding habitat?

The soil profile is another consideration within a habitat. Have you ever witnessed a freshly exposed or dug area of soil? Did you notice differences in color and texture? There is a story that happens in each of these layers, or horizons.

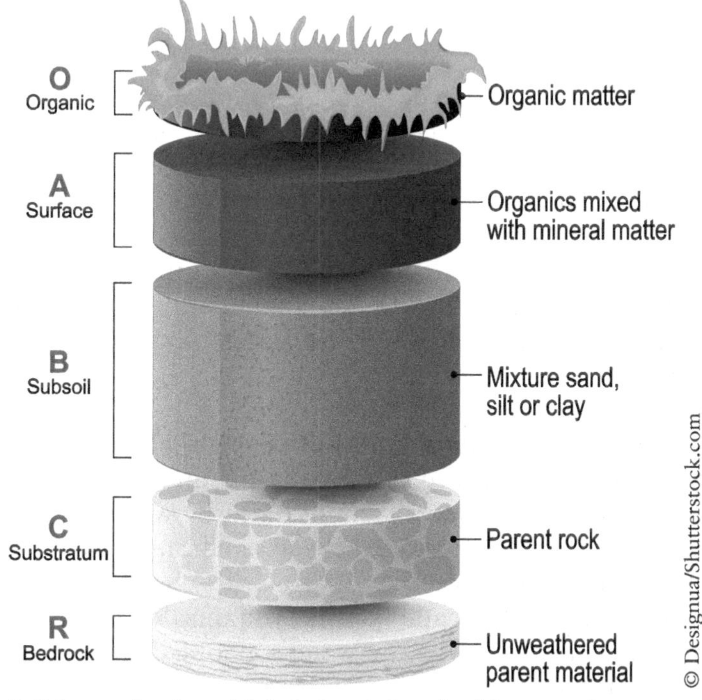

Soil layers (horizons) labeled and described in general terms.

Look at the diagram. In the top O horizon, you may have freshly fallen or decomposed leaves, twigs, animal waste, and fungi. This is called the **leaf litter** or undecomposed organic matter. The A horizon is the topsoil with porous, partially decomposed matter (**humus**) that is dark in color and crucial for plant growth. Together, the O and A horizons have the most biodiversity. Roots and most of the decomposers are typically found here in this area subjected to changes in soil temperature and moisture.

The B horizon, or subsoil is mostly inorganic (abiotic) with nutrients from A and broken-down minerals from C. In layer C, you will find low or no nutrients and it consists of weathered bedrock or parent rock, and the R horizon is solid rock with the source of the minerals in the area. Throughout the Great Lakes region, the bedrock is mostly buried by glacial deposits.

Soil Organisms. Life living in soil has value. Those animals that burrow create openings in the soil for the necessary distribution of oxygen and water. This is called **aeration.** Decomposition (see Chapter 4), which is due to living things, allows nutrients to recycle throughout the area. The leaf litter quality along with the conditions of the physical environment (temperature and moisture influences) will affect the rate and quality of decomposition. This process involves all consumers, especially bacteria and fungi.

Considerations. Soil quality can degrade for various reasons. **Erosion** is a major problem. This is when soil particles and nutrients are removed by wind or water action. It can also be attributed to farming practices, logging, overgrazing, and recreational vehicles. To create new O and A horizons, it can take 200–1,000 years to build 1 in. (2.54 cm). Is this a **renewable resource**? A renewable resource is a natural feature that humans depend upon and can be replenished within a person's lifetime. Erosion coupled with rapid population growth is a crisis for the future of agriculture. What types of soil and agricultural laws have been enacted where you live?

Terrestrial ecosystems are closely tied to aquatic ecosystems. What happens on land affects the water. The water that passes through the terrestrial environment will carry with it the nutrients needed by organisms, and pick up other substances like minerals and soil and transport them into nearby waterways. Runoff of soil into water (**sedimentation**) will pollute the water and cause disruptions in aquatic ecosystems. These are global concerns.

When trying to interpret aquatic systems, you must consider the surrounding landscape to truly understand why the water's abiotic conditions exist and why certain biota inhabit the area.

3.5 Freshwater Ecosystems

There is a direct connection between land and water. The nutrients and salts from terrestrial habitats are resources carried by water that flows across the varied landscape to ponds, lakes, and oceans. In this section, you will start by trying to understand properties of water—the vector of nutrients or home to many organisms. This will help you interpret aquatic ecosystems. Knowing the characteristics of the unique water molecule leads the naturalist to realize why certain conditions or species relationships exist in aquatic environments. The foundational knowledge can be used to predict consequences of an altered system and derive potential remedies or ways to prevent issues from arising.

> ***Nature Journal:*** First, what do you already know about water? Consider its properties and where you might find it. Next, think about a time you spent near a lake, river, wetland, or ocean. Discuss what you remember about the water and the surrounding landscape. What might the naturalist be interested in observing or studying within an aquatic ecosystem?

Water Properties and Nature

The quantity and quality of water concerns everyone on the planet. Conserving and preserving any of our water sources should be a priority for our health and sustainability of this precious natural resource. In terms of interpreting nature, recognizing water in two of its three states is usually obvious—as a liquid and solid state (ice). The third state, gas, is present when water evaporates and forms water vapor (see Chapter 4 to learn about how the water molecule cycles through nature). Of these three states, water in the gaseous state is the least dense due to the absence of hydrogen bonds holding molecules together. The **molecular arrangement** of liquid water is the densest of the three and becomes less dense as its bonds align while expanding the water molecules as it freezes into the solid state of ice.

Another phenomenon in relation to water's density at varying temperatures occurs when water reaches its densest at a single temperature. As previously discussed, water becomes less dense as water warms and as it cools to freezing at 0 °C. At 0 °C, water changes to ice that floats as its density decreases. Ice is a poor conductor (in other words, a good insulator) and reduces heat loss from below. Only shallow ponds ever freeze solid. When water reaches 4 °C, water molecules are close together and become the densest form of water. This can become the driving force of **lake turnover.** This occurs in regions where surface temperatures of water could reach 4 °C in the fall and spring seasons. The dense

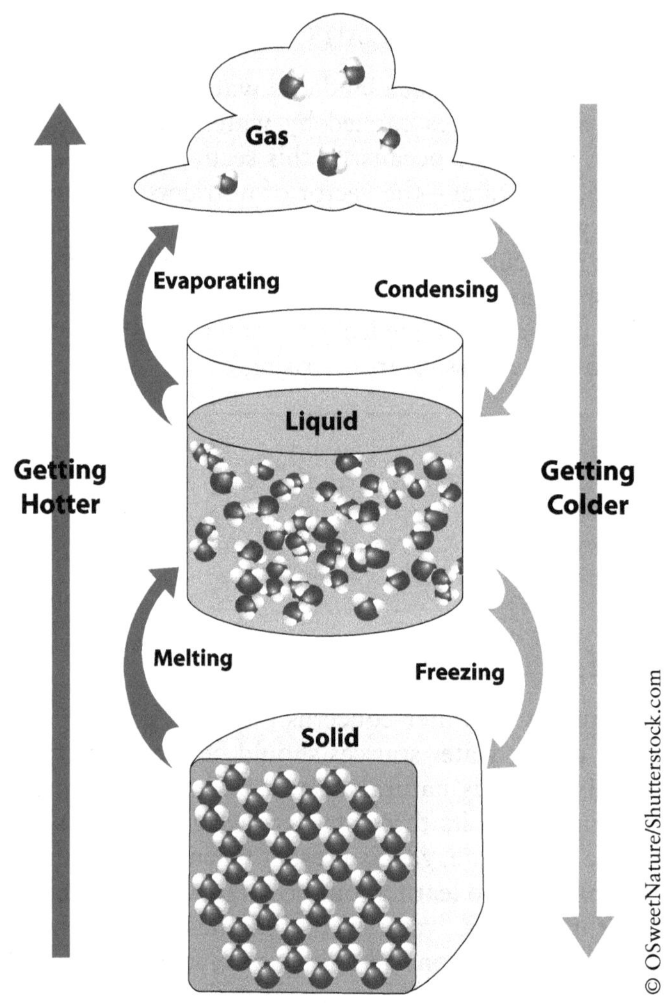

Note the molecular arrangement and temperature of water in all three of its states—solid, liquid, and gas.

water causes a displacement of the water temperature layers beneath it that results in changes in the dynamics of the ecosystem. More details are discussed later in this chapter.

In winter when surface water reaches 0 °C, the molecules align and expand into solid water that floats, or ice. This permits life to still exist in the underlying liquid water in lakes and rivers. Additionally, water serves as an **insulator** in all three of its states. Life in aquatic habitats can flourish because of this property. Plants and animals are seldom affected by sudden temperature changes in the air

Grapes growing near a large lake.

due to its **high specific heat**. It takes a large amount of energy just to raise the temperature of water by one degree Celsius. This is especially the case in deep water. Temperatures there change even more slowly. Moreover, the climate of the landscape surrounding large bodies of water, such as the Great Lakes or ocean is buffered by water's insulating property. Water releases energy from these large water bodies when the surrounding area is colder than the water. Large lakes or the sea will absorb and release heat much more slowly than the air. Therefore, the near-shore landscape climates stay warmer into the fall season and stay cooler longer into the spring. This climate is ideal for orchards and vineyards. ***Activity:*** Use a map and other resources to locate where you would find these fruit-growing farms—are they by a large body of water?

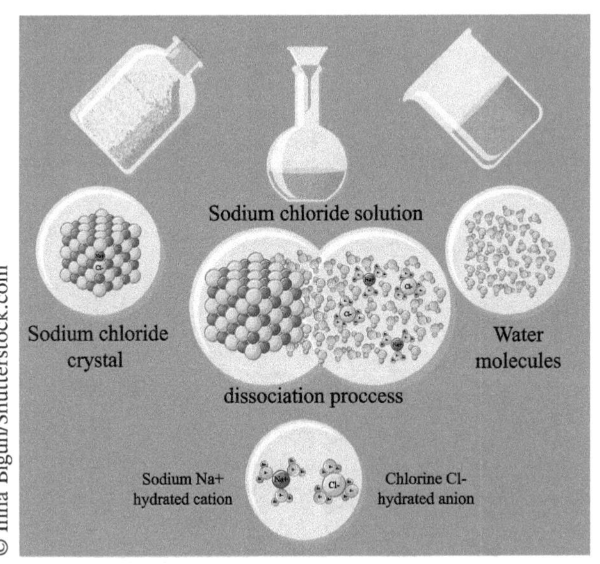

Sodium chloride, or table salt will dissociate or dissolve in water. The opposite charged ions from each molecule will attract one another forming a sodium chloride solution (salty water).

Another relevant property of water pertains to the actual water molecule. It has an uneven distribution of electrons causing a slight charge at each end. Water is made up of two parts hydrogen (a positive charge) and one part oxygen (a negative charge). Water will attract positively or negatively charged ions—like a magnet—opposites attracting one another. Because of this phenomenon, water is often referred to as the greatest solvent in the world.

The uneven distribution of electrons has another great significance in nature. It allows for plants to acquire their necessary nutrients in the soil. The water molecule will form strong bonds with many things and will separate molecules into individual ions. The compounds attracted to water are called **hydrophilic** ("water loving"), and the phenomenon of **adhesion** allows plants to absorb the separate minerals. Those certain minerals are **nutrients**, necessary for nourishing plants and animals, and required for survival by helping build cells and tissues. Floating plants take minerals from water, rooted aquatic plants from the lake or river bottom, and animals acquire minerals from the plants and animals consumed (think about it—animals are more than half water!). Minerals also get released from plants and animals as they die and decay in the water. This keeps those important minerals cycled in the ecosystem.

Cohesion is another water property that has a strong bond formed, but this time it is between water molecules (water "sticking" to water). Pour water in a glass and you will see this property in action—no separation observed in your liquid! After you drink your water, you will notice remains of water on the sides of your glass—an example of adhesion. The cohesive and adhesive properties of water are also at work in nature.

Beads of water on swan feathers demonstrate the water's high surface tension.

Consider how water moves through plants. With the presence of the sun and warm temperatures, evaporation of water occurs on the surface of plants (and of course other things). Evaporation pulls the cohesive water molecules from the plant roots, and along with adhesive properties, the water moves up through the plant xylem and out its openings (stomata)—this is called evapotranspiration (see Chapters 4 and 6 for discussions on the water cycle and plant characteristics). ***Activity:*** Have you ever put food coloring in water you gave a cut flower in a vase? Or how about putting a clear plastic bag around a plant or leaves of a tree branch to capture this evapotranspiration process? These things would demonstrate the phenomena of cohesion and adhesion properties.

Cohesion and adhesion are also at work when considering water's **high surface tension.** This creates a habitat for plants and animals and allows water to "bead." Each water molecule is pulled equally in every direction except at the surface where there is an interface between the water and the air. The strongest pull occurs between water molecules resulting in an elastic-like film that forms at the surface. What have you observed that demonstrates water's high surface tension?

In some cases, molecules will "fear" water or simply not bond to it. These molecules are termed **hydrophobic**. Oils are a perfect example. Pour some cooking oil into your glass of water and you will notice it remains intact and does not mix with the water to form a solution. The oil on feathers, especially those found on waterfowl allows water to bead and not soak into the feathers. In the bead of water, the water bonds strongly together, while the water molecules form more closely at the surface because of a weaker attraction to the air. Again, a high surface tension is demonstrated. Petroleum-based oil from oil spills in our natural

A barrier to contain and collect the oil pollution in the water.

surface water causes concern as that oil clogs respiratory passages in plants and animals and destroys tissue. However, due to the oil's hydrophobic property, environments can be saved. Oil can be harnessed from water since it does not go into solution. While still harmful to the aquatic ecosystem, the oil can sometimes be removed by professionals and disposed of by proper means and standards that do not pollute the environment further.

Each of us will have a different visual of what a clean, or good quality aquatic environment looks like. By the way, never drink any water you find in nature—no matter how clean it seems. Good quality aquatic ecosystems will have diverse food webs with balanced nutrients. Through plants photosynthesizing, the respiration of plants and animals, and decomposition, the necessary oxygen and carbon dioxide gases will cycle through the environment.

Oxygen is required by aquatic plants and animals, but can be limited in some types of water. Although soluble in water, oxygen diffuses very slowly into it. Therefore, lower levels of oxygen exist in aquatic environments than in the air. When comparing standing water to moving water (e.g. lakes versus rivers), less oxygen exists. So, natural currents, or wind and wave action will speed oxygen diffusion into water. Also, cooler water will hold more dissolved oxygen than warm water. Sensitive species like trout need high oxygen. If the environment warms, then these fish will leave the area or potentially die. In contrast, carbon dioxide is more soluble in water than oxygen. Carbon dioxide, which is necessary for photosynthesis, dissolves from the atmosphere into water systems easily, becoming available to plants. Other sources of this gas include decay of organic material, plant and animal respiration, and groundwater. More details of biogeochemical cycling are found in Chapter 4.

While considering the contents of water, you might recognize differences in clarity. Another property of water is that it is transparent. In aquatic ecosystems, light needs to penetrate the water to allow for photosynthesis. If the water is cloudy or **turbid** that indicates suspended materials are absorbing most of the light and solar energy. This tends to occur where there is runoff and after heavy rains or snowmelt. Fewer plants will likely result because of low light energy, thereby decreasing the levels of the necessary oxygen in the system. Additionally, the water will become warmer because of the additional contents in the water column, resulting in less oxygen.

Does your lake look turbid? What is the material suspended in the water making it cloudy? Look around the perimeter of the lake or along the river. Do you notice any bare dirt? Erosion of that sediment into aquatic ecosystems is a major concern. Not only will the temperature of the water warm and less oxygen made available, but the destruction of habitats and fish spawning areas will also occur. The easiest remedy for this situation is to plant native shrubs and trees. The way to prevent erosion and runoff into aquatic ecosystems is to keep the native vegetation along shorelines.

Nature Journal: Now, think about your water use. What are the ways you use water? On average, how much water do you think you use in a day? There are insightful water use calculators online to help you determine the amount. Use this method to determine your water use footprint. What are your thoughts about the results? Why should people be concerned about water use? List ways to minimize your water usage. Where does all your water go after you use it? What types of things does that water attach to as it moves through your landscape? How does people's water use affect your Nature Area?

Lake Environments

Have you ever spent time at a lake or pond? What types of things do you remember observing? How did it make you feel? What do you know about what happens under the water? In this section, we will make connections within watersheds to help understand lake ecosystems. As discussed in Chapter 2, a watershed pertains to a drainage basin or catchment area where all the surrounding area "sheds" or drains water to a common low-lying body of water—like a lake, river, or ocean.

Given the location in which this book covers, we must include the Great Lakes. These lakes dictate much of what exists within its basin—the climate, weather patterns, habitats, and people. The interconnected Great Lakes contain almost 20% of the world's supply of fresh surface water. These massive North American inland freshwater lakes—Huron, Ontario, Michigan, Erie, and Superior (HOMES) border eight states and one province. They contain 23 quadrillion liters of water and the deepest point exists in Lake Superior where the water sinks to depths of 406 meters. Thousands of species live in and around these incredible natural, aquatic features. Some of the unique features you would encounter traveling along the Great Lakes would include the world's largest freshwater dunes, fossilized coral, glacial and volcanic geology, heavy forests, petrified trees, and the dramatic Niagara Falls plunging within this system before the waters eventually travel out to the Atlantic Ocean.

Beyond the magnificence, the Great Lakes have had a long history of concerns but also these immense wonders have benefits as well. In recent years, governments and partnerships between the United States and Canada have formally

Nature Journal: Choose one of the Great Lakes nearest to you, one you have visited, or one that you would like to visit someday. Research the Great Lake in terms of its history on geological, cultural, industrial, ecological, and personal levels.

The Great Lakes and the neighboring states and provinces.

recognized the tremendous value of this natural resource. Significant efforts and funding have helped combat pollution, protect coastal habitats, and prevent or minimize the spread of invasive species. Protecting the Great Lakes will require the continued collaboration, but the effort is critical so that we can preserve the wonder and beauty of this precious ecosystem.

While learning about aquatic systems, you will want to have a familiarity with the associated terminology. **Aquatic ecology** refers to the study of the interactions between living and nonliving things within these water systems. **Limnology** focuses on the physical, chemical, and biological aspects of inland aquatic environments (**oceanography** refers to this type of study for the sea, a large, saltwater environment). Reference to still waters, or lakes and ponds is considered **lentic** environments, whereas **lotic** conditions refer to moving or flowing waters such as rivers and streams.

In general terms, some people define lakes and ponds differently. Naturalists may identify the main difference between these lentic systems based on size. Lakes will be larger and deeper than ponds with a **fetch**, or greatest distance across the surface greater than five acres. With the larger span, wave action may be greater in lakes allowing more oxygen to dissolve into the system than in ponds. Both lakes and ponds have standing water, but some ponds may lose its water for about eight months. Due to its greater depth, lakes will have a varied vertical temperature compared with the uniform temperature of most ponds. The deeper lake environments have fewer rooted plants than ponds. Ponds tend to have shallow water that allows rooted plants to grow throughout. You will find organisms adapted to any of these lentic conditions.

Environmentally, you are starting to understand water properties and characteristics of lakes to help you interpret an ecosystem. Realize that lake turnover can circulate possible pollutants that had settled into the sediments that could affect the ecology, health of the lake, and humans. Rivers may carry toxins into lakes, and since lakes may have relatively low oxygen, the pollutants may persist and have a negative effect on the system. What else is important to know about a lake environment? Making good observations in and around a lake will help you understand the system and identify any concerns.

If you can make connections between things happening within an environment, like cause and effect relationships, then you are demonstrating systems thinking. This is ecology. When you add humans to the scenario, then it becomes social-ecology. Do you spend time thinking about what you observe in nature? Revisit Chapters 1 for ways to raise your nature awareness, especially if the aquatic environment is new to your nature study. Thinking about potential interactions that could occur in and around the lake can lead to the preservation and management of lentic systems. What else can indicate the environmental quality of a lake? Knowing the complexity of the food webs that exist within it can provide the relative quality of the lake. Identifying **indicator species** in the area, or those organisms sensitive to change in the environment can signify favorable conditions and potentially good water quality. What is the highest **trophic level**? If a healthy top predator exists in the system, then niches are filled, and a balanced ecosystem may result. Knowing these and other details of the lake, like its water chemistry, requires monitoring of the lake to get a good representation of its trends. Monitoring, or close observation also reveal if exotic or invasive species appear. How could the lake be affected if a new species becomes established? How are humans impacted by the presence of invasive species? Refer to the section on biodiversity loss in Chapter 4 for further insight on this topic to help you interpret your surroundings.

Often times, changes to the lakes happen due to human actions. Altering the surrounding landscape, in particular, eliminating native shoreline vegetation will increase runoff of sediment potentially creating turbid waters and increasing nutrients in the lake. The result of the excess nutrients, especially nitrogen and phosphorus will increase photosynthesis in the water. Typically, algal blooms will result and cause an altered ecosystem—an environment of poorer quality due to less oxygen available in the water and decreased transparency.

Types of Lakes. The origin of lakes vary. In the Great Lakes and Upper Midwest region, most lakes exist due to glaciation over 9,000 years ago. Modern examples of lake formation through these means can be seen when ice blocks melt, or the development of moraines cause a dam effect (see Chapter 2 for further details on glaciers). Rivers can create lakes when the outer edges of meanders become naturally dammed or sediment may build up at the mouth of a river. Beavers create lake or pond environments from dam building on streams. Landslides and tectonic plates may also create new lakes. Often times, people will form lakes for farm animals' water source, irrigation purposes, stocking fish, or aesthetics. Healthy aquatic ecosystems will attract a variety of wildlife and can bring beauty and pleasure to a landscape.

The productivity or level of nutrients found within lakes can be classified by various means. For the purpose of this book, the three types of lakes are found at opposite ends of the nutrient spectrum or in between.

Oligotrophic and eutrophic lakes are interpreted as having few nutrients and many nutrients respectively, and mesotrophic with a relatively, medium-level of nutrients. By knowing the features of each, the naturalist can make some deductions and predictions about what may live in the area.

Oligotrophic lakes will have a deep, steep-sided basin due to a similar shaped watershed or surrounding area. This small drainage area provides limited nutrients to the lake so a low algal biomass likely exists, and its steep sides do not allow many shoreline plants. Light penetration can shine deep into the lake because of these conditions. The lack of plants and the lake's great depth contributes to clear, cooler temperatures and low biomass of species dwelling on the bottom of the lake, or **benthos**. Organisms living in oligotrophic lakes will have sensitivities or adaptations to the abiotic factors.

In **eutrophic** lakes, a large watershed contributes to this shallow ecosystem. The high level of nutrients (i.e. nitrogen and phosphorus) found in the water likely originates from the runoff and adds to high algal biomass. The shallow lake allows for rooted plants in a broad shoreline area. With an abundance of life, significant death tolls occur, which builds up the bottom of the lake making it shallower. The light and oxygen are lower in this environment than in oligotrophic conditions. Water temperatures tend to also be warmer and turbid. Animals found here will tolerate these conditions, such as carp or suckers. The benthos will have a high biomass.

If a lake does not seem to exhibit pure oligotrophic or eutrophic conditions, then often it is acceptable to consider it as **mesotrophic**. The lake will have, nutrient levels and other characteristics falling somewhere between the other lakes. Mesotrophic lakes tend to have what anglers might consider great fishing with smallmouth bass, perch, walleye, and northern pike (see Chapter 8 to learn about fish and to peruse a field guide of fish in the region). Mesotrophic lakes also tend to exhibit complete turnover in the fall and spring months which offers a thorough mixing of the nutrients and disperses oxygen.

Lake Habitats. Imagine standing on the shore of a lake looking out at the water. Where do you think you would find the most biodiversity around you? It is likely (or hopefully) right where you stand. You may have terrestrial plants at your back and you could notice plants growing out of the water nearby. This is a great place to view wildlife. The plants in the water are rooted in the benthic and exist in the very diverse **littoral zone**, or nearshore area where plants grow. You may find plants **emerged** (rooted with leaves growing above the water line), **floating** (rooted with leaves laying on top of the surface) or **submerged** (all photosynthesizing parts are below the water surface). Panfish, like northern bluegill and pumpkinseed will live among the cattails (emergent plants) and lily pads (floating plants). Also, in the **benthic zone** you will find organisms who are mostly scavengers or decomposers like scuds or midges (see Chapter 7 for more examples of freshwater invertebrates). Beyond the littoral zone, you will notice open water, or the **limnetic zone.** Here you may discover algae, cyanobacteria, diatoms, protozoans, zooplankton, and fish like walleye. Due to water's high surface tension, a habitat at the water's surface exists called the **neuston**. You might see water striders, duckweed, and mosquito larva. Water's level of transparency at various depths will also create different habitats for animals. They can find refuge in the darker regions of the lake or take advantage of visual opportunities in the light for seeking food sources in this **euphotic zone.** ***Activity:*** Draw a diagram of the lake habitats described in this section. Be sure to label the littoral, benthic, limnetic, neuston, and euphotic zones and the characteristics and organisms within each area.

JOURNAL

Nature Journal: Look at a map. How close is the nearest **inland lake** (a lake other than one of the Great Lakes) to your Nature Area? A **topographical map** would show the **contour lines** around that lake to indicate watershed size and the steepness of the landscape. Other maps may show relief or relative changes in elevation to help determine the same things. Research the lake's **bathymetry** or depth contours. This could help reveal habitats or some potential fishing areas! Make some predictions about how that lake might look in terms of the surrounding land, the littoral zone, and its transparency. Based on your observations, do you think it is oligo-, eu-, or mesotrophic? Make plans to visit the lake to build on your reflections. Complete a Sit Spot Routine [as described in Chapter 1] while you are at the lake. Leave the area a better place than when you found it. Bring a trash bag to pick up litter you may find.

In the summer months, many lakes deep enough will stratify—or become layered into distinct temperature bands. These layers of varying temperatures create different densities to drive the stratification phenomenon. This results in different environmental conditions going vertically through the water column. Mesotrophic lakes will often show this **thermal stratification** feature.

In a stratified lake near the surface, or **epilimnion**, the warmest and best-mixed water with the most light penetration exists. Going deeper, the **metalimnion** layer experiences cooler temperatures, and higher oxygen levels with abundant small fish.

Within this layer exists the **thermocline**. Perhaps you have experienced the thermocline when you swam deeper into a lake and it suddenly became colder! The thermocline is where the temperature drops significantly at a single point in depth and prevents further mixing of the water in summer. The water continues to cool and reaches **isothermic** conditions, or it is of the same temperature for the remaining depth. This deepest layer is called the **hypolimnion**. Typically, no light penetration exists, yet you will find high oxygen levels in this colder water but will deplete as you reach closer to the bottom where decomposers utilize it resulting in elevated carbon dioxide. Only certain animals adapted to these conditions are found here. ***Activity:*** Draw the vertical profile of a stratified lake. Label the epilimnion, metalimnion, thermocline, and hypolimnion. Indicate the change in temperature at the various layer depths.

We learned in the previous section on water that lakes in some regions and situations will "turnover" in fall and spring due to the change in density of water at different temperatures. As some of our summer lakes cool in the fall and winds increase, surface temperatures may reach water's densest of 4 °C. At this point, the surface water begins to sink and displaces those temperature layers of the summer as was just discussed. The result is a cycling of the water, which often will churn up bottom sediments and cause the displacement of algae, zooplankton, and fish that had once existed in somewhat predictable places in those stratified temperature layers of summer. Toxins, pollutants, or litter that had settled to the bottom through the year can also be churned up and dispersed during these times of turnover.

Changes to Lakes. We also learned previously that nature is always changing, and that succession is inevitable. We learned that a terrestrial landscape will go through a somewhat predictable change in community structure from a field to a forest progression. Succession is also a natural process in aquatic ecosystems. Lakes will naturally fill in over time and may become a terrestrial environment unless they have a constant source of incoming water like a spring or stream.

Other occurrences that can alter aquatic ecosystems is cultural eutrophication, acidification, and the presence of invasive species. **Cultural eutrophication** occurs when humans speed up the process of aquatic succession by increasing nutrients in water from runoff, fertilizers, sewage, etc. This happens with increased urban sprawl. **Urban sprawl** is common these days where people who once lived in a city seek a more remote or removed living space away from highly populated areas. People find a more nature-centered area, and many want property with a view or access to water so they will build along an inland lake. If the property owner is not aware of their impact on nature, erosion and runoff increase into the lake. Cultural eutrophication leads to algal blooms. This will decrease water transparency, thereby causing the loss of submerged plants and reducing the lake's

Aquatic succession occurs by the accumulation of sediment from runoff and the buildup of dead organic matter in a lake. Eventually a lake can transform into a terrestrial habitat.

Notice the manicured lawn leading to the lake. Lawns have shallow, tight roots that prevent easy infiltration of water. Fertilizers, pesticides, and pet waste can easily runoff to the lake.

habitat structure. This creates fewer places for spawning, maturation, and general survival of aquatic animals. Due to the fast reproduction rate and short life span of the algae, massive biomass of dead organisms settles into the sediment. Oxygen concentrations in the lake will lower because of the bacteria needing it to decompose this material. The lake the people once knew is no more. The appearance, inhabitants, and smell may be less attractive.

Acidification of a lake will make the appearance of water quite clear, but will cause aquatic wildlife to decline. This issue exists throughout the region at varying levels. Although the natural release of sulfur from volcanoes or soils may seem like a distant occurrence for some of us, the effects are carried by winds and released in precipitation all around us. What is a more obvious source for acidification is fossil fuel combustion at industrial facilities. The sulfur dioxide emitted combines with water to create sulfuric acid, which has a very low pH, meaning highly acidic. This will also contribute to acid rain. Where does the rain go? It goes everywhere on land and into our water. If a neutralizing rock is exposed in the lake, like limestone or other carbonate, then that may help neutralize the water. Fish and other life forms cannot survive in a pH lower than 6–5 (7 is neutral on the logarithmic scale of 14). Lakes with a high transparency can indicate acidification, or it can also suggest the presence of an invasive species (i.e., quagga mussel, *Dreissena rostriformis*) filtering the water at an accelerated rate compared with native species. However, realize that a transparent lake may just indicate oligotrophic conditions.

An **invasive species** will heavily colonize a habitat because of its fast reproductive rate or absence of a predator. It can be a native or nonnative (exotic) species and can be found in the water or on land. A **native species** is one found in a region before the colonization of Europeans in North America. This is sometimes referred to as **pre-European settlement**. For our purposes, based on what officials working to combat these species refer to them as, we will classify the problematic nonnative species as invasives or invasive species. These plants or animals may adversely affect the habitats they invade economically, environmentally, or ecologically by displacing the native species of which many organisms may depend upon as a resource.

Some nonnative species were introduced to habitats with an intent to stock as game or food species populations, or to prey upon another nonnative species. Unintentional introductions have occurred from the ballast waters of freighters, as part of shipping crates, or releases by people who may not know the consequences that could result. Invasive species will affect native species by competing for prey or habitat, predation, habitat alteration, hybridization, or the introduction of harmful diseases and parasites. Their presence may also change the native species' size and age structure, distribution, density, population growth, or may result in extinction.

You will learn more about these issues and what can be done to prevent or combat invasives in Chapter 4 and in later chapters. An effective and efficient approach for combating invasive species is by encouraging a diverse habitat of

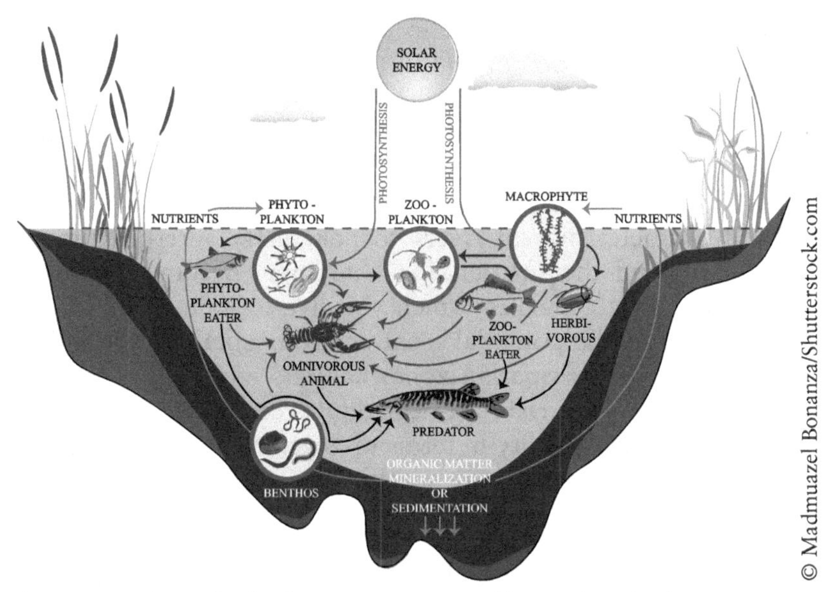

A cross-section of a freshwater ecosystem illustrating the interactions that can occur. Describe relationships between biotic factors, and between the biotic and abiotic factors. Identify what habitat each organism is found in by using the terms from this chapter.

native vegetation to provide a complex structure for native wildlife. If possible, create this scenario along waterways to promote a healthy aquatic ecosystem. Regardless of where you live in the Great Lakes basin or Upper Midwest region, you will likely have some familiarity with aquatic ecosystems of various kinds. Read further for more insight on these systems.

River Characteristics

Rivers are dynamic and show tremendous variation through time and space. The shape, size, and content of a river are constantly changing, forming a close and mutual interdependence between the river and the land it traverses. Imagine following one river from a small trickle in the hills, over rapids and waterfalls, through areas of peaceful, nearly motionless water, all the way to its meshing with the mighty waves of a Great Lake. On a grander scale, a river may carve through majestic mountain ranges, create deep gorges and canyons, gentle valleys, lush meadows, and mighty plains, all the while providing the setting for a diversity of biological communities. Over time it may change drastically from a roaring, overflowing force in the spring, to a still, icy-cold mass in the winter.

The surface water of rivers has significant importance to people. Although rivers only have a small fraction of ~0.0001% for all the water in the world, they carry vital nutrients and water while draining the land's surface. Rivers are of immense importance geologically, biologically, historically, and culturally. They are critical components of the water cycle by moving or draining surface water that accounts for nearly 75% of the Earth's land surface.

Understand characteristics of rivers to help make interpretations of these aquatic ecosystems and the surrounding landscape. Become familiar with the social-ecology of the watershed in which you live to gain a deep awareness of your surroundings and the potential local environmental concerns. Some types of land use existing in this **riparian zone** (habitats and communities along the river) may include farming, urban areas, logging, residences, factories, and varying degrees of streamside stability in terms of vegetation. Revisit Chapter 2 for a refresher on watersheds and for more specific types of things to consider regarding any watershed.

Importance of Rivers. Rivers provide habitat, nourishment, and means of transport to countless organisms. If you have ever observed a river valley firsthand, you will marvel at their powerful forces that create the surrounding majestic scenery. Rivers provide travel routes for exploration, commerce, and recreation. As the water flows, they leave valuable deposits of sediments, such as sand and gravel from which humans extract as a natural resource. Rivers also form vast floodplains where many of our cities are built, and the flow from some rivers provides the electrical energy for use in everyday lives.

Rivers as Natural Watercourses. Rivers are waterways, flowing over the land surface in extended hollow formations (**channels**). The channels drain discrete areas of mainland with its natural gradient. In basic terms, the existence of a river depends on three things: the availability of surface water, a channel in the ground, and an inclined surface to allow the flow of water. In this sense the term, "river" includes all kinds of watercourses, from the tiniest of brooks to the largest of rivers (the term stream is used interchangeably with river when describing characteristics of the watercourse). Essentially, a river represents the excess of precipitation over evaporation for a certain land area (see Chapter 4 for a discussion on the hydrologic, or water cycle).

A Close Mutual Interdependence. Rivers and streams are characterized by flowing waters and are called lotic systems (as opposed to lentic systems, such as lakes). Lotic environments have been described as having four dimensions: a longitudinal dimension that includes the length of the river within which has a pronounced zonation of chemical, physical, and biological factors (comparable to the

Nature Journal: Visit a nearby stream, one that is safe to wade into with the water below your knees. Always let somebody know you will be spending time in or near the water. At any point while in the water or along the edge, identify and discuss the four dimensions of that lotic system. Include a labeled sketch of these areas. Choose a stream dimension and create a field investigation question to conduct within it. Determine methods for collecting data. If feasible, collect and record your data. Discuss your study in terms of the area's social-ecology. See Chapter 2 for further insight on field investigations.

vertical stratification in lakes); a lateral dimension involving exchanges of organic matter, nutrients and biota between a location in the stream channel and the adjacent floodplain; a vertical dimension consisting of a hydraulic connection linking the river channel with groundwater; and a fourth dimension of time that pertains to the velocity of the water flow.

Rivers are very dependent on climate. Their characteristics are closely related to the precipitation and evaporation regimes in their drainage areas. Three main types of rivers have been distinguished. **Perennial or permanent rivers** have a constant flow of water (although there may be considerable seasonal variation in the amount of flow) and occur in regions where precipitation generally exceeds evaporation. This would describe many of our larger rivers that are constantly lotic or flowing. **Periodic** rivers may run dry occasionally but have streamflow during regular periods of variable duration. These occur in regions where evaporation exceeds precipitation on an annual average but periodically precipitation is greater. You would find small amounts of water flowing through here much of the time except in the spring after snowmelt and high rain events where these streams often swell over the banks. **Episodic** rivers only rarely and fleetingly have water in their channels. These occur in very arid climates, such as desert regions where flash floods occur.

The Greek philosopher, Heraclitus said, "You cannot step into the same river twice." River systems are very dynamic and exist in a constant state of change. No matter how many times you visit the same river in the same place, that river will be different each time. Due to the constant flow, it will bring new things to a given point continuously—debris, downed trees, runoff, and organisms.

Size Classification of Streams. Stream ecologists and **anglers** (people who fish) can have a universal language in terms of stream classification by using a scale of "orders" for the different characteristics, starting with the first stream order. Remember, streams and rivers are essentially the same. Typically, people use "stream" to indicate a relatively smaller river. The number given to **stream orders**

have nothing to do with rapids. They pertain to having year-round water and describe the relative size and relationship it has with other branches of the stream. It correlates with drainage area, but it is also controlled by topography and geology. It is helpful for comparing conditions to similar rivers and making predictions about potential stream changes under the influence of different types of situations.

The smaller the order number, the closer it is to the **headwaters** (origin) and the narrower it is—this affects what is contributing to it and what could live there. Predictions can be made by figuring out the order classification. A stream order number of 1 indicates a stream flowing directly from its headwaters. This could lead you to a desirable location for fishing. When you have two of the same stream orders convening, then you can claim it to be the next (higher) number for its order. Stream order helps to conceptually organize the streams in a watershed. As streams increase in order, they also increase in length, exponentially. Worldwide, about 70–75% of stream miles occur as headwater (first-order) streams.

Stream Features. A common feature along a river is a **floodplain**. A floodplain is a flat area immediately adjacent to a stream. It may cover with water during periods of high discharge (floods). Not the ideal place to build a house or business or set up a camp during a rain! However, it is a great place to get close to a river to sit, go for a walk, or fish.

The **meanders** of rivers refer to the complex dynamics occurring as the river moves back and forth within the channel walls creating a range of habitats for the river **biota** (life). The current will be powerful as it hits the outer meander of a riverbank. It erodes the side and bottom of the bank. Typically, this outer part of the meander will be twice the average depth of the river. In the diagram that follows, find the meandering stream. Notice the sediment getting deposited on the opposite side of that riverbank. This is from the eroded sediment from the upper meander.

All streams exhibit **sinuosity**. Sinuosity is the degree of the meanders that varies from stream to stream based on stream gradient, local geology, hydrology, and land use. You will typically find big meanders where there is the least amount of energy from the river current. It has a low gradient and a fine-grained substrate or stream bottom. This type of sinuosity is often near the mouth or end of a river. Small or tight meanders are found in high energy areas and will occur where there is a high gradient, often near the headwaters, and has coarse substrates like cobbles and boulders.

Perhaps a lake you are familiar with was not always a lake but part of a river. You can usually easily determine this by referring to a map or exploring the area. An **oxbow lake** forms when a river meanders too much and almost becomes a complete loop. The resulting lake is when the river creates a new channel between the beginning and end of the loop and cuts off the meander core.

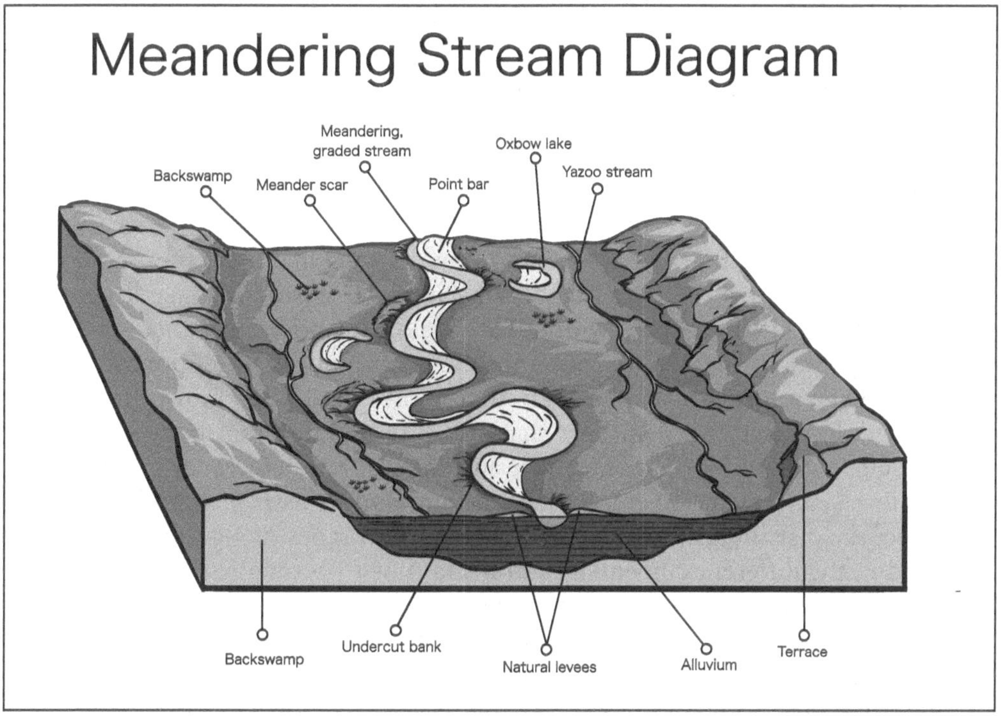

Potential features of a river system. Examine each feature and try to determine how it formed and its effect on the system. In which direction does the water flow in this diagram? Where would you expect to find deeper water in the river? Where is the floodplain?

All the characteristics you can observe about a stream creates unique habitats within and along that stream for plants and animals. Animals, especially benthic **macroinvertebrates**, those small aquatic organisms attached to the substrate and debris that you can actually see, will only be adapted for one of these habitats. If you know a species' natural history, then you will know where to find them. Knowing where they are located can help you find those that prey upon them, like fish. Within a stream you may observe these features that also serve as habitats: **riffles** (fast, shallow current typically with a rocky bottom), **pools** (twice the average depth of still water), **runs** (flat, unbroken surface), **cascades** (falling water), and **plunge pools** (eroded pool at the base of a cascade). Habitats along a river can include: wetlands, woody debris, organic debris, vegetation, undercut banks, and boulders.

The river current shapes the make-up of the substrate, delivers nutrients and food downstream, and is a force that must be faced or adapted to by organisms. It plays a large role in determining the distribution, **morphological** adaptations (what an organism or certain features look like), and the river current can even dictate behavior of stream organisms. Due to this strong influence, most stream organisms appear adapted to fast currents or slow currents, but not both. For example, body shapes of stream dwelling organisms offer clues to specific adaptations that allow organisms to cope with the current and attachment devices can serve to anchor invertebrates to the substrate.

Body shapes that are essentially flattened allow the organisms to minimize the force the river current exerts on them as with the the mayfly larva (Ephemeroptera) attached to rocks in the riffle zone. Simply by looking at the body of a fish, you can deduce where you might find that species. The northern bluegill (*Lepomis macrochirus*) is taller than its width. This makes it easier to maneuver around the rooted plants in the **littoral** (nearshore) areas found in lakes, ponds, and rivers. Trout have a torpedo-shaped body that allows the fish to move through fast waters with less energy expenditure than a fish with a different body shape.

The current plays a significant role in the river's transparency, temperature, oxygen levels, and water chemistry. The current and its associated abiotic factors will affect the distribution of algae and plants. What types of algae and plants can be found in the different currents? **Diatoms** (a type of brown algae made of silica found in microscopic, yet intricate patterns) will be found in slow- or fast-moving water. Green algae will be found attached to hard substrates in fast currents or floating in slow water. Rooted plants will thrive in slow water with soft sediments in the littoral zone.

The substrate of a river can also affect the organisms that live in that stretch. Look at the bottom of a river. If it looks uniform (the same), and especially if it has fine sediment, that is an indicator of little biodiversity. A variety of substrate material provides many places for organisms to live or find food, and makes streams complex and dynamic. Diversity and density of invertebrates also often increases with the presence of organic matter as part of substrate. Optimal fish spawning habitat will have a varied substrate. Most river fish need large rocks to mixed gravel to spawn on. This is another reason to reduce erosion. The substrate can be critical habitat for certain species.

Sediment distribution can be relatively predictable along the length of the stream. Starting at the headwaters, the mean gradient, or slope of the river is highest and will decrease (or become less sloped) near the mouth. In areas of a high mean gradient, the sediment in the stream will have large boulders and cobbles. As the gradient becomes less steep, the river current cannot carry heavy rocks, so you find finer grain materials as you move downstream. Streams will meander less when the gradient is steep because of the high energy flowing through the riverbanks. Whereas downstream the energy is less, and the stream takes the path of least resistance in the substrate and banks. Organisms will be adapted for certain regions of the stream. Knowing the natural history of invertebrate and vertebrate animals, and aquatic plants will help you predict where you might find them on a map and in nature.

> ***Nature Journal:*** Find your Nature Area on a topographical map. A topo map will provide the type of detail needed to find the drainage divides located around you. However, on most maps, you can still find the nearest river or stream, and can follow it from its headwaters to the mouth of where it drains into another body of water. What stream order level would you classify that stream at the nearest point to your Nature Area? How close is the next nearest stream? Make plans to visit the stream and find a Sit Spot there to take notes on your observations, questions, and sketch things of interest. What do you notice about any meanders you see, potential habitats for river organisms to live, what is the substrate like, what do you think the oxygen level is in relation to the current? See Chapters 1 and 2 for help with making observations, getting to know your "place", and more information on watersheds.

Effects of Dams. Dams play a role in the infrastructure of many communities. With any human manipulations to nature, there will be consequences. There are advantages and disadvantages of damming rivers. Damming a river may offer power generation for the people living and working in the area, it may have

A generalized diagram of a dam. In which direction is the flow of water? Notice the difference with the volume of water upstream compared with the water downstream. How would the river ecology change if the dam were removed?

irrigation purposes, prevent nearby floods, divert water for various reasons, create more waterfront property for homes, or provide additional recreational opportunities (you might actually be swimming in the backwaters of a dammed river when you thought it was a lake!). These advantages of dams will also benefit local economies.

The disadvantage of dams, is that the river flow is obstructed and a myriad of changes to the ecosystem results. With the flow obstruction, temperature increases in the backwaters causing dissolved oxygen in the water to be greatly reduced. Living things need oxygen, and some need higher levels than others (like trout). Entire food webs are altered in this scenario, especially trout species who need to migrate up specific streams to spawn. Species intolerant to the changed conditions will die. Sediment builds up in this unnatural lake environment and more **anaerobic** conditions take over, meaning no or little oxygen is available, so more bacteria populate the system. Although many people enjoy the backwaters of dams because there is more surface area to recreate, there is a balance to the system that was lost. The drainage of the entire surrounding landscape is slowed, the quality of the water is lowered, and the entire ecosystem functions in a different, and in some scenarios an unhealthy way. Lastly, entire towns, homes, properties, and livelihoods may be at stake if a dam breaks. For the sake of safety and the environment, many watershed management plans are requiring the assessment of dams and the local ecology. Through inventories and monitoring of these altered systems, removal of hazardous and environmentally degrading dams can restore balance to the ecosystem.

Stream Macroinvertebrates. A way to monitor a river system is to inventory the "bugs". Many terrestrial insects that fly will hatch in nearby water. Adults lay eggs on the surface of the water or on aquatic plants (depending on the insect species). The insect egg hatches, and the next stage will spend time in the water before emerging, drying off its wings, and allowing "blood" or plasma to pump through the veins in the wings before flying away. You can volunteer to become a stream monitor and collect these amazing creatures while they still reside in the water. They are **bioindicators** that tell us the relative quality of the water. Contact your local watershed council or natural resources agency to learn more details.

These macroinvertebrates, or animals without backbones that are visible to the naked eye, live mainly on the stream bottom or on other substrates like logs or leaves in the water. For this reason, you may also refer to them as benthic macroinvertebrates. As important links in the food chain, they not only serve as fish food but also help recycle nutrients in the system. They are good bioindicators of stream quality because they remain essentially sedentary, and are susceptible to changes in the system. Due to their high reproductive rate and short life span, problems or trends in the water quality can easily be detected if a change in the species assemblage is observed from monitoring a site. This can be a very rewarding and fun community science opportunity to get involved in.

Wetland Wonders

The last aquatic ecosystem we will cover are wetlands. You will learn what makes a wetland classified as a wetland, the four main types of wetland, the benefits of wetlands in terms of ecological services, the threats to the health of wetlands, and who you can contact to help protect aquatic systems. Identifying wetland characteristics will help you make interpretations within the ecosystem. Understanding wetland services will lead you to appreciate the benefits of wetlands provided to nature and humans.

What Makes a Wetland a Wetland?

For a wetland to exist, certain things need to happen. The soils would have all pore spaces between the grains waterlogged or full of water, thereby being classified as **hydric soils.** Typically, the wetland would arise in a low-lying area where rain and runoff would keep the soils saturated and create an area covered with shallow water. You would expect the wetland to support water-loving or **hydrophytic** plants and animals adapted to living in this aquatic environment. Groundwater is often at or near the ground's surface allowing the wetland to remain wet and constantly "fed" from below.

You might find wetlands next to rivers or lakes that regularly overflow. People might also create wetlands intentionally such as state agencies flooding an area for waterfowl breeding, or on private land for hunting, fishing, or aesthetics. Unintentionally, construction may block the natural flow of water causing a stream or groundwater flow to back up and overflow. Beavers as wetland builders will dam streams and flood large areas. They can turn meadows into marshes or parts of forests into swampland. These varied ways wetlands come into existence among other natural variables will determine what kind of wetland results.

Major Wetland Types

Many wetland varieties exist, and often they have mixed qualities of what will be explained in this section. By knowing these four main types, you will be able to make some quick, yet fairly confident claims in your classification. The marsh, swamp, bog, and fen are all wetlands yet have unique features that set them apart from each other.

Marshes. Of the four wetland types, marshes will typically be the most productive or diverse due to their water depth, flow, and plant life. They will have shallow water (<3 ft.; 1 m deep), but in some instances may be well over or less than this amount. They will remain wet and not stagnant through the year, and are common along rivers, lakes, and the sea. Marshes have gained the recognition as one of the most productive habitats on earth teeming with wildlife. Some common marsh inhabitants (who will also be found in other wetland types but can be found in a marsh) include cattails, water lilies, bulrushes, arrowhead, reeds, pickerel weed, some grasses, rushes, sedges, and wild rice. The plant roots and lower stems remain submerged in rich, wet soil, and the upper foliage is emerged in the air. The rate of photosynthesis is very high and plant growth is very fast and abundant. Fish, ducks, frogs, water snakes, insects, etc. raise young in marshes and predators will be found hunting in this diverse area.

Swamps. Identified by trees standing in the water—dead or alive, swamps make great places for birds to perch or nest. The isolation of the trees by the water makes them somewhat protected from predation. Swamps have open surface water, but a tree canopy in some places are usually present. You find trees or shrubs (evergreen or deciduous) growing in swamps sometimes resembling a wooded wetland. Silver and red maple, aspen, willow, cedar, tamarack, and black spruce may grow here where smaller animals abound and a few predators persist. Woodpeckers, swallows, flying squirrels, some ducks, and great blue herons use the dead standing trees. Look for skunk cabbage in the early to mid-spring along the margin of swamps (and other wet areas). Bears love skunk cabbage and will feast on it as they emerge from their winter slumber.

Bogs. If you were to jump up and down on the ground in a bog, you would see the shrubs and trees move! You would be on the **peat**, a thick mat of partially decomposed matter that floats on the water below. The peat is often several meters thick, approximately 12 m or 40-ft. thick. Bogs are poorly drained with usually no outflow, and commonly found in, but not restricted to the boreal forest and tundra. They have stagnant, brown, highly acidic water, with little or no dissolved oxygen.

If you tested the water's pH, you would discover that it reads acidic due to no inflow or outflow of the system and the collection of leaves and conifer needles that contribute their tannins to the water. Because of these conditions, bogs generally do not have high species diversity. You will find, however, unique species adapted to the environment.

A surface carpet of mosses—especially the presence of *Sphagnum* sp. only grows in bogs. Other than *Sphagnum*, you will discover other plants like leatherleaf, blueberries, cranberries, sedges, and orchids. Even more, there are carnivorous plants, such as the bladderworts, sundews, and pitcher plants that can be found in bogs. These meat-eating plants will trap and "digest" insects as their source of nutrients (especially nitrogen).

Another fact, bogs have buried treasures! They provide us with a treasure trove of information. Dead plant and animal matter do not decompose fully, and few raw nutrients are available for new plant growth. The bog conditions preserve remains of plants and animals that have allowed us to interpret past landscapes. The slow decomposition rate preserves the organisms that fall into bogs for thousands of years. Ancient pollen, leaves, animal bodies (with hair, muscles, and teeth intact) have been recovered. Other discoveries include buried sediments from radioactive fallouts and releases of heavy metals in the environment. Studying layered bog remains, in particular from core samples can help recreate ancient climates and help make interpretations of how vegetation and the landscape has changed over time.

Fens. This last wetland category will have some flow and have a bottom of limestone (which will make the water more neutral to basic or alkaline). You could still "shake" the ground by jumping on the peat, but you will not find *Sphagnum* anywhere because those plants need acidic conditions. Dominated by grasses and sedges, fens resemble a flooded field of hay with a slow flow of freshwater.

Wetlands at Work

Learn about the ecosystem benefits wetlands provide wildlife and people.

Wetlands give the world many "free services." In terms of wildlife, wetlands serve as a migration vacation or a place for migrating animals (especially birds) to replenish with rich food supplies before resuming their travels between winter and summer homes. Wetlands also provide hundreds of plant and animal species vital habitat to live. The health of our nation's waterfowl population is directly tied to these areas. Seventy-five percent of all waterfowl, our ducks and geese breed only in wetlands. For many animals, wetlands offer a natural nursery to raise young. The convenient food supply and thick vegetation provides good hiding for the growing families. Lastly, wetlands will reveal a haven for rare species. Nearly half of threatened and endangered species either live or depend on wetlands for survival.

If those reasons do not convince you of why wetlands need protection, then let us shift our focus to a personal level. Wetlands benefit people in many capacities. They act as sponges by absorbing excess water in an area. This flood prevention thereby reduces damage to the surrounding natural and human environments. Wetlands will also filter contaminants such as excess nutrients, heavy metals, or other toxic chemicals by adhering to vegetation and sediment. This prevents these

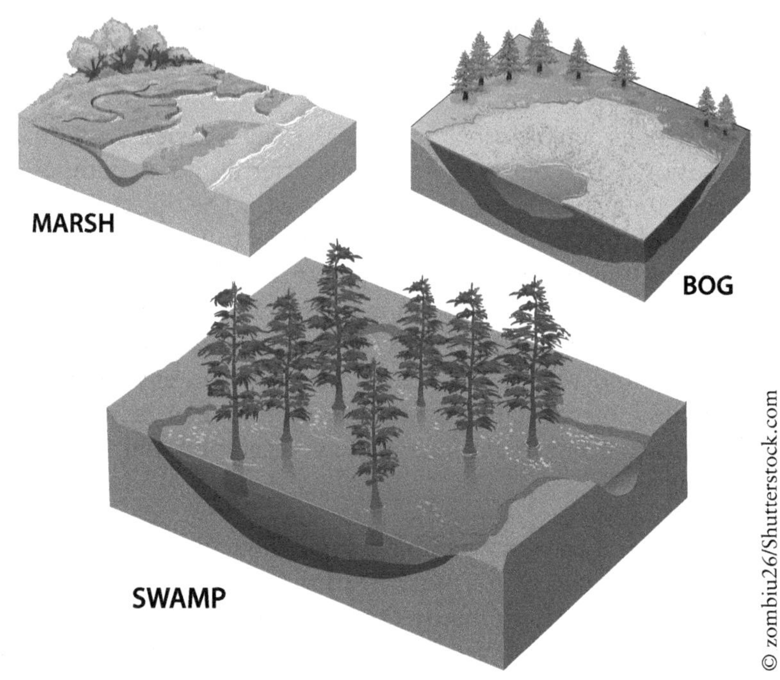

The swamp, marsh, and bog are generalized in this picture. What are the major differences with the features? In nature, at a glance, fens may resemble bogs but knowing plant types and conducting a pH test would be determining factors between the two wetlands.

concerns from traveling further in the watershed and this ultimately protects our drinking water supply.

As a similar benefit, wetlands serve as silt trappers. As wetlands slow potential flood waters, the runoff of silt and other sediments in the water will settle out among roots and stems of wetland plants. This protects streams and lakes downstream from build-up of sediment that could stifle the growth of aquatic plants and animals. Related to floods, storms that reach coastal wetlands along a large lake or sea will act as a buffer to the effects of strong winds and waves on shoreline communities of people and wildlife.

Additionally, with added water to the system, the holding capacity of wetlands will serve as groundwater rechargers. Water migrates downward through wetlands to maintain the groundwater levels. In this so-called holding tank of water, wetlands support the commercial fishing industry because in some places they are used for stocking and provide fish and shellfish with food and a place for breeding and raising young. These food sources are then used for commercial purposes and sold to grocers and restaurants. Another important aspect of wetlands is for a multitude of recreation opportunities for people. The biodiversity attracts bird-watchers, photographers, hunters, anglers, and anyone who can appreciate the wetland habitat and species that abound.

Wetland Loss

Since pre-European settlement, or the 17th century, the Great Lakes basin has lost around 70% of its wetlands, most of which has been on private land. Unfortunately, the perception of wetlands has been viewed as "wastelands" and "just swamps." The ignorance can cause great damage to the services these habitats can provide. Economic incentives over the years for development and urban sprawl has resulted in the loss of half of spawning grounds for fish, waterfowl habitat, flood control capability, and loss of erosion control and sediment-trapping capability in those areas.

The major causes of wetland loss and degradation is often a combination of **biological, chemical, and physical alterations** to the water. In the following examples, you will see the relationship between the alteration to the water and what is happening to the surrounding land. You will also be able to make connections between two or more types of these alterations. Biological alterations can include the introduction of nonnative plants that would need to be controlled through biological, chemical, or mechanical means or prevented by proper land management, such as minimizing the disturbance to the landscape. However, the removal of vegetation is also a problem. The plants play a significant role in all ecosystem services previously described. Chemical alterations to the habitats result from the release of pollutants and toxic chemicals in the ecosystem and runs off into the water causing an increase in nutrient levels. Physical destruction or degradation of wetlands include the filling, draining, dredging, damming, destructive recreation of all-terrain vehicles, and peat mining.

What is so significant about peat? **Peat** is a natural resource and coveted by many. Its uses are as a natural fertilizer and soil conditioner for gardens and farms, and supports local economies with its mining operations from the northern reaches of our region. Although valued, it is a resource that will take a long time to renew, and in the meantime, destroys the ecosystem services of the habitat. What are alternatives to peatland crop production for sod, carrots, onions, celery, potatoes, lettuce, cranberries, mint, radishes; or the horticultural use for containerized seedlings used in the forest industry? Think about utilizing local farms, or growing and preserving your own food sources using sustainable practices.

Nature Journal: Find a wetland near where you live. How would you classify it—marsh, swamp, bog, or fen? What types of observations led you to that deduction? Notice the topography around you. Does the wetland have a basin shape? Discuss any types of potential degradation to the wetland. Find a place to sit comfortably where you have a view of the wetland. Spend time listening and looking for birds in the area. Reflect on your experiences and observations (see Chapter 1 for reminders of what to include for a thorough reflection).

What To Do

Many education materials exist to raise awareness about wetlands. Read local brochures, landowner guides, watch videos, attend workshops, find school curriculum support materials, and discover wetlands in children's literature to share with youth. Contacting local organizations or agencies with wetland education to see how you can get involved in stewardship activities will help them move their mission forward to protect our natural resources while also offering you rewarding stewardship experiences.

3.6 Reflecting on Chapter 3—Parts of Nature

This chapter offered a plethora of useful information to help you decipher parts of nature in terrestrial and aquatic habitats. For the land environments, you learned how to classify and interpret habitats, the characteristics of succession, how to recognize major terrestrial ecosystems, and basic geology and soil types. To help you think about freshwater ecosystems, you uncovered the mysteries of water properties in nature, discovered lake environments, river features, and wetland wonders.

By knowing and understanding this material, you can start to make meaningful interpretations of what you encounter in nature. You will have created a foundation to build upon. By keeping good notes in your Nature Journal, the interests you have in nature will emerge and questions for you to ponder and research will develop. Rather than viewing any part of nature as unproductive, unimportant, or worthless, shift mindsets through education to replace those words with productive, vital, and irreplaceable. Create awareness. Look at a map and make plans of where to find adventure nearby. Make it a goal to explore as many habitats you can within a five-mile radius! Next, seek out each of the habitats discussed in this chapter. Keep record of your experiences.

REVIEW

Review Questions

1. Describe the regional landscape where you live.
2. What is a habitat? How would you provide a classification, or name for a habitat?
3. What does an animal in a habitat tell you?
4. Explain habitat signs and stimuli (structure, vertical layers, horizontal zones, complexity, patchiness, edge, size, special features, and other organisms).
5. Define succession. Discuss the difference between primary and secondary succession. What are the seres? What is degradative succession? Why is it considered a type of primary succession?
6. Discuss the field to forest habitat types.
7. What are the main goals of forest management?
8. What is the difference between a rock and a mineral? What are characteristics of a mineral that will help you identify it? Discuss the three rock types. Explain the rock cycle. What are fossils?
9. Describe soil formation, characteristics of soil, soil layers, soil organisms, and considerations of soil.
10. Discuss the properties of water.
11. Describe lake types, habitats, aquatic succession, and cultural eutrophication.
12. Discuss the importance of rivers, as natural watercourses, stream orders, stream features, effects of dams, and macroinvertebrates.
13. What makes a wetland a wetland?
14. Discuss the four major wetland types.
15. Explain the benefits wetlands provide wildlife and people.
16. What threatens wetlands? What can people do to help?

PART II:

The Science of Nature Study

Look deep into nature, and then
you will understand everything better.
— Albert Einstein

Naturalists learn basic ecology by spending time in nature and making observations. In Part II, The Science of Nature Study, you will read about the essential ecological concepts that naturalists should know. You can apply this information to your nature observations (or vice versa) to give deeper meaning to those experiences.

Each ecological concept is separated into its own section in the only chapter found in Part II. Chapter 4 covers the scales of life found within an ecosystem; evolution basics; spheres of earth and energy flow; biogeochemical cycles; and biodiversity, conservation, and management.

4 Ecology Essentials

Nature is all about connections. Connections exist between wildlife and their environment, within an environment, and involving people. As introduced in earlier chapters, basic ecological concepts will help the naturalist with interpreting their observations made in nature. ***Ecology*** *is the study of the interactions living things have with each other and interactions living things have with their environment. Ecology also provides information about ecosystem benefits, and how we can use natural resources in ways that can leave the environment healthy for future generations.*

An ecologist studies these interactive aspects of nature. An ecologist is not to be confused with an environmentalist. An environmentalist is a person who is aware of and concerned with the protection of the environment (all biotic and abiotic components of nature). They advocate for or against issues regarding the environment. Issues may relate to a plant, animal, or the quality of other natural resources like water or air. What people decide to do with information is personal, while the data and claims shared by ecologists and naturalists are supported by empirical evidence and logical reasoning (see Chapter 2 on scientific thinking).

4.1 Scales of Life: Populations and Communities

In Chapter 2, the concept of the scales of life was introduced from the level of the organism (with the green frog as an example) to its population, community, ecosystem, through to the biosphere. Here, you will think about it a little more deeply.

Starting with the simplest of the levels, the **organism** is an individual living system such as an animal, plant, fungus, or microorganism. For something to be considered an organism, it must follow all the criteria of reacting to stimuli, reproduction, growth, and maintenance of life. If something only has one of these characteristics, it is not an organism. For research on an organism, the field ecologist or naturalist might pursue the descriptive field investigation question regarding the behavior of the one male white-tailed deer (*Odocoileus virginianus*) in the herd that frequents the

POPULATIONS

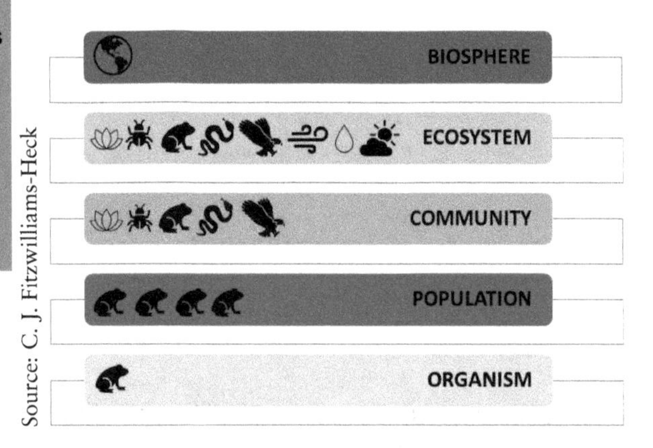

Source: C. J. Fitzwilliams-Heck

Nature Area in the morning hours during October (see Chapter 2 for more details on conducting field investigations). The population ecologist would be interested in all the white-tailed deer in the region, whereas, the community ecologist would want to know how those deer interact with the populations of other species – plant or animal. Studying deer on the ecosystem level would need to factor in how the abiotic and biotic conditions impact the deer in a given region. On a biosphere level, how would this deer species and its relationships compare with others of the same species in different locations on the planet.

Population Dynamics

A **population** includes all members of a *single* species living together in a specified geographical area. A **species** is a group of organisms capable of interbreeding and producing fertile offspring. In any given room, contemplate the number of populations present. Perhaps you counted one for the human population in the room, or a higher number for all the different types of pets that live with you. Did you consider bacteria, yeast, your gut flora, insects, arachnids, or any infections? Many populations live around us and we may not realize it.

To start thinking about the populations around you in nature, consider simply becoming aware of what types of species live near you by doing the activities in Chapter 1 to help you become a naturalist. From there, you could conduct a field investigation as discussed in Chapter 2. A field investigation focused on populations might simply be how many American goldfinches (*Spinus tristis*) are active in the Nature Area in the evening hours from late April through mid-May? To think about the population dynamics further, how do those numbers fluctuate through the day or year, or during different weather conditions, what is their behavior during these times?

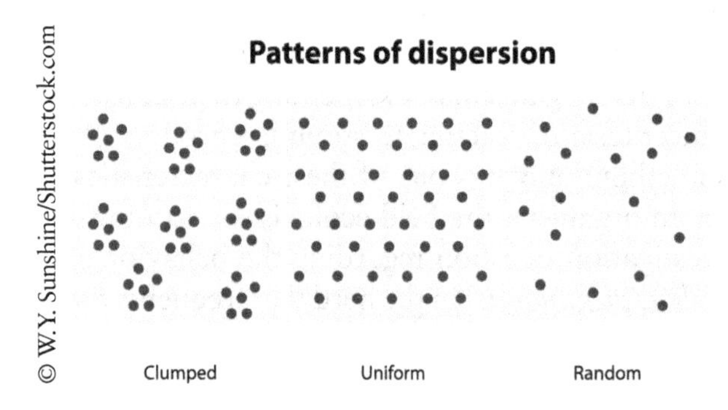

© W.Y. Sunshine/Shutterstock.com

The population ecologist and naturalist may want to determine population size or density. This information can inform whether changes in numbers occurred over time indicating a potential concern, how much territory is needed to sustain a population, or how much recovery time is needed for a population after a disturbance. To make any comparisons, the investigator needs to have ongoing monitoring or conduct inventories before and after an occurrence. It is difficult, if not impossible to count every species so a sample will provide insight needed to make calculations and predictions. Chapter 2 gave some examples of sampling methods, such as using quadrats, sample plots, and transects.

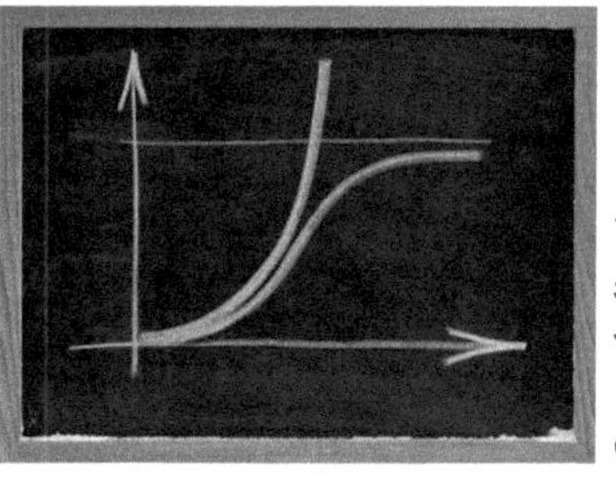

Two population growth curves over time: exponential and logistic growth, or J- and S-curves, respectively. The population limits of logistic growth corresponds to limited resources regulated by the carrying capacity of the habitat.

Understanding the behavior or tendencies of plants and animals in terms of general distribution patterns can lead to better sampling techniques. Three main types of patterns occur—clumped, uniform, and random. Clumped distribution is the most common arrangement observed in nature. You will find plants or animals grouped around optimal conditions such as nutrients, food sources, temperature, and finding a potential mate. For small animals, the cluster can provide protection or strength in numbers. Uniform patterns are observed in territorial organisms. A hawk population in an area may have relatively even dispersion because they defend the area in which they hunt and breed. The perfect example of a random pattern is the growth of dandelions, a generalist that can grow almost anywhere and does not tend to have a pattern to where it is distributed.

You will observe three modes of population growth patterns in nature—exponential, logistic, or complex growth. **Exponential growth** is represented by a J-shaped curve where population numbers start to increase slowly, then multiply in a continuing upward J-shaped trend over time. This will be exhibited by generalists or organisms characterized by a fast reproductive rate and generation time. The **logistic growth** of a population exhibits a similar starting pattern but the numbers level over time at the **carrying capacity** of the habitat. The carrying capacity of a population reflects how many individuals an area can sustain. **Complex growth** resembles the S-shaped, logistic growth pattern but the population fluctuates above and below the carrying capacity in a more realistic pattern.

The population growth curve for any organism results from the interaction between the **environmental resistance** acting on the population and the **biotic potential** of the organisms. The types of things affecting population numbers and preventing a continual growth pattern is environmental resistance, or also referred to as **limiting factors**. These outside forces that can limit population growth can

© Ole Schoener/Shutterstock.com

Purple loosestrife (*Lythrum salicaria*), a nonnative herbaceous plant has a fast reproductive rate as do other *r*-selected species. This invasive plant can outcompete and displace native plants.

include predators, disease, competitors, space to live, available breeding partners, or disaster. The other consideration is biotic potential, or the ability and success of reproduction.

Two main reproductive strategies of species exist, but most species fall somewhere in between the two types of strategies if it were on a spectrum. Having an understanding of this concept helps scientists study natural histories and evolutionary biology of populations. Populations limited by reproduction rate are classified as *r*-selected species, and those populations limited by the area's carrying capacity are considered K-selected species (K standing for carrying capacity). The *r*-selected species are **density independent** and relatively unstable populations showing a J-shaped curve then plummeting, only to accelerate and fall repeatedly. The size of the population does not affect the trend. Typically, these organisms are opportunists, smaller, short-lived, produce many offspring, and provide no care for the offspring. The K-selected species hover around the carrying capacity or at an equilibrium with their population size. They are **density dependent** and the number of individuals will play a significant role in limiting growth, but this makes for a relatively stable population (S-shaped growth curve). These organisms are larger than *r*-selected species, long-lived, produce fewer offspring, and provide greater care for those offspring. ***Activity:*** Create graphs to illustrate the population trends for *r*- and K-selected species. Be sure to label the axes and provide a title and descriptive caption for each.

© Nicholas Hunter/Shutterstock.com

Black bear (*Ursus americanus*) exhibits population growth patterns of K-selected species.

The **generation time** is another factor contributing to the classification of population growth projections. This pertains to the time between the birth of one generation and the next. The more closely an organism is to exhibiting *r*-selected species characteristics, the shorter its generation time. Bacteria have some of the quickest times between generations. Lastly, **survivorship**, or when death is likely to occur relates to the reproductive strategy of a species and is considered in population studies.

Community Relationships

Broadening our scope, we now think about the scale of life beyond the population level, the **community**. Communities are all species that potentially interact with each other and living in a given region. Something a community ecologist or a naturalist might investigate are the interactions red-winged blackbirds (*Agelaius phoeniceus*) have with other animals and plants in a wetland community. Random samples in a habitat using one of various methods can reveal relative species abundance, species richness, and species diversity (or biodiversity). The data you collect in a study on ecological communities can be analyzed to help determine what lives in an area, their approximate numbers, and the potential relationships could be interpreted.

Relative species abundance provides a percent expression of the proportion of a species or species' group representing the total of all species counted in an area.

© isabela66/Shutterstock.com

The keystone, found at the midpoint of an arch, will support the entire archway. This holds true for keystone species whose presence or absence will have a significant effect on the habitat.

© miroslav chytil/Shutterstock.com

© A. J. Gallant/Shutterstock.com

© Procy/Shutterstock.com

The keystone species, gray wolf (*Canis lupus*) and American beaver (*Castor canadensis*). What types of potential interactions or population trends can their presence and absence have in a habitat?

Species richness provides the count of the individuals of each species counted. Both of which can be used for the biodiversity determination of the area. **Biodiversity** tells the complexity of the species composition of the community. With increased diversity comes more complexity. More complexity in an ecosystem creates more stability, which translates into increasing the odds of overcoming adverse conditions such as disease or a change in the environment. Chapter 2 gives details on types of field investigations that could be done to collect and calculate these values. Realize various types of diversity existing in the community. There is not only species diversity, but geographic and genetic diversity as well. You will find high biodiversity in areas in which species are widely distributed and when great variety is found in a species gene pool. The latter is the key to a species' survival (read the next section to learn more on evolution). Disturbances in a habitat, such as a fallen tree or selective cutting can create more structure to the area resulting with increased biodiversity. However, if the habitat becomes too fragmented it can cause a decline in biodiversity (the last section of Chapter 4 discusses this further).

Each organism has a role in the community. Understanding an organism's **niche** in the community will also help decipher the potential interactions taking place in any community. The niche is how any organism makes a living or does its job in nature. It describes how it can respond to or influence surrounding resources and competitors. Two species cannot coexist stably if they occupy identical niches. Examples can include the food the organism eats, its predators, temperature tolerances, and the time when they are the most active. These roles often influence how the ecosystem functions.

Plants also serve important niches. They are a food and shelter source for animals, affect the carbon and oxygen cycles through photosynthesis, and they also can stabilize shorelines to prevent erosion and runoff thereby protecting water quality. The roles organisms provide often influences how the ecosystem functions. The next time you encounter a plant or animal in nature, pause to think about that organism's niche it serves.

Another aspect of a community to recognize are any ecological dominants present. An **ecological dominant** refers to a few species especially abundant in a habitat. Looking across a landscape, do you notice many of one type of plant or tree? If so, that indicates the plant or tree species has the most **biomass**, or organisms that comprise the greatest amount of mass in the area. How can this impact the habitat?

Some species can have a tremendous impact on a community. Some maintain the ecological community structure. These are classified as **keystone species**. Their impact is greater than expected based on abundance of that organism. Its absence or presence will bring significant change to the community. Usually the keystone species is a top predator. The role may become more apparent once removed or enters a habitat either by natural forces or introduction.

The interactions in a community have different perspectives dependent on who they relate with and the resulting response. The first type of interaction is **intraspecific**, indicating relationships between the *same* species. Intraspecific interaction or competition is a contest within the species for food, prey, nutrients, space to live, or a mate. This can happen within any species. For example, interactions among birds such as, blue jays (*Cyanocitta cristata*) occur daily for various reasons.

It might be easy to recognize interactions between animals, but think about what intraspecific competition might look like in the plant kingdom. When you have two of the same plant species growing next to each other, consider their needs and their responses if those needs are not met. How can plants compete if they cannot move? See Chapter 6 to help you think about plant more critically.

If there are two *different* species involved, then that is an example of **interspecific** competition. You can witness blue jays squawking at Eastern fox squirrels, or the squirrels chattering at the jays, or the birds could get more aggressive to drive away the squirrels. This interaction could be for resources or space to live. The results of an interspecific relationship can vary depending on the species and the situation.

When two populations compete for the same limited, vital resource, only one population will outcompete the other and potentially bring about a local extinction. This is the **competition exclusion principle** as described by a Soviet biologist, G. F. Gause in the 20th century from observing two species of *Paramecium* in a controlled laboratory experiment where they were given the same food while living in the confines of a test tube. His articulation of this scientific law would be refined over the years to recognize the application to the natural world. In nature, many variables act on an individual at any given time. One of the two species would outcompete the other if the two species had the exact same survival requirements, were constantly exposed to the same conditions, the same food choices were available, and lived in a confining space.

Gause had further described another phenomenon from the experiment after he changed one of the *Paramecium* species. With this new species combination, he did not observe exclusion but witnessed coexistence. **Resource partitioning** is the dividing up of scarce resources among species that have similar requirements and both populations can survive at the carrying capacity. We observe this in nature more than competition exclusion. This is not to say that there is never a so-called winner species when describing an interspecific interaction.

Predator–prey relationships, is an interspecific interaction many of us know. The **predator** is one who seeks out and sometimes needs to hunt for its food. The **prey** is the target species of the predator, a food source. This complex occurrence not only happens between animals but can also be found between animals and plants as they prey upon or eat the vegetation. Either way, populations are linked and often cyclical changes in abundance or decline can be observed. A high prey population may indicate low numbers of the predator population in the area. Whereas, when predator populations are high, prey species tend to be low. Of course, overlap of high and low numbers of each species can occur and other trends observed due to varying types of environmental resistance acting on one or both populations.

Another term we need to know as naturalists is symbiosis. This helps define many types of relationships in a community. **Symbiosis** describes an association between two or more dissimilar organisms who live together in a dependent or close association. Predation would fit the definition; however, you find in some readings that since death is an immediate result of one species then it does not fit the classification. Our interpretation of the predator-prey cycle fits the description of a close and dependent association. Other examples of symbiosis include **parasitism**, **commensalism**, and **mutualism**. All these symbiotic examples describe a close and often long-term interaction between different species.

Parasitism is a type of symbiotic relationship between organisms of different species. The **parasite** benefits at the expense of its host. This is a form of predation, but with parasitism, the **host** usually does not die immediately. The parasite will acquire the benefits from the host (i.e. a place to live or as a food source) while it remains alive. The host's immune system and general health will likely eventually decline, resulting in death.

The braconid wasp infects the living, tomato hornworm (*Manduca quinquemaculata*) caterpillar with its eggs. The wasp immobilizes the caterpillar to lay its eggs, then eventually the wasp larvae will feed on their hornworm host.

The honeybee (*Apis* sp.) on the flowerhead of a New England aster (*Symphyotrichum novae-angliae*). What are the potential mutualistic benefits? Describe more examples of mutualism!

American robin (*Turdus migratorius*) sitting on its nest in a tree as an example of commensalism. What are other examples of commensalism?

Mutualism describes a relationship where both species benefit from their relationship. The bee and a flower is a classic example. The bee obtains a food source from the flower it lands upon. The flower may pass along its pollen. The bee visits another flower of the same species nearby and the pollen that collected on its body may transfer to the new flower. Fertilization of the flower may result because of these interactions. The bee benefited because it obtained food and the flower benefited because it was fertilized. Both species have a positive result.

Commensalism, another type of symbiosis is when one species benefits while the other is unaffected—harmed nor benefited. An example can include most birds who build a nest in a tree. The bird benefits by finding a place to build a home for its young, and in most instances, the tree goes unaffected. Some bird species' **defecation** (fecal discharge from the body) will result in killing the tree especially if they nest there recurrently. This is the case with the great blue heron (*Ardea herodias*).

Mimicry can also be considered symbiotic. To mimic means to copy, and this type of development requires natural selection for it to persist in a species. Different types of mimicry exist in nature. Two that occur are Batesian and Mullerian mimicry. **Batesian mimicry** is where one species assumes the form of another in terms of its coloration or behaviors. This offers superior protective capability. In this situation, the **model** has harmful characteristics (e.g. stings, bitter tasting, etc.) and the **mimic** looks and behaves the same way but does not

Not harmful, but resembles a stinging insect. Yet with the aposematic coloration, the clear-winged moth (from Order Lepidoptera) is avoided.

possess any of the damaging qualities as the model. The protection comes from the **aposematic** coloration that exists between the model and the mimic. Aposematic colors are those who signal a warning in nature. Yellow, orange, or red and often in combination with black will indicate potential problems resulting from coming near or eating the individual. Therefore, both the model and mimic are mostly left alone by predators. In **Mullerian mimicry**, similar colors and markings exist between two different species but both species have harmful properties to potential predators. Previously thought as an example of Batesian mimicry, viceroy butterflies (*Limenitis archippus*) are also distasteful to predators like its lookalike the monarch butterfly (*Danaus plexippus*). Therefore, this is an example of Mullerian characteristics.

Another type of survival tactic dealing with color includes **cryptic coloration.** This is the potential of being hidden or camouflaged in its environment. In other words, **camouflage** is the result of cryptic coloration. You may be familiar with camouflage clothing that military, hunters, or naturalists wear to help them blend in with their surroundings. If you wore an outfit in your house that had the colors and patterns of a forest from head to toe, you would likely not be camouflaged but you would have cryptic coloration or the ability to be hidden in the right habitat.

Spend time thinking about potential relationships occurring around you in nature. Become familiar with ways to classify these relationships and how different variables can act upon the species involved.

If you took this katydid (*Microcentrum rhombifolium*) away from the green vegetation, it would no longer appear camouflaged.

Nature Journal: Select an animal species or a plant species that exists within or near your Nature Area (or right outside your home) to focus your attention and apply some of the concepts learned in this section. What is its typical distribution pattern that you would expect based on your research and observations? Would you classify it as an *r*- or K-selected species? Why? What are its limiting factors? Discuss the intra- and interspecific interactions that it faces within your Nature Area.

Review Questions

1. What is ecology? Discuss the difference between the terms environment and environmentalism.
2. Define the scales of life for organism, population, and community.
3. Describe the types of things a population ecologist tries to answer.
4. Draw the three population patterns: uniform, random, clumped. Which is the most common? What could cause these patterns?
5. Explain the population growth curves: exponential, logistic, and complex.
6. Define an *r*-selected species. What does their population growth curve look like? What is K? What are the strategies of K-selected species and what does their population growth curve look like?
7. Define generation time. What is survivorship?
8. What types of questions do community ecologists try to answer?
9. Define the terms ecological dominant and keystone species. Give examples.
10. Define the term niche and give examples.
11. What is the difference between intraspecific and interspecific interactions?
12. Describe the competition exclusion principle. Give an example.
13. Explain resource partitioning. Give an example.
14. What is symbiosis? Describe the dynamics of predator and prey relationships. What is a parasite? How could predation and parasitism be of value to a community?
15. What is the difference between Batesian and Mullerian mimicry?
16. Compare mutualism and commensalism.

4.2 Evolution Basics

Have you ever noticed that the plants and animals around you seem to fit perfectly where they live? What is it about certain plants that can grow in dry, sandy soil? Is it just by chance that bird beaks seem ideal for the food sources in the area?

When you start to recognize features of nature on a deeper level, it may indicate that you are considering its ecology. You may be studying the various ways organisms interact with each other and with the elements in the environment. To really start to understand these relationships and help identify trends or themes in nature, the naturalist needs to consider the history of the environment—how it has changed over time. Chapter 3 identified parts of nature that helped you interpret some possibilities of how an ecosystem can change, but another fundamental way of viewing the world is through the lens of evolution.

Ecology has existed since life began. Bacteria, (**cyanobacteria**) some 3.5 billion years ago, began to capture sunlight and released oxygen through photosynthesis which made evolution of other life forms possible. The ozone layer was also formed through this released oxygen which protected life from the sun's intense ultraviolet radiation. Life emerged from optimal environmental conditions and life helped create environments conducive for other life. Organisms adapted to conditions and developed niches that have influenced the environment. Essentially, evolution could be viewed as ecology over a very lengthy time scale.

Evolution is the foundation of biology. **Evolution** is simply the change in characteristics (or traits) of organisms over many generations. We cannot say evolution occurs on an individual or organismal scale of life. However, when new traits persist through generations and permeate a population, then we may say evolution has occurred. The length of time for this to occur depends on the life strategy of the species as discussed previously in this chapter.

Having at least a basic understanding of evolution will help with your nature study and may improve your appreciation of living things. Without going too far into genetics, one needs to realize a few things—all individuals in a population vary from one another (**variation**), some traits of parents are passed onto offspring (**inheritance**), species with traits best suited for a particular habitat or condition have a greater likelihood of survival and reproduction (**natural selection**), and for any of this to occur usually takes a considerable amount of time (change can take thousands of years).

Background

Charles Darwin, a British naturalist in the mid-1800s was the first to articulate his findings for evolution on the Galapagos Islands and suggested how it occurs. He documented evidence of a variety of finches (birds) on the island with different beaks and realized they likely descended from a single species that had flown there from South America. Each finch species evolved a unique beak to feed on a certain type of food.

Basic Genetics

In the time of Darwin, what is known about genetics now did not exist. He did not realize features or changes observed in a species' traits occurred from changes in **genes**. Genes serve as the information needed for determining looks, behavior, and many aspects about an individual. Plants and animals consist of thousands of genes.

All our genes are found in the nucleus in each of the trillions of cells in our body. Cells are organized into tissues (i.e. skin) muscle, bone, and has blueprint information for that particular cell. This genetic information is carried within an organism's DNA. DNA (deoxyribonucleic acid) is a long molecule tightly wound into a package called a **chromosome**. Humans have two sets of chromosomes with 23 in each cell. One set is inherited from each parent.

Chromosomes are copied when the cells divide. A new trait happens when something causes a change with the copying of the DNA that makes up the gene. This could have no effect, minimal effect, or a big effect on an individual and ultimately the population. A new species seldom shows a radical difference from the original species, but a new feature of a species can persist over generations in a response to a changed environment. More details about the theory of evolution, how it occurs, and its evidence is found in the remainder of this section.

Just like all branches of science, knowing the terminology and incorporating scientific thinking will help us understand concepts more fully. As discussed in Chapter 2, scientific thinking supports claims made with empirical evidence and logic, while keeping an open mind to new possibilities of evidence.

Scientific Law and Scientific Theory

Let's now distinguish the fundamental difference between a scientific law and a scientific theory. The **scientific law** is something simple, true, universal, and absolute. It will often be expressed in a mathematical equation to help model and predict outcomes. Examples include the law of gravity, laws of motion, laws of thermodynamics, gas laws, conservation of mass and energy law, and the law of elasticity. **Scientific theory** pertains to a set of related observations and events based upon proven hypotheses. A **hypothesis** is an educated guess or testable prediction based upon observation. The hypothesis gains support or is refuted by experimentation or continued observation. In the case of a scientific theory, the results have multiple verifications by many researchers. One scientist cannot create a theory, only a hypothesis. A theory is much more complex and dynamic than a law.

The scientific law and scientific theory have been simplified in a comparison of a slingshot and a car, respectively. The slingshot (a handheld y-shaped structure, usually made of wood, metal, or plastic with an elastic band tied to the top prongs to be used to fling an object toward a target) has only one moving part, an object to be slung, and the speed and trajectory is predictable through a calculation. The car is very complex, with many moving parts all working together to carry out a common task (transportation). Improvements are constantly made, but the overall function remains the same. Theory components are improved upon through time, but without changing the truths of the theory.

Scientific theories are well documented and proven beyond reasonable doubt. Scientists derive new hypotheses of each theory based on empirical evidence to make the theory more elegant, concise, or all-encompassing. Theories can be modified, but they are seldom, if ever, entirely replaced. Examples include the theory of relativity, atomic theory, quantum theory, and the theory of evolution.

How Evolution Works

The facts of evolution are the observed changes (empirical evidence) in populations over time. The theory component of evolution involves the current scientific explanations of how the changes occur. Some ways evolution happen are through **mutations, gene flow, genetic drift**, and **natural selection.**

A mutated gene will occur randomly in an individual's DNA. Genetic **mutations** occur naturally and affect gene sequences. The results in the offspring may or may not be inherited beyond that individual. Inherited variation will help increase the gene pool. The more variety in the gene pool the better. With more gene varieties present within a population, the greater the chances of them overcoming adverse conditions or adapting to a changed environment.

Gene flow, or the immigration of new individuals into a population may result in an increase in the frequency of certain genes within the population. This can also be a positive thing for the gene pool's diversity. **Genetic drift**, on the other hand, is random. This is when certain genes occur more frequently in a population by chance when members reproduce. Realize that natural populations are genetically variable, and a range of physical and behavioral traits exist.

With **natural selection**, some genetic traits are more suited for an environment or conditions. These favorable traits are transmitted to the next generation at a higher frequency. Over time, those with the best-suited traits will persist in the population. A measure of **fitness** indicates reproductive success, not strength. Meaning, the fittest individuals will have the most descendants.

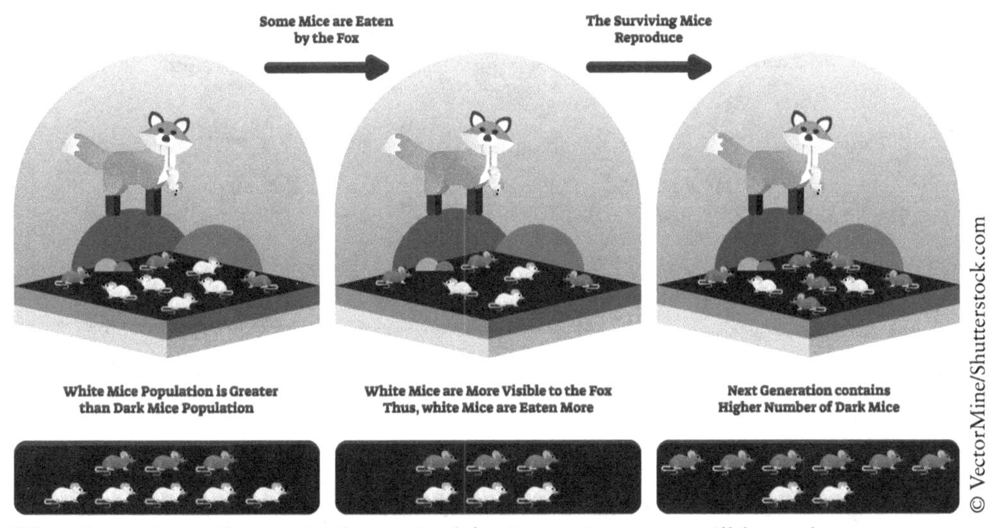

The adaptations of a species best suited for its environment will have the greatest probability of being passed on to the next generation. Which mice were favored by the fox? Why? What was the result?

As previously introduced, **adaptations** are an inheritable change in organisms allowing them to live successfully in an environment. Adaptations enable organisms to cope with environmental stresses and pressures and can be morphological (how they appear), behavioral (how they act), or physiological (how they function). For the feature to be considered an adaptation, it needs to endure through many generations. This is a gradual process caused by natural selection but can occur rapidly with *r*-selected species.

Traits are not developed because of a need. Existing characteristics enabled them to survive and take advantage over others in the environment. This allows for greater chances of producing more offspring. The result is what you may have heard of as the, "survival of the fittest." This is natural selection, and the driving force for evolution. It is the main process whereby organisms better adapted to their environment tend to survive and reproduce.

The process of natural selection leading to evolution begins with environmental resistance or limiting factors acting on a population. Based on the biotic potential, competition, and inheritable variation of the species will determine if traits

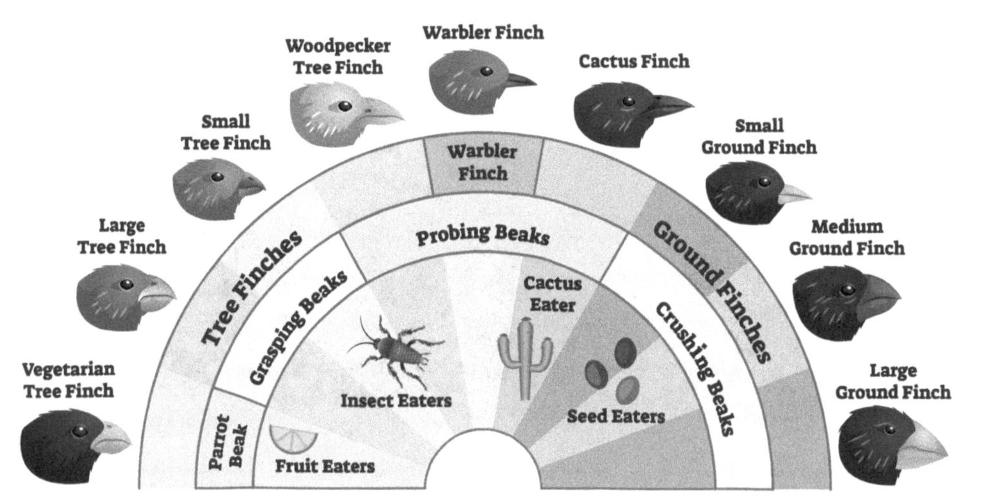

Competition for resources within and between species plays an important role in evolution through natural selection. The finches Darwin observed are thought to have evolved through adaptive radiation that caused diversified beak shapes adapted for different food sources. This process can result in speciation.

will persist in the population supposing the environment still favors certain genes. Evolution can be said to occur if a change is witnessed in the traits of a population through many generations.

Evidence for Evolution

We can interpret observations to understand evolutionary relationships. Evidence of evolution is observed with microorganisms, **comparative anatomy** and morphology, comparative embryology, paleontology and stratigraphy, biogeography, biochemistry and molecular biology.

Microorganisms. Bacteria and viruses are *r*-selected species and entities that can rapidly adapt to a changing environment. These observable changes often evolve into new strains altogether. Therefore, evolution can be said to have occurred when the changed trait persists through generations. This can take a short amount of time to witness. Immunities to pharmaceuticals can be a result of this phenomenon.

Comparative anatomy and morphology. When considering the anatomical evidence for evolution, **homologous structures** help scientists find relatedness

HOMOLOGOUS STRUCTURES

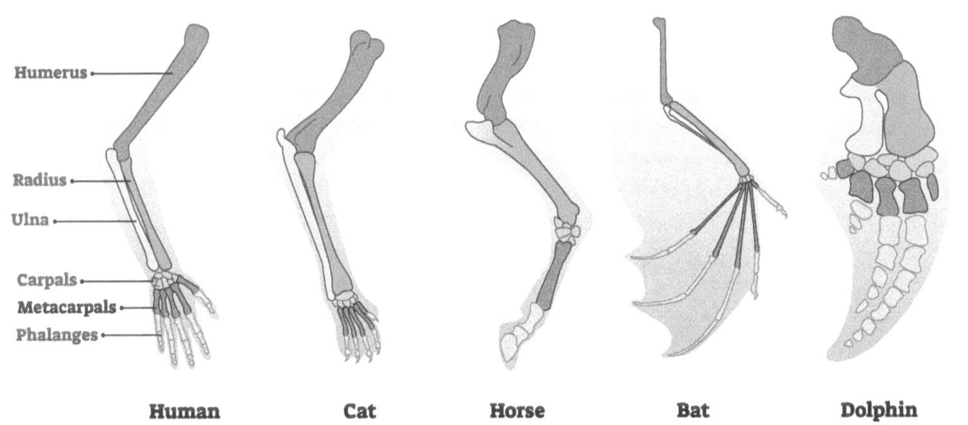

The bone structures among these species' forearms have a basic similarity (homologous) which can be used to classify organisms.

among species. Anatomical structures in different species originated by heredity from a structure in the most recent common ancestor of the species. For example, the bone structure of a pterodactyl, a bat, and a bird have some basic similarities that suggests they had a common ancestor with a similar forearm structure. In terms of bone structure, these animals differ with, for example, the bat has a wing between its fingers and the bird has feathers all along the forearm. These are **analogous structures** because even though they have closely related functions they do not have a common ancestor for that function. Each of these species evolved their ability to fly from a different ancestral line.

Comparative Embryology. Comparative embryology is also used as an indicator for the evidence of evolution. A resemblance exists between the development of the embryos in different species. The explanation is that they share a great degree of genetic and DNA similarity, therefore their early development will share the same qualities of having a primitive eye, gill slits, a neurochord, and a tail.

Paleontology. Paleontology is the study of fossils in sedimentary rock that shows similarities and differences to living species. **Stratigraphy** is the layering of the rock in which the rocks and fossils can be aged through relative age dating or the Principle of Superposition. These concepts can be applied by showing where the fossils were found in relation to one another. Older rocks are found below newer rock. This can be used to determine how species assemblages have changed over time or in what ways a particular species may have evolved.

EVOLUTION

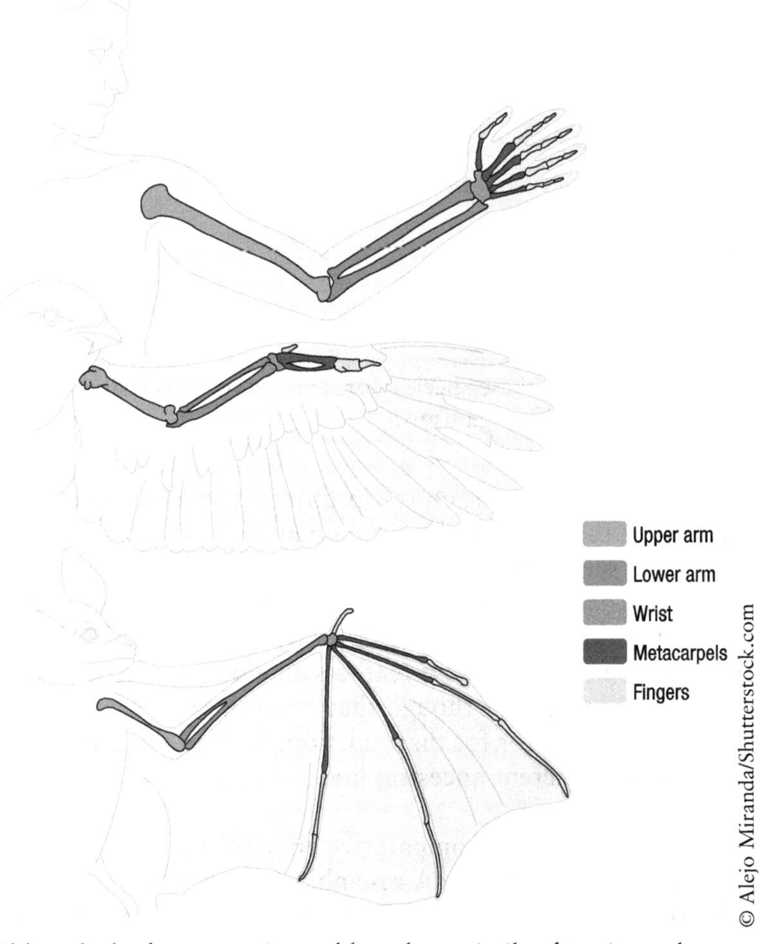

Although the forearms pictured here have similar functions, the structures differ to indicate different descendants from different times (analogous).

Biogeography. Biogeography also provides evidence of evolution through the distribution of species spatially and temporally. Characteristics of some organisms were a result of continental drift when the Earth's plates shifted the land masses apart. A stage of evolution was reached when the land became isolated and moved into a new area or latitude causing a shift in environmental conditions.

If two organisms are closely related, their chemical processes and DNA will be very similar. This refers to the comparison of **biochemistry** and **molecular biology** between species. Because of these scientific advances in the sample analyses, scientists can determine the degree of relatedness. **Speciation** is an evolutionary process of when new species arise. It requires selective mating, which results in a reduced gene flow. Selective mating can be the result of geographic isolation, behavioral isolation, or temporal isolation. For example, a change in the physical environment is **geographic isolation** by an extrinsic barrier. A change in camouflage available in the habitat could result in **behavioral isolation.** And, a shift in mating times due to seasonal variation or habitat changes can result in **temporal isolation.** One species of deer could shift location and therefore changes its "rut" or timing for breeding readiness. Regardless of mechanisms or evidence for evolution, the fact is that changes in population adaptations are observable and speciation can eventually occur.

Review Questions

1. Describe the difference between a scientific law and a scientific theory.
2. Define evolution.
3. What parts of the study of evolution are facts and what are theories?
4. Explain these mechanisms for evolution: mutations, gene flow, genetic drift, and natural selection.
5. Discuss the meaning of an adaptation and give examples.
6. Explain the meaning for the phrase, "survival of the fittest".
7. What are the lines of evidence for evolution? Describe the evidence.

4.3 The Four Spheres of Earth and Energy Flow

Widening our perspective and moving further up the scale of life, the **ecosystem** is the fundamental unit of ecology. An ecosystem has a self-sustaining community of organisms and includes their physical environment in which they interact. It encompasses the biotic and abiotic factors.

For an ecosystem, the sun is the ultimate energy source. There is a one-way flow of energy through the ecosystem while all matter is cycled. Matter occupies space within the habitat and has mass. Nutrients and water are taken up by living organisms and cycled back to an abiotic component of the ecosystem. As an ecosystem researcher, you may want to investigate connections between the macroinvertebrate assemblages within a stream and the surrounding abiotic conditions.

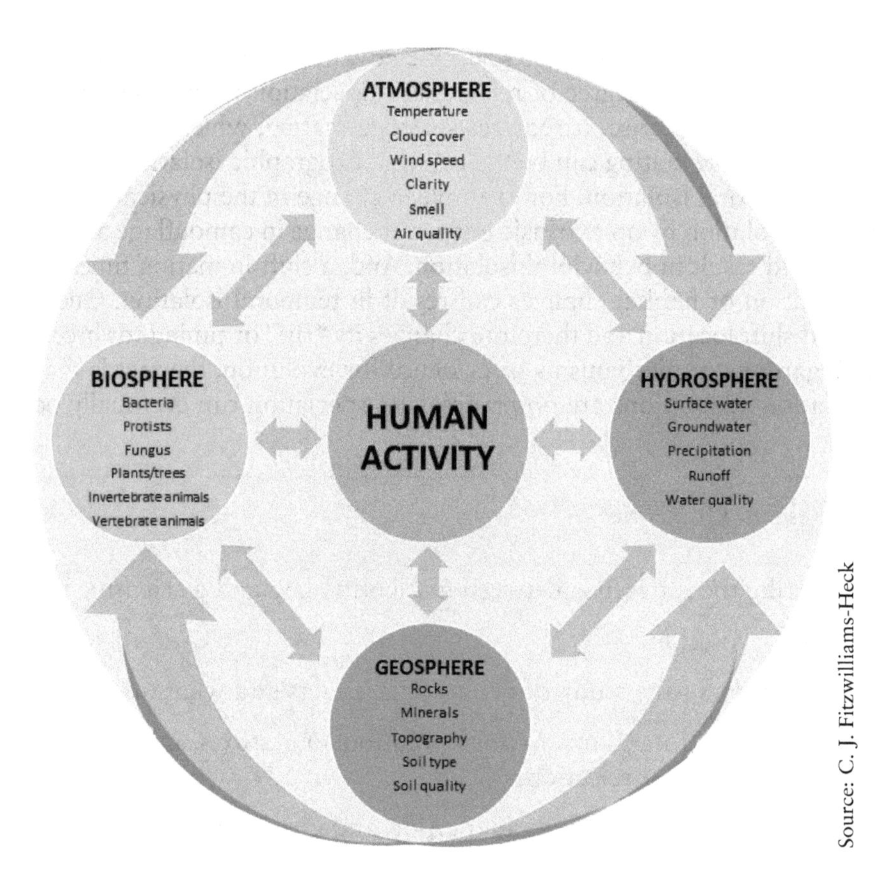

Source: C. J. Fitzwilliams-Heck

The Four Systems Form an Ecosystem

When you stand at any location, take a moment to reflect on the four systems that affect the living and nonliving things around you. These four systems are the atmosphere, hydrosphere, geosphere, and biosphere. As a naturalist, addressing these four systems or spheres will help you focus on more than what you first observe. It helps hone your observational and interpretation skills. So, what's with the "-spheres"? A sphere resembles a three-dimensional ball. With no beginning and no end in sight, these four systems act just the same with a continual recycling of matter.

The **atmosphere** acts as the Earth's shield or envelope. Approximately 45% of solar energy reaches the earth's surface, while about 55% is reflected back into space. The gases in the atmosphere of biological importance include nitrogen which comprises about 78% of the atmosphere, oxygen with about 20%, water vapor occurs at approximately 1%, and carbon dioxide is around 0.035%. Observations of the atmosphere can include how it feels which can indicate temperature, and wind direction or intensity. You can also describe how the air and sky looks and how it smells. This can be an indicator of air quality.

The **hydrosphere** references the water found in, on, and around the planet. It is found as groundwater, in lakes, rivers, streams, and the moisture component of the atmosphere. The **geosphere** is the nonliving part underfoot such as rocks, minerals, and soil. The **biosphere** makes up the living components in the ecosystem. Realize that energy flows in one direction in the ecosystem while matter will move between these four spheres. Each system is linked. Connections within the earth system involve atmospheric chemistry and temperature affecting organism behavior (biosphere), weathering (atmosphere or hydrosphere) of the geosphere, and temperature can affect evaporation rate (hydrosphere). Plants (biosphere) affect the atmospheric carbon dioxide level, aid in the physical and chemical weathering of rocks (geosphere), and can control the water transfer (hydrosphere) from the soil (geosphere) to the air (atmosphere). Rainfall and runoff (hydrosphere) erodes the land surface (geosphere). Soil (geosphere) and water (hydrosphere) limits plant growth which affects the animals in the area (biosphere). You get the idea. The possibilities are almost endless.

Nature Journal: List the four spheres. Identify where each sphere exists in your Nature Area. Labeling the spheres on your base map of the area can be useful in visualizing and interpreting the connections. Diagram these connections on your map or in a separate section of your journal. Explain at least one connection between each sphere you identified. Once you start recognizing these relationships, you may find it fun to identify as many connections as you can. So, keep listing them until you run out of ideas.

Parts of the Biosphere

All parts of the biosphere are made up of organisms that can be classified into 1 of 3 groups. They are either a **producer** (also known as photosynthesizers, autotrophs, or plants), **consumer** (i.e. heterotrophs, herbivores, carnivores, omnivores, predators, prey, parasites, scavengers, or detritivores), or a **decomposer.**

The energy flows through ecosystems from the sun to the producers. Plants have the ability to change the solar energy into a usable food source through the process of photosynthesis (see Chapter 6 for further details). This process of "self-feeding" classifies green plants as **autotrophs.** From there, energy passes to consumers who need to ingest another organism for nutrients. This would also be called a **heterotroph**. Various types of heterotrophs exist. A plant eater is an herbivore, whereas a carnivore is somebody whose diet is mainly meat, and an omnivore's diet consists of both. **Scavengers** are those consumers who feed off mainly dead animals (**carrion**) and do not rely on hunting as a predator. **Detritivores** are a special group of consumers that feed on **detritus** (dead organisms or cast-off material from living organisms like exoskeletons, antlers, leaves, twigs, etc.). These organisms break down once living, **organic matter** (tissues that still retain the carbon that exists in all living things). Worms and beetles are good examples of detritivores. All that dies will get broken down into its inorganic components to be recycled through the ecosystem. This is done by **decomposers**. Fungus and bacteria have the ability to decompose matter.

JOURNAL

Nature Journal: To help you have an ecosystem perspective, choose an organism found within your Nature Area. Determine: Where and how does your organism acquire its nutrients? How is its food source sustained? What affects their food source's survival in your Nature Area? Who within the area depends on them and in what capacity?

For another fun activity at your Nature Area or other place of interest, find a log and spend time discovering the different stages of decomposition on top of it. Make as many observations as possible. Gently roll the log to examine what lies beneath it and replace it the way you found it when finished. Do you notice other logs around you? Compare logs of differing stages of degradation. Are there any snags in the area? These are fun to investigate as well. Take caution when around these decaying, standing trees; they may topple easily. Chapter 2 can help you think about organizing a field investigation.

The various feeding relationships define the **trophic levels** in a **food chain** and **food web**. A food chain illustrates the flow of energy through the ecosystem with the focus on one, linear feeding relationship possibility. The base of food chains begins with the first trophic level, the plants or the producers of the ecosystem. Without the plants, the herbivores and carnivores would not exist in the area. The second trophic level would be the plant predator, herbivore, or also known as a **primary consumer**. The **secondary consumer** appears as the third trophic level and preys upon the herbivore. The **tertiary consumer** (fourth trophic level) gains energy from the secondary consumer—and so forth. Ecosystems typically do not support trophic levels beyond the fifth. Creating a food web will display the complex interconnectedness among food chains. It can help the naturalist and scientist realize what could happen if one of the "strings" of the web was

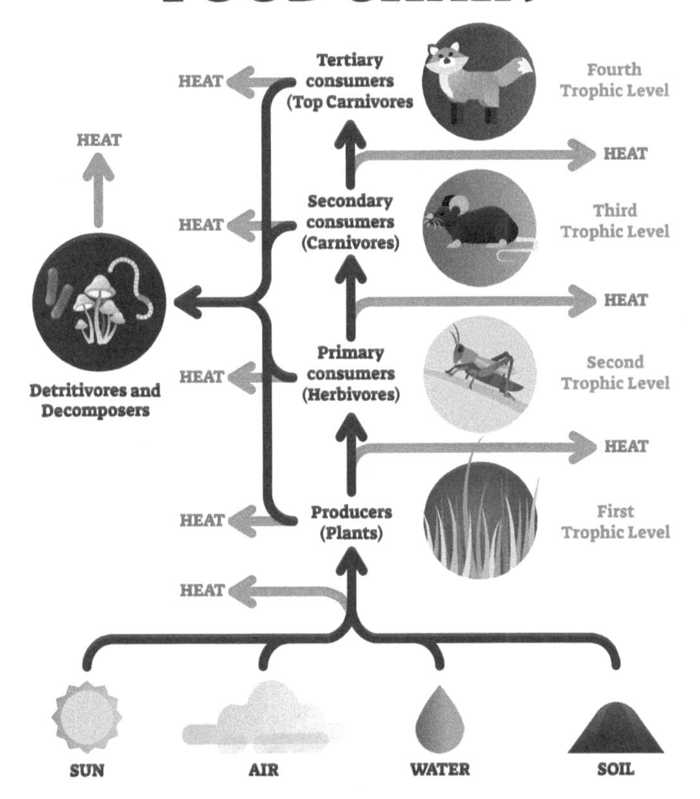

Explain what is occurring at each step of this food chain. Start from the bottom of the diagram. The abiotic factors will influence the growth of the vegetation in an ecosystem. The producers provide the energy for the subsequent trophic levels of consumers. Note the heat loss from every activity in the food chain, and all living things will eventually decompose to allow for the recycling of matter in the system.

removed, meaning the loss of one organism can have a detrimental effect on the whole ecosystem.

It is also important to realize that it is more energy-efficient to eat plants of the ecosystem instead of animals. Each trophic level has a fraction of the energy that is available in the level below it.

There is more energy available at the producer level than at the consumer level. A greater amount of biomass exists with the plants in the ecosystem. However, a small fraction of what organisms consume is transformed into actual tissue mass. Most nutrients retained are used for metabolism and other daily activities. We see the laws of thermodynamics within a food chain.

In this food web, you can trace the energy flow from the plants to the various animals in the ecosystem. The more species you can identify, such as for the plants and other primary producers, the more complex the web will appear and show the potential relationships and effects that can occur if something were gone from the system.

Energy loss occurs at each step in the food chain with the lowest energy level available at the top with the predator. Only about 10% energy is utilized at any trophic level with approximately 90% of the energy lost due to heat production and **homeostasis.** The percent will vary dependent on the animals and caloric content. All molecules produce heat as an unusable byproduct, and homeostasis is the natural regulation of the organism's internal environment to maintain a stable condition (i.e. breathing, sleeping, blood circulation, etc.).

Nature Journal: Create a food chain by starting with an organism that could live in your Nature Area. Depending on what type of organism it is, draw arrows from or to it as an indication of energy flow through your ecosystem. Write names of plausible organisms to include in your food chain and link them to your starting organism with arrows showing energy flow. You should have a linear display of the food chain (or a circular diagram to indicate the decomposers releasing nutrients into the soil for plants to grow). Label the trophic levels. Identify the producer, primary consumer, secondary consumer, tertiary consumer (if present), and decomposers. Choose one of the organisms in your food chain to branch off from to create another chain with the result of a realistic food web for your Nature Area. Do not try to label the trophic levels or consumers in this new food chain you created, because their trophic status will likely change in the different interconnected chains of the web.

To understand energy within an ecosystem further, knowing a few things about the **Laws of Thermodynamics** comes into play. These scientific laws define the energy conversions in the universe. The aspects of these laws most relevant for our nature study include the first and second laws.

The first law states that energy is neither created nor destroyed, but changes form. The sun's energy changes form as the plants utilize it in photosynthesis to create chemical energy. The universe has a finite or limited amount of energy due to the Earth being considered mainly a closed system in terms of matter cycling. This is a major reason why sustainable energy use is critical to ecosystem quality.

The second law of thermodynamics states energy spontaneously flows in only one direction. Hot temperatures naturally transfer to cooler temperatures. Have you watched warm surface water of a lake evaporate into the cool air on a late summer's morning? Heat cannot flow from a colder body to a warmer body. Also, high energy of any kind cannot be sustained and results in a lower energy state. In the Earth's closed system, matter moves freely and disorder (**entropy**) increases. All the energy transformations occurring in an ecosystem will have heat loss. Temperature, pressure, and density differences will eventually distribute evenly or dissipate in the ecosystem.

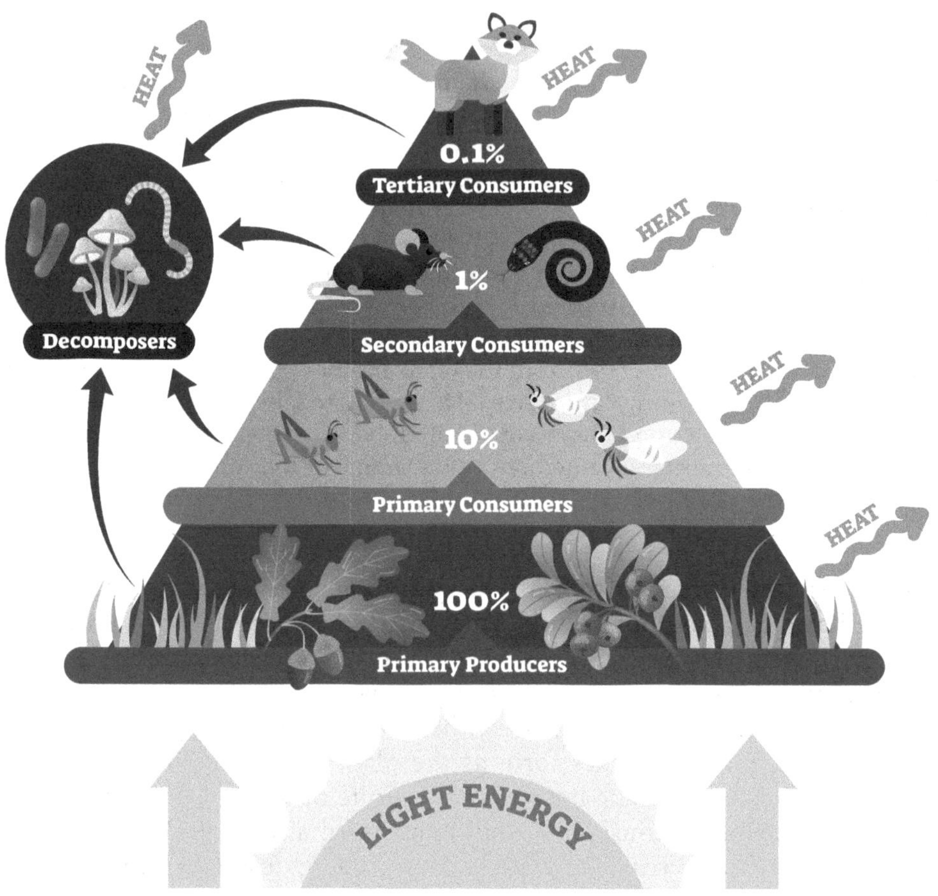

In the **biomass pyramid** or **energy pyramid**, the relative mass of the trophic levels is displayed with the primary producers at its base signifying the greatest amount of energy available for the ecosystem. In the generalized example, of the 100% of the total plant biomass in an ecosystem, only 10% of its energy is transferred to the herbivores of the community, which can only support 1% biomass of secondary consumers, and only 0.1% biomass of the ecosystem as a predator. Animals have a greater energy cost than plants. It takes a considerable amount of biomass in an area to support a tertiary consumer. A habitat cannot support very many predators because of this fact. What are the other types of things to consider in terms of what supports organisms in an ecosystem?

Review Questions

1. Describe the four systems of Earth that form an ecosystem.
2. What is the difference between a food chain and food web? How does the energy flow? Why can it be important to construct and study a food web?
3. Describe the biomass or energy pyramid. How much energy is lost with every transfer? What two things are lost with each energy transfer?
4. Why do animals have a greater energy cost than plants?

4.4 Biogeochemical Cycles

How do the carbon, nitrogen, and water cycles affect you? These cycles are critical to living organisms including the cycling of phosphorus and sulfur. In this section we will think about the cycling of some abiotic components of an ecosystem.

Abiotic components exist as one of two types. One type is the **conditions** of the ecosystem that include the setting or surroundings, and the local weather (i.e. temperature, cloud cover, wind speed, relative humidity). The other category of abiotic factors are the **resources**, like water and nutrients that exist in the habitat to help the biota survive. Ninety-two elements exist yet only 30 are vital to life—these are classified as **nutrients**. The movement of water and nutrients between the abiotic and biotic components of ecosystems is **biogeochemical cycling**. With a fixed amount of resources within Earth's relatively self-contained system, these elements recycle at varying rates. The scientific study of how living systems influence and are controlled by the geology and chemistry of the Earth is referred to as **biogeochemistry**. Most major environmental problems like global warming, acid rain, pollution, and excess greenhouse gases are analyzed using biogeochemical principles and tools.

Carbon Cycle

Why is the carbon cycle so relevant to nature study? It is happening everywhere around us. **Carbon** is stored in living systems. Carbon is considered the "building block" for life; it is necessary for all things to exist! When referring to the biomass of an ecosystem, this is a measure of the carbon (C) stored in living systems. Carbon is linked with O_2 to form CO_2. Carbon dioxide (CO_2), an essential gas to life, comprises approximately 0.035% of atmospheric gas. The O_2 and CO_2 gases cycle between all four spheres and are interdependent.

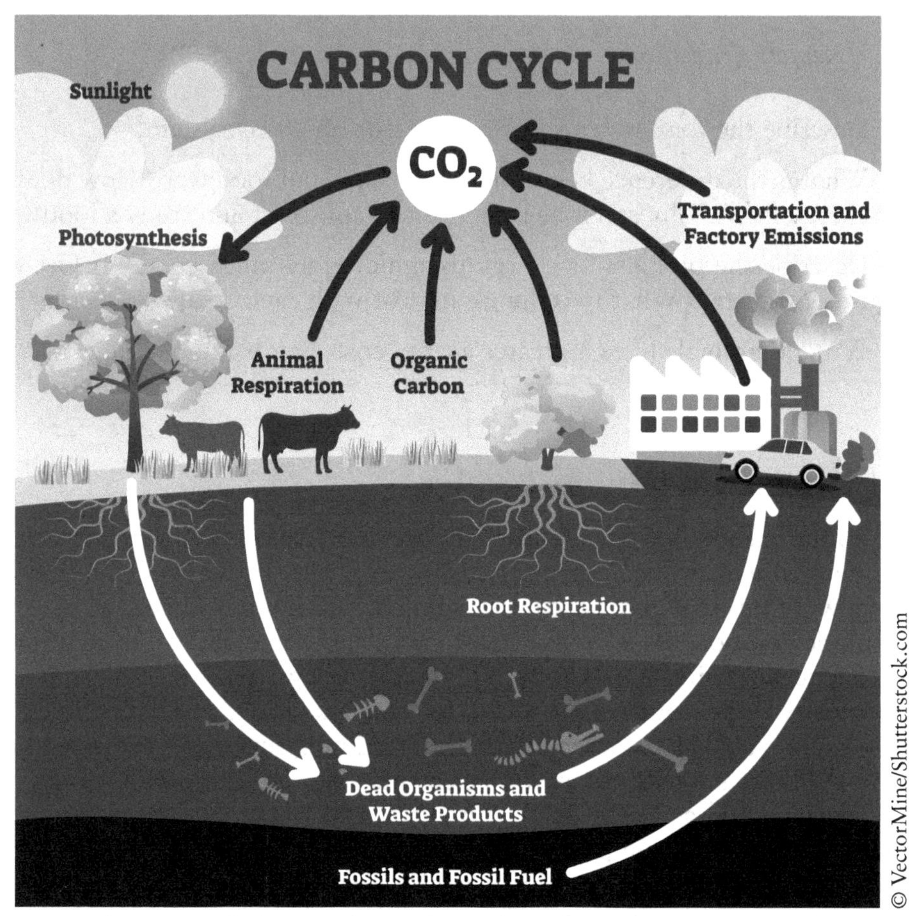

Photosynthesis, from all plants and green algae, takes in and assimilates carbon dioxide from the atmosphere. This is the greatest amount of carbon transferred in the ecosystem. Plants and animals expel carbon dioxide in respiration. Animals will acquire carbon from the plants and animals they eat. Carbon will also transfer to the soil through decomposition. Over time and pressure from the burial of those fossils, its extraction and burning of it releases carbon dioxide and is a form of energy (**fossil fuels** such as coal or gas).

The Greenhouse Effect and Global Warming

Before reading any further, do you think the greenhouse effect is a negative thing for the planet? No, we need it for life to exist ... but not too much of it. So, it is a good thing for us on earth. When there is an excess of the greenhouse effect, then we experience global warming – this is not a good thing.

Greenhouse Effect

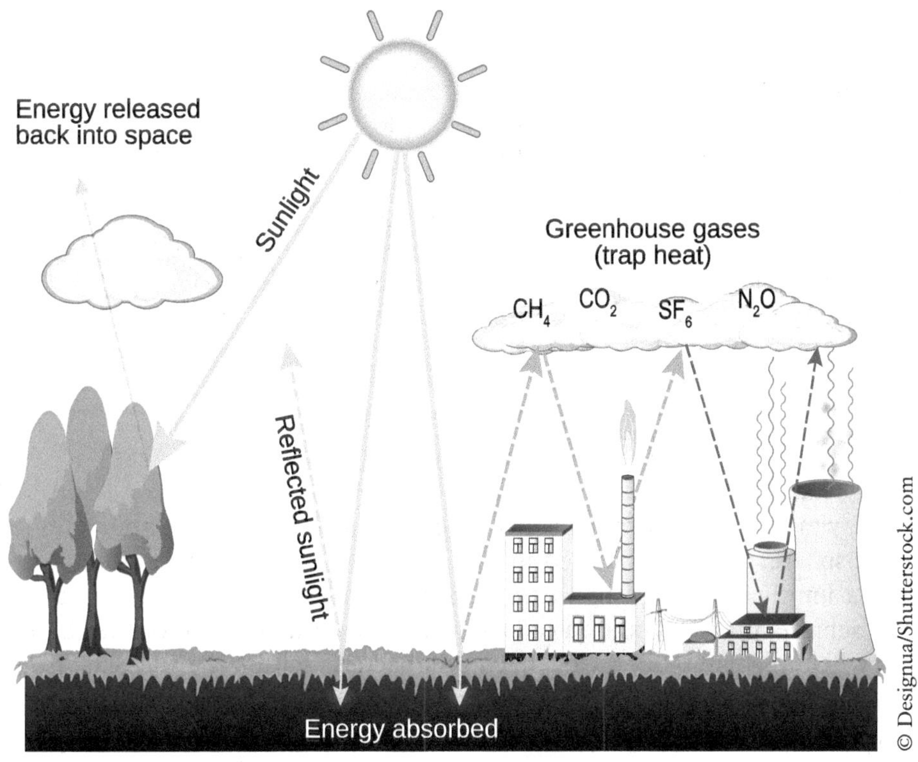

Trace the path of sunlight and energy as the earth absorbs it and becomes trapped by greenhouse gasses. The greenhouse effect and global warming are both illustrated here.

First, what is the **greenhouse effect**? Actual greenhouses help grow plants in cold temperatures. The glass of greenhouses lets in light but traps heat. In the atmosphere, greenhouse gases act as the glass panes in a greenhouse (i.e., water vapor, carbon dioxide, nitrous oxide, and methane). Without their existence, the Earth would be 60 °F (16 °C) colder.

The sunlight enters the Earth's atmosphere passing through a blanket of greenhouse gases - the metaphorical glass panes. Land, water, and the biosphere absorb sunlight's energy. Once absorbed, energy is sent back into the atmosphere allowing some to escape back into space, while most of the energy stays trapped in the atmosphere by greenhouse gases. This is how the greenhouse effect works, but if the blanket of these gases is too thick, then less energy escapes and it will cause the earth to heat up beyond what is natural—this is **global warming**.

> ***Nature Journal:*** Imagine if all the trees, shrubs, and herbaceous plants were removed for as far as you could see around your Nature Area. How would this affect the carbon cycle? What if burning fossil fuels continued or increased? How would an increase in temperatures through the year affect the abiotic factors (conditions and resources) nearby? Discuss how those changes could impact the living things within and near your Nature Area. What could be done with your Nature Area to ensure a healthy cycling of carbon? Apply your answers above to the biosphere level. Research how global warming leads to climate change and explain it. Why is this a negative thing for the planet? What can be done on a global scale to minimize and stop climate change?

Nitrogen Cycle

Another vital element to life is nitrogen. Nitrogen is necessary for DNA, RNA, proteins, vitamins, and hormones. Although nitrogen is the most abundant gas in the atmosphere, it cannot be used in this form by organisms. Therefore, it is considered a limiting factor to life. Atmospheric nitrogen needs to be "fixed". How can we achieve **nitrogen fixation** or change it into a usable form for life? Lightning strikes will immediately fix nitrogen, certain bacteria within the soil will also fix it, and the synthetic fertilizer humans use adds nitrogen to crops, gardens, and lawns to help the plants turn greener and grow faster.

The phenomenon of soil bacteria fixing nitrogen often begins with a symbiotic relationship between the roots of **legumes** (plants from the "pea or bean" family, such as clover) and nitrogen-fixing bacteria growing on them. The roots provide a home for the bacteria, and the bacteria provides essential nitrogen to the plants, which in turn allows the cycling of this nutrient in the ecosystem. Plants absorb nitrogen as nitrate or ammonium salts dissolved in soil water. Animals get nitrogen from eating plants and animals.

> ***Nature Journal:*** As discussed in the section of Chapter 3 about aquatic ecosystems, excess nitrogen can wreak havoc on these systems due to what water carries with it in runoff. Research the potential ways humans have affected the nitrogen cycle near your Nature Area and within your community. Where do these nitrogen sources, if any, exist within your subwatershed: fertilizer use, fertilizer production, burning fossil fuels, animal waste, sewage, erosion, and runoff into waterways. What could be done to minimize the impact of these activities?

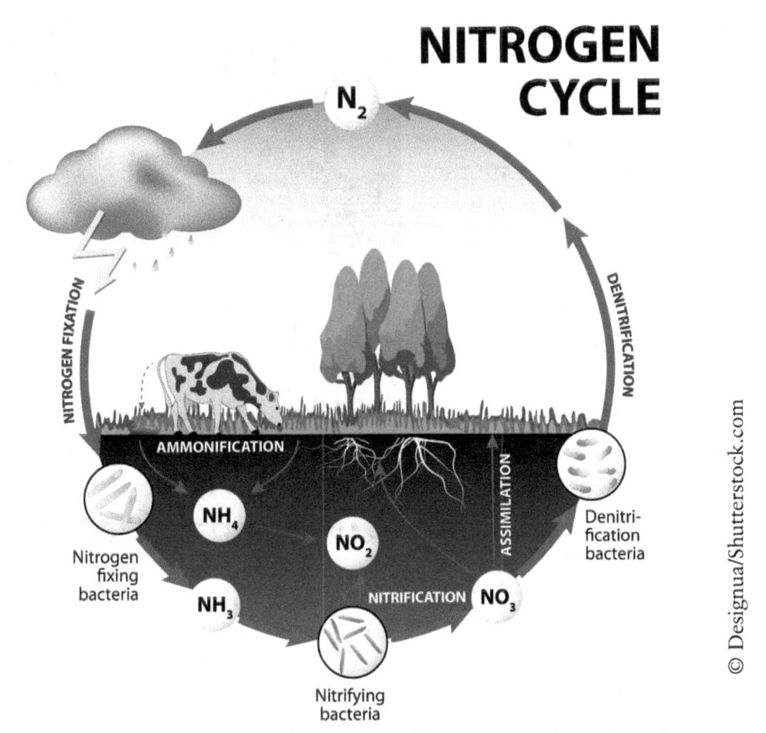

© Designua/Shutterstock.com

This nitrogen cycle shows lightening and bacteria in the roles of nitrogen fixation. Note the various forms nitrogen takes in the ecosystem.

Water Cycle

We need water. Water and its ecosystems are essential to most life, but humans are threatening this natural resource through overuse and misuse. All the water we have on or around our planet is constantly being recycled or is unusable.

Ponder this ... all the water we have now in our hydrosphere is all the water there ever was and ever will be. Perhaps the rain that just fell on your head was once part of a dinosaur! Remember the rules of ecology, there is no away in nature—everything must go someplace. What are the possible fates of a water droplet? Suppose you begin in a large lake. Most **evaporation** that occurs in our region happens from the Great Lakes' surface, otherwise the most evaporation on the planet occurs from the oceans.

Evaporation is when water in its liquid form transforms into its gaseous state due to heating of the water (think about how global warming affects the rate of evaporation and its consequences!). This water accumulates as water droplets in cloud formations, or **condensation**. When the cloud reaches its water-holding capacity, it

CYCLES

forms **precipitation** such as rain or snow varieties. As that water droplet, you may precipitate onto water or land. On the land, you may experience **infiltration** and penetrate the ground to recharge the **groundwater** that will either collect in natural **aquifers** or flow to a nearby lake or stream. If this is not possible, then you will either collect on the surface or leave the area via **surface runoff** and travel downslope. You may evaporate from the surface if warm enough, or you could be taken up through the roots of plants only to exit their stomata by means of **transpiration.** The water will continue to move as a cyclical process. All of the fates of a water droplet are possible because of the unique properties of the water molecule as discussed in Chapter 3.

Water covers more than 75% of the Earth's surface and is the most abundant compound in nearly all living organisms. With 97.5% of the water as ocean, the usability decreases for humans. The 2.5% of freshwater is mainly frozen which leaves a small percent (<1%) as fresh, available water. Clean water is even scarcer.

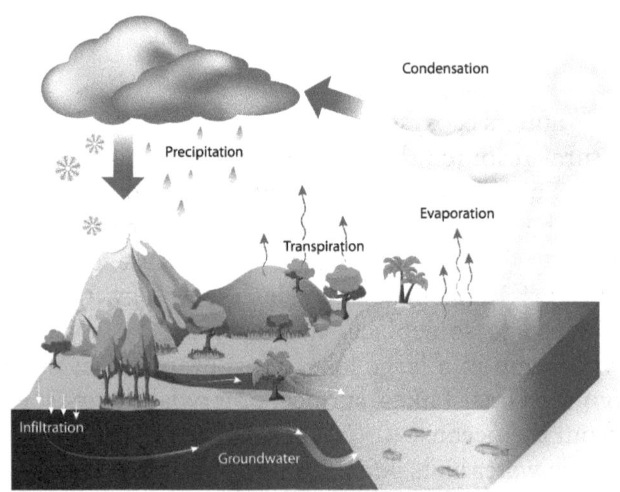

The sun drives the water cycle. It heats water to the point of evaporation and enters the air as water vapor only to eventually condense into precipitation. What are all the possible fates of the water after it falls to the earth?

Nature Journal: Suppose more pavement and buildings (**impermeable surfaces,** or "**hardscapes**") existed than green spaces. These areas do not allow water to infiltrate into the ground. How could this have a negative impact on the hydrosphere?

You may be amazed to learn how much water volume runs off a rooftop in a rain event. Consider a roof of 1,000 sq. ft. (92 sq. m) during an event that produces 1 in. (2.5 cm) of water. How many gallons (cubic meters) do you think runs off? About 600 gallons (2 cu. m) runs off in just that one rain event! Think of how much your garden could be watered in a season if you could harness that runoff in a rain barrel, or direct that flow into a garden, or plant a garden or native wild area of grasses, flowers, shrubs, and trees.

Take inventory of the hardscapes around where you live (i.e. rooftops, driveway, sidewalks, etc.). Map it out on paper. Calculate the surface area for all your hardscapes (convert to inches or cm). Multiply those dimensions by the number of inches (cm) of rainfall (suppose 1 in. or 2.5 cm if this is a hypothetical situation). Divide by 231 to the number of gallons (because 1 gallon is 231 cu. in.; or 1,000 because 1 cu. m. is 1,000 L).

The number you get in the calculation is the amount of runoff that occurs in the area after a rain event. Where is that water going? Where are the biggest problem areas? What could be done around where you live or work to reduce the amount of water unable to infiltrate in the ground?

Review Questions

1. Describe what is meant by biogeochemical cycling. What is its significance to nature.
2. Diagram and explain the key points of the carbon cycle, nitrogen cycle, and water cycle.
3. How do humans benefit from each cycle (carbon, nitrogen, and water)? Discuss how humans negatively impact each cycle. How can that negative impact affect humans? What can people do to fix or prevent the degradation to the ecosystem from those acts you described?

4.5 Biodiversity, Conservation, and Management

Try to imagine a world with no regulations on how we interact with nature. Is there anything you do in a day that does not involve or depend on the environment? Improving biodiversity through management can help conserve our precious natural resources. A variety of living things can create a more beautiful and exciting place to live.

BIODIVERSITY

"Biological diversity" or **biodiversity** describes the varieties of organisms and the complex ecological relationships that gives the biosphere its unique, productive, and balanced characteristics. Nearly 2 million species have been identified, yet taxonomists estimate 3–100 million different species alive. Why so many unknown species? Think about the variety of hard to reach locations on the planet, extreme conditions that exist where living things could survive, the minute size of some organisms, and often science for the environment gets eliminated for federal funding.

Biodiversity exists on several levels to preserve ecological systems and functions. As discussed in Chapter 2's section on field investigation methods, there are different factors of **species diversity** to consider when addressing the variety of species in an area, such as species richness, species abundance, and relative species abundance. Other considerations are geographical, genetic, and ecological diversity. **Geographical diversity** considers how species are distributed across a region. Having a high geographic diversity means that the population is potentially spread across a wide variety of ecosystems. This can ensure a population persists even if a portion of the organisms were subjected to adverse conditions, like flooding, and were eliminated in just that part of the area. A population with a high **genetic diversity** would help ensure the potential for a population to adjust and eventually adapt to a changed environment. **Ecological diversity** can be used to encompass all these aspects of biodiversity. It incorporates the richness and complexity of a biological community. You can view this broad perspective of examining the ecosystem by the number of niches occupied, how many trophic levels that exist, the ecological processes that capture energy, the variety of food webs sustained, and the efficiency of the recycling of matter within an ecosystem. Realize that not all habitats have an efficient or diverse system. In those instances, the area may continually degrade causing further issues unless something is done to restore its balance.

Conservation Biology. Conservation biology is a scientific discipline devoted to understanding factors, forces, and processes that influence the loss, protection, and restoration of biological diversity within and among ecosystems. This branch of science is applied and goal-oriented to prevent extinction. E. O. Wilson, an American scientist, naturalist, and writer is not only recognized as an expert in nature and ants, but also is considered the seminal figure of biodiversity and conservation. In the 1970–1980s, Wilson and other biologists grew very alarmed by the degradation of natural systems they had spent their lives studying. Since then, conservation biology has been focusing on the analysis of ecosystem inventories to help protect areas of concern where biodiversity is or could become threatened.

Wildlife Management. Wildlife management is closely tied to both biodiversity and conservation. It applies ecological knowledge to plant and animal populations and their communities to strike a balance between the needs of those populations

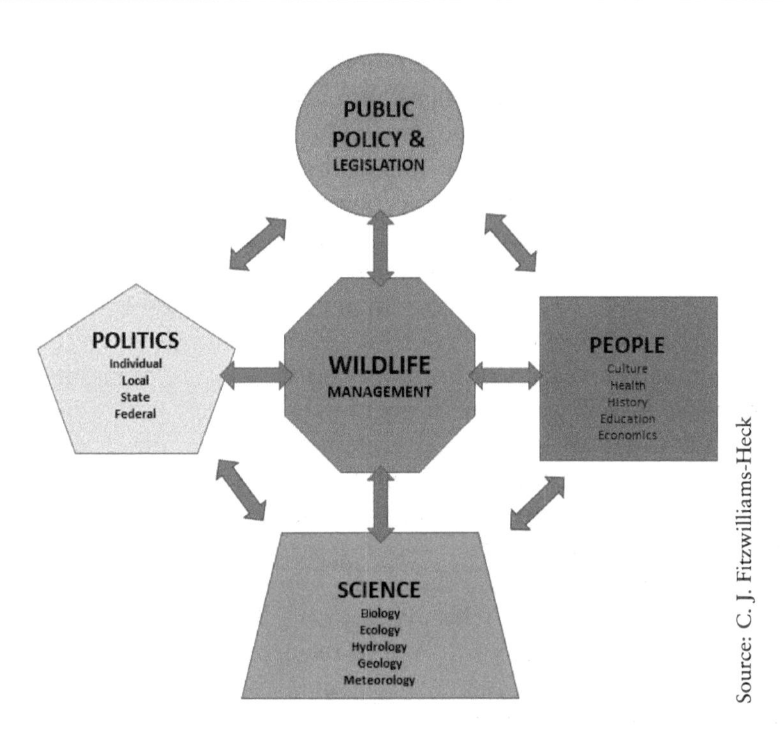

Source: C. J. Fitzwilliams-Heck

and the needs of people while maintaining ecosystem biodiversity. Wildlife managers use scientific data to help communicate and balance the needs of the people in a region to make informed recommendations for public policy regarding natural resource management. Prior to decisions or policy making, public forums may be offered or surveys sent to citizens for input. This is the people's chance to voice concerns and offer informed opinions on what they value in nature.

Benefits of Biodiversity

Why care about protecting biodiversity? As a naturalist, you may need to discuss these points to help others realize and understand the importance. Most people you ask would hopefully agree that just knowing wildlife exists around them is important. Bird watching, wildlife viewing, and nature photography are a few examples of how people enjoy the outdoors. Humans can benefit from biodiversity in many ways. Without biodiversity, conservation, and management these activities would not exist, and neither would the deeper, social–economic–ecological connections. Human benefits of biodiversity include increases in food, medicine, economy, ecological complexity, and aesthetics.

There are nearly 100,000 types of edible wild plants that could be utilized by people. About half of our food crops depend upon the diversity of insects for pollination. Genetic diversity of our agriculture can also enhance and provide insurance against losses of prevalent strains of our staple crops. For medicines, more than half of all prescription drugs contain some natural products. At least a dozen of the top 25 prescription medicine comes directly from wild plants, while the rest scientists developed from the chemical analysis of wild species.

Biodiversity can also improve local economies. Another tie to insect diversity for pollination of crops is the amount of revenue the agricultural industry generates with a multibillion-dollar income. Alternatively, small operational **organic farms** with crop production intended for the local communities can help raise the local economy. They too depend on biodiversity. The organic practices eliminate the use of harmful pesticides and synthetic fertilizers or hormones on the crops. Among their many sustainable farming approaches, organic farmers may also use a diversity of plants to promote pollinators, prevent plant predation, and generate nitrogen fixation. With large-scale, commercial agriculture practices, the land is cleared to farm year-round. With these practices, biodiversity is lost in all of the Earth's spheres and often, workers may not be compensated for their work. Another aspect of organic farming is the fair and equal treatment given to those who work to plant, care, and harvest the crops. ***Activity:*** Choose a favorite food item you eat. Trace all the fossil fuels used to produce, package, and transport your food to reach you. Buying local organic food has many advantages. Research where you can connect with a small farm near you!

Another connection to the environment and economic growth is travel. Ecotourism is created by increased biodiversity. Tourists will pay money to enter parks, camp, tour, and experience novel natural communities and protected ecosystems. Places like national or provincial parks also employ thousands of people and generate billions of dollars each year. Having these natural attractions can boost local economies, but the downside can be from people disrupting or destroying habitats and could introduce nonnative species.

Promoting biodiversity preserves ecosystem services, and directly provides things of pragmatic value to us. The ecological benefits of biodiversity include food, fuel, and fiber; shelter and building materials; air and water purification; waste decomposition, climate stabilization and moderation, nutrient cycling, soil fertility, pollination, pest control, and genetic resources. The total value of these ecological services is a multitrillion dollar per year benefit. Increased diversity may help ecosystems withstand environmental stress better and recover more quickly than those communities with fewer species. Although we try, we do not fully understand the complex interrelatedness between organisms. Sometimes we may be surprised at the effects of losing seemingly insignificant members of biological communities. Ninety-five percent of potential pests and disease-carrying organisms in the world are controlled by natural predators and competitors. Maintaining biodiversity is essential to preserving these ecological services.

All of this is great, but what can biodiversity do for me personally? Aesthetics and cultural benefits can improve by having high biodiversity. **Aesthetics**, or the overall beauty perceived of an area will attract people outdoors to enjoy hunting, fishing, camping, hiking, wildlife watching, creating art, and other nature-based activities. Nature can also motivate people to exercise more. It can offer psychological and emotional restoration. In some cases, cultures depend on the existence of beautiful or unique natural places for meditation, worship, or to collect certain forms of wildlife. Nature can carry spiritual connotations, and to some people, a religious or moral significance. If we lose biodiversity, we lose these special places that can offer so much to us.

Levels of Species Loss

Many of us have heard or read about extinct or endangered animals, but what do these, and other classifications mean? **Extinction** means that the very last member of a species has died and the entire species has vanished forever from Earth. You can have local extinctions, called an **extirpation** where they may have disappeared from a population locally, but not the entire species globally. If a species is labeled as **endangered**, that indicates their ability to survive and reproduce is jeopardized [by human activity] and is in danger of becoming extinct or extirpated. **Threatened** species have very low numbers and breeding pairs and are likely to become endangered soon. Species of **special concern** have usually experienced changes to their preferred habitats and the likelihood of their population decreasing has increased. The number of individuals for each classification varies depending on the biotic potential of the species. Special monitoring and restoration may be implemented in any of these situations but would be most effective when the situation is less severe.

What Causes Biodiversity Loss?

HIPPO! That is: *h*abitat alterations, *i*nvasive species, *p*ollution, *p*opulation growth, and *o*verexploitation cause biodiversity loss. The acronym will help you remember the types of things acting against biodiversity. As a naturalist, you need to be able to recognize and discuss these aspects. With this familiarity, you can start to identify areas around you that may be at risk of losing species, thereby losing ecosystem services to benefit you and the functioning of nature. Realize that species loss is a natural process. Studies indicate on average, one species goes extinct naturally every 500–1,000 years. Approximately, 99% of all species that ever lived are now extinct. In undisturbed areas, data indicates one species per decade will become extinct or extirpated. The concern lies with the growing human population and their impacts on ecosystems has increased extinction rates by a factor of 1,000.

As you examine the causes of biodiversity loss, you will realize how everything is connected and human actions are essentially the reason. If you were to identify the biggest concern that leads to the other causes of loss, then you would find habitat alterations as the greatest cause of extinction today. Billions of hectares of forests, woodlands, and grasslands have been converted to commercial use, croplands, grazing lands, logging, and urbanization. Those organisms adapted to their preferred habitats that are now fragmented or gone, has caused population declines of birds and mammals of about 85% worldwide.

> ***Nature Journal:*** Look around your Nature Area. How much, if any of it remains in its original state? How do you suspect the land was used over time? Is there evidence of fire, logging, or wetland draining? Chapter 2 can help you think further on interpreting your "place." Where is the nearest place you could visit that has been untrammeled by humans?

Habitat Alterations. The habitat alteration most recognizable and impactful to biodiversity is **fragmentation** of the landscape. This is the reduction of habitat into smaller and smaller, more scattered patches. It reduces biodiversity because many species (especially large mammals, like bear) require large territories to subsist. Some species require the interior of a habitat to live, such as birds who need deep forests to reproduce successfully. Fragmentation divides populations into isolated groups and makes species vulnerable to catastrophic events like storms or disease. Very small populations may not have enough breeding adults to be viable even under normal circumstances. As habitats get increasingly fragmented, immigration becomes more impossible which leads to local extirpations of species. Hopefully, you may start to notice in some cities and around suburban housing developments, **green spaces** have increased. These are areas of vegetation with minimal disturbance by the **built environment** or construction by humans. Natural, constructed, or remediated **corridors** have been designated to help with the connection of green spaces to promote immigration and migration of species—and to provide people a place to walk and observe nature.

Invasive Species. As discussed in Chapter 3, **invasive species** may become aggressive and crowd out or outcompete native species that are essential food sources for wildlife, especially pollinators and migratory birds. The accidental or intentional introduction of nonnative species are freed from predators, parasites, pathogens, and competition in their new territory and will overrun an area quickly. Over 140 nonnative, invasive species have been established in or around the Great Lakes since the 1800s and the species count continues to rise. Remember from Chapter 3, nonnative species did not previously exist in a place and quickly becomes invasive. The reference to invasive species in terms of conservation pertains to nonnative species. Invasives are virtually impossible to eradicate because by the time you notice them, they have already become established. Help stop the spread of invasives by learning about its natural history, and how it can be properly identified and eliminated.

In terrestrial environments, landscape with native plants especially anywhere away from your home where you might not monitor it closely enough to prevent spreading, do not move firewood beyond where you find it, learn to recognize and eliminate invasives properly to not cause more damage to the environment, and report any invasive species observed where you discover its existence. In aquatic systems, remember to clean, drain, and dry. Clean off all visible parts of anything that encountered the water. Drain water from your boat from the bilge, motor, bait bucket, boots, etc. Lastly, dry everything completely for days before entering another aquatic system. The larvae of invasive mussels or parts of plants can go undetected and you may unknowingly transport and introduce them to another aquatic ecosystem. Currently, you can report invasives electronically on the website or app of the Midwest Invasive Species Information Network (**MISIN**). There is also the North American Invasive Species Network who has a platform for sharing invasive species information, the Global Invasive Species Information Network.

Many nonnative species have no serious ecological impact, but introduction of a single key species can, as in the example of the sea lamprey, quagga mussel, or emerald ash borer, cause a sudden and dramatic shift in an entire ecosystem's structure. New species can significantly change interactions between existing species and between those species and their nonliving part of the environment. Ecosystems will change with the presence of nonnative species—they will still function, but in a new way. If species diversity is low in an ecosystem, then the greater the chances of exotics taking over. Either there is a lack of competitors or predators or there are niches available in an ecosystem to allow the nonnatives to become established. The question comes down to, what do we want our landscape to look like? Invasive species is perhaps the second worst threat to native biota.

Nature Journal: Research an invasive species near your Nature Area. Use the Natural History Journal Pages at the end of this chapter as a guide.

Pollution. **Pollution** in all four spheres also threatens biodiversity. Air and water pollution, agricultural runoff, industrial chemicals, etc. can cause death at all trophic levels. Areas of pristine habitats may also experience these issues due to prevailing winds and being carried by water. Alternative energy sources (i.e. solar, wind, geothermal), buying local, and creating more green spaces can help minimize pollution. These remedies make pollution less of a threat to biodiversity.

Population. Human **population** growth, on the other hand, exacerbates environmental problems. More people usually means more habitat change, more invasive species, more pollution, and more overexploitation of natural resources. This is the ultimate reason behind proximate threats to biodiversity.

Overexploitation. **Overexploitation** has two considerations, overharvesting and overconsumption. Overharvesting means the depletion or extinction of species from the wild from too much hunting and fishing. This was observed with the passenger pigeon, American bison, and Atlantic cod. This is also observed with the commercial trade of animal artifacts and live specimens, as well as eliminating predators perceived as pests and not as part of the balance of the ecosystem. Overconsumption of resources is in the form of too much timber cutting, fossil fuel use, etc. that may result in species extinction. Current amphibian declines are attributed to a complex combination of chemical contamination, disease transmission, habitat loss, ozone depletion, ultraviolet penetrance, and climate change. As you can see, much overlap exists between the HIPPO factors of species loss and it is a synergistic interaction of these factors.

Nature Journal: To explore species diversity at your Nature Area, consider reading Chapter 2 for a discussion and examples of field investigation methods to help you expand your nature study. Taking inventory of what exists around you and analyzing and synthesizing the data can provide baseline information for you to compare future monitoring of the area. This will help you realize if a change has occurred and if there is a need for concern.

Conservation Approaches

To help with the preservation of biodiversity, conservation biologists have various methods. Species of concern can be identified through the inventories or monitoring of habitats by scientists or community, or also known as citizen scientists. If the habitat is also potentially fragile, the species can be labeled as an **umbrella species.** The preservation of an entire designated habitat will protect one species, and in turn results in protecting other species. Another method of conservation is the Endangered Species Act of 1973 that restricts actions on, forbids trade of, and prevents extinction of species. The protection of this act also protects our ecosystem services.

Other methods of protecting biodiversity include **captive breeding** where endangered species are bred in zoos to boost populations to reintroduce into the wild, and **cloning** in which molecular techniques are used to clone endangered or extirpated species. Both efforts are futile if the habitat in which they prefer no longer exists or is threatened. International treaties also exist to help protect biodiversity. The Convention on International Trade in Endangered Species of Wild Fauna and Flora, 1973 (CITES) bans international trade and transport of body parts of endangered organisms. The Convention on Biological Diversity, 1992 (CBD) promotes biodiversity conservation, to use it sustainably, and ensures fair distribution of its benefits. Many success stories of saving wildlife exists. Some include the bald eagle and gray wolf. ***Activity:*** Research a conservation success story where you live!

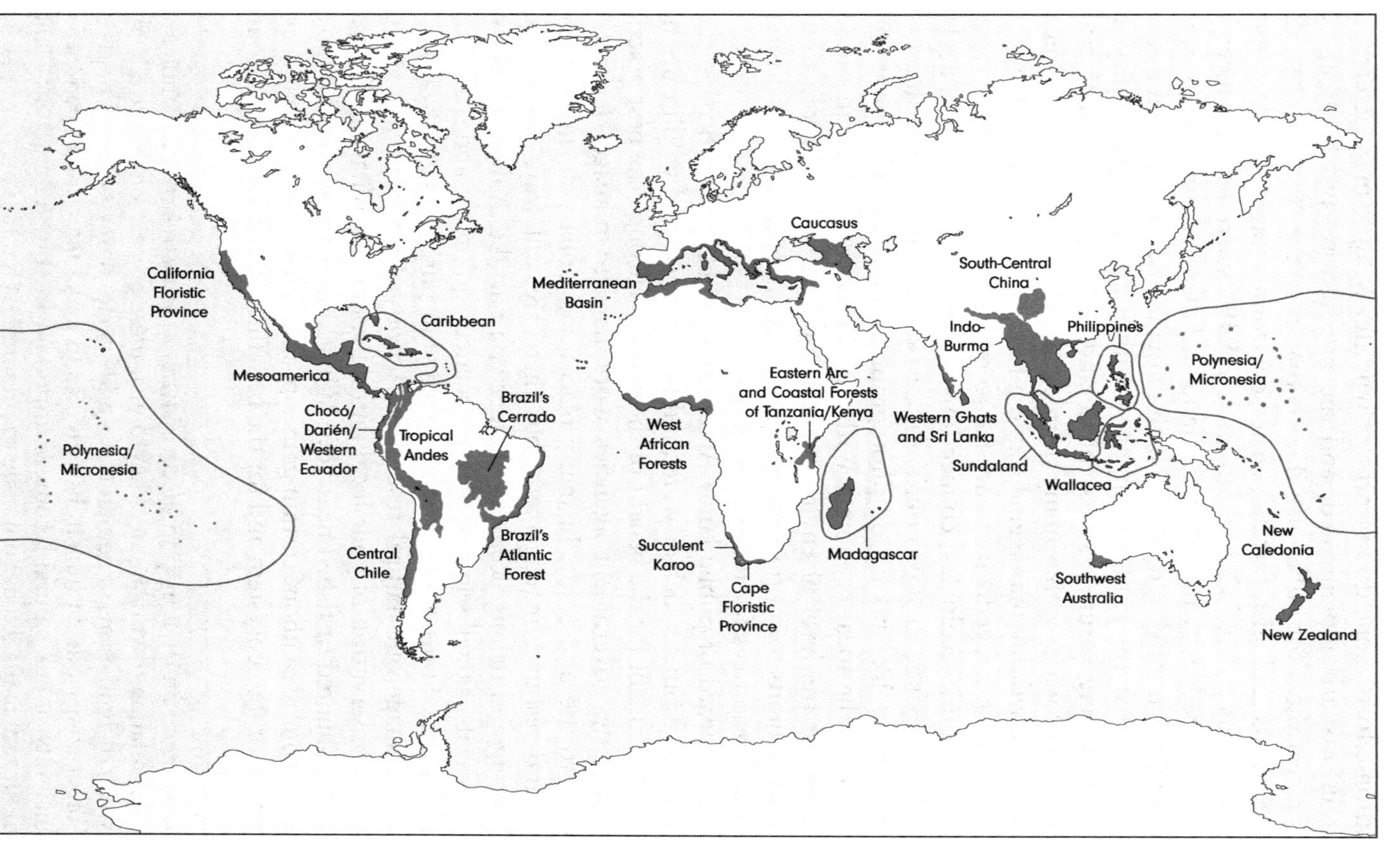

Biodiversity hotspots, highlighted in red, are areas that support especially high number of species **endemic** to the area (found nowhere else in the world). Hotspots are mostly near the equator, especially tropical rain forests and coral reefs. The areas are isolated by water, deserts, or mountain ranges. Many conservation initiatives occur in these secluded locations.

Conservation efforts alone do not protect wildlife. Wildlife management also ensures those efforts are implemented, and the populations and landscape are regulated to protect them for public use. This is thanks to the "father" of wildlife management, a 1930s ecologist, forester, writer, environmentalist—Aldo Leopold. His seminal work was *A Sand County Almanac* (1949) where he introduced the "land ethic." He helped us realize that the land includes everything in, on, and around it from all the nonliving to all the living organisms, and we need to think about and treat all that the land encompasses with a moral responsibility. The biggest issue with letting nature take its course and the biggest issue with trying to manage the land or water is that many people may not understand how nature works. This is where wildlife management and a land ethic come into play.

Conservation and management are needed to maintain a biodiverse system to support social–economic–ecological connections. The premise is that if a healthy, balanced ecosystem exists, then ecotourism increases in that location. Increased visitors for recreation like biking or photography, and sport like hunting or fishing will increase the local economy. With more people involved in outdoor activities, the greater the level of environmental appreciation and protection to maintain healthy habitats to support the ecosystem services they offer of clean air, water, and natural resources for other human use.

Have you ever heard of public land? American's Public Trust Doctrine states that certain natural resources such as water, public land, fish, and wildlife are held in trust by the government for the benefit of the people. Through the best practices mentioned previously, government agencies work hard to manage our natural resources for us to use and enjoy. In many places, tax dollars do not currently go toward natural resource management, but there are still ways we can help. Purchasing hunting and fishing licenses, whether you partake in them or not will help the environment - as will some of their associated equipment and special conservation license plates. Some of the proceeds from your purchases go toward helping protect and manage our natural resources. ***Activity:*** Research ways you can help support the conservation of your local natural resources. What are the **game species** that can be hunted or fished near you? What kinds of laws, regulations, or concerns exist in your area about certain game species? Their management is critical for the balance of the ecosystem and for the benefits of the people as a resource.

Nature Journal: Study a map to discover where the nearest public lands are located in relation to your home. Plan to visit these areas soon. Make a comparison between them in terms of habitat types and species observed. See Chapters 1 and 2 for organizing your nature study activities to maximize your experiences. Is the sustainability of our natural resources important to you? Research how the land and wildlife are managed in your state or province. What do you want nature to look like in the future?

Why Should Nature Be Managed?

Habitats are impacted by agricultural land development, urban sprawl, subdivision and sale of land, invasive species, climate change, and other pressures. These issues are increasingly degrading and destroying wildlife habitat. This causes the decline of some wildlife populations and challenges the sustainability of our natural resources. What can we do?

By managing *your* land for wildlife, you can address these issues and make a difference around your apartment complex, your yard, the 60 acres you own, or the landscaping around your workplace. Increasing native green spaces will serve as a continuum of habitat. Landowners are critical stewards of our land. Taking a preservationist's outlook, or letting nature takes its course, may not be the best method of action. In a relatively short amount of time, humans have changed the natural balance of ecosystems that had taken flora and fauna millennia to adapt and evolve to live in. Without our intervention, the changes that we keep incurring will result in the native species demise and ecosystem disruption.

Contact your local conservation district or state/provincial natural resource agency to learn how you can manage the land around you. By thoughtfully managing your land for wildlife, you are not only providing habitat and promoting biodiversity, you are also contributing to clean air and water, and the health of our environment and economy. What do you want nature to look like seven generations from now? Live life now in the way that would create the world you want in the future.

Review Questions

1. Discuss the types of biodiversity.
2. Distinguish between extinct, extirpated, endangered, threatened.
3. How do we benefit from biodiversity? Give examples of each benefit.
4. Describe how humans threaten biodiversity. What does HIPPO stand for? Other than humans in general, what specifically is considered the greatest threat to biodiversity?
5. Define and discuss invasive species, how they possibly impact an ecosystem, and the methods to control them.
6. How do we try to protect biodiversity? Explain in terms of conservation and management.

Invasive Species—Natural History Journal Pages

Select an **exotic, nonnative invasive** species currently found near your Nature Area. Your species should be nonnative with respect to it not originally found in pre-European settlement in North America, *and* invasive in terms of it having characteristics to outcompete other species in the area. Mostly, these types of organisms are referred to as just invasives. Your selected species can be terrestrial or aquatic, and a plant or animal (remember, that includes insects!). Consider species problematic to ecosystems and are recognized by natural resource management as a threat to the area.

Invasive species could have negative ecological, economic, social, and public health impacts. Nonnative species are of special concern to conservation biologists, invasion biologists, and ecologists in general. They have been directly, or indirectly involved in up to half of the known extinctions and are threatening thousands of other species with extinction. These invasive species are also costing us billions of dollars per year in terms of the cost of damage to natural and managed ecosystems, the cost to control these invaders, and the loss of commercial fisheries and agriculture.

Use reputable websites and resources for your research about a local invasive species. Consider starting with your state or province's natural resources agency's website. Track where you find the information you use, and add to the references section (website name, page title, and website address).

1. Select a current nonnative, invasive species that could potentially live within/ near your Nature Area.

   Common Name: ______________________________

   Kingdom: ______________________________

   Phylum: ______________________________

   Class: ______________________________

   Order: ______________________________

   Family: ______________________________

   Genus: ______________________________

   species: ______________________________

2. Create a sketch of your invasive with key features labeled to help with identification (i.e. size, colors, stripes, or other markings, etc.).

3. List important/interesting characteristics and behaviors related to its life cycle or existence.

4. a. Is the invasive species considered terrestrial or aquatic?

   b. List details/features of this invasive organism's preferred habitat.

5. The distribution range of the invasive
   a. Draw your state/province. Shade the map where the invasive is found.

b. Shade the map of North America to show where the invasive species inhabits.

c. Where in the world did the species come from? Shade the invasive's region/country of origin.

d. How did the invasive get to your region?

e. How does this species continue to spread?

6. What specific problems might this nonnative, invasive species have on native species and to its community?

7. How might this exotic species affect humans, either directly or indirectly?

8. What can people do to help prevent the spread of this organism?

9. List all resources used (**books:** author's last name, first initial, year, and book title; **websites:** name of site, date retrieved, and web address):

PART III:

Natural Histories and Field Guides

"Reading about nature is fine, but if a person walks in the woods and listens carefully, he can learn more than what is in books . . ."

—George Washington Carver

In Part III, you will focus on the **natural history** of organisms. You will learn about the general characteristics of fungus, plants, and invertebrate and vertebrate animals. This practical knowledge can help you understand and relate to what you encounter in nature. To further develop your understanding of species around you and to identify them, a field guide for each group of organisms found in the region is provided at the end of the chapters. You will also find Natural History Journal Pages there to guide your research on local species of interest. Refer to Chapters 5–8 along with other sections of the book to enhance your nature study.

5 Fungus

*Fungus are everywhere! Approximately 25% of earth's biomass is fungus. As a novice or expert naturalist, you may have stopped to marvel at fungus in nature. Amazingly, only about 200,000 species of fungi have been described of the estimated 1–1.5 million species. Most fungi will exist underground without making an appearance (or waiting a very long time to show itself), or they may exist on a small scale. In a tropical rainforest, studies have revealed that even very small samples of fresh and decayed leaves from the forest floor can harbor over 145 different species of **microfungi**! Fungus may be interesting to look at, but we also learned in Chapter 4 about its function of decomposition in the ecosystem. What other characteristics or benefits does it offer us?*

5.1 Importance

The study of these incredible organisms called fungi is called **mycology**. They have ecological and economic importance without which we cannot imagine life. In terms of ecological services, fungus breaks down dead organic material to continue the cycling of nutrients through ecosystems as described in Chapter 4. Most plants could not grow without symbiotic fungi that inhabit plant roots and supply essential nutrients.

Consider human health. The first mass-produced antibiotic was derived from fungus, *Penicillium*. The future of lifesaving medicine could be found with further fungal discoveries and uses. We also use fungus in a variety of our foods. For example, yeast for the rising of bread, in soy sauce or cola, and we use it in the fermentation process of making alcohol. On a level of concern, fungus can also cause problems or diseases in humans, such as ringworm, athlete's foot, yeast infections, and so on. Ecologically, plant diseases arise in the form of rusts, smuts, leaf/root/stem rots, and crop damage from fungus species. However, scientists learn a great deal from these r-selected species. Yeast and other fungi are often termed "model organisms" because of their fast growth and evolution, thereby helping making advances in genetic and molecular biology.

5.2 Defining Fungus

First, let's define fungus by what they are not. Fungi are NOT PLANTS!!!! They do not have a cell wall made of cellulose. Fungi do NOT photosynthesize. They do not have green chlorophyll and do not use sunlight to produce food. How are they defined? They are sessile organisms with spores and a cell wall of chitin. Fungus are defined by how they obtain food. All fungi are **absorptive heterotrophs**. They secrete enzymes into their food source, digest it externally, and then absorb the nutrients.

There are two types of absorptive heterotrophs—saprophytic and symbiotic. **Saprobes** grow on dead things such as decaying tree trunks on the ground from which they acquire their nutrients. Otherwise, fungus form symbiotic relationships with other living things, with whom fungus lives in a long-lasting or permanent association. Types of symbiosis a species might exhibit are parasitic (athlete's foot or ear fungus), commensalistic (living harmlessly in the guts of animals), or mutualistic relationships (fungus growing in soil that forms a **mycorrhizal** or plant root symbiosis). Symbiosis is discussed in more detail in Chapter 4.

5.3 Fungus-like Organisms

As discussed in Chapter 1, in terms of its taxonomy or classification, fungus belongs to the Domain Eukarya. Prior to the appearance of the species as Kingdom Fungi, fungus-like organisms classified as Kingdom Protista existed. You can still find these protists today growing just about anywhere, but you will find them in mostly mesic forests especially in the spring after the rains. Look on logs, wood chips, leaf litter, or at the base of trees. They are absorptive heterotrophs like fungus, but these fungus-like organisms can move slowly as it spreads across the surface of where it absorbs its nutrients via its plasmodium (a lumpy mass of **protoplasm**). The most common species in the Great Lakes and Upper Midwest region is a slime mold called dog vomit or *Fuligo septica*. The colors range and will vary slightly dependent on temperature, pH, and its substrate. Can you guess where this species gets its name? Yes, it looks like dog vomit. Sometimes the slime mold takes a different shape and fans out in a network of veins. Slime molds do not harm plants in the form of disease but can block leaves from photosynthesizing if it starts to grow on the green parts of plants. If you feel the need to rid a slime mold from your property, you can break it apart and let it dry out or rake up its pieces on mulch or lawn clippings. Alternatively, slime molds can reveal a fascinating nature study. Observe them up close with a magnifier and monitor them over time.

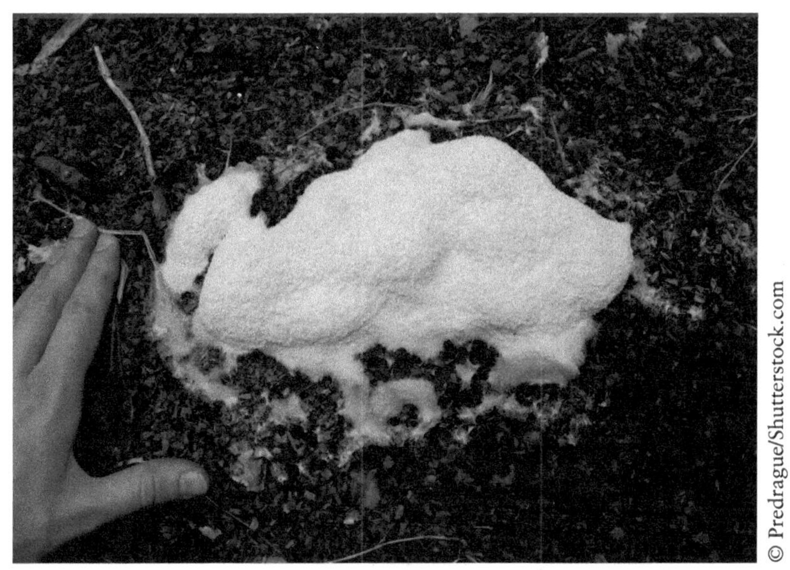

The fungus-like slime mold from Kingdom Protista, dog vomit or *Fuligo septica*. It comes in various sizes and shapes, and the colors can be yellow, white, orange, or red.

5.4 Kingdom Fungi

All fungi do not have the "mushroom shape." Mushrooms are technically the reproductive structures of just one particular kind of fungus. Referring to the visible part of a fungus as a "mushroom" has become acceptable among many naturalists. There are many, many kinds of fungus (and beyond the scope of this book!). Rather than using the taxonomic level of "phylum," fungi are classified into "divisions." Divisions are distinguished by the spore-producing structures they form. The four main fungus divisions are Zygomycota, Basidiomycota, Ascomycota, and Deuteromycota.

Fungus Division	Common Name	Examples
Zygomycota	Zygote Fungi	bread mold
Basidiomycota*	Club Fungi	puffballs, earthstars, gilled mushrooms, shelf-fungus
Ascomycota*	Sac Fungi	truffles, morels, ergot, stinkhorns, rusts
Deuteromycota	Imperfect Fungi	yeast, athlete's foot, and in soy sauce and cola

*Most commonly found in nature.

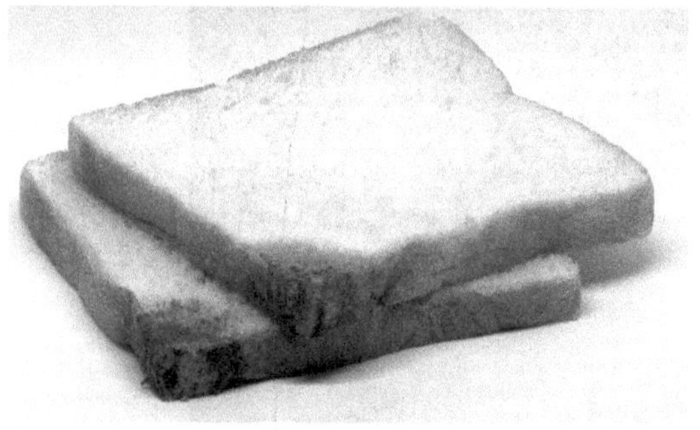

© Kumpanat Phewphong/Shutterstock.com

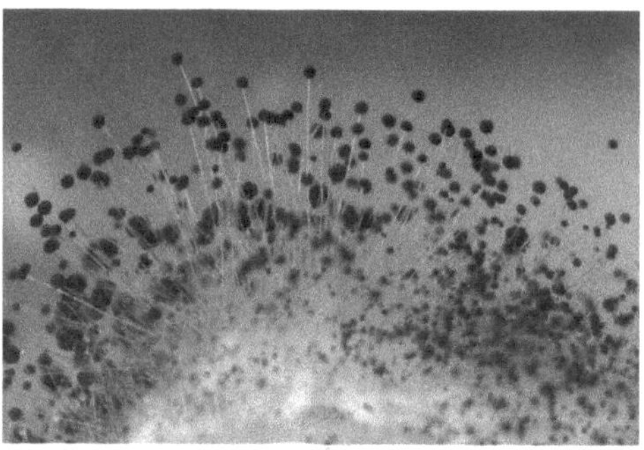

© Rattiya Thongdumhyu/ Shutterstock.com

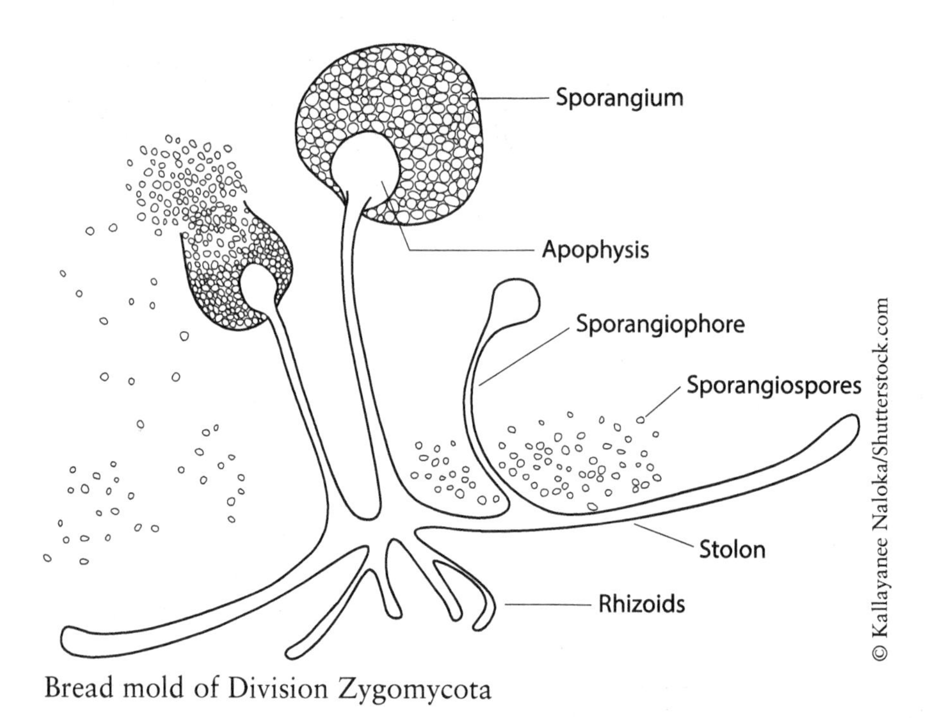

© Kallayanee Naloka/Shutterstock.com

Bread mold of Division Zygomycota

5.5 Spore Dispersal

Fungi are **sessile** or immobile. They have two ways to extend their range. Depending on the species, they can either grow into adjoining areas or disperse its spores.

Puffball species releasing its spores.

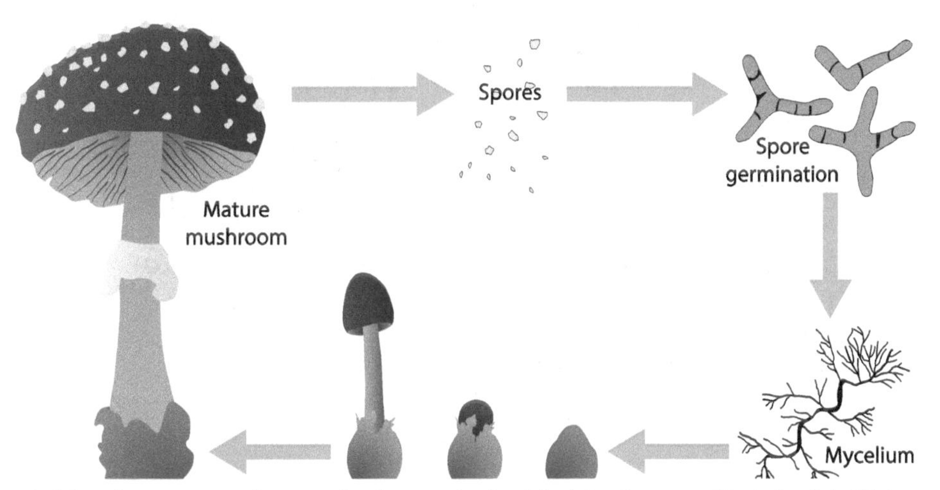

The visible, above-ground part of a fungus is the "fruiting" organ (**fruit body**). This part is often referred to as the mushroom, regardless of its division classification. The fruit body is where spores are produced for reproduction means. The spores disperse and will germinate in conditions favorable for growth. The **mycelium** are the "roots" of the fungus and can spread for miles underground.

Passive mechanisms for spore dispersal could involve wind, water, or animals. Spores release from under or within the cap after having a disturbance act against it. Active spore-release mechanisms for dispersal exist in other species, like with the means of a bursting cell, rounding-off, or basidiospore discharge. These seemingly spontaneous spore releases need favorable conditions such as temperature, moisture, oxygen, and nitrogen necessary for fungus to germinate and continue to grow.

Wind Dispersal

Once the fungus is ready to release its spores, they will be carried by the wind for dispersal.

Giant Puffball, *Calvatia gigantea*, will dry out and crack to allow wind to carry away its spores. This is not energy-efficient. It produces trillions of spores because the chance of a spore landing in a suitable habitat is extremely small. Most spores will land close to the parent fungus.

Water Dispersal

Spores will be released by water and travel in or on the water flow.

Earthstars, *Astraeus hygrometricus*, rely on raindrops to depress the sack to release spores, then the water helps carry them away.

Animal Dispersal

The animal dispersal mechanism of spreading spores greatly improves the chance spores will be deposited in a site favorable for germination and growth. It allows fungi that rely on animals for germination to produce fewer spores because each has a greater chance of success.

Truffles, *Tuber* sp., are produced below ground and must be unearthed to be dispersed. At maturity, they produce an aroma that attracts animals. Animals will dig them up for food. The spores are not digested and will eventually pass through the animal at some distance from where the truffle was dug up.

Stinkhorns, *Phallus* sp., smells like rotten meat and will attract flies that then get coated with its spores.

Bursting-cell Dispersal

In the method of bursting-cell dispersal, spore release is explosive because of a weak point at the tip of the ascus that ruptures suddenly. Spores have a better chance of being dispersed farther if the air is turbulent.

Pilobolus sp. will have a sticky mass of spores discharged at 35 ft./s (10.5 m); up to 6 ft. (~2 m) high; and lands as far away as 8 ft. (~2.5 m).

Rounding-off Dispersal

Rounding-off is a rapid motion dispersal mechanism that results when a rapid motion will result when the force keeping a deformed surface under stress is suddenly released in a catapulting reaction.

Sphaerobolus sp. (cannonball fungus)

Basidiospore Discharge

Spores cover the surface of gills or pores on the underside of a mushroom's cap. Discharge range is related to gill spacing or tube (asidiospore house) diameter.

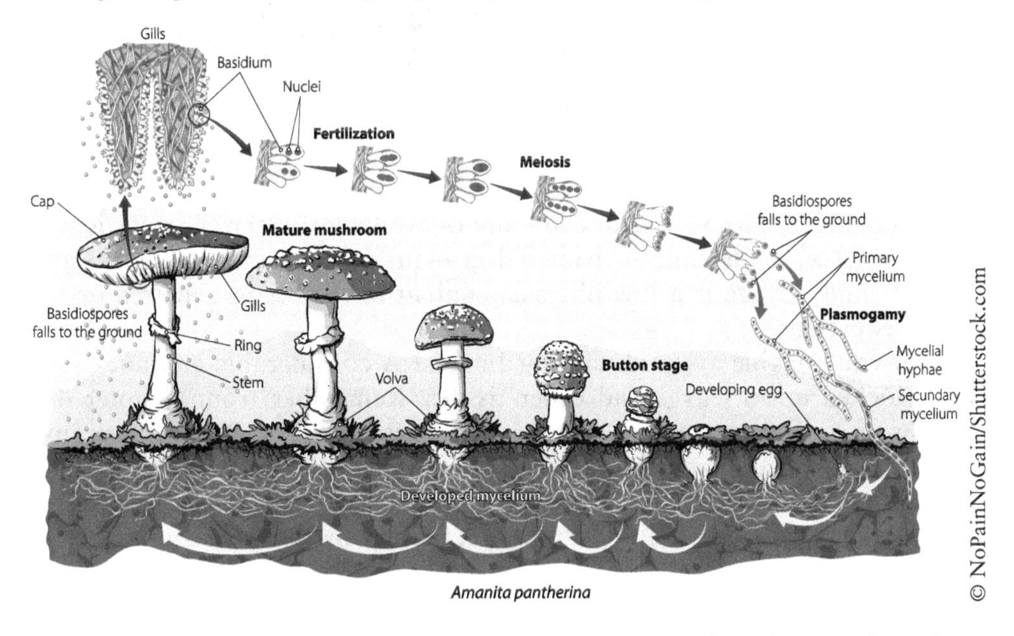

Amanita pantherina

The "mushrooms' spore discharge" is sometimes still debated. Regardless, fewer spores are produced by this fungal group if this is the main method of dispersal.

5.6 Ecological Role of Fungus

As discussed in Chapter 4, fungi are major decomposers in most terrestrial ecosystems (along with Domain Bacteria). Decomposition is the reduction of the body of a formerly living organism into simpler forms of organic and especially inorganic matter. Fungus have a critical role in biogeochemical cycles and food webs that enable essential nutrients to be recycled in the ecosystem.

Fungus growing on a log representing the decomposition process.

The rate of decomposition depends on temperature, moisture, chemical composition of the organic matter, and amount of available oxygen. The rate is slowed if there is slow or no fungi or bacterial growth, due to temperature too low or too high, and if there is a low nitrogen content in the organic matter, or if low oxygen is present.

Can you imagine a world without fungus? A considerable amount of nitrogen is locked up in leaves and other tissues. Remember from Chapter 4 that nitrogen is a vital element of life yet is a limiting factor. Without fungus for decomposition, there would not exist enough nitrogen for plants to make new leaves, stems, and wood. The ground would be buried by dead leaves and wood lying forever where they fell. Furthermore, the mycorrhiza, the mutualistic relationship between plant roots closely associated with fungal structures (mainly hyphae), would not exist. Plants give the fungus nutrients for the plants made from photosynthesis (sugars and carbohydrates), while the fungus gets protection from predation. Fungus provides the plant moisture and mineral nutrients (phosphorus and nitrogen) while also protecting the plant against certain diseases and pathogens. Over 90% of plant species are dependent on mycorrhizae for survival. In these underground partnerships, such as aspen tree roots and fungal hyphae, the plant roots have increased surface area. How is this a good thing for the plants?

Some fungi compete with other organisms or directly infect them. Beneficial fungus restrict, and sometimes eliminate, populations of noxious organisms like pest insects, mites, weeds, nematodes, and other fungi that kill plants. Fungus also forms another mutualistic symbiosis in the associations present in lichen.

Lichen

Lichens are not a single organism, but are a partnership between a fungus and algae or a **cyanobacteria** (photosynthetic bacteria formerly classified as blue-green algae). The lichen body consists of fungal filaments (**hyphae**) surrounding cells of green algae/cyanobacteria. Algae/cyanobacteria have chlorophyll (green pigment to photosynthesize) to help provide the fungus some of the algae's sugars or carbohydrates it produces as a food source and may also provide the fungus with other nutrients. The fungus component of lichen protects its partner(s) from drying out and shades them from strong sunlight by enclosing the photosynthetic partners within the body of the lichen. Lichen is classified as a member of Kingdom Fungi because the fungus has the greatest influence on the final form of the lichen's body shape and its flexibility.

Lichen has a remarkable resistance to drought and will persist for a long time. A dry lichen can absorb 3–35 times its weight in water. It has the ability to absorb moisture from dew, fog, or humid air. The process of drying out slowly makes it possible for photosynthetic partner(s) to make food for as long as possible. It also makes it feasible for lichens to live in harsh environments (deserts or polar regions), and on exposed surfaces (bare rock, roofs, tree branches).

Foliose: flat leaf-like lichens

Crustose: crust-like lichens

Fruticose: mini shrub-like lichens

Squamulose

Lichen shapes or characteristics that demonstrate the four basic varieties found in an ecosystem.

Lichen spread mostly by small pieces of their body getting blown around. All reproductive structures are present in the fragment, so growth can begin immediately (spore dispersal is rare). Most lichen grow slowly, probably because their environments have limited water supply. They tend to live for many years and can be hundreds of years old (used to date rock surfaces on which they grow). Lichens can be used as bioindicators for air pollution with the fruticose and foliose varieties indicating relatively good air quality. Lichen also has antiviral and antibacterial properties that have been used or replicated in medications. They have been used in scented soaps and perfumes, and as a type of dye. Furthermore, they are soil builders in an ecosystem and an important food source for animals especially in extreme environments where nutrients may be limited.

Nature Journal: Go for a walk nearby to look for lichen. Try to classify them into one of the four basic shapes described in this section. Is it found in the field guide section of this chapter? Note where you found it—the habitat type and its substrate. Make a drawing with its color and measurements. Are there more of its type around? Or different varieties? Does there seem to be a trend for where the lichen are found? Consider a field investigation revolving around lichen (see Chapter 2).

5.7 Mushroom Hunting

Naturalists love to discover and observe fungus they encounter while on their Nature Strolls. Mushrooms often surprise us when we find them because they may not have been there the last time we visited a location. Although, due to the extensive **mycelium** of the fungus, the organism had likely lived subterranean for a while (decades or even centuries!) until conditions were favorable for its fruiting body ("mushroom") to emerge.

Nature Journal: If you conduct regular field inventories of your Nature Area, you may find mushroom surprises from time to time. Consider heading out after a warm, steady rain. Due to the variety of species and appearances, you will observe them either camouflaged among where they grow or stand out among the leaf litter or logs. It can sometimes feel like a scavenger hunt when the conditions are favorable for them to emerge. Note the conditions leading to their emergence and sketch and map where you found them. Monitor the mushroom(s) daily. Does it dry out? Do you notice evidence of something eating it? How does it look as it dies? Does it seem to refresh with additional rains or humidity? Use this book and other resources to make an accurate identification and learn more about its natural history.

Other than mushroom hunting for pure naturalist purposes, some people prefer to forage the mushrooms to eat. For example, the morel (*Morchella* spp.) is a coveted species by many. Morels can motivate people to get outside and focus their attention. Collectors are so passionate that festivals are held, and favorite spots are closely guarded secrets. Species can be found from mid-April through mid-June depending on latitude and local weather conditions. Be sure you are collecting true morels and not the toxic, false morels (*Gyromitra* spp.). Other desired wild mushrooms foraged include oyster mushrooms (*Pleurotus* spp.), inky caps (*Coprinoid* group), giant puffball (*Calvatia gigantea*), boletes (*Boletus* spp.), and hen of the woods (*Grifola frondosa*) to name a few. This book is not to be used as a guide for foraging edible mushrooms. The author advises against collecting or foraging unless you are in the presence of an expert. Do not assume if a mushroom was eaten by an animal that it is safe for humans to eat—this is not true!

Many mushroom species are poisonous and are responsible for numerous cases of sickness and death every year! However, learning to forage mushrooms can be rewarding. You need to be absolutely positive in the fungus identification before ever experimenting to eat one! Consult many resources and attend a field workshop on mushroom hunting and identification. Some methods that can help identify poisonous mushrooms do not always hold true but can serve as a guide. Here are some features to help you recognize mushrooms to avoid: warts or scales on the cap or top of mushroom; gills that are white or light-colored, not brown, on the underside of the mushroom; gills that look like thin, leaf-like plates underneath the mushroom; an upper ring around the upper part of the stem; a lower ring around the lower part of the stem; or the base of the stem looks like a bulb (or cup).

What are the symptoms of mushroom poisoning? Everyone may experience symptoms differently. Early symptoms may include feeling sick, stomach cramps, vomiting, and watery or bloody diarrhea. If you think a mushroom caused a reaction when ingested, collect samples of the mushroom that was eaten. Carefully dig up a few mushrooms, complete with underground parts, to help with the identification. Call a physician, local poison control center, or hospital emergency room. Bring mushroom samples of what was eaten to the emergency room.

To enjoy mushrooms safely, get excited to simply find and observe them in nature. Learn to identify the ones that grow near you by referring to the field guide section of this chapter. If you find that you want to become more of an expert in mycology, start with the books listed in the reference section for this chapter. Happy mushrooming!

Review Questions

1. What is the term for the study of fungus?
2. Discuss how fungi are beneficial to the environment and to humans. How can fungus be harmful?
3. Explain why fungus is not classified as a plant. How is an absorptive heterotroph defined?
4. Describe two methods of how fungi obtain their food.
5. Briefly discuss the fungal relationships: parasitism, commensalism, and mutualism.
6. What kingdom are fungi categorized under? Do all fungi have the "mushroom shape"? What is the next taxonomic level in which fungi belong (a more specific category than kingdom). How are these fungi distinguished? List the four basic types and give examples of each.

7. Draw and explain the basic mushroom structure. What is the largest part of a fungus? When does the "mushroom" appear?
8. What are the two ways fungi can extend their range? What are the means for passive mechanisms and active-release mechanisms for spore dispersal? What has to be right for germination and growth? Discuss and give examples for each method of dispersal. Which is the most energy-efficient method for the fungus? The least? Which method produces fewer spores?
9. Fungi play many important ecological roles in nature. Describe its role as a decomposer. How is the rate of decomposition affected? Explain how the world would be different if there were no decomposers. Why are mycorrhizae important to plants? Discuss how fungi can impact other organisms and their populations. Describe the ecological role of lichen. What are the main characteristics of lichen?
10. What are possible characteristics to look at to determine whether a fungus is poisonous? What was the advice given in the chapter for mushroom hunting?

5.8 Field Guide: Fungus

Fungus diversity and natural history is intriguing. You can find all colors and a variety of shapes in places that you never observed them before! In this section, you will see some common **macrofungi** that have large fruiting bodies (the "mushroom" stage) and can be found at least in the Great Lakes basin and Upper Midwest.

Nature Journal: A natural history template appears at the end of this section to help focus your nature studies on fungus you encounter or would like to study. Start paying closer attention to the fungus you encounter and keep track of what you find. Create a natural history entry for each one.

It is beyond the scope of this field guide to provide you with the information necessary to make a confident identification for edibility purposes. Please use multiple resources to accomplish this and refer to an expert's opinion for training. DO NOT EAT any mushroom unless you are absolutely certain of its identity. If you don't know, don't eat it.

How to Identify Fungus

Read through the following section to become familiar with the aspects you should consider when trying to identify a fungus you encounter. When recording your fungal finds in your Nature Journal, these would be key things to observe.

Anatomy

Know the basic anatomy of a fungus. Learn the terms associated with each feature. You will see and use these terms in communication about all fungal species. These are also the structures that will help you discern between species and make an accurate identification. Fungus are also grouped by where they are typically found and how they produce spores.

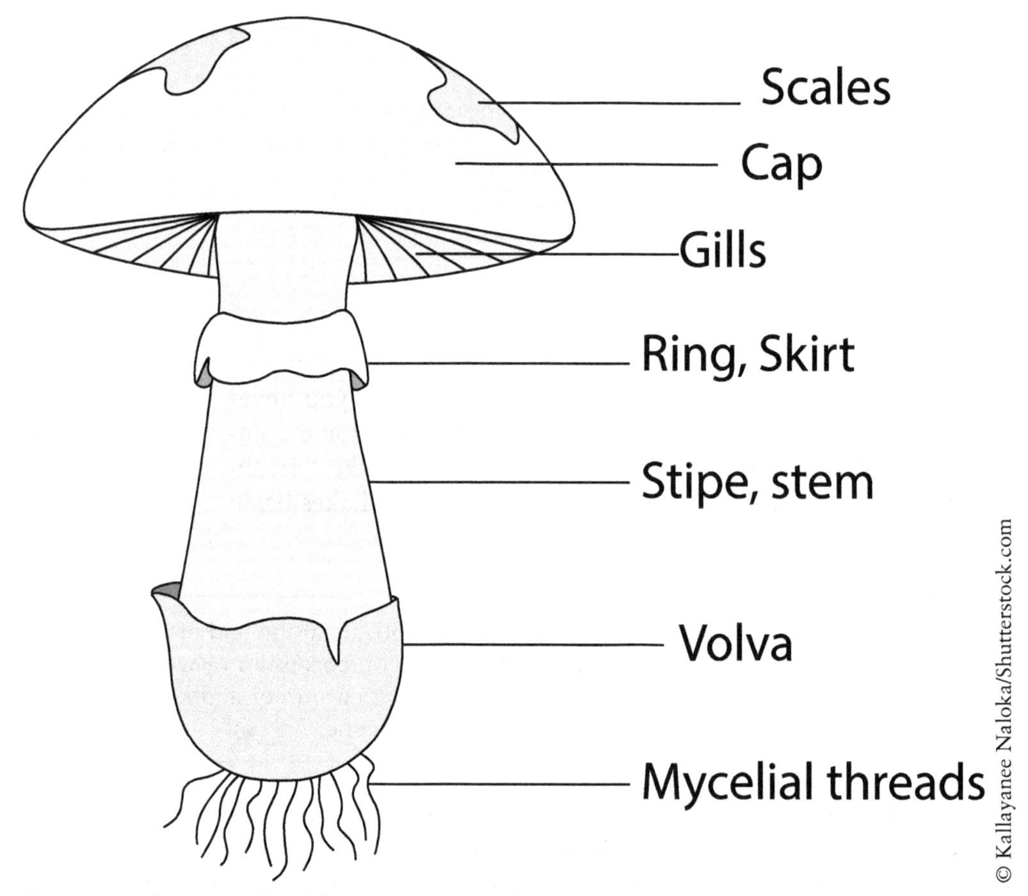

Structure of a generalized fungus species.

Location

Describe the habitat where you found the fungus. Was it growing in the lawn, through the leaf litter, or wood chips? Was it growing on a log or on the side of tree? Was the tree living or dead? Saprobes will usually be found growing on decaying wood. Mycorrhizal fungus will grow out of the ground or may grow from the base of a tree. Parasitic fungus grow on living trees and can cause damage.

Overall Shape or Cap Type

The spores are housed in the fruiting body or "mushroom" part of the organism. Learning to examine this part critically will help you make an identification. What is the shape of the overall mushroom? Is it spherical with no stem? Or shelf-like with a small or no stem visible? If it has a cap with a stem, what is the cap shape? Are there scales or warts (or bumps) on the cap? What do the edges of the cap look like? How big is the mushroom?

Below the Cap

Look below the cap—does it have gills or seem spongy and porous? Some mushrooms seem to have gills, but upon closer inspection you will find they are actually *folds* that are extensions of the stipe and are more plate-like. Mushrooms with folds also do not have a true cap, but the entire mushroom is a continuous structure like with the chanterelles (*Cantharellus* spp.). If the mushroom cap had gills, how are they arranged? You may need to cut it lengthwise to observe the gill attachment to the stem. They may or may not be directly attached to the stem, or the gills could have notches, or run down the length of the stem.

Stem

Examine the stipe or stem. Look for the presence of a ring, or a skirt. Examine the base of the fungus—is there a volva? Does it have a bulbous base? Any patterns or scales on the stipe?

Color

Often a species will grow in a variety of colors, so always read the possible descriptions of a species. The location and age will alter its color. Those growing in sunshine will likely show as a paler hue than those growing in the shade. Sometimes it is necessary to streak the tissue to observe if it stains or if it changes color if bruised when you pinch it.

Spore Color

Spores are invisible to the naked eye. However, their color can help with identification. Place the cap with the rounded side up on a piece of white paper and cover it with a bowl for a few hours. You should see a dusting of spores under the cap. If not, the spores are likely white and will show up if you touch the paper under the cap.

Fungus Species Accounts

Based on your answers to the above questions on identifying your fungus specimen, you can look for the organism in an orderly way. Use the quick classification system described below of the overall mushroom appearances. Your specimen will be classified as one of the following shapes or groups: cap, atypical, shelf, spherical, cup, coral, miscellaneous, lichen, or protist (slime molds). ***Read the description to find a match to your species in question. Then go to the corresponding numbered category to review the examples.*** If you do not find an exact match, often the picture will lead you to an idea of what genus classification your specimen may be under. Use additional resources for further assistance in identification.

1. ***Caps*** with the traditional stem structure and obvious cap with underside as either gills or folds.
2. ***Caps*** with the traditional stem structure and obvious cap with underside of fine pores or spongy.
3. ***Atypical mushrooms*** without a regular cap with gills or pores but do have a stem.
4. ***Shelf fungus*** typically growing on trees or logs (with gills underside). They resemble an actual shelf and lack an obvious stem.
5. ***Shelf fungus*** typically growing on trees or logs (with fine pores or spongy underside). They resemble an actual shelf and lack an obvious stem.
6. ***Spherical*** or ball-like mushrooms without a distinctive stem.
7. ***Cup-shaped*** mushrooms resembling a cup or may be flattened.
8. ***Coral*** fungus tending to grow low to the ground in a grouping that resembles coral with its irregular branches or stick-like structures pointing upward.
9. ***Miscellaneous*** mushrooms are those species that do not fit the other categories.
10. ***Lichen*** are typically flattened with various degrees of scales or tufts. Only most common varieties are listed.
11. ***Slime molds*** are fungus-like organisms from Kingdom Protista that are often mistaken for actual fungus. They have a round, wet blob–like appearance.

1. Cap (with stem and underside gills)

Amanita spp. are extremely poisonous, and deaths are reported from eating them due to confusing their identity with edible mushrooms. Although the variety of species can look very different from one another, there are some common characteristics. The young button stage is surrounded by a veil and resembles a miniature puffball until the parts grow and expand leaving a cup at the base and the skirt on the stem. All are mycorrhizal and have a symbiotic relationship between their mycelium and with living tree roots.

© Ari N/Shutterstock.com

Destroying Angel (*Amanita virosa*)

Description: smooth, white cap; gills free from stem; stem white and collared near top; white bulbous base with cup
Size: 1.5–5 in. wide; stem 2–10 in. tall
Habitat: deciduous woods (oaks preferred); often under a single living tree
Season: summer through fall
Spore Print: white
Note: Most deadly mushroom known—one cap can kill an adult.

© Jon Paul Photo/Shutterstock.com

Fly Agaric (*Amanita muscaria*)

Description: cap yellowish to orange with white warts (unless washed away); gills free from stem, white to yellow; white stem with white or cream collar toward top with whitish bulbous base with scales
Size: 2–7 in. (up to 18 cm) wide; stem 3–7 in. (up to 18 cm) tall
Habitat: deciduous and coniferous forests; usually under a single tree
Season: summer through fall
Spore Print: white
Note: Toxic, generally not lethal; can cause delirium and coma-like conditions for hours.

© Randy Bjorklund/ Shutterstock.com

Bleeding Mycena (*Mycena haematopus*)

Description: small, dainty, conical caps (often fringed edge) with red-brown center and red-gray at edge; gills white or grayish-red; stem brown (shows red when cut); grows in large numbers
Size: ½–1.5 in. (up to 4 cm) wide; stem 1–3 in. (up to 8 cm) tall
Habitat: decaying logs and organic matter
Season: spring through fall; species-dependent
Spore Print: white
Note: Belongs to the ***Mycena* spp.** all seemingly delicate with smooth or pleated caps, but species colors and odors may vary.

© Johannes Dag Mayer/ Shutterstock.com

Meadow Mushroom (*Agaricus campestris*)

Description: cap white or buff, sometimes brownish scaled or fringed; gills free and pinkish (darken with age); stem white with thin collar near top and will disappear
Size: ¾–4 in. (up to 10 cm) wide; stem 1–3 (up to 8 cm) in. tall
Habitat: grows singly in grassy areas
Season: late summer through early fall
Spore Print: black or dark purple to brown
Note: ***Agaricus* spp.** are saprobe meadow mushroom–types generally sold in stores; however, they resemble deadly amanitas and should not be eaten until verified by an expert.

© Henri Koskinen/ Shutterstock.com

Pungent Russula or the Sickener (*Russula emetica*)

Description: slimy, bright red cap (fades with age) depressed in center with ridged edges; white gills attached to stem; white stem
Size: 2–4.5 in. (up to 12 cm) wide; stem 2–4.5 in. (up to 12 cm) tall
Habitat: near conifers; in bogs
Season: mid-summer through mid-fall
Spore Print: white
Note: Toxic, causes intestinal distress; ***Russula* sp.** have toxic and edible varieties that vary in color; mycorrhizal.

© Henrik Larsson/ Shutterstock.com

Shaggy Mane (*Coprinus comatus*)

Description: columnar cap, white with brown shaggy scales; cap edges flare and turn inky with age; gills white; white stem with collar and bulbous base
Size: ¾–2.5 in. (up to 7 cm) wide; stem 3–8 in. (up to 20 cm) tall
Habitat: large clusters found scattered along roadsides, lawns, and pastures
Season: early summer through fall
Spore Print: black
Note: Of the **Inky Caps (*Copinroid* spp.)**; caps deteriorate quickly after picking.

© Ivan Marjanovic/ Shutterstock.com

Jack O'Lantern (*Omphalotus illudens*)

Description: orange-yellow cap, gills, and stem; gills run down stem
Size: 2–5 in. (up to 13 cm) wide; stem 2–8 in. (up to 20 cm) tall
Habitat: on stumps in large clusters or on buried roots (oaks preferred)
Season: summer through late fall
Spore Print: white or cream
Note: Luminescent in dark (greenish glow to gills); contain potent toxins causing intestinal, respiratory, and circulatory issues; resembles some edible mushrooms.

© Jesus Cobaleda/ Shutterstock.com

Wood Blewit (*Lepista nuda* or *Clitocybe nuda*)

Description: violet cap (fading to purplish-gray to brown); pale violet gills partially attached (fading to light brown with age); pale violet stem with fine white hairs; base brownish
Size: 1.5–6 in. (up to 15 cm) wide; stem 1.25–4 in. (up to 10 cm) tall
Habitat: deciduous or coniferous forests; compost piles; wood chips; hedgerows
Season: late summer to late fall
Spore Print: light pink
Note: Blewits are saprobes and have a purplish color fading to tan with age; resembles toxic species.

© FotograFFF/Shutterstock.com

Yellow Chanterelle (*Cantharellus cibarius*)

Description: of the **Chanterelles (*Cantharellus* spp.)**; frilly-edged cap and trumpet shape; cap yellow to orange (other species exist as black, orange, red, purple, cream); orange buff gills or folds run continually upward from top of stem into cap
Size: ¾–6 in. (up to 15 cm) wide; stem 1–3 in. (up to 8 cm) tall
Habitat: grows in soil in woods (oak and coniferous), roadsides
Season: summer to early fall
Spore Print: white or pale buff
Note: Coveted by foragers, fruity aroma, mild pepper flavor; compare to toxic jack o'lantern (orange, similar shape but with true gills).

2. Cap (with stem and underside pores)

© Henri Koskinen/ Shutterstock.com

Aspen Bolete (*Leccinum aurantiacum*)

Description: cap orange-red; underside porous white; stem whitish with brownish scabers that darken with age
Size: 4–8 in. (up to 20 cm) wide; stem 4–7 in. (up to 18 cm) tall
Habitat: associated with aspen trees and grassy areas
Season: summer through fall
Spore Print: yellow to olive-brown
Note: Scaber Stalks (*Leccinum* spp.) all have raised scales or scabers on stem, caps smooth, and underside is porous or spongy; species come in a variety of colored caps; mycorrhizal; targeted by maggots; can cause intestinal distress.

© Ari N/Shutterstock.com

King Bolete (*Boletus edulis*)

Description: cap pale brown; underside white with small pores or spongey; white stem near top transitioning to tan below with netlike vertical ridges; bulbous base
Size: 3–10 in. (up to 25 cm) wide; stem 4–7 in. (up to 18 cm) tall (larger west of region)
Habitat: mycorrhizal in coniferous forests and sometimes near oaks and other deciduous trees
Season: summer through very early fall
Spore Print: brown
Note: Favorite of insects and people.

© muuraa/Shutterstock.com

Old Man of the Woods (*Strobilomyces floccopus*)

Description: gray to black cap shaggy or scaly, young may have fringed edges; underside white; porous; stem shaggy
Size: 1.5–6 (up to 15 cm) in. wide; stem 2–6 in. (up to 15 cm) tall
Habitat: under deciduous trees (mainly oaks)
Season: summer through fall
Spore Print: black
Note: Bruises red at first then black.

3. Atypical Caps (with stem)

© Henri Koskinen/ Shutterstock.com

Common Stinkhorns (*Phallus impudicus*)
Description: dark, pitted, tapered head with whitish circular tip; rough white stem and volva
Size: ½ in. wide; stem 3–8 in. (up to 20 cm) tall
Habitat: wood chips, clustered on soil, forests, fields
Season: summer through fall
Spore Print: difficult to achieve; brownish
Note: *Phallus* and *Mutinus* spp. are saprobes that resembles a puffball when young but swells with water and lengthens rapidly within a few hours. Mature tops have a foul odor with slime-containing spores. Attracts flies and beetles to transport spores.

© Henri Koskinen/ Shutterstock.com

Fluted White Helvella (*Helvella crispa*)
Description: white, thin-fleshed cap folded and saddle-like; ribbed stem and hollow; flesh brittle
Size: ½–2 in. (up to 5 cm) wide; stem 1.5–5 in. (up to 13 cm) tall
Habitat: in ground of deciduous or coniferous forest; sometimes found on decaying wood; disturbed areas
Season: summer through fall
Spore Print: white
Note: Many **Elfin Saddles (*Helvella* spp.)** grow in the region; colors vary.

© LapailrKrapai/ Shutterstock.com

False Morel (*Gyromitra esculenta*)
Description: brown to red-brown cap wrinkled and folded irregularly with no pits; stem whitish (not hollow)
Size: 1.5–4 in. (up to 10 cm) wide; stem 1–3 in. (up to 8 cm) tall
Habitat: coniferous forest scattered under trees
Season: spring
Spore Print: not visible
Note: *Gyromitra* spp. are toxic, sometimes even lethal; look very similar with varying colors and habitat; mistaken for true morels.

© Tomasz Czadowski/ Shutterstock.com

Morel (*Morchella escuenta*)
Description: pitted, honeycomb and spongy cap. Stem completely hollow with a continuous cavity lengthwise.
Size: 1–2 in. (up to 5 cm) wide; 1.5–4 in. (up to 10 cm) tall
Habitat: old orchards; deciduous forests; grassy areas; near recently dead elms; after a fire
Season: spring
Spore print: cream, white to yellow
Note: Not to be confused with toxic look-alike species.

4. Shelf (with gills)

Split Gill (*Schizophyllum commune*)

Description: cap fuzzy, tough, fan-shaped, white to gray (brownish when wet); gills gray and radiating outward from point of attachment; edges grooved; not stem
Size: ½–1 in. (up to 2.54 cm) wide
Habitat: usually clustered on logs, stumps, sticks of deciduous trees
Season: found year-round
Spore Print: pinkish
Note: Some claim it is the most common and widespread mushroom in world; being studied for medicinal purposes.

5. Shelf (with pores)

© Nikolay Kurzenko/ Shutterstock.com

Artist's Conk (*Gandoderma applanatum*)

Description: semi-circular shelf with irregular edges; stemless but thick at attachment then thinning toward edge; topside gray or brown with concentric banding sometimes and very hard; underside whitish with fine pores
Size: 2–25 in. (up to 63 cm) across
Habitat: on living but mostly dead trees (mostly deciduous, some conifers)
Season: year-round, perennial
Spore Print: brown
Note: Causes white rot (parasite and decomposer roles); due to darkening when underside is scratched, often used for scratch art; if cut from top to bottom, pore layers show each year's growth.

© Randy Bjorklund/ Shutterstock.com

Birch Polypore (*Piptoporus betulinus*)

Description: half-dome or "pillowy" growth with dull white top surface (smooth when young then cracks) and thick, rounded edge surrounding pore surface; underside porous, whitish (may become toothy with age); no stem but may taper at attachment point
Size: up to 10 in. (up to 25 cm) across (usually smaller)
Habitat: grows only on birch trees, stumps, or logs (mostly dead wood)
Season: spring through summer (dark in winter and through following year)
Spore Print: white
Note: Cause brown rot; both parasite and decomposer.

© john paul slinger/ Shutterstock.com

Chicken of the Woods (*Laetiporus* spp.)

Description: orange to yellow topside; underside yellow and porous or sponge-like
Size: 20 in. (up to 50 cm) wide
Habitat: deciduous and coniferous forests found on living or dead trees
Season: late spring through fall
Spore Print: white
Note: With proper identification, young outer edges edible (meaty texture, chicken taste); toxic lookalike, jack o'lantern species (gills present).

© Julie Vader/ Shutterstock.com

Turkey Tail (*Trametes versicolor*)

Description: thin, overlapping colorfully striped (cream-colored, tan, dark brown), fan-shaped caps; banded colors typically alternate from smooth to hairy (view under magnifier); wavy edges; underside whitish with very fine pores; no stem but narrow to a neck of attachment
Size: ¾–3 in. (up to 8 cm) wide (single cap)
Habitat: grows on dead wood, mostly of deciduous trees (sometimes conifers)
Season: white to pale yellow
Spore Print: spring through fall
Note: Being studied for anticancer properties.

© Aleksander Bolbot/ Shutterstock.com

Hen of the Woods (*Grifola frondosa*)

Description: large cluster overlapping fan-shaped caps; gray, tan, cream top; white or yellow below with small pores; short, thick, branched stem
Size: 8–25 in. (up to 63 cm) wide
Habitat: near base of stumps or trunks of deciduous trees
Season: early through late fall
Spore Print: white
Note: Grows in the same location yearly.

6. Coral and club fungi

© igor.kramar.shots/Shutterstock.com

Crown-tipped Coral (*Artomyces pyxidatus*)

Description: large white to yellowish-pink cluster with several branches that curve away from base; darkens to tan; each stem with shallow, tiny cup making it appear crown-like
Size: 1.5–4 in. (up to 10 cm) wide; 4–6 in. (up to 15 cm) tall
Habitat: grows on decaying wood of deciduous trees (willow, aspen, maple preferred)
Season: early summer through fall
Spore Print: white
Notes: Can be confused with toxic yellow-tipped coral fungus.

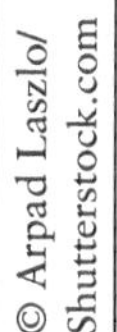

© Arpad Laszlo/Shutterstock.com

Dead Man's Fingers (*Xylaria polymorpha*)

Description: finger-, or club-shaped; outside black at maturity with white tough tissue inside; rough skin; immature stage is pale
Size: ¾–4 in. (up to 10 cm) tall
Habitat: found in deciduous forests; in loose groupings on decaying wood
Season: spring (immature); summer through fall (mature)
Spore Print: black
Note: Saprobes.

© Khun Ta/Shutterstock.com

White Worm Coral (*Clavaria vermicularis*)

Description: white club-like appearance, looks like mung sprouts growing up from ground (yellows with age); generally unbranching or has very few
Size: 2–6 in. (up to 15 cm) tall; up to ¼ in. (up to 1 cm) wide
Habitat: ground in woodlands; grassy or mossy spots
Season: summer through fall
Spore Print: white
Note: Worm Coral (*Clavaria* and *Clavulinopsis* spp.), similar looking with various colors; brittle texture and breaks easily.

7. Spherical Mushrooms

© Tomasz Czadowski/ Shutterstock.com

Common Puffball (*Lycoperdon perlatum*)

Description: white to tan in color, with spherical top when young with slightly rough surface, ages to brown and powdery; elongated bases (no true stem)
Size: <3 in. (up to 8 cm) diameter
Habitat: grows from ground in groups in woods or grassy areas
Season: early summer through fall
Spore Print: brown
Note: Releases spores from top splitting.

© EdwardsMediaOnline/ Shutterstock.com

Rounded Earthstar (*Geastrum saccatum*)

Description: small, smooth tan sphere when emerging; skin splits and unfolds to star-shaped base (darkens with age)
Size: 2–4 in. (up to 10 cm) across
Habitat: ground near deciduous and coniferous trees; often near stumps
Season: present year-round; fruits summer through late fall
Spore Print: brown
Note: Earthstars (*Geastrum* spp.) are saprobes; if sac is pressed (mainly from raindrops) will likely cause spore release.

© Oxford_shot/ Shutterstock.com

Giant Puffball (*Calvatia gigantea*)

Description: white spherical fungus (turns yellow to brown with age at center then outward); short, nonvisible stem or point of attachment at base
Size: 6–20 in. (up to 50 cm) wide
Habitat: woodland edges; open grassy areas
Season: early summer through fall
Spore Print: brown
Note: Perfectly round and pure white throughout indicates a healthy puffball; anything other can be toxic (Common Earthball is black on the inside).

8. Cup-shaped Mushrooms

© Jolanda Aalbers/Shutterstock.com

Wood Ear (*Auricularia auricula*)

Description: irregular shape, flattened to cupped, appears ear-like; thin and rubbery; smooth, gray-tan to red-brown spore-producing surface; gray-brown non–spore-producing surface (facing substrate) sometimes violet, downy with vein-like wrinkles
Size: 1–6 in. (up to 15 cm) wide
Habitat: found on decaying wood of deciduous and coniferous trees, logs, and branches
Season: spring through fall (mostly fall)
Spore Print: white or cream
Note: Saprobes; an anticoagulant.

© Edita Medeina/Shutterstock.com

Orange Peel Fungus (*Aleuria aurantia*)

Description: bright orange, shallow cups often with wavy edges; no stem
Size: up to 4 in. (up to 10 cm) wide
Habitat: bare, hard surface; disturbed areas; gardens; new lawns
Season: summer through fall
Spore Print: white
Note: Orange and Red Cup Fungus (*Aleuria* and *Sarchscypha* spp.) have similar appearances, most inedible; saprobes.

9. Miscellaneous Mushrooms

© iwciagr/Shutterstock.com

White-egg Bird's Nest Fungus (*Crucibulum laeve*)

Description: mature stage looks like tiny nests with tinier, flattened spheres inside (spore producers); young stages have a dome cover appearing like a cushion
Size: <½ in. (up to 1.27 cm) (nest); <1/16 in. "eggs"
Habitat: grows singly or clusters on decaying branches or wood
Season: spring through fall
Note: Bird's Nest Fungus (*Crucibulum* and *Cyathus* spp.); similar appearances with variations in nest and eggs; nests often referred to as splash cup; rain triggers spores to eject; "nest and eggs" best viewed under a magnifier.

10. Lichen

© jessicahyde/ Shutterstock.com

British Soldier Lichen (*Cladonia cristatella*)

Description: fruticose (upright and bushy); green-gray branching stalks with bright red spore-producing tips
Size: 1–1.5 in. (up to 3 cm) tall
Habitat: decaying wood; base of trees; mossy logs; sometimes soils and shaded fields
Note: *Cladonia spp.* resemble structure but without red tips and may have distinctive cups.

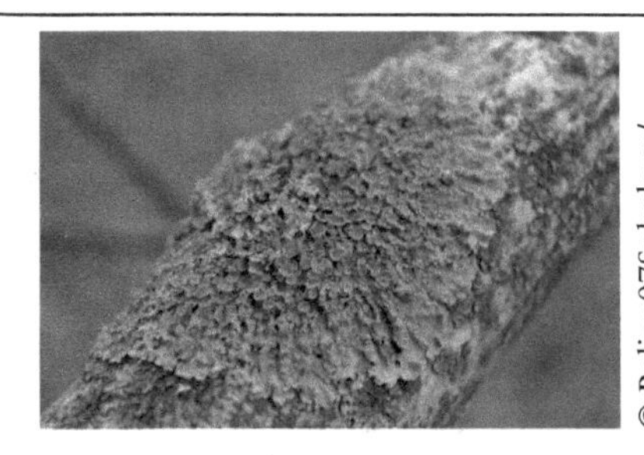

© Polina07Sokolova/ Shutterstock.com

Common Orange Lichen (*Xanthoria parietina*)

Description: bright orange (full sun; duller in shade); foliose (flat and leaflike)
Size: variable
Habitat: exposed rock; branches of deciduous trees; disturbed areas
Note: Easily peeled away, if not then likely a crustose (surface hugging) lichen genus.

© Arnon Polin/ Shutterstock.com

Green Shield Lichen (*Flavoparmelia caperata*)

Description: foliose (flat and leaflike); bright green when hydrated (green-yellow when dry); round leafy lobes with wrinkled center
Size: variable
Habitat: grows on tree trunks and branches; widespread
Note: Indicator of clean air.

© ArtEvent ET/ Shutterstock.com

Reindeer Lichen (*Cladonia rangiferina*)

Description: fruticose (upright and bushy); coarse, gray-white branched tufts with fine hairs
Size: up to 4 in. (up to 10 cm) tall; large mats
Habitat: widespread in northern parts of region; coniferous forests with sandy soil, bogs, or tundra
Note: Other species vary in color; favorite food of caribou.

© Rasa Gruzdiene/ Shutterstock.com

Ring Lichen (*Evernia mesomorpha*)

Description: fruticose (upright and bushy); greenish blue to yellow-green (when dry); tuft-like with granular surface
Size: variable patches
Habitat: often found growing on trees (especially where bark has fallen off)
Note: *Evernia* spp. are "bushy" lichens.

11. Kingdom Protista (fungus-like organisms)

© Predrague/Shutterstock.com

Dog Vomit or Scrambled Egg Slime (*Fuligo septica*)

Description: yellow, amorphous mass, usually with a wet appearance
Size: variable
Habitat: deciduous forests, growing on decaying wood
Season: spring through fall
Note: Has the ability to move and absorbs nutrients.

Fungus—Natural History Journal Pages

Choose a fungus found in your Nature Area or from the field guide that interests you. Research the following information in this book and in other reputable sources.

1. Common Name: Morel
   Scientific Name: Morchella escuenta
   Kingdom: Fungi
   Division: Ascomycota
2. Description: pitted, honeycomb, spongy cap, hollow stem
   Size: 1-2in wide, 1.5-4in tall
   Habitat: deciduous forests, grassy areas, recently dead elms
   Season: Spring
   Spore Print: cream, white to yellow
3. Sketch the overall shape of fungus with colors and special features labeled.

4. Shade the map of North America to show the range where the fungus is found.

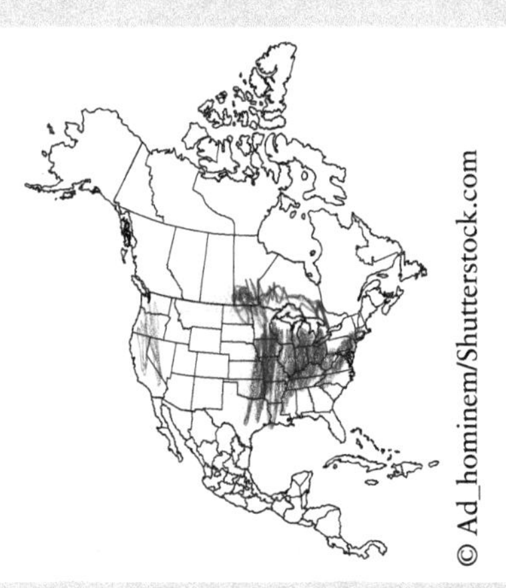

5. How does the fungus obtain its nutrients? Mycorrhizal relationships

6. What is its role in its ecosystem? Decomposer

7. Are there any precautions needing to be taken in regards to this fungus? If so, what should be considered?

Not to be confused with its toxic look-alike, the false morel

8. Discuss additional facts about this species.

Great tasting mushroom, very sought after in the midwest during the springtime.

9. List all resources used (**books:** author's last name, first initial, year, and book title; **websites:** name of site, date retrieved, and web address):

Fitzwilliams-Heck, C.J., 2021, A Practical guide to Nature Study

real Simple, Feb 25, www.realsimple.com

6 Plants

We would not exist without plants. As a naturalist, we need to have an awareness and basic understanding of plants. Recognizing the differences and similarities, and observing and researching their uses in nature will reveal some of the so-called secrets of an ecosystem.

Plants occupy important niches. They provide food, shelter, and oxygen to many animals. We have considered how a habitat is classified by its plant types, the features of different vegetative stages of succession, and parts of a forest. We have also learned about the roles plants play in biogeochemical cycling and in each of the four spheres. We also recognized the value of plants in the stabilization of soil to prevent erosion. Planting native plants, especially native trees, are the best remedy and greatest economical investment to improve the environmental quality of ecosystems. Elaborate on these concepts by reviewing past chapters and conducting your own research to learn more specifics about botanical features or plant physiology. Chapter 6 will focus on basic ***botany****, or the study of plants. Topics covered will include general characteristics of plants, how they function, the four main plant divisions, review questions, an introduction to making identifications, field guides to common plants and trees in the Great Lakes and Upper Midwest region, and templates for your own natural history studies.*

6.1 Plant Basics

All plants are sessile organisms, meaning they cannot move independently. However, plants still need to obtain energy for growth, reproduction, and tissue repair. Like animals, food (nutrients) serves as the source of this energy. All green plants manufacture their own food (**autotrophs**) through **photosynthesis** ("photo-", meaning light and "-synthesis" to make). Most plants in Kingdom Plantae carry out this process along with the **phytoplankton**, or algae in Kingdom Protista, and **cyanobacteria** in Domain Bacteria. Solar energy is necessary for this process; therefore, the sun is the ultimate source of energy for living things. Plants produce more food than they utilize, thereby comprising the greatest amount of biomass in an ecosystem that becomes, directly or indirectly, the food supply for all animals in the habitat.

Photosynthesis converts the sun's solar energy into chemical energy through the action of light with the help of **chlorophyll.** Water and carbon dioxide are the raw materials needed for this process. The result or products of photosynthesis are sugar, oxygen, and water. Chlorophyll, the green pigment within the cell bodies of **chloroplasts,** absorb the solar energy to change and transfer it until the energy is stored as a sugar molecule (various sugar types are formed in different circumstances; all are a type of carbohydrate). Temperature, light intensity, water availability, and carbon dioxide levels affect the rate of photosynthesis.

During the daytime, most photosynthesis takes place in the green leaves within the chloroplasts. However, the epidermis (upper and lower layers) does not contain chlorophyll but serves as a protective layer called the **cuticle,** which is waxy to prevent excessive water loss. The **mesophyll** layers in between, with the exclusion of the veins where the **vascular tissues (phloem and xylem)** exist, contain chlorophyll and help make food. In the diagram, note the difference in mesophyll types from the upper and lower areas of the leaf. Gas exchange occurs through the spongy layer. The **stomates** occur abundantly in the lower epidermis (less abundant in the upper layer) to allow gas exchange of oxygen and carbon dioxide. Two **guard cells** surround each stoma, contain chlorophyll, and control the size of the opening based on the amount of water available (open when water is plentiful and closed when it is not).

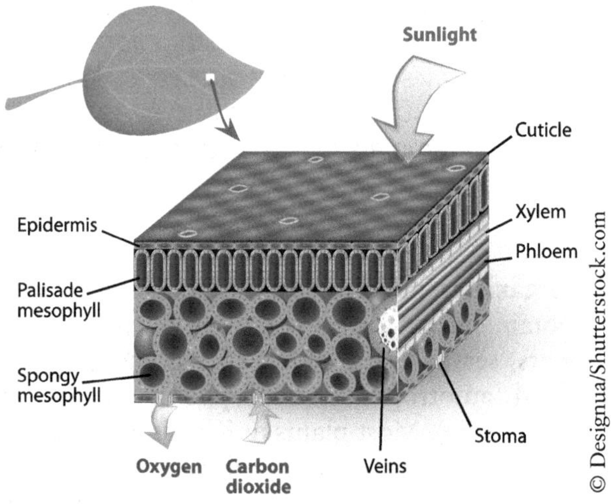

Leaf anatomy (cross-section view) showing the process of photosynthesis (see the text description for further details).

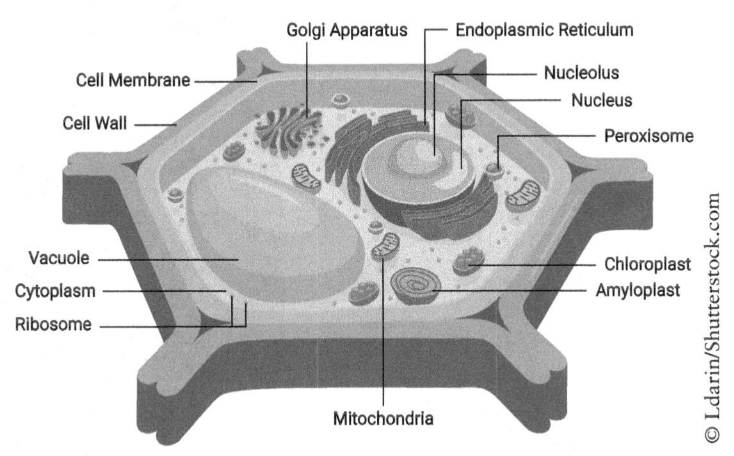

Plant cell anatomy. Throughout the plant, you will find multiple cells (the hexagonal structure pictured here) comprised of these structures or organelles, each having essential functions. Plant cells differ from animal cells. A few obvious differences include the presence of an outer cell wall in plants (made of cellulose) to provide protection and support for the organism, the angular appearance of the cells, and the presence of chloroplasts.

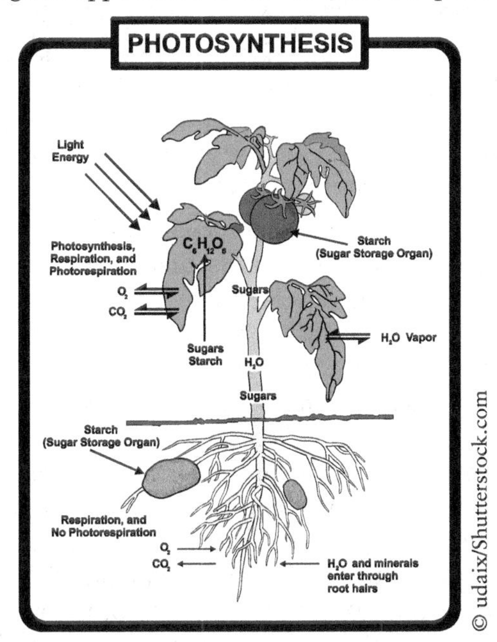

The chemical equation for photosynthesis is: $6CO_2 + 6H_2O \rightarrow C_6H_{12}O_6 + O_2$. Plants use light energy to combine carbon dioxide with water to produce glucose (sugar) and oxygen. The sugars can be stored as starch and used as energy reserves. (See text for further details.)

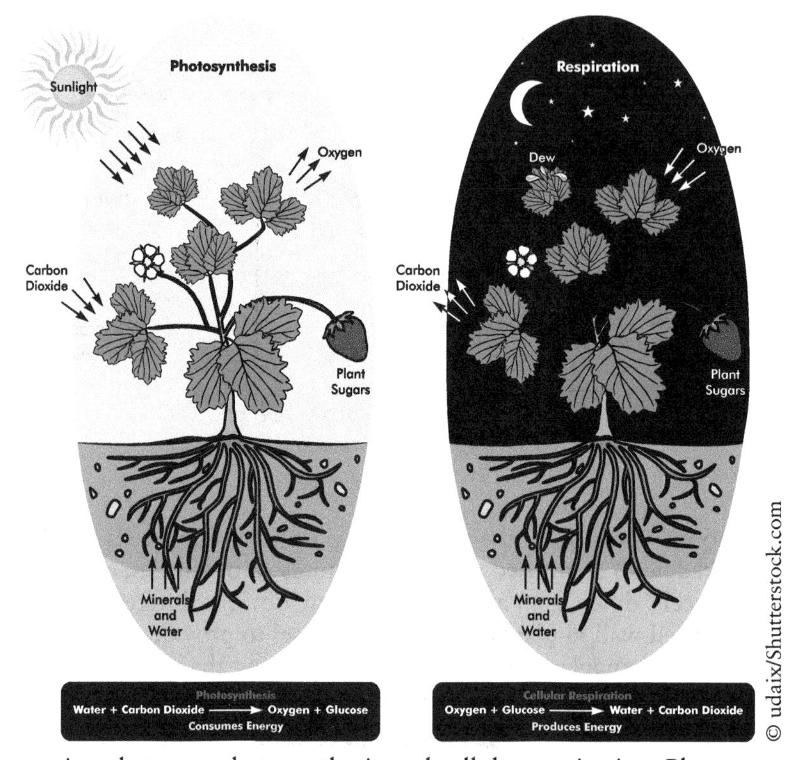

Comparison between photosynthesis and cellular respiration. Plants cannot photosynthesize at night due to the absence of solar energy. During cellular respiration, the plant takes in glucose and oxygen, and gives out carbon dioxide and water and releases energy.

It may come as a surprise that some plants "eat" animals to supplement their photoautotroph nutrition. Plants like the **pitcher plant** (***Sarracenia*** **sp.**) or **sundew** (***Drosera*** **sp.**) can trap insects and other small invertebrates and digest them. From the digested tissues, the plants obtain usable nitrogen and other chemicals that are likely deficient in their preferred habitat, such as bogs. View the pitcher plant in the field guide of herbaceous plants section later in this chapter. Notice that the leaves at its base are vase shaped; this is where water collects and is mixed with digestive enzymes. Hair-like structures pointing downward within the hollowed area prevent trapped insects from escaping. The animal will drown and become digested. Bacteria within the liquid may also aid in the digestive process. Sundews have leaves with sensitive hair-like structures tipped with shiny sticky globules containing the digestive enzymes. When touched, hairs bend over and restrain the insect or other prey while the juices digest it. The plant resumes its normal position upon completion of the digestion process.

In other cases of plants acquiring nutrients, they may exhibit other heterotrophic behavior. Some plants lack chlorophyll so they do not photosynthesize. However, these species still possess other characteristics of plants; so they remain within the same kingdom—cell walls of cellulose rather than chitin as in fungus. **Ghost plant** (*Monotropa* **sp.**) and **coral root orchid** (*Corallorrhiza* **sp.**) are assisted by mycorrhizal fungi in which the plant roots tap into, absorbing water and minerals in the rich forest humus where they grow. These plants are parasitizing the fungi while the fungi exist in a mutualistic relationship with nearby trees. Other heterotroph plants that are parasitic include **dodder,** ***Cuscuta*** **sp.** and the partially parasitic plant **mistletoe,** ***Phoradendron*** **sp.**

The plant world is diverse and has many fascinating aspects to learn. From the basic concepts you just read, you may start to appreciate their capabilities and roles in ecosystems. What variety of plants can you find in nature? Recognize their differences and similarities. Use the field guides at the end of this chapter to inspire further investigations and to learn about what you might have discovered on your Nature Strolls. Now, let us investigate the ways in which you can classify the plants you encounter.

6.2 Plant Divisions

As we look around outside, we hope to see green. Herbaceous plants, shrubs, and trees are hopefully visible. If not, perhaps you can change that by doing some native plantings!

As you become more observant of plant life, you will start wanting to know their niche - what role they serve in their ecosystem. You also might start to wonder about their taxonomy. To make this process simpler, start with the basics. Let's focus on four broad groups in which you can easily classify almost any plant (which of course, includes trees) that you may encounter. The four main varieties of plants this book focuses on are: Bryophyta (i.e. moss), Pterophyta (i.e. ferns), Coniferophyta (i.e. conifers or evergreens), and Anthophyta (flowering or broadleaf plants). The organisms listed in parentheses make up the largest portion of each division. Although most references to these groups in this book will use the common examples, you should also know the actual division name and remember that other representatives exist. In a temperate forest, you will likely find the presence of all four of these plant divisions. In other areas or habitats, these divisions might not seem obvious at all - or you might just notice one type.

Activity: Make it a goal to find the four plant divisions in any habitat you visit. At the very least, get to know your plants and trees within your Nature Area. Keep reading to learn the basics about each division. This will help you understand and appreciate all the plants you encounter in nature.

The basis of modern plants is rooted in a common ancestor, green algae. Green algae are plant-like, single- or multi-celled protists mostly found in aquatic ecosystems (from Kingdom Protista—see Chapter 1 for further information of the basic taxonomy). These organisms photosynthesize but typically do not possess xylem, phloem, stomata, leaves, or root structures like you would find in vascular plants. Algae also have the ability to move via cilia or flagella. These filamentous or hair-like structures may cover the entire surface of the organism (**cilia**) or appear in a single or paired whip-like **flagella**, but, as plankton, their movements are limited and are based on the water flow.

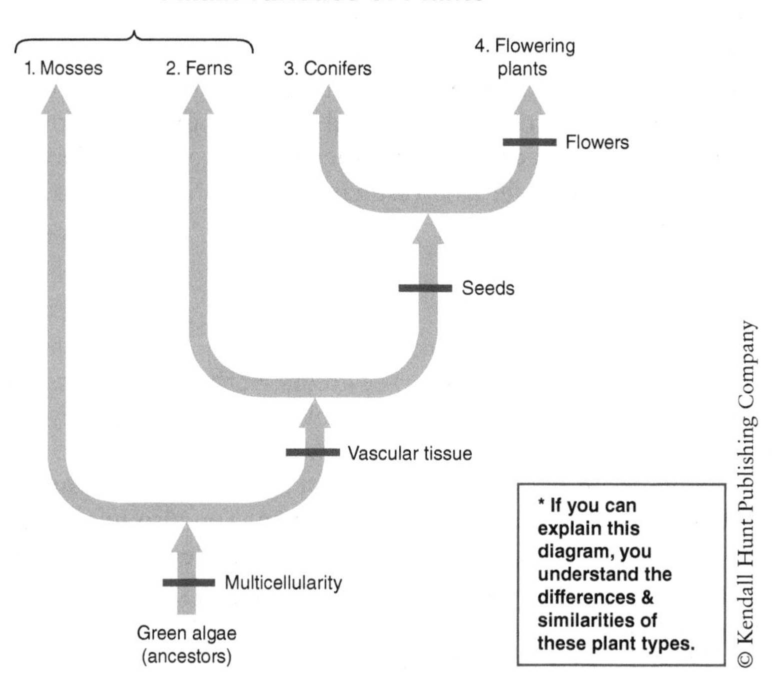

The green algae gave rise to Kingdom Plantae, and we will focus on the four main divisions within it. All plants are multicellular, made of multiple cells, unlike most algae. Starting with the most primitive plants and moving toward an increasingly more complex perspective of the plant kingdom, we essentially advance in the evolutionary time scale. We will begin with moss, then progressing to ferns, conifers, and then ultimately flowering plants. The features they possess help classify them—vascular tissue, seeds, and flowers are the basic structures they may or may not possess. Moss lacks all of them. These four plant divisions still exist today (including the protist, green algae).

Moss

One of the most ancient plant groups is Division Bryophyta with the common example of mosses. Other examples include liverworts and hornworts. They are considered amphibians of the plant world—able to live in dry environments yet require water for its life cycle. A dried moss colony can resurrect after a rain or

© Hanahstocks/Shutterstock.com

a dewy morning. Moist (mesic) environments will have the most luxuriant and diverse moss growing. Moss exists everywhere, from the tropics to the poles. Mosses are essential in soil building, growing on and spreading over bare surfaces. As the most primitive of the plant divisions, it will grow usually no more than an inch from the ground, has no true roots, and lacks vascular tissue (xylem and phloem) to transport water and nutrients.

The life cycle of mosses begins with the carpet of green plants you see in the image. The next opportunity you have, examine the moss under a magnifying glass to observe the individual plants. Individual male and female plants exist at the tips of separate male and female gametophytes (see the diagram for the life cycle of moss). **Gametophytes** are where the **haploid** sex cells (housing of half the number of chromosomes—**gametes**, or genetic makeup for a future plant) are held within the sex organs—**antheridia** (male) and **archegonia** (female). When enough water exists, the male's sperm swims across the neighboring moss. Success of fertilizing the female's egg is left up to chance. If fertilization occurs, then the female moss plant will sprout a filamentous structure that houses the spores within a capsule at the tip. This is termed a **sporophyte** and has the complete number of chromosomes (**diploid**). Hundreds of spores will eventually erupt. The **spore** is haploid and will form a new gametophyte—in other words, it has the potential to sprout a new moss plant on damp soil. The spore has no protective covering, so the weather and other elements will affect its survivorship significantly and determine if the plant will persist and if the cycle will continue. Spore cases help with identification, and serve as a source of entertainment! Pinched at just the right time of ripeness, spores will erupt in a dust cloud and float away!

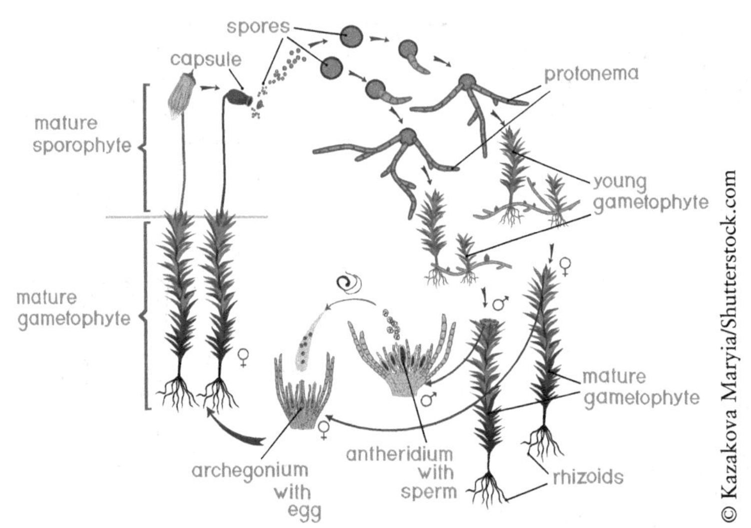

Life cycle of moss. Follow the path of the spores as they develop into male or female plants and see how fertilization occurs. When does the plant have the complete set of chromosomes (diploid)? See more details in the text.

Ferns

Division Pterophyta, or ferns, for the purposes of this book, appeared next in the fossil record and possess a true vascular system. Ferns have varied habitats, from dry to moist, but are typically found in mesic soil in the shade of a forest. Consider a common, mature, fern frond (a petiole stalk with an entire or typically compound

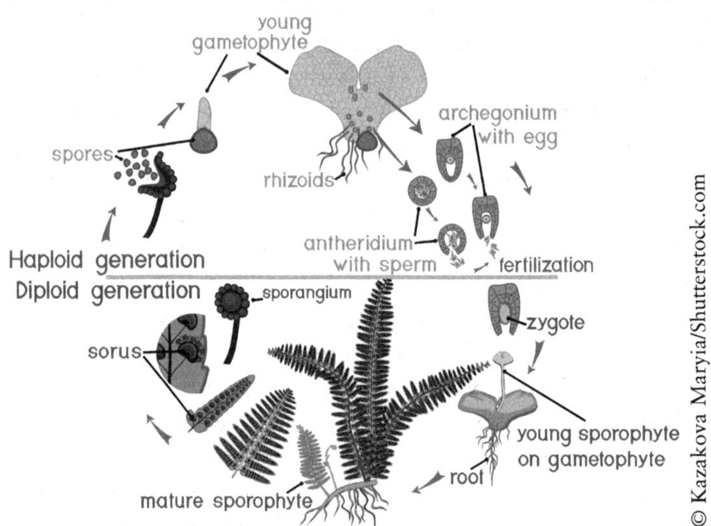

Life cycle of a fern. Start with the fern frond, find where the spores are held, then follow the path to the generation of a new fern. Explain the process. See the text for further information.

blade). The spores are generally held within **sporangium** on its underside; otherwise, spores may exist in a stalk-like structure. The spores will be released and scatter. Under ideal conditions for the species, spores will grow a prostrate, heart-shaped leaf-like structure that contains the male and female sex organs. The male's sperm requires water for it to swim and, by chance, fertilize the female's egg. Once fertilization occurs, a "fiddlehead"—or young, curled fern frond—grows, and the cycle repeats again.

> ***Nature Journal:*** Visit your Nature Area and search for representatives of each of these plant divisions. For the mosses, use a magnifier to look closely at the individual plants and discover the parts diagramed for its life cycle. If summertime, how many species of ferns can you find? Try to identify them. Examine the specimens for their reproductive means. Sketch and discuss your discoveries.

Conifers

The seed producers appear next in the plant division evolutionary progression—conifers (Division Coniferophyta) and flowering plants (Division Anthophyta). The conifers are perennial woody plants – mostly trees with some shrubs which retain their needle-like leaves through the year (with some exceptions). They are referred to

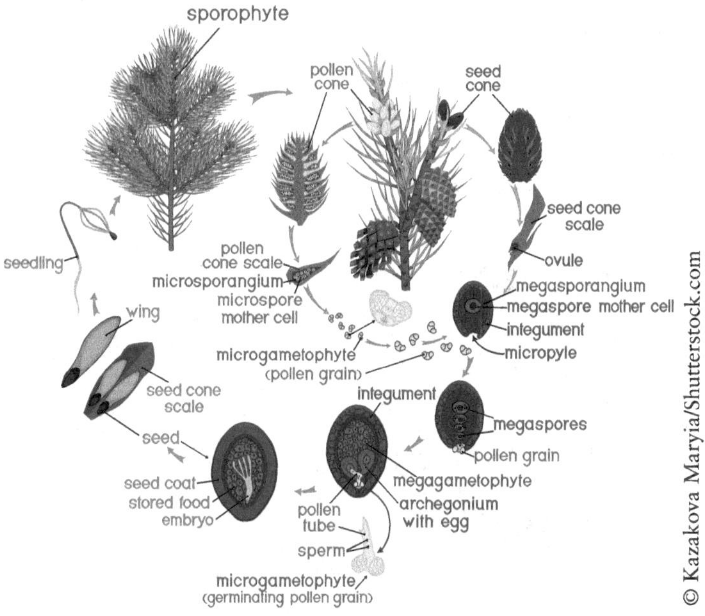

Life cycle of a conifer. On the tree branch, you will find its cones. The smaller male cones have the pollen that will be windblown to the open, female cones. Follow the path from fertilization of the pollen grain to the mature tree (sporophyte). See the text for further descriptions. Some terminology in the diagram will help those who want to go deeper into their nature study.

as the softwood trees, and cone-bearing seed plants. Common examples include the pines, spruces, hemlock, firs, and cedars. **Seeds** are different from spores since they have a young plant (diploid embryo) growing within the protective coat of the seed along with stored food for the embryo's nourishment. The conifers are considered gymnosperms ("gymno"- means naked and -"sperm" means seed). After fertilization, the exposed seeds will eventually fall from the cones in which they are housed.

The other development evolved in the plant world is the presence of **pollen**. The pollen is how sperm (haploid) from the male cones are ferried through the wind to potentially fertilize the female's egg (haploid) within its open cone (not all trees have both sex organs). The result is a seed that may eventually fall to the ground or get eaten by an animal to potentially germinate in the ground into a seedling; then, the mature tree starts the process over. View the generalized diagram of a conifer's life cycle and follow the development of the seed (diploid).

Flowering Plants

In this most recent division of plants to evolve, Division Anthophyta, we see the most diverse and complex group of plants with very efficient pollination. Wind-pollinated flowering plants exist, but the presence of flowers will also attract potential pollinators. Nectar, a sugar-rich liquid produced by flowers, will entice insects, hummingbirds, and some bats. The smell, colors, patterns, and shape of the flower will also lure different potential pollinators. Grasses, wildflowers, and hardwood deciduous shrubs and trees are examples of flowering plants. Some grasses and woody plants may not produce showy flowers, but will have structures more suitable for wind pollination rather than insect pollination.

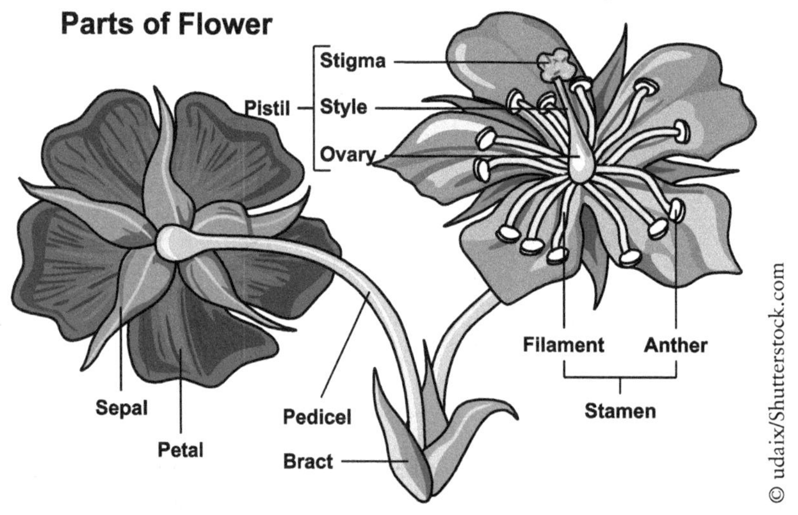

The illustration provides examples of common features of flowering plants. Each family will have variations of what you see here. Learning these parts will help with understanding the plant's natural history and to make quicker identifications.

Activity: If you do not have flowers, consider planting native plants near your home or in a pot for your porch, windowsill, or desktop. What types of effects does growing and caring for flowers or plants have on you?

Flowering plants are considered angiosperms, or those with protected seeds. Think about that apple you ate recently. Where were the seeds? Protected. The seeds are encased within the flesh of the carpel, the reaction of a fertilized egg within the plant's ovary that swelled.

In the example of the apple tree, it produces showy flowers with both male and female parts (not all trees have both parts). The pollen, held at the tip of the male's stamen within the anther will erupt or get disrupted. Upon release, the sperm held within it will potentially adhere to the top of the female's pistil where it may travel down to its ovary for fertilization. The resulting fruit that forms protects the seeds yet lures animals to eat it. This is another way for the tree to spread once the fruit is eaten. The protective seed coat may allow the seeds (and the growing plant within) to remain viable within the animal's digestive tract. Not all flowering plants produce a showy fruit, but will flower even if as a subtle display. **Herbaceous** (vascular and nonwoody) and woody varieties of this division exist (deciduous shrubs and trees), therefore making this the most abundant plant division in nature.

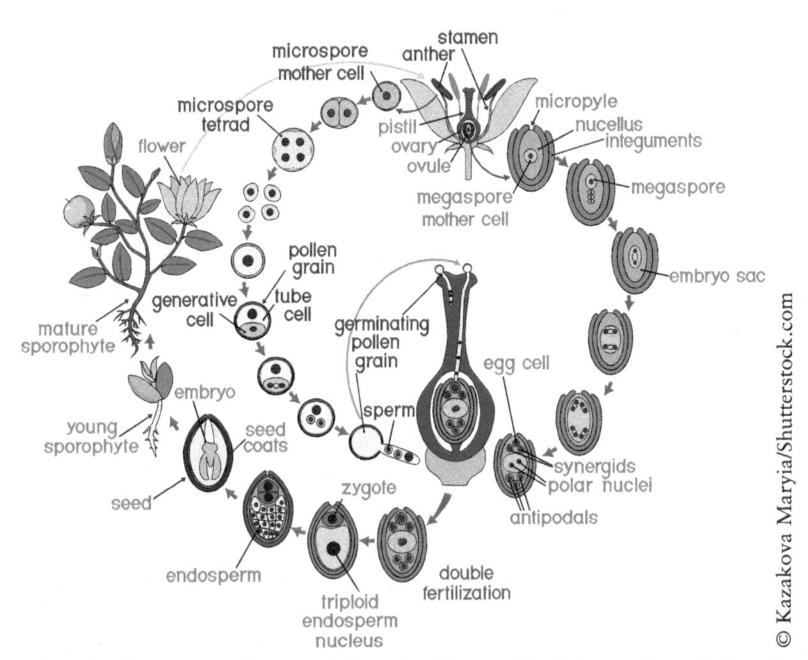

Life cycle of a flowering plant. Within the flower, find the male and female structures. The pollen's sperm will fertilize the female's egg in fertilization. See the text for more details. The additional terminology in this diagram will help you go deeper into your nature study.

*Is it a **monocot** or **dicot**?* More than likely, the plant in question is a dicotyledon, or dicot with over 200,000 species globally compared with about 50,000 monocotyledon, or monocot species. Dicot plant embryos have two seed leaves (**cotyledons**), flower parts of fours or fives, and vascular tissue arranged in ringed bundles surrounding a central pith. Examining the leaves would show netlike, or branching vein patterns. Examples of dicots include deciduous trees and shrubs, many flowering perennials, and most vegetables grown in your garden. Monocots have one seed leaf, narrow leaves with parallel veins (this often can help with making a quick deduction that a plant is a monocot), scattered vascular bundles in the stem, and no cambium. Flower parts are in threes or multiples of threes (petals, sepals, stamens). Examples of monocots include grasses, grains, orchids, lilies, rushes, and palms.

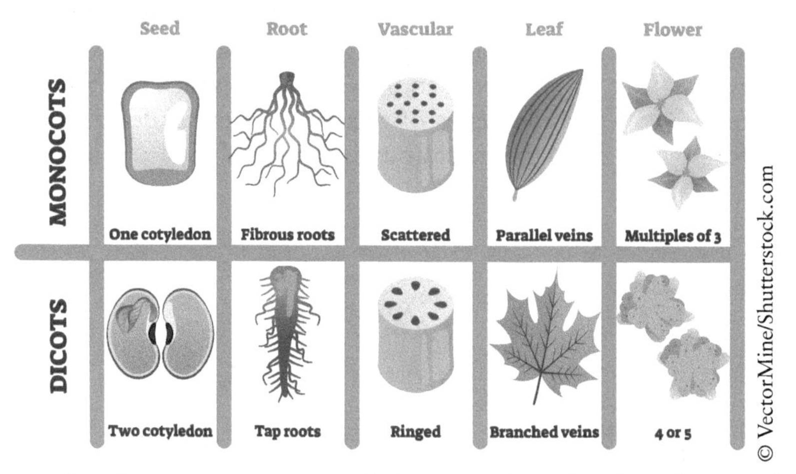

A comparison between monocots and dicots. Looking for these features forces you to look closely at the plant and can be one of the first things you consider when making a species identification.

Types of Growth Variations. You may have read or heard of **perennial** plants, but do you know what this classification indicates? Simply put, its roots and underground stems persist for many years, and send up shoots every spring that produce flowers and seeds every year. The survival of the plant varies by species. **Annuals**, you may have planted as young plants or seeds along your walkway or in a planter. These plants grow from seeds and flower the first year. They will also produce seeds, but then die. Whereas, if you have a **biennial**, it will produce leaves the first year, but will flower and seed the second and then die. What growth variations do the flowers around you exhibit?

> *Nature Journal:* When spring arrives, look for the first sign of flowers blooming. Look at it closely and describe and sketch its details. Include all parts of the flower and leaf arrangement and shape. Is it a monocot or dicot? Complete Natural History Journal Pages for that plant (see the template at the end of this chapter).

6.3 The Woody Structures

Trees are a valuable resource for the ecosystem services they provide and personal benefits they offer humans. Supposedly, most people can use the equivalence to 1 tree per year that is 100 ft. (30.5 m) tall and 18 in. (46 cm) diameter to meet their personal everyday needs—excluding the ecological benefits of clean air and erosion control. Most trees are a renewable resource, but knowing where your products come from and whether they are from sustainable means is a good practice to ensure those ecological benefits are served.

Activity: For your personal use, think about how you use trees. Consider its derivatives from residues, fibers, fruits, and chemicals. Have you used any of these today? Coffee, toothpaste, shampoo, soap, cosmetics, vitamins, pharmaceuticals, cooking utensils, syrup, juice, cardboard, chewing gum, plastic, tires, paper, ink,

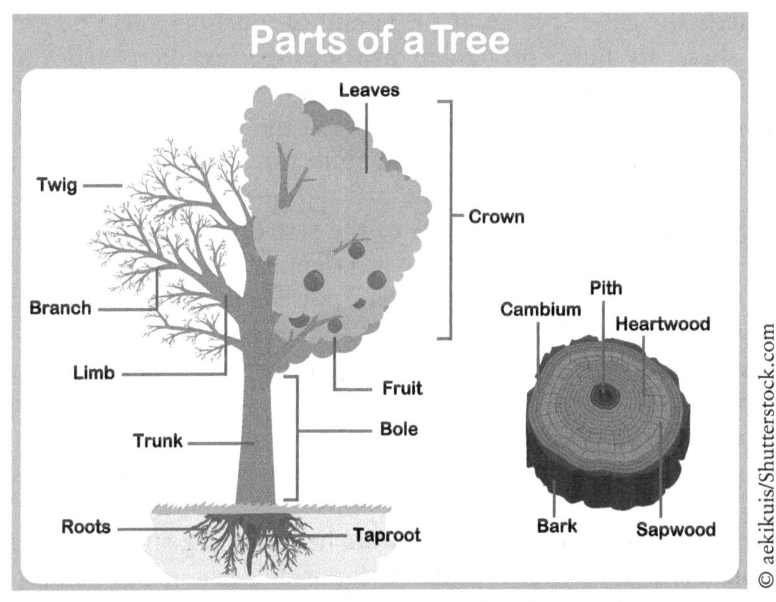

When discussing parts of a tree, use the terminology here to best communicate to other naturalists. Looking at the diagram, what do you suppose the difference is between the roots and taproot? How do the trunk and bole differ? Notice the "tree cookie," or the slice of the trunk and its various layers—what is the significance of each section (see text for further details)?

mushrooms, ice cream, bowling pins, snowshoes, baseball bat .... What part of a tree do these things come from? Can you think of more examples?

A further definition of a tree: it is a woody plant with a single trunk that is usually not branched at the base. At maturity they will reach taller than 15 ft. (4 m) with a distinct crown, trunk, and be at least 3 in. (8 cm) in diameter at chest height. Shrubs are also a woody plant but have many branches near its base and are less than 15 ft. (4 m) at maturity. When looking across a landscape, notice whether the woody plants have a distinctive trunk, or if multiple trunks or branches exist at its base.

Like any plants, the growth of trees and shrubs are influenced by the abiotic conditions within its habitat such as by the soil, moisture, wind, and crowding. Some will grow taller than usual or remain stunted for its lifetime. With crowding, it can also be considered a biotic factor acting upon it, such as with intraspecific or interspecific competition (see Chapter 4 for details). However, it is the abiotic elements for which they compete. The age of the tree or shrub influences its growth, as does the presence of any damage from fire, insects, or parasitic fungus. All these things will affect the growth of the tree.

The bark of woody plants is also important in healthy growth. It is a multi-layer of protective tissues on the outside of the tree comprised primarily of cork cells. **Lenticels** for gas exchange (CO_2 in and O_2 out) may also be present in the bark, and these are more prominent in some species like cherries and locusts. Moving inward, you find the **cambium**. This is the part of the wood that is alive! It contains the cell-dividing parts surrounding the trunk, limbs, branches, twigs, and roots. The outer portion of the cambium gives rise to the **phloem**, which is found just behind the bark and has the critical role of transporting food and other materials from the crown downward to the rest of the plant. Toward the inside of the tree, the cambium gives rise to the **xylem**, which takes up water and other trace minerals up from the roots and transports it to the rest of the tree. This is classified as the **sapwood** (outer wood). The **heartwood** is near the center of the trunk where xylem cells have become plugged with resin and other compounds, creating the densest layers in the trunk. The **pith** is at the center and makes up a small portion of the woody mass, but in twigs it can sometimes prove useful in the identification of a tree, such as the black walnut having a dark pith and its close relative butternut having a buff-colored pith.

The growth of the tree is evident in its height and girth or thickening of the trunk. From the perspective of photosynthesis and acquiring the necessary solar energy, the tree has a **primary growth** pattern to lengthen limbs and branches. **Secondary growth** will result when needs for photosynthesis have been met and the tree begins to thicken. This process leaves radial growth layers patterned in the wood. These are the annual rings you can count to determine a tree's age and to distinguish between good and poor growing seasons.

6.4 What Happens to Leaves in the Fall?

The broad leaves in the Great Lakes and Upper Midwest region cannot survive the winter. There is not much sunlight for the plants to nourish themselves, and it gets very cold. The green leaves of summer rely on chlorophyll for their color and the food-making process of photosynthesis. With less light in the fall, photosynthesis eventually stops, and the green chlorophyll disintegrates. The colors revealed in the fall are mainly due to decreased light rather than temperature, and the result of chemical substances once suppressed or newly manufactured by the chlorophyll. These substances are pigments that create the appearance of different colors in the leaf.

One such pigment is xanthophyll, which creates a yellow color observed in birch and poplar, and can also be seen in corn. Carotenes appear as gold, orange, and red in maple species and gives carrots their orange color. Anthocyanin turns pink, scarlet, purple to black in sweet gum, dogwood, and sassafras; and also gives blueberries and cherries their color. Oaks produce tannins, resulting in brown leaves. These examples given for a tree's leaf color in the fall is what they would mostly display, but realize that different tree species have varying proportions of these pigments with different amounts of chlorophyll left in a given autumn.

Other variables to consider for leaf color are temperature and precipitation. When we see a bright display of fall colors, it is usually a result of the area experiencing colder temperatures than average yet still above freezing. This condition triggers the production of anthocyanin, and bright, vibrant colors result. However, if there is an early frost, this pigment is weakened or does not get created. Drought in an area causes very little color change, and leaves drop earlier than usual. When we experience an autumn with drab, dull colors, then we probably had mostly warm and cloudy days. Lastly, consider forest diversity if you are looking for a pretty autumn scene. If there is a variety of trees in an area, then the color display will also be varied and have a longer duration in the season compared to a forest with fewer tree types.

The other process in the fall to understand is the preparation for leaf drop. An **abscission layer** of a cork-like nature forms across the point of leaf attachment. Once completely grown, the leaf's stem (**petiole**) drops from the twig. The abscission layer prevents water loss in the tree during the winter months. In the winter, look for buds on the twigs of deciduous trees. These are next year's leaves developing inside those moisture-conserving scales. Those trees are not dead in the winter! Most conifers are evergreens and retain their leaves through the winter. Their needles are small and covered with a waxy covering to prevent water loss. In our region, winter is a time of drought. Water is usually frozen and unavailable. Photosynthesis in trees requires considerable water, so these trees will not likely produce food (or produce very little) during this season.

Timing is everything. **Photoperiodism** refers to the responses any organism has to changes in day length or amount of light exposure. The varying amounts of daylight affects the chemicals in trees. Increasing light in spring prepares trees for leaf emergence, whereas decreasing light signals trees to drop their leaves in the fall. It is a relatively timed breakdown of chemicals within their cells that affects leaf emergence or leaf drop. This phenomenon will in turn affect the behavior of the animals dependent upon these trees.

6.5 Ecological Importance of Trees

Trees provide a large component of our necessary oxygen in the atmosphere. Trees provide homes and serve as a food source for many different animals. Roots bind soil and host a myriad of beneficial fungi. Moss and lichen find a substrate on tree trunks, which serve as a source of nesting material, shelter, and food for animals. Cavities of trees are used by birds, snakes, and squirrels as a home or refuge. The layers of a tree support a diversity of life, with the canopy hosting an array of seemingly hidden butterfly, bird, and insect species. The species and age of trees within a forest controls what other species exist in the habitat. The remains of trees also provide special features and complexity necessary for biodiversity to exist (see Chapter 3). Humans also benefit from ecological services provided by trees—that of improving air quality, stabilizing the soil, protecting aquatic ecosystems from sedimentation and thermal pollution, and supplying us with a multitude of products that can be harvested or manufactured from them, especially the maple sugar collected from sugar maples (*Acer saccharum*)!

With the fundamentals of plant biology, the ability to recognize and understand the basic plant divisions, and the ecological importance of plants, you are ready to delve deeper and learn how to make identifications.

Review Questions

1. Discuss reasons why humans need plants.
2. How can plants obtain nutrients?
3. Describe characteristics of the Kingdom Plantae.
4. What are the four main varieties of plants? Discuss how they differ and how they are similar. Give characteristics of each division of plants.
5. Compare and contrast the structure and purpose of spores and seeds.
6. Discuss how fruit forms and the benefits to flowering plants.
7. Describe influences on tree and shrub growth.

8. Define the function of bark and the wood's components of cambium, phloem, xylem, sapwood, and heartwood.
9. Discuss what happens to trees in the fall and the winter. What is the significance of photoperiodism?
10. How are trees beneficial to an ecosystem?
11. Discuss the types of things you would examine to make an accurate identification of a plant or tree (next section).

6.6 Field Guide: Primitive Plants

Before delving into the field guide for herbaceous plants, here are a few words about and photos of other photosynthesizers of potential interest you may encounter. Peruse this section to familiarize yourself with these common species.

Kingdom Protista—The Phytoplankton, Green Algae

Division Eukarya, Kingdom Protisa. As discussed in this chapter previously, algae are typically classified as a protist but will sometimes still be found in other reference material as the plant kingdom. Regardless, there are distinctive differences, yet both photosynthesize and exist as the foundation of their food web. As a naturalist, you may have wondered about algae observed in aquatic environments.

This is a sight you may find on the surface of standing water. If you take samples, use rubber gloves and wash hands thoroughly with soap and water after coming in contact with the water.

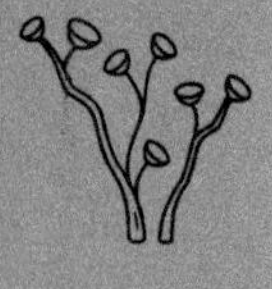

Activity: Taking a sample and using a magnifier (preferably a pocket or laboratory microscope), you could perhaps identify the species, or at least see some identifying structures to determine its taxonomic division or simply to just enjoy seeing a new environment up close. Identification of these organisms are beyond the scope of this guide. However, you are encouraged to find resources to help you go deeper into understanding this fascinating realm of nature.

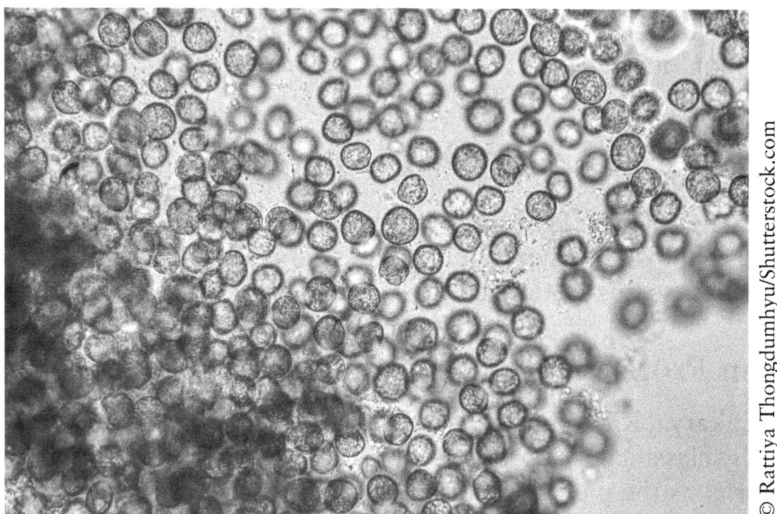

Chlorella sp. occurs in clumps or individual cells among other algae; they are found in organic-rich or polluted waters; often have musty odor.

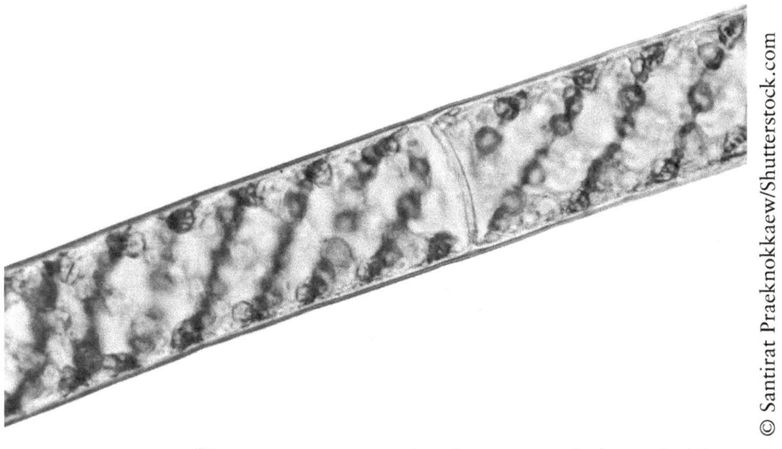

Spirogyra sp., a common filamentous green alga has a spiral-shaped chloroplast in each cell; it is found in dense colonies on pond surfaces in spring.

Green Algae. Many fascinating species of algae exist. For this algal group, upon close inspection you would observe bright green pigments in chloroplasts (or plastids) and a nucleus that is well defined. The structure varies among the genus and can occur as single cells, or as round, flat, or filamentous forms. These are most abundant in standing water (lentic systems).

Kingdom Plantae—Division Bryophyta, The "Mosses"

Many species of bryophytes (or "mosses") exist. Because of their lack of a vascular system, they grow close to the ground. Therefore, the naturalist needs to simply get close to the specimen to examine their leaves and spore cases. Research online for good identification resources. Remember, because they have no vascular system, these plants are not classified as herbaceous. As a naturalist, you will see these organisms and will want to know something about classifying them. Examples of bryophytes include hornworts, liverworts, and moss—each encompassing hundreds of species. Common examples of each group are shown below to help you get started with recognizing key features and making connections.

Hornworts: Develop a long (0.4-0.9 in.; 10–22 mm) horn-like sporophyte from their gametophyte. In the gametophyte phase, you will find a flat, green-bodied plant in damp environments, garden soils, or on the tree bark. They will split lengthwise to release spores. These are not as common as the other bryophyte examples in the region.

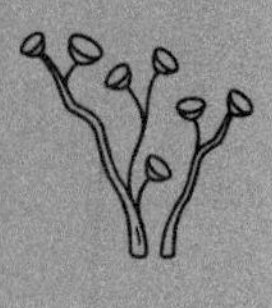

Liverworts: Exist as extremely small, variable plants with flattened stems and indistinguishable leaves with small rhizoids. Upon magnification, membrane-bound oil bodies within their cells are apparent, which do not exist in other bryophytes.

The Marchantiophyta (hepatics or liverworts) are common with many species and difficult to discern due to the small structures.

Liverwort or umbrella liverwort, *Marchantia polymorpha,* with its obvious (yet still small-sized, up to 0.4 in.; 10 cm) gametospores as umbrella-like features.

Mosses: The most abundant and speciose of the bryophytes will persist as green, clumpy, conspicuous plants found in moist environments and out of direct sunlight. The small leaves attached to a stem are used for water and nutrient transportation. Mosses can absorb a large amount of water and, after rains, will often feel wet and spongy.

Haircap Moss (*Polytrichum commune*)—2 to 10 cm

Pincushion Moss (*Leucobryum glaucum*)—5 to 9 mm

6.7 Field Guide: Herbaceous Plants

Plants are [hopefully] part of your everyday life. If not, think of how you could change that situation. If you want to learn more about species other than trees, this section will get you familiar with some common herbaceous plants in the region. Herbaceous plants, the nonwoody vascular varieties including pterophytes (ferns and their allies) and anthophytes (flowering plants), have other considerations for identification. The unique features and terminology are explained within the field guide.

Ferns and Their Relatives

The plant divisions provided here will offer insight to what you may encounter on a Nature Stroll in the summer months in the Great Lakes basin and Upper Midwest region.

Division Pterophyta (*ferns*).

Read about characteristics of this group earlier in Chapter 6. What you discover in this section will help you identify ferns of different families. Each example provided is somewhat common in the region and exhibits a different spore-producing structure.

Royal Ferns (Osmundaceae): The 2 ft. (<1 m) fronds in this family arch outward and often form a mound from which it grows; spores in brownish-black upright clusters
Cinnamon Fern (*Osmunda cinnamomea*)

© Todd Boland/Shutterstock.com

Tree Ferns (Cyatheaceae): spreading fronds divided into leaflets (compound) and subleaflets; spores are found on underside in sori
Bracken (*Pteridium aquilinum*)—likely the oldest fern type, and it has a worldwide distribution; petiole 3 ft. (1 m) tall, 3 ft. (1 m) wide triangular compound leaf

© Furiarossa/Shutterstock.com

Spleenworts (Aspleniaceae)—diverse fern family; widely distributed; tends to hybridize within family
Lady Fern (*Athyrium filix-femina*)—delicate, minutely-toothed, 3 ft. (1 m) tall, drooping tip

Sporangia shown on the underside of lady fern

Sensitive Fern (*Onoclea sensibilis*)—up to 2 ft. (0.5 m) tall; yellow-green; coarsely textured

Sporophyte rachis (brown, beaded) of last year's sensitive fern found in a wetland, "bead fern"

Division Sphenophyta

Horsetails (Equisetaceae). Jointed stems, leaves (when present) in whorls; stems contain silica ("scouring rush" name given for its abrasiveness and usage); tends to grow near water
Horsetail (*Equisetum hyemale*)—May be referred to as snakegrass or puzzlegrass, but are not a grass! This type of common name causes confusion; using the general classification of horsetail or scientific name is preferred. Up to 3 ft. (1 m)

Wood Horsetail (*Equisetum sylvaticum*)—30–60 cm tall

Division Lycophyta

Club Moss or Ground Pine Family (Lycopodiaceae)—not mosses or pines, but often mistaken for both! Using *Lycopodium* as a classification is preferred. These primitive plants have needle-like leaves covering all stems and branching thickly; spore cases arranged as cone-like at tip of upright stems.
Lycopodium clavatum—2–6 in. (5–15 cm) high, but grows much longer prostrate (along the ground) up to 3 ft. (1 m)

Wildflowers

In this section, you will find wild, herbaceous plants that flower, Division Anthophyta (no woody plants here!). The key and descriptions revolve around the presence of a visible flower when trying to make an identification. The botanical terms in this practical guide have been kept to a minimum to help with comprehension of concepts. If not explained in context, you may find definitions previously discussed in this chapter. Identifying wildflowers can be fun for anyone, and different methods for keying them out exists. This guide will direct you to an identification by way of flower color.

How people would classify a color for a flower can vary. Often, many shades can fall between two major color groups, or color names mean something different to people. If you cannot find what you need in one section, try another section to see if the color could be interpreted differently. Regardless of how you determine your wildflower's classification, certain features of the plant also need consideration: (1) *height* of the plant; (2) *flower type*, if it is regular (radially symmetrical) or irregular; (3) *leaf arrangement*, if any are present, how are they arranged along the length of the stem (absent, basal, alternate, or opposite); (4) *leaf type*, if they are entire, lobed, or divided and what do the veins (parallel or intersecting) and margins look like (smooth or toothed); and (5) *habitat*, this will help you decide on a species or direct you to finding a specimen of interest. This type of information is included in the description of the plants in the field guide. Take note of the family names so that you can start making connections among the genus and species of that group. Use additional resources if necessary to make more specific or accurate identifications.

Plant Families to Know

As you may have gathered by now, species from the same family will often present similar characteristics. This helps the naturalist make comparisons among groups and can reduce the time it takes to key out a species. Some of the most common plant families are aster, grass, lily, mint, mustard, parsley, pea, and rose.

At the very least, you could identify many plants to fit into one of these eight groups. Here is what to remember for these common families:

Aster (Asteraceae): *composite flower* of many small flowers attached to center disc (petals radiating from a central point); each flower 5 petals; numerous stamens (i.e. sunflowers)

Grass (Poaceae): monocots, *parallel venation*; stem round, hollow with sheathing; inflorescence terminal and axillary spikes

Lily (Liliaceae): monocots, parallel venation, *floral parts in multiples of 3s*, 3 petals and 3 tepals (sepals identical to petal), 6 stamens, 3-parted pistil (i.e. onions, tulips)

Mint (Lamiaceae, formerly Labiatae): *stem square* in cross-section; flowers grouped in leaf axils or terminal spikes; fused sepals; 2-lipped corolla; 2 or 4 stamens

Mustard or Crucifers or Cabbages (Brassicaceae, formerly Cruciferae): *4 petals*, 6 stamens (4 tall and 2 short), 4 sepals, 1 pistil; small, regular, bisexual flowers; mostly annuals (i.e. broccoli, cabbage, cauliflower)

Parsley (Apiaceae): *umbrella-like umbel* (several); filigree-like, pinnately compound leaves

Pea or Legumes (Fabaceae): *irregular flowers* with 1 large "banner petal," 2 "wing petals," and a "keel" (i.e. beans, peas)

Rose (Rosaceae): *5 petals*, sepals with many stamens (i.e. edible fruits such as apples, pears, cherries, raspberries, strawberries; trees, shrubs)

Activity: Look for examples of these families throughout the following field guides and especially when you discover plants in nature! Complete a Natural History page for representatives of each family. At the very least, sketch the flower, leaves, and other identifying features. By completing these activities, you will remember the key characteristics of eight plant families. Just from a quick observation of a plant, you could recognize if it fit one of these families. Further identification will be made easier from that point.

Identification Key

Flower Color

Use the descriptions of each color category to help you make a decision on what section to search for your flower and plant name.

1. **White:** flowers in this section will appear pure white, cream, off white, pearly-white, yellowish-white, whitish-green, and those mostly white but with various colored markings
2. **Yellow to Orange:** you will find flowers described as yellowish-green, yellow-orange, orange-yellow, and red-orange
3. **Pink to Red:** pink, dusky pink, crimson, rose, red, and bi-colored flowers with pink or red
4. **Lavender to Purple:** colors range from lavender to violet, magenta, red-purple, and dark purple
5. **Blue:** blue or violet-blue flowers are found in this category
6. **Green and Brown:** green, green with purple, greenish-brown, brown, or brownish-purple

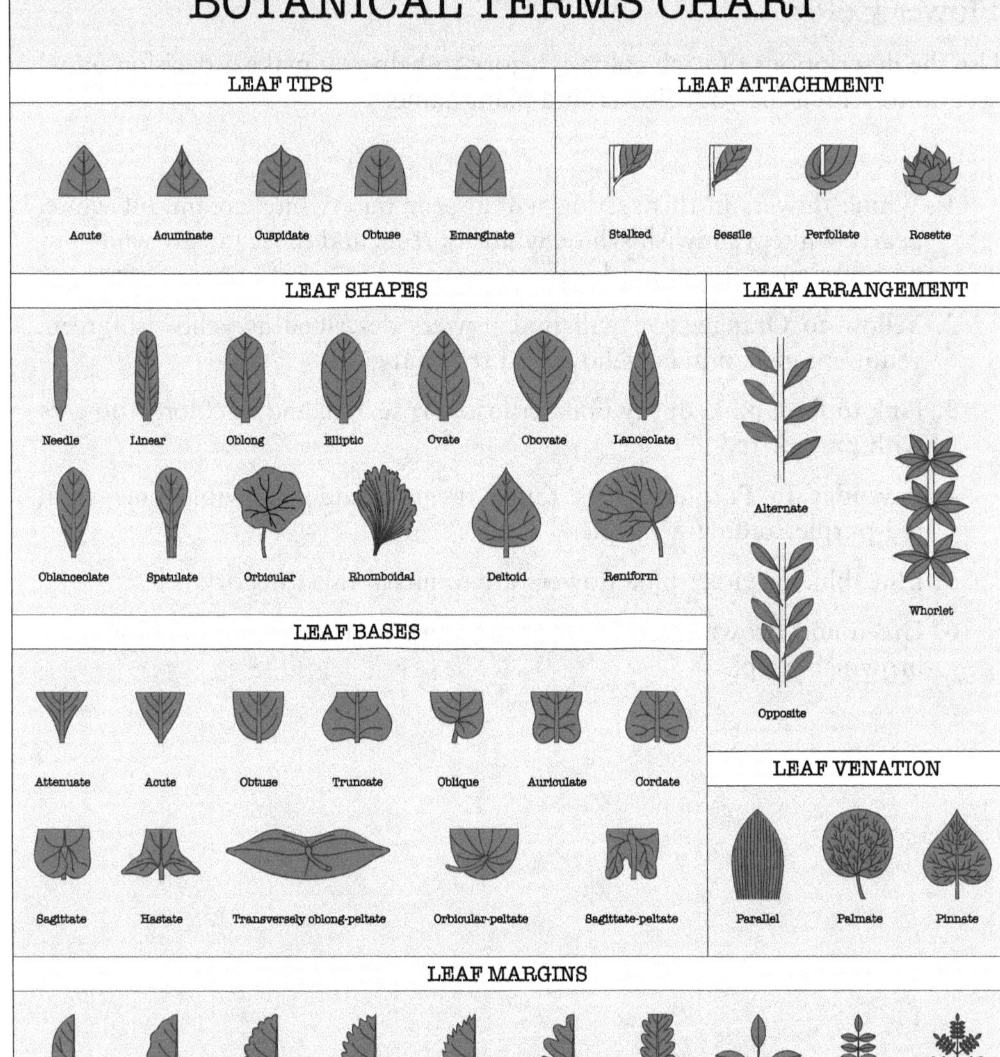
BOTANICAL TERMS CHART
LEAF TIPS
Acute
Acuminate
Cuspidate
Obtuse
Emarginate
LEAF ATTACHMENT
Stalked
Sessile
Perfoliate
Rosette
LEAF SHAPES
Needle
Linear
Oblong
Elliptic
Ovate
Obovate
Lanceolate
Oblanceolate
Spatulate
Orbicular
Rhomboidal
Deltoid
Reniform
LEAF ARRANGEMENT
Alternate
Whorlet
Opposite
LEAF BASES
Attenuate
Acute
Obtuse
Truncate
Oblique
Auriculate
Cordate
Sagittate
Hastate
Transversely oblong-peltate
Orbicular-peltate
Sagittate-peltate
LEAF VENATION
Parallel
Palmate
Pinnate

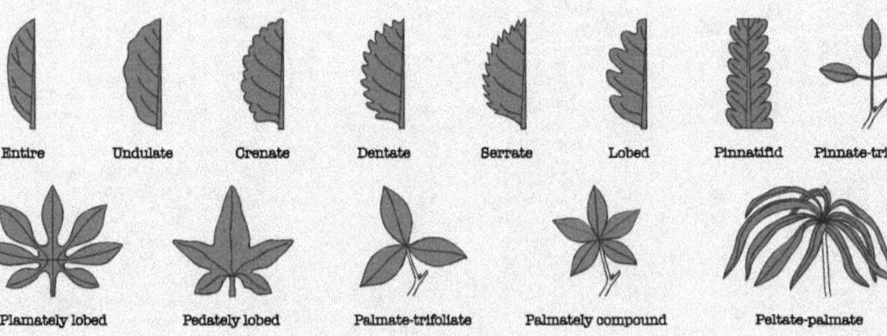
LEAF MARGINS
Entire
Undulate
Crenate
Dentate
Serrate
Lobed
Pinnatifid
Pinnate-trifoliate
Pinnate
Bipinnate
Plamately lobed
Pedately lobed
Palmate-trifoliate
Palmately compound
Peltate-palmate
Tendrils
Stipulate

Habitat Descriptions

Within the color-coded sections that follow, the flowers are listed in a somewhat pattern of seasonal blooming sequence and then grouped by botanical relatedness or similarities. Most **habitat descriptions** have been condensed to these classifications:

Woods: mainly deciduous forest; bloom times are typically spring before leaf-out of trees

Dry woods: oaks and conifers dominate these areas with well-drained soil and bracken fern

Wet woods: conifer, moss, ferns, liverwort associations within moist sites; common in northern part of the region

Meadow: open area with minimal woody cover; disturbed sites—roadsides, abandoned fields, ditches, fencerows

Dunes: sandy, coastal areas along the Great Lakes

Swamps: wetland with stagnant water; trees present

Bogs: wetland dense vegetation mat with various water-tolerant plants

Take Pictures Not Petals

Make all your plant observations where the specimen grows … without picking it. If you need to look more closely or spend more time, simply take a photograph. Plant collections, or herbariums can be a very useful resource, but the intent of this practical guide is to encourage making thorough observations in the field and documenting those experiences in the Nature Journal. If an herbarium is desired, then proper techniques should be researched before collecting. Never harvest all of a species in an area. Consider creating photo journal for the plants you collect. Take close-up photographs of the parts of the plant to help with identification.

Although many people forage wild plants as a food source or for fun, this practical guide does not inform or encourage that any of the plants or plant parts can be eaten or used medically without further research or consultation with an expert.

White

Wildflower Species Accounts

1. White

© Kyle Selcer/Shutterstock.com

May Apple or Mandrake (***Podophyllum peltatum***)
Family: Barberry (Berberidaceae)
Height: 12–18 in. (30–46 cm)
Flowers: spring; fragrant; nodding on short, fuzzy stalk from fork of two leaves; 1 per plant; stamens yellow surrounding pistil (yellowish-green); fruit large, apple shaped
Leaves: 2, deeply lobed, covering nodding blossom; immature plants have one umbrella-like leaf; grows in colonies; seeds, leaves, roots toxic
Habitat: open woods, wetland margins

© Nikolay Kurzenko/Shutterstock.com

Trillium (*Trillium grandiflorum*)
Family: Lily (Liliaceae)
Height: 12–18 in. (30–50 cm)
Flowers: spring; large 3 petals per stem; 3 prominent pointed green sepals below and between longer petals; turns pink with age.
Leaves: 3 broad, whorled narrowing to a pointed tip; veins conspicuous; 1 set per plant
Habitat: woods

© Gerry Bishop/Shutterstock.com

Foamflower or False Mitrewort (***Tiarella cordifolia***)
Family: Saxifrage (Saxifragaceae)
Height: 6–12 in. (15–30 cm)
Flowers: spring; delicate, short stems on fuzzy central stem; long stamens extend beyond petals
Leaves: basal, 3–5 lobes, indented at base; long, fuzzy stems
Habitat: woods

© Mike Truchon/Shutterstock.com

Dutchman's Breeches (*Dicentra cucullaria*)—similar to ***D. canadensis***
Family: Fumitory (Fumariaceae)
Height: 6–12 in. (15–30 cm)
Flowers: spring; waxy, inflated petals dangle from stem like pantaloons; 2 spurs spread upwards forming "legs of pants"; yellow-tipped
Leaves: delicate, finely divided, long stalked; arise from base
Habitat: woods

© LesPalenik/Shutterstock.com

Bloodroot (*Sanguinaria canadensis*)

Family: Poppy (Papaveraceae)
Height: 2–6 in. (5–15 cm)
Flowers: spring; single, 8 (sometimes 10, rarely up to 16) white petals with yellow filaments in center; 4 petals longer than others to give a squarish appearance; closed in cold or cloudy weather
Leaves: base of plant; rounded; margin wavy or coarsely toothed; up to 4 in. (10 cm) during flowering (larger afterwards); stem smooth, erect, leafless, contains red juice that can cause a rash
Habitat: moist woods

© MargareeClareys/Shutterstock.com

Wild Lily-of-the-Valley or Canada Mayflower **(*Maianthemum canadense*)**

Family: Lily (Liliaceae)
Height: 2–8 in. (5–20 cm)
Flowers: spring; small, individual clusters at tip of stem; sweet smelling; 2 petals and 2 tepals (4-part flower); 4 protruding stamens; fruit is cluster of green berries with red speckles
Leaves: usually 2 (1 or 3) heart-shaped base; partially surrounds stem; veins parallel from base to tip
Habitat: woods

© WildFarkas/Shutterstock.com

Starflower (*Trientalis borealis*)

Family: Primrose (Primulaceae)
Height: 4–10 in. (10–25 cm)
Flowers: late spring, early summer; 5–9 (usually 7) sharp-pointed petals; flower pairs at tip (vary 1–4); stamens prominent
Leaves: 5–10 long, tapered, shiny; whorled below flower; 2–4 in. (4–10 cm) long; [mostly] smooth margin
Habitat: woods, swamps, bogs

© Andrea J Smith/Shutterstock.com

Bunchberry (*Cornus canadensis*)

Family: Dogwood (Cornaceae)
Height: 8–10 in. (20–26 cm)
Flowers: late spring, early summer; "petals," actually 4 modified leaves surrounding greenish-cream true flowers; fruit, clustered red berries in center
Leaves: dull green above shiny beneath; prominent parallel veins; rounded ending in tip; whorled of 6 near top
Habitat: woods, swamps

White

© Sigur/Shutterstock.com

Bearberry or Kinnikinick (***Arctostaphylos uva-ursi***)
Family: Heath (Ericaceae)
Height: up to 12 in. (30 cm)
Flowers: spring, early summer; white or white with pink; oval, constricted at opening then flaring into 5 open lobes; clusters of 5–10 at tip of branches
Leaves: smooth, leathery, shiny at the top; evergreen; woody, creeping
Habitat: dunes, rocky areas

© ChWeiss/Shutterstock.com

Wild Strawberry (*Fragaria virginiana*)
Family: Rose (Rosaceae)
Height: 3–6 in. (8–15 cm)
Flowers: early summer; up to 1 in. (2.5 cm) across; 5 rounded petals; many yellow stamens; separate from leaves; arise near ground; stalks usually shorter than leafstalks; fruit is red berry, fragrant, seeds in sunken pits covering surface (woodland strawberry smaller with raised seeds)
Leaves: compound with 3 blunt-toothed leaflets; stems and underside hairy
Habitat: meadows, dry woods

© Hanaa Ghobrial/Shutterstock.com

Fleabane (*Erigeron strigosus*)
Family: Aster (Asteraceae)
Height: 2–4 ft. (6–12 dm)
Flowers: spring to fall; aster-like; 50–100 "petals" (more numerous than asters)
Leaves: narrow; smooth to toothed margin; basal leaves longer than broad; stem leaves smaller and stalkless; stems sparsely leafy; branches frequently near top; stem and leaves very hairy (especially lower on plant)
Habitat: meadows

© Robert Mutch/Shutterstock.com

False Solomon's Seal (*Smilacina racemosa*)
Family: Lily (Liliaceae)
Height: 16–36 in. (4–6 dm)
Flowers: spring, early summer; clusters of branched, numerous, tiny flowers at tip of stem; fruit, white berry with brown turning red with purple
Leaves: large, flat plane, row on both sides of stem; oval with pointed tip; 4–8 in. (up to 20 cm) long 1–3 in. (up to 8 cm) wide; veins parallel and prominent; leafstalks short; stems arched with hairs
Habitat: woods

© Natalia Rezvova/Shutterstock.com

Smooth Solomon's Seal (*Polygonatum biflorum*)
Family: Lily (Liliaceae)
Height: 18–24 in. (4–6 dm)
Flowers: late spring; tubular, up to 1 in. (20 mm) long; dangles usually in pairs from slender axil stalks
Leaves: alternate, stalkless (sometimes clasping); parallel veins; not hairy; smooth margin; stem unbranched and usually arching
Habitat: woods

© Nikki Yancey/Shutterstock.com

Thimbleberry (*Rubus parviflorus*)
Family: Rose (Rosaceae)
Height: 3–6 ft. (1–2 m)
Flowers: early summer; showy, 1–2 in. (up to 5 cm) across; 5 oval petals; numerous yellow stamens; fruits are thimble-shaped, red, many seeded
Leaves: large, 4–8 in. (up to 20 cm) across; 5 palmate lobes (resembles a maple leaf); coarse, irregularly toothed; veins prominent
Habitat: open woods, meadows

© RealityImages/Shutterstock.com

Field Bindweed (*Convolvulus arvensis*)
Family: Morning Glory (Convolvulaceae)
Height: grows along ground; stems trailing or twining around vegetation; 15 ft. (4.5 m)
Flowers: summer; may be pink or tinged pink; funnel-shaped; 1 in. (2–3 cm) across
Leaves: opposite; slightly downy; arrow-shaped attached to short leafstalk
Habitat: meadows, dunes

© Jeff Holcombe/Shutterstock.com

Partridgeberry (*Mitchella repens*)
Family: Madder (Rubiaceae)
Height: grows along ground; stems creeping (roots may be visible)
Flowers: summer; paired, flaring, tubular with 4 hairy petals on the inside; red berries
Leaves: rounded pairs; ½ in. (12 mm) long; evergreen
Habitat: dry woods

White

© mizy/Shutterstock.com

Ox-eye Daisy (*Chrysanthemum vulgare*)
Family: Aster (Asteraceae)
Height: 1–3 ft. (up to 1 m)
Flowers: summer; showy 2 in. (5 cm) wide, 15–30 "petals" with notched tips; yellow, depressed center disk
Leaves: smooth; sparsely hairy; up to 2 in. (5 cm) long; narrow; dark green; coarsely toothed margins; base leaf clasps stem
Habitat: meadows; nonnative

© Robert Mutch/Shutterstock.com

Cow Parsnip (*Heracleum lanatum*)
Family: Parsley (Umbelliferae)
Height: up to 10 ft. (3 m)
Flowers: summer; flat-topped cluster up to 8 in. (20 cm) across; petals notched deeply at tip; larger petals on outer edge than inner
Leaves: compound up to 18 in. (45 cm) across with 3 coarsely-toothed leaflets; hairy underside; clasp stem; ridged stem and hollow
Habitat: meadows

© Olha Solodenko/Shutterstock.com

Water-hemlock (*Cicuta maculata*)
Family: Parsley (Umbelliferae)
Height: 3–6 ft. (up to 2 m)
Flowers: summer, fall; starburst clusters up to 4 in. (10 cm) across of tiny flowers
Leaves: 2–3 times pinnately compound; long, narrow, sharp pointed leaflets with sawtooth margins; all parts of plants poisonous when eaten
Habitat: meadows, streambanks, swamps

© Martien van Gaalen/Shutterstock.com

Wild Carrot or Queen Anne's Lace **(*Daucus carota*)**
Family: Parsley (Umbelliferae)
Height: 2–3 ft. (6–9 dm)
Flowers: summer, fall; flat-topped cluster of small flowers; some dark flowers may be in the center
Leaves: immediately below flower several 3–5 narrow bracts; leaves deeply and finely cut; stalk bristly; strong carrot odor; skin rash can occur if handled
Habitat: meadows; nonnative

© mizy/Shutterstock.com

Hoary Alyssum (*Berteroa incana*)
Family: Mustard (Brassicaceae)
Height: 1–2 ft. (3–6 dm)
Flowers: summer, early fall; tiny, elongated clustered at tip; 4 petals deeply notched (looks like 8); sepals and stalk hairy
Leaves: stalkless; longer than broad; tapers at ends; margin smooth
Habitat: meadows; nonnative

© Iva Villi/Shutterstock.com

Wintergreen or Checkerberry **(*Gaultheria procumbens*)**
Family: Heath (Ericaceae)
Height: 3–7 in. (8–18 cm)
Flowers: summer; small, waxy, egg-shaped, nodding, single blossom; fruit red, clustered at center
Leaves: thick, oval, shiny, dark green; evergreen
Habitat: dry woods

© Ezume Images/Shutterstock.com

Ghost Plant or Ghost Pipe **(*Monotropa uniflora*)**
Family: Heath (Ericaceae)
Height: 5–10 in. (13–26 cm)
Flowers: summer, early fall; single, nodding, tubular flower with same colored stem
Leaves: barely visible; scale-like; white or pinkish; stem smooth and waxy
Habitat: woods, swamps

© MisterStock/Shutterstock.com

Poison Ivy (*Rhus radicans*)
Family: Cashew (Anacardiaceae)
Height: as an individual plant 2–8 in. (up to 19 cm); may grow up trees or fences as a thick, fibrous vine; mostly forms unified colony over the ground surface
Flowers: early summer; (not pictured) inconspicuous 5 petals clustered in leaf axils; fruits cluster of white, shiny berries
Leaves: 3 leaflets from a common point; oblong with pointed tips; irregular lobes on one or more sides of the leaflets (highly variable shapes); all parts of plant are poisonous to the touch and if burned; "leaflets of three, let it be!"
Habitat: meadows, woods, dunes

White

© Martin Fowler/Shutterstock.com

Water Lily (*Nymphaea tuberosa*)
Family: Water Lily (Nymphaeaceae)
Height: floating; long stalk below water surface
Flowers: concentric rows of tapered petals; larger petals in the outer ring compared to inner; yellow disk at center
Leaves: large, flat, round with deep notched base; underside greenish-purple
Habitat: aquatic

© Anest/Shutterstock.com

White Campion (*Silene latifolia*)
Family: Pink (Caryophyllaceae)
Height: up 48 in. (12 dm)
Flowers: summer, early fall; 5 deeply notched petals displayed at tip of inflated tubular sac (hairy); tend to open in evening
Leaves: opposite; hairy; up to 1.5 in. (4 cm) long; margins entire; stem hollow
Habitat: meadows; nonnnative

© Gerry Bishop/Shutterstock.com

Boneset (*Eupatorium perfoliatum*)
Family: Aster (Asteraceae)
Height: 18 in.–5 ft. (up to 15 dm)
Flowers: summer, fall; flat-topped 9–23 blossoms in each head; individuals tuft-like with no apparent petals
Leaves: opposite pairs; broad at base, surrounds stem (appears stem pierces leaves); hairier on bottom than top; stem hairy
Habitat: swamps, bogs, meadows

© Marta Jonina/Shutterstock.com

Yarrow (*Achillea millefolium*)
Family: Aster (Asteraceae)
Height: 1–3 ft. (3–9 dm)
Flowers: summer, fall; flat-topped cluster of many, small 5-petaled blossoms
Leaves: finely cut divisions; larger leaves at bottom; stems smooth with cottony fuzz; fragrant
Habitat: meadows; nonnative

© Sann von Mai/Shutterstock.com

White Clover (*Trifolium repens*)
Family: Pea (Fabaceae)
Height: 3 in. (7 cm)
Flowers: summer; 0.8 in. (2 cm) heads at end of stalk; visited by bumblebees and honeybees
Leaves: trifoliate, smooth margin, elliptic with long petioles
Habitat: meadow; nonnative

© Robert Biedermann/Shutterstock.com

Garlic Mustard (*Alliaria petiolata*)
Family: Mustard (Brassicaceae)
Height: 12–40 in. (30–100 cm)
Flowers: spring, summer; small, 4 petals arranged in a cross shape; fruit 4-sided slender capsule
Leaves: stalked, triangular to heart shaped 4–6 in. (10–15 cm) long; coarsely toothed margin
Habitat: woods; nonnative, very aggressive and invasive

© Gl0ck/Shutterstock.com

© Arcaion/Shutterstock.com

Japanese Knotweed (*Polygonum cuspidatum*)
Family: Knotweed (Polygonaceae)
Height: 10–13 ft. (3–4 m)
Flowers: late summer; early autumn; small, erect 6 in. (15 cm) long
Leaves: bamboo-like stem (raised nodes); broad, oval, truncated base; margin entire
Habitat: meadows, wetlands; nonnative, very aggressive invasive

Yellow to Orange

2. Yellow to Orange

© Nadezhda Nesterova/Shutterstock.com

Marsh Marigold (*Caltha palustris*)
Family: Crowfoot (Ranunculaceae)
Height: 1–2 ft. (3–6 dm)
Flowers: spring; conspicuous up to 1.5 in. (38 mm) across with 5–9 "petals"
Leaves: large, smooth, shiny, rounded; indented at base; margins saw-toothed; stem short, straight, hollow
Habitat: swamps, streamside

© LifeCollectionPhotography/Shutterstock.com

Buttercup (*Ranunculus acris*)
Family: Crowfoot (Ranunculaceae)
Height: 2–3 ft. (6–9 dm)
Flowers: spring to fall; 5–7 glossy, overlapping petals; bushy stamens prominent in center
Leaves: divided on lower stem into 3 segments; deeply and sharply lobed; upper leaves small; hairy, mostly on underside
Habitat: wet meadows

© Pitofotos/Shutterstock.com

Trout Lily or Adder's Tongue (***Erythronium americanum***)
Family: Lily (Liliaceae)
Height: 6–10 in. (15–26 cm)
Flowers: spring; nodding, one per plant; 6 backward curving petals (can vary); back is purplish-brown; 6 prominent red or yellow anthers beyond bell-shaped head
Leaves: 2 at plant base; mottled brown; plants in colonies with many sterile plants (nonblossoming)
Habitat: woods

© Hana Stepanikova/Shutterstock.com

Dandelion (*Taraxacum officinale*)
Family: Aster (Asteraceae)
Height: 2–18 in. (5–46 cm)
Flowers: spring to fall; 2 in. (5 cm) wide head of many tiny florets at end of erect stalk; seeds form in conspicuous round fuzzy ball
Leaves: basal rosette, deeply cleft and pointed lobes; stem leafless, hollow, white juice when cut
Habitat: meadows; nonnative

© Manfred Ruckszio/Shutterstock.com

Yellow Wood Sorrel (*Oxalis stricta*)
Family: Wood-sorrel (Oxalidaceae)
Height: prostrate up to 18 in. (46 cm)
Flowers: late spring to fall; 5 petals ¼–½ in. (6–13 mm) across; petal base may be reddish; flowerstalk longer than leafstalk
Leaves: clover-like, 3 radiating leaflets from central point; distinct notch at tip
Habitat: meadows

© oroch/Shutterstock.com

Yellow Rocket (*Barbarea vulgaris*)
Family: Mustard (Brassicaceae)
Height: 1–2 ft. (3–6 dm)
Flowers: spring, summer; 4 petals; elongated clusters at tips of flowering branches
Leaves: deeply clefted on lower leaves into one large, terminal lobe and many other small lobes; upper leaves coarsely toothed; base clasps stem
Habitat: meadows; nonnative

© Le Do/Shutterstock.com

Common Cinquefoil (*Potentilla simplex*)
Family: Rose (Rosaceae)
Height: 12–24 in. (30–60 cm)
Flowers: early spring–June; 5 yellow petals
Leaves: 5 leaflets palmately pinnate; prostrate stems that root at nodes; eaten by many caterpillars of Lepidoptera; many species of *Potentilla* with variations of leaves
Habitat: woodlands, fields

© Brian Lasenby/Shutterstock.com

Wood Lily (*Lilium philadelphicum*)
Family: Lily (Liliaceae)
Height: 1–3 ft. (3–9 dm)
Flowers: early summer; cup-like, upward facing; reddish-orange to yellow; 6 petals with purple spots at base; stamens project beyond petals; sometimes >1 bloom per stem
Leaves: 4–8 whorled around stem; long, narrow, pointed; stem white and powdery, upright with flower at tip
Habitat: meadows, dunes

© K Hanley CHDPhoto/Shutterstock.com

© melissamn/Shutterstock.com

Goat's Beard (*Tragopogon dubius*)
Family: Aster (Asteraceae)
Height: 1–3 ft. (3–9 dm)
Flowers: summer, fall; single, yellow at stem tip; outer 5-notched "petals"; flowerstalk inflated below blossom and hollow; seed head large, round, feathery
Leaves: alternate, grass-like, clasps stem; smooth stem, upright, milky juice
Habitat: meadows

© Mark Herreid/Shutterstock.com

Butterfly Weed (*Asclepias tuberosa*)
Family: Milkweed (Asclepiacaceae)
Height: 1–2 ft. (3–6 dm)
Flowers: early summer to fall; clusters at top of stem; bright orange; lower parts turn backwards
Leaves: narrow, long, soft hairs; margins entire
Habitat: meadows, dunes

© Sharon Wills/Shutterstock.com

Black-eyed Susan (*Rudbeckia hirta*)
Family: Aster (Asteraceae)
Height: 1–3 ft. (3–9 dm)
Flowers: summer to fall; large, conspicuous yellow "petals," central disk brown, domed
Leaves: longer than broad, very hairy; stems hairy, erect
Habitat: meadows

© Iva Vagnerova/Shutterstock.com

Orange Hawkweed (*Hieracium aurantiacum*)
Family: Aster (Asteraceae)
Height: 8–24 in. (2–6 dm)
Flowers: summer; orangish to red up to ¾ in. (2 cm) across in clusters at top of stem
Leaves: basal, 1–2 small leaves on stem; very hairy; stem milky juice
Habitat: summer; nonnative

© Angel L/Shutterstock.com

Sundew (*Drosera rotundifolia*)
Family: Sundew (Droseraceae)
Height: up to 9 in. (23 cm)
Flowers: summer; tiny ¼ in. (7 mm) across; 3–15 in loose spike
Leaves: most conspicuous part of plant; basal, round blades on long, flat leafstalks; bristly red hairs with clear, sticky exudate on tips to aid in attracting, trapping, and digesting insects
Habitat: bogs, swamps

© M Rose/Shutterstock.com

Jewelweed or Touch-me-not (***Impatiens capensis***)
Family: Touch-me-not (Balsaminaceae)
Height: 2–5 ft. (6–15 dm)
Flowers: summer; single, dangling, orange blossom, flaring reddish-brown spotted petals; prominent, curling spur at rear; 1 in. (2.5 cm) across; mature seed pods burst when touched
Leaves: smooth, thin, long stalks, oval shaped with rounded teeth
Habitat: wet meadows; streamsides

© dragunov/Shutterstock.com

Yellow Pond Lily (*Nuphar variegatum*)
Family: Water Lily (Nymphaeaceae)
Height: grows just above water surface
Flowers: summer; cup-like yellow blossom; large, yellow disk in center
Leaves: large, floating, narrow notch at base
Habitat: aquatic (lakes, ponds)

© Maren Winter/Shutterstock.com

Evening Primrose (*Oenothera bennis*)
Family: Evening Primrose (Onagraceae)
Height: 2–5 ft. (6–15 dm)
Flowers: summer to fall; 4 broad, yellow petals; cross-shaped stigma in center; opens late in day, wilts next day; 4 turned-back sepals appear as 2
Leaves: 4–8 in. (10–20 cm) long, narrow, stalkless, smooth to slightly hairy; margins smooth to occasional small teeth
Habitat: meadows, dry woods, dunes

© Irina Borsuchenko/Shutterstock.com

Common St. John's Wort (*Hypericum perforatum*)
Family: St. John's Wort (Hypericaceae)
Height: 12–30 in. (3–7.5 dm)
Flowers: summer to early fall; yellow, numerous, 5 petals 1 in. (2.5 cm) across; prominent, bushy stamens in center; black dots along petal margins
Leaves: opposite pairs, stalkless, smooth; stems smooth, branched
Habitat: meadows; nonnative

© simona pavan/Shutterstock.com

Butter-and-Eggs (*Linaria vulgaris*)
Family: Figwort (Scrophulariaceae)
Height: 12–32 in. (3–8 dm)
Flowers: summer to fall; upright spike; yellow with orange marking near center; long-pointed spur at base; 1 in. (2.5 cm) wide
Leaves: numerous, light green, long, narrow, pointed on each end
Habitat: meadows, dunes; nonnative

© fedsax/Shutterstock.com

Common Mullein or Hunter's Toilet Paper (***Verbascum thapsus***)
Family: Figwort (Scrophulariaceae)
Height: 3–6 ft. (9–18 dm)
Flowers: summer to early fall; long, upright spike of yellow 5-rounded petaled blossoms (variable blooms on same plant)
Leaves: thick, woolly, largest at base; stem tall, erect, woolly
Habitat: meadows; nonnative

Horsemint (*Monarda punctate*)
Family: Mint (Lamiaceae)
Height: 1–3 ft. (3–9 dm)
Flowers: summer to fall; yellow with purple, tube-shaped flower, whorled, 2-lipped, >1 per stem
Leaves: up to 3 in. (8 cm) long, narrow, slightly fuzzy, pairs with smaller leaves at base
Habitat: meadows, dunes

Canada Goldenrod (*Solidago canadensis*)
Family: Aster (Asteraceae)
Height: 1–5 ft. (3–15 dm)
Flowers: summer to fall; feathery plume, tiny yellow blooms at stem tip
Leaves: long, narrow, smooth, coarsely saw-toothed; stem downy upper part
Habitat: meadows

Pink to Red

3. Pink to Red

© rsev97/Shutterstock.com

Spring Beauty (*Claytonia virginica*)
Family: Purslane (Portulacaceae)
Height: 3–7 in. (7–18 cm)
Flowers: spring; 5 petals, pink or white with dark pink veins; 3-part style; 2 green sepals; branches off from stem; blooms for 3 days
Leaves: thick, single narrow pair, 1–3 in. (2–8 cm) long, leafstalk definite
Habitat: woods

© Beth Beebe/Shutterstock.com

Wild Columbine (*Aquilegia canadensis*)
Family: Crowfoot (Ranunculaceae)
Height: up to 3 ft. (1 m)
Flowers: spring to early summer; reddish with yellow center; nodding at end of filamentous stem; 5 petals with long tubes extending upward with 5 curving spurs
Leaves: divided basal leaves, and divided into 3 segments; leaflets of 3; stem leaves similar with no stalk and smaller
Habitat: dry woods, meadows, bogs

© jadimages/Shutterstock.com

Pink Lady's Slipper (*Cypripedium acaule*)
Family: Orchid (Orchidaceae)
Height: 8–18 in. (20–46 cm)
Flowers: late spring to early summer; dark to light pink; nodding, inflated pouch (the "slipper"), darker veins
Leaves: 2 broad, basal leaves (sometimes 3); parallel veins; hairy
Habitat: dry woods, swamps, bogs; never pick

© EQRoy/Shutterstock.com

Wild Geranium (*Geranium maculatum*)
Family: Geranium (Geraniaceae)
Height: 1–2 ft. (3–6 dm)
Flowers: spring to early summer; 5 pink, rose, purple, blue, white petals with dark veins (usually); single or few flowers per stem; 1.5 in. (4 cm) across
Leaves: palmately compound, 3–5 lobes; irregular and coarsely toothed at tips; long stalked near base and shorter toward top; stems hairy, erect
Habitat: meadows, woods

© ggw/Shutterstock.com

© Julie Beynon Burnett/Shutterstock.com

© demamiel62/Shutterstock.com

Pitcher Plant (*Sarracenia purpurea*)
Family: Pitcher Plant (Sarraceniaceae)
Height: 12–18 in. (3–5 dm)
Flowers: late spring to early summer; dark reddish-purple; single, nodding; globe-shaped with 5 petals folded inward
Leaves: basal, tubular, hollow; usually filled with water and dead insects; greenish-red with yellow blotches
Habitat: bogs, marshes, swamps

© Ruud Morijn Photographer/Shutterstock.com

Everlasting Sweet Pea (*Lathyrus latifolius*)
Family: Pea (Fabaceae)
Height: up to 7 ft. (2 m); vine
Flowers: late spring, summer; pink (white to purple range); 4–10 showy blossoms clustered near tip; flowerstalk up to 7 in. (18 cm)
Leaves: 2 leaflets; 1- to many-forked tendrils; 2-lobed, uneven stipules
Habitat: meadows; nonnative, invasive

© LFRabanedo/Shutterstock.com

Red Clover (*Trifolium pratense*)
Family: Pea (Fabaceae)
Height: up to 24 in. (6 dm)
Flowers: spring through fall; dark reddish-purple globular bloom at end of branch; color varies to pale pink or white; fragrant
Leaves: palmately compound, 3 leaflets with v-shaped design; leafstalk hairy
Habitat: meadows; nonnative

Pink to Red

© Pavlo Lys/Shutterstock.com

Dame's Rocket (*Hesperis matronalis*)

Family: Mustard (Brassicaceae)
Height: 3–4 ft. (9–12 dm)
Flowers: spring to midsummer; pink, white, or purple open clusters of 4-petaled stalked blossoms 1 in. (2.5 cm) across; hairy sepals; fragrant; fruit is cylindrical capsule; do not confuse for native, wood phlox (*Phlox divaricata*) with 5 petals and opposite leaves
Leaves: alternate, short or no stalks; not indented at base; finely hairy; smooth to fine-toothed margins; blade longer than wide
Habitat: woods edges, meadows; nonnative, invasive

© guentermanaus/Shutterstock.com

Wild Bergamot (*Monarda fistulosa*)

Family: Mint (Lamiaceae)
Height: 2–4 ft. (6–12 dm)
Flowers: summer; cluster of tufted pink to purple blossoms at stem tip
Leaves: paired, narrow, smooth to slightly hairy, margins toothed
Habitat: meadows

© Ihor Hvozdetskyi/Shutterstock.com

Common Milkweed (*Asclepias syriaca*)

Family: Milkweed (Apocynaceae)
Height: 2–4 ft. (8–12 dm)
Flowers: summer; light pink to purple; spherical flower cluster; fragrant; seed pods long pointed, bumpy, tipped with long, silk-like hairs
Leaves: thick, 4–6 in. (10–15 cm) oblong; hairy underside; milky juice when stems cut
Habitat: meadows, dunes

© Evannovostro/Shutterstock.com

Teasel (*Dipsacus sylvestris*)

Family: Teasel (Dipsacaceae)
Height: 2–6 ft. (6–18 dm)
Flowers: summer to fall; numerous, tiny, pink (or lavender) tubular florets with sharp pointed bracts between each; seed heads persist through winter
Leaves: long and wide at base, paired and sheath stems; scattered spines; stems rigid, upright, grooved, spiny
Habitat: meadows; nonnative

© PhotoChur/Shutterstock.com

Bull Thistle (*Cirsium vulgare*)
Family: Aster (Asteraceae)
Height: up to 7 ft. (2 m)
Flowers: summer to fall; single (up to 3) heads; pink to purple; 2 in. (5 cm) across; stiff; yellow tipped spines over bracts
Leaves: deeply cut; coarsely toothed; sharp pointed spine at tip; margins with short spines; woolly underside; stem coarse and prickly
Habitat: meadows; nonnative

© Stephen Bonk/Shutterstock.com

Joe Pye Weed (*Eupatorium maculatum*)
Family: Aster (Asteraceae)
Height: up to 6 ft. (18 dm)
Flowers: summer to fall; flat-topped clusters; pinkish-purple; 8–20 blossoms per head
Leaves: whorls of 4–5 long, narrow, pointed at both ends; veined; sawtooth margin
Habitat: swamps, streamsides, bogs

© James W. Thompson/Shutterstock.com

Spotted Knapweed (*Centaurea maculosa*)
Family: Aster (Asteraceae)
Height: 1–4 ft. (3–12 dm)
Flowers: summer to fall; shaggy pinkish-purple blossom at tip of stem; consists of small individual, tubular flowers; overlapping, black-tipped bracts underside
Leaves: deeply cut, rough, grayish-green; few; stems rough and grayish-green
Habitat: meadows; nonnative, invasive

© Nikolay Kurzenko/Shutterstock.com

Cardinal Flower (*Lobelia cardinalis*)
Family: Lobelia (Lobeliaceae)
Height: 1–5 ft. (4–15 dm)
Flowers: summer to early fall; bright red blossoms on upright spike; up to 2 in. (5 cm) long
Leaves: thin, long, narrow; 6 in. (15 cm) long; upper leaves small; small toothed margins; stems may be slightly hairy
Habitat: meadows, wet margins

4. Lavender to Purple

© Michael Benard/Shutterstock.com

Jack-in-the-Pulpit (*Arisaema triphyllum*)
Family: Arum (Araceae)
Height: 1–3 ft. (3–9 dm)
Flowers: spring; tiny, inconspicuous; purplish-green, brown canopy and covering surround flowers
Leaves: 1–3, divided into 3 leaflets at top of upright; stem
Habitat: woods, swamps

© Evgeniy Eivo/Shutterstock.com

Hairy Vetch (*Vicia villosa*)
Family: Pea (Fabaceae)
Height: 2–3 ft. (6–9 dm)
Flowers: late spring, summer; tiny pea flowers in tight clusters on one side of spike; lavender to violet
Leaves: pinnately compound, 10–20 leaflets; fuzzy; thread-like tendrils; stems upright to reclined, mostly trailing, fuzzy
Habitat: meadows

© Tikhomirov Sergey/Shutterstock.com

Self-Heal or Heal-All **(*Prunella vulgaris*)**
Family: Mint (Lamiaceae)
Height: up to 12 in. (3 dm)
Flowers: spring through fall; variable colors of purple; 2-lipped with upper hood-shaped and lower with 3 lobes; arise from between green, leafy bracts at the top of stem
Leaves: opposite, oval and tipped, finely-toothed, entire; stems 4-sided, erect, slightly fuzzy
Habitat: meadows

© ID1974/Shutterstock.com

Blazing Star (*Liatris spicata*)
Family: Aster (Asteraceae)
Height: up to 6 ft. (1.8 m)
Flowers: late summer; many clusters of feathery blooms along spike
Leaves: long lower leaves, size decreases toward top; very narrow; smooth margin
Habitat: meadows, streambanks

Lavender to Purple

© Chris Hill/Shutterstock.com

© Milla V Krivdina/Shutterstock.com

Common Burdock (*Arctium minus*)

Family: Aster (Asteracea)
Height: 2–6 ft. (6–18 dm)
Flowers: summer to fall; purplish tubular florets at tip of green, spiny bur; burs cling to clothing
Leaves: large lower leaves 1 ft. (30 cm); upper heart-shaped without notch; leafstalks hollow; margins wavy to finely toothed
Habitat: meadows

© Peter Turner Photography/Shutterstock.com

New England Aster (*Symphyotrichum novae-angliae*)

Family: Aster (Asteraceae)
Height: 3–7 ft. (9–21 dm)
Flowers: late summer to fall; lavender to purple; showy (1–2 in.; 2.5–5 cm); 45–100 "petals"; head somewhat sticky
Leaves: narrow, entire, smooth margin; clasps stem
Habitat: meadows, marshes

© Flower_Garden/Shutterstock.com

Purple Loosestrife (*Lythrum salicaria*)

Family: Loosestrife (Lythraceae)
Height: 2–5 ft. (6–15 dm)
Flowers: summer to fall; purple flowers on spike; 6 petals
Leaves: narrow, up to 4 in. (10 cm); usually opposite; stems upright, smooth to fuzzy
Habitat: swamps, streamsides; nonnative, aggressive invasive

© dabjola/Shutterstock.com

Nightshade (*Solanum dulcamara*)
Family: Nightshade (Solanaceae)
Height: 2–12 ft. (6 dm-4 m)
Flowers: summer to fall; 5 purple to blue petals (backward); yellow center cone protruding; fruit green, then red cluster; toxic
Leaves: 3 lobed, oval with center leaflet largest (varies on stem); stems woody at base, hairy, weak, vine typical
Habitat: meadows, swamps

© Elena Elisseeva/Shutterstock.com

Purple Coneflower (*Echinacea purpurea*)
Family: Aster (Asteraceae)
Height: 2–5 ft. (6–15 dm)
Flowers: summer to fall; reddish purple rays drooping (7–40); showy up to 4 in. (10 cm); bristly central orangish disk
Leaves: toothed, egg-shaped, long stalked lower on stem
Habitat: meadow

© Curioso.Photography/Shutterstock.com

Phragmites or Common Reed (***Phragmites australis***)
Family: Grass (Poaceae)
Height: 6–13 ft. (2–4.5 m)
Flowers: summer to fall; dense, branched clusters at end of stem; feathery deep purple at maturity
Leaves: flat, smooth, green to grayish; hollow stem
Habitat: wetlands; nonnative, aggressive invasive

5. Blue

© Kimberly Boyles/Shutterstock.com

Common Blue Violet (*Viola papilionacea*)
Family: Violet (Violaceae)
Height: 3–8 in. (8–20 cm)
Flowers: spring; 5 petaled blue to dark purple; lower petal extends backward into spur; 2 side petals bearded; dark colored veins
Leaves: heart-shaped; course-toothed margin; "stemless"
Habitat: meadows, woods

© Anton Zhuk/Shutterstock.com

Periwinkle or Myrtle **(*Vinca minor*)**
Family: Dogbane (Apocynaceae)
Height: trailing plant, ground hugging; 1–2 in. (2.5–5 cm)
Flowers: spring; blue-violet, funnel-shaped, 5 flaring lobes
Leaves: evergreen, shiny, dark green, oval, pointed
Habitat: meadows, woods; nonnative, invasive

© Shebeko/Shutterstock.com

Wild Lupine (*Lupinus perennis*)
Family: Pea (Fabaceae)
Height: 1–2 ft. (3–6 dm)
Flowers: spring to early summer; petals blue to violet or pinkish-white; short stalks from central stem (up to 10 in.; 25 cm) with many blossoms; fruit in hairy pods
Leaves: palmately compound, 7–11 leaflets (2 in.; 5 cm) long; stems erect, fuzzy, branched
Habitat: meadows

© Carmen Rieb/Shutterstock.com

Blue Flag Iris (*Iris versicolor*)
Family: Iris (Iridaceae)
Height: 2–3 ft. (6–9 dm)
Flowers: late spring to early summer; blue petals with darkened veins stand upright and sepals of same color and pattern are longer and open downward; base yellowish
Leaves: long, narrow, sword-like; mainly at base of plant
Habitat: swamps

Blue

© Miroslav Hlavko/ Shutterstock.com

Ground Ivy (*Glechoma hederacea)*
Family: Mint (Lamiaceae)
Height: 2–20 in. (5–50 cm)
Flowers: spring through summer; blue, funnel shaped; opposite clusters of 2–3 flowers bilaterally symmetrical
Leaves: opposite; rounded, lobed; attached to stem at center; hairy topside; stems square
Habitat: meadows, woods; nonnative, invasive

© Natalya Erofeeva/Shutterstock.com

Forget-me-not (*Myosotis scorpioides*)
Family: Borage (Boraginaceae)
Height: 6–24 in. (1.5–6 dm)
Flowers: spring to fall; small (1.4 in.; 6 mm) pale blue, 5 petals, center yellow and star-like
Leaves: alternate; longer than broad, hairy; stems hairy, upright then prostrate
Habitat: moist areas; nonnative

© marineke thissen/Shutterstock.com

Pickerelweed (*Pontederia cordata*)
Family: Pickerelweed (Pontederiaceae)
Height: 1–3 ft. (3–9 dm)
Flowers: summer to fall; blue, tiny flowers on spikes (4 in.; 10 cm) long; 2-lipped petals with 3 lobes
Leaves: heart-shaped, broad, smooth margin
Habitat: ponds, shallow water

© Dudakova Elena/Shutterstock.com

Chicory (*Cichorium intybus*)
Family: Aster (Asteraceae)
Height: up to 5 ft. (15 dm)
Flowers: summer to fall; showy blue blossoms, petals blunt and fringed at tip; stalkless
Leaves: jagged, sharp-pointed lobes lower leaves; stem leaves few and small; hairy; clasps stem; stem tough, shallow grooved with few spines
Habitat: meadows; nonnnative

© Jeff Holcombe/ Shutterstock.com

Blue Vervain (*Verbena hastata*)
Family: Vervain (Verbenaceae)
Height: 18 in.–5 ft. (4.5–15 dm)
Flowers: summer to early fall; many branched flower head of spikes with clusters of small, blue, 5-petaled flowers; flowering begins at base of spikes then progresses toward tips
Leaves: opposite; long, narrow, pointed, toothed margins; stems 4-sided, hairy
Habitat: meadows, marshes

6. Green and Brown

© Sara Koivisto/ Shutterstock.com

© Andrew Sabai/ Shutterstock.com

Skunk Cabbage (*Symplocarpus foetidus*)
Family: Arum (Araceae)
Height: up to 5 in. (13 cm)
Flowers: early spring (can emerge while snow-covered); stemless, ground level; mottled brown to greenish-purple hood with rolled edges; center knob with tiny, yellow flowers
Leaves: leaves develop may or may not be visible during flowering; later season, very large up to 2 ft. (6 dm) long, somewhat heart-shaped; margins smooth or wavy
Habitat: swamps

© Adam Jan Figel/Shutterstock.com

Stinging Nettle or Common Nettle or Nettle Leaf **(*Urtica dioica*)**
Family: Nettle (Urticaceae)
Height: 3–7 ft. (1–2 m)
Flowers: spring to fall; small, greenish-brown; numerous in dense axillary inflorescences
Leaves: narrow heart-shape; 1–6 in. (3–15 cm); very hairy; strongly serrated margin; plant stem is erect, green, hairy; stinging hairs may come off when touched and inject several chemicals causing a painful sting
Habitat: disturbed land; near buildings; understory in wet environments; meadows

© Bildagentur Zoonar GmbH/ Shutterstock.com

Curled Dock (*Rumex crispus*)
Family: Smartweed (Polygonaceae)
Height: 1–4 ft. (3–12 dm)
Flowers: summer to fall; small greenish-brown, inconspicuous, whorls on stalks
Leaves: long (up to 1 ft.; 30 cm), strongly veined; wavy margin
Habitat: meadows

© Hachi888/Shutterstock.com

Common Cattail or Bulrush **(*Typha latifolia*)**
Family: Cattail (Typhaceae)
Height: 3–9 ft. (1–3 m)
Flowers: spring to summer; compact cluster at tip, brown flowers with no petals or sepals; female separate from male flowers—bottom and top of each other, respectively, with no gap
Leaves: sword-like, extending up the length of plant, ½ in. (1.27 cm); flat, bases clasp stems; nonnative, narrow-leaved cattail (*T. angustifolia*) not as tall and leaves thinner with a prominent gap between flower types
Habitat: wet open areas

© IvanaStevanoski/Shutterstock.com

Bur Reed (*Sparganium* spp.)
Family: Bur Reed (Typhaceae)
Height: up to 5 ft. (1.5 m)
Flowers: summer; ball-like head of small greenish to white flowers
Leaves: very narrow, up to 1 ft. (30 cm) long
Habitat: wet areas

© Mala Iryna/Shutterstock.com

Ragweed (*Ambrosia artemisiifolia*)
Family: Aster (Asteraceae)
Height: up to 4 ft. (1.2 m)
Flowers: summer to fall; inconspicuous, green; males on spikes at stem tips, flat, downward, heavy with yellow pollen; females fewer, below male in leaf axil
Leaves: deeply cut, many lobes with deep cuts; margin smooth; stem hairy
Habitat: meadows

© Oksana Shevchenko/ Shutterstock.com

Leafy Spurge (*Euphorbia esula*)
Family: Spurge (Euphorbiaceae)
Height: 1–2 ft. (3–6 dm)
Flowers: spring to fall; small, greenish-yellow, inconspicuous but surrounded by prominent yellowish-green rounded bract with pointed tips
Leaves: long, narrow, smooth margin; scattered on lower stem but numerous toward top; stem smooth
Habitat: meadows, dunes; nonnative

6.8 Field Guide: Identifying Trees

© Andrei Kim/Shutterstock.com

Activity: Can you recognize and name at least five trees? To help you get started, spend time getting close with trees around you and just ask yourself some questions about those trees to help build your naturalist intelligence. In the beginning, do not worry about trying to identify the species; just focus on familiarizing yourself with the characteristics and how you would describe them. Use this section to learn the types of questions to ask when presented with a tree for identification.

Questions to Ask Yourself

When you meet a tree, can you determine if the tree is a conifer or deciduous tree? A **conifer** is cone bearing and typically evergreen or has needle-like leaves, whereas a **deciduous** tree usually has broad leaves and will shed them in the fall. Now, can you recognize the difference between the two types of trees? This is the first question the naturalist asks, and this will become second nature once you familiarize yourself with trees.

Conifers

Find a **conifer** nearby—any will do. Observe the following features about the tree:

- *Silhouette:* What is the overall shape of the tree? Is it rounded or more pyramid shaped?
- *Bark:* What is the color and texture of the tree bark? Mature bark is a truer texture of the species than young bark.
- *Leaves:* If possible, examine the tree's needles (technically, modified leaves). How would you describe them? Single needles attached to the twig? Or are they clustered at the base of the needles and attached together on the woody twig? Are they round, square, or flat in cross-section? Or is the needle arrangement simply flat and scaly?
    - Spruces: single needles attached directly to the twig; square in cross-section
    - Pines: clustered in groups on the twig 2–5 per bundle
    - Firs: single needles arranged on top and sides of twig; flat
    - Hemlock: single arrangement on twig, short, and flat
    - Cedars or juniper: flat and scaly
- *Cones:* Are there any cones growing on the tree? How about cones that have fallen to the ground around the tree? Describe their size and shape.

Deciduous Trees

Find a nearby **deciduous** tree. Observe these characteristics of the tree:

- *Silhouette:* Notice the **silhouette**, or overall shape of the tree. Is it round or oval?
- *Bark:* Describe the color and texture of the bark (especially on mature trees). This can serve as a quick way to identify the tree, especially in the winter or if the branches are too high. Some trees have very unique bark.

- *Branching Pattern:* This is an important feature in the identification of the tree, and can be determined in any season. Look closely at the branching pattern of the twigs, buds, leaves, and leaf scars. Do these features exist directly across from each other along the length of the woody twig or branches (opposite branching)? Or do they look offset down its length (alternate branching)?
  - More alternate branching species exist than opposite
  - Learn the general leaf shapes of opposite branching trees for quick identification; all are very different from one another; draw each leaf type to help you learn and remember the tree types
    - To remember opposite branching trees, remember this acronym (or make up your own): MADCapHorseCat
      - **M**aple—palmately simple leaf
      - **A**sh—pinnately compound leaf
      - **D**ogwood—simple, entire leaf with parallel veins to the margin
      - **Cap**rifoliacea—many shrubs; Honeysuckle Family—simple/compound, entire/palmate, berries; vibernums
      - **Horse**chestnut—palmately compound
      - **Cat**alpa—simple, entire (large heart-shape)
- *Leaves:* If leaves are present, what is the shape? Is there a single leaf attached to the woody twig? Or does there seem to be a network of leaflets connected to it? For terminology, refer to the *Botantical Terms Chart* prior to the *Wildflower Species Accounts*.
- *Other Features:* Does the tree have any other identifying characteristics? Like flowers, seeds, or fruits? How about the presence of thorns or hair-like structures? Are the buds visible? Describe all of these features.
- *Common Deciduous Trees:* The six most common deciduous trees in the region are maple, ash, oak, beech, birch, and aspen. Unfortunately, non-native invasive insects are infesting or threatening many of these trees. Research the invasive species in your nearby forests.

Nature Journal: Discover the woody plants in your Nature Area. What do you observe about the conifers and deciduous trees? Use a magnifying glass to study the buds up close and examine the twig features, and then use binoculars to look high on the branches to make observations. Start with the most common tree at your site to learn everything you can about it, then the second most abundant, or choose a tree of interest to study. Use the Meeting-a-Tree Chart on the following page and the Natural History Journal Pages at the end of this chapter to help guide your study of trees.

After spending time getting acquainted with the trees nearby, learn to identify those trees based on your observations—and recognize them by name. To use this section efficiently and make proper identifications, it is imperative to make detailed observations of key botanical features. Knowing plant and tree anatomy along with its corresponding terminology will help. Use the diagrams here and throughout Chapter 6 to help you learn the features and terms quickly.

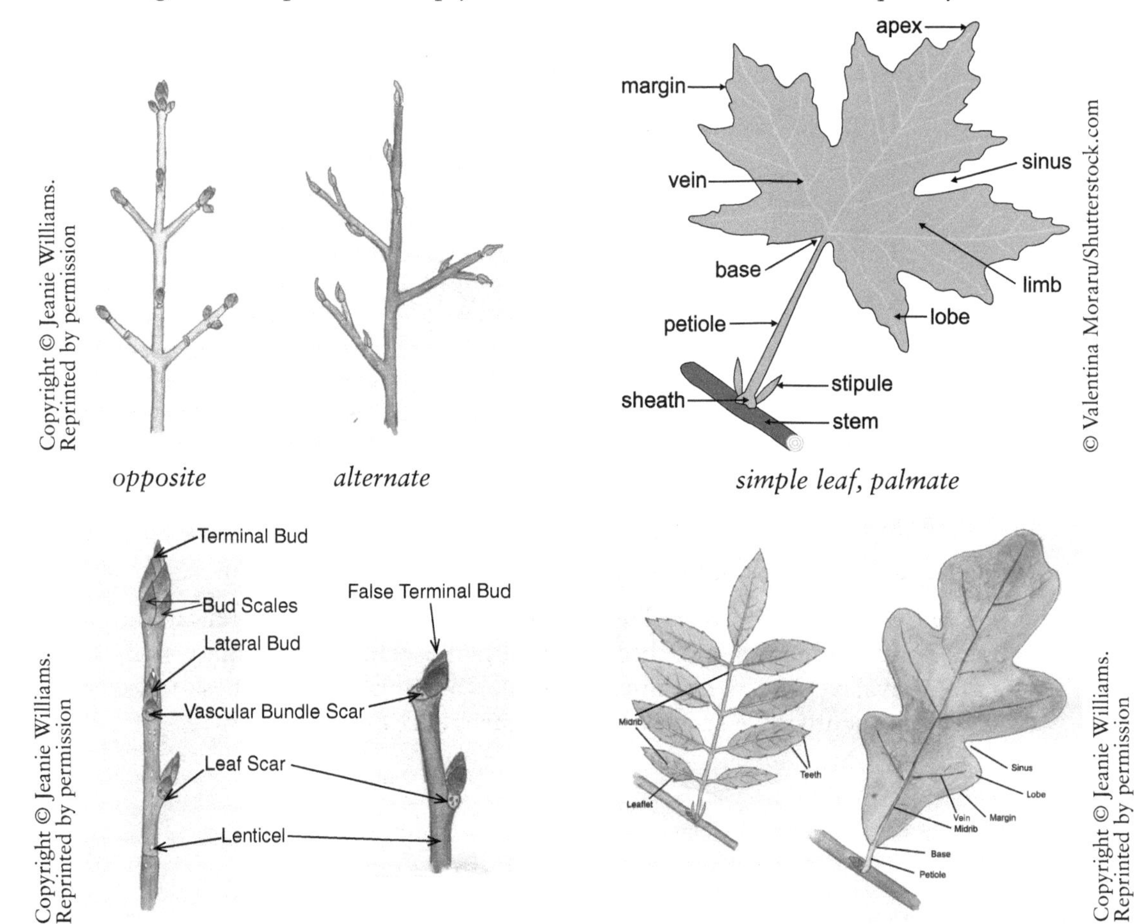

opposite *alternate*

simple leaf, palmate

compound leaf, pinnate *simple leaf, pinnate*

Meeting-a-Tree Chart

Completing this tree chart will lead you to the trees' identification.

Spend time making observations of the trees in your Nature Area and other places where trees catch your attention. Making good observations will make identification easier. Fill in the information on the chart below to help you learn the trees. ***When presented with a tree to identify, consider all these characteristics.*** *Use the field guide that follows to help learn its scientific name. Other books, the internet, or tree identification apps can also help. Some trees used in landscaping are cultivars and may be difficult to identify to species.*

Is your tree a Conifer (C)/ Deciduous tree (D)?	**If C, are needles in clusters?/ If D, what is the branching pattern (opposite/alternate)?**	**If C, are needles single, or how many needles per cluster, or are they flat and scaly?/ If D, leaves simple or compound?**	**Draw needle for C / or leaf for D (include size and point out key characteristics)**	**Describe Bark (i.e., color, texture, draw pattern if relevant, etc.)**	**Draw/describe cones (C), seeds (C/D), buds (D), flowers (D)**	**Full Common Name (i.e., Eastern White Pine)**	**Other notes (page no., habitat, location, height, overall tree shape, tips for identifying, fun fact, etc.)**
1.							
2.							
3.							
4.							
5.							

Tree Identification Keys

As a naturalist, you will likely start to collect field guides to help with making species identifications and to learn more about natural histories. You will discover that each field guide will have its own method for displaying and organizing the information. You might find a pictorial guide, descriptive keys for orders or families, perhaps a guide that is simply alphabetical, or sorted by habitat, or it could have a dichotomous key.

A dichotomy indicates a decision needing to be made. In nature study, a **dichotomous key** is an efficient tool designed to distinguish the differences among a set of organisms, in this case, a group of trees. The key separates trees into various categories, based on physical characteristics, until there are only two species remaining—similar to a process of elimination.

The use of keys requires careful observation and usually involves the learning of some basic terminology learned in the previous pages or elsewhere. At first, patience is often needed until you become proficient at using the key and using the field guide section of this book. Using dichotomous keys forces you to look closely at your organism in question, and ultimately you gain further knowledge about that species.

How to Use a Dichotomous Key. *Start at number one and read both options. From there, go to the number that pertained to your specimen, and continue the same process of reading both options for the number and going where you were directed to until you have a name for your tree. Find the tree's picture and description in the field guide to determine if you made an accurate identification. Realize that not all organisms in the region will be found in this guide, but you can hopefully determine the genus or family of your specimen in question.*

Which Tree Key Should I Use?

Start here to get oriented to using the appropriate key for identifying your woody plant. Ask yourself the following questions to determine which section to search for your tree or shrub in question. Will it be coniferous or deciduous?

Use this dichotomous key to help you determine what to do next. Is your specimen in question a:

1a. Coniferous tree or shrub - the leaves are needlelike (2)

1b. Deciduous tree or shrub - the leaves are flat and broad (3)

2a. Use the corresponding conifer tree key after this section to help you identify your specimen to the species level.

2b. Flip through the first section of the woody plants to view examples and descriptions of conifers.

3a. Use the dichotomous keys provided if your tree or shrub currently has leaves. The deciduous tree keys are separated by the tree's branching pattern. This approach will at least take your search to the family level, and in some cases the genus and species (4)

3b. If no leaves are present, determine the branching pattern of the deciduous tree to search in the appropriate section of the woody plants descriptions (opposite or alternate).

4a. Look at the point of leaf attachment, leaf scars, or buds down the length of the twig. Use the appropriate key for *opposite* branching.

4b. Look at the point of leaf attachment, leaf scars, or buds down the length of the twig. Use the appropriate key for *alternate* branching.

Go to the section on tree and shrub descriptions to read further about your potential specimen.

Coniferous Tree Key

Use this key for conifers only!

(conifer, evergreen, needleleaf)

1a. Needles in bundles, clusters, or groups (2)

1b. Needles single; or flattened and scaly (3)

2a. Needles in clusters of more than 5 needles (Tamarack, *Larix laricina*)

2b. Needles 2 to 5 per bundle (Family Pinaceae; Pine species, *Pinus* spp.; see options a–d)
   a. Five needles per bundle (Eastern White Pine, *Pinus strobus*)
   b. Needles in pairs, 4–7 in., brittle; bark reddish, blocky (Red Pine, *Pinus resinosa*)

c. Needles in pairs, 1.5–3 in., spirally twisted; bark orange, peely near top (Scotch Pine, *Pinus sylvestrus*)
d. Needles in pairs, ¾–1.5 in.; straight to slightly twisted (Jack Pine, *Pinus banksiana*)

3a. Needles not single or bundled, but are scaly and flattened (4)

3b. Needles single on the twig (5)

4a. Has cones; needles are scaly, flat, and fan-like; bark has narrow vertical strips (Family Cupressaceae Northern White Cedar, *Thuja occidentalis*)

4b. Has berries; needles may be scaly and prickly, scaly and rounded (Family Cupressaceae; Eastern Red Cedar, *Juniperus virginiana*)

5a. Needles single; flat (6)

5b. Needles single, square in cross-section, stiff, sharp (Family Pinaceae; Spruce species, *Picea* spp.; see options a–d)
a. Needles ½–¾ in. long; twigs hairless (White Spruce, *Picea glauca*)
b. Needles ½–¾ in. long; twigs hair; columnar shaped tree with dense crown (Black Spruce, *Picea mariana*)
c. Needles ¾–1.25 in. long, blue-green to silver-green and sharply incurved (Blue or Colorado Spruce, *Picea pungens*)
d. Needles 1–1.5 in. long; twigs bright orange-brown; numerous drooping branches (Norway Spruce, *Picea abies*)

6a. Needles ½ in. long with short petiole (stem) (Family Pinaceae; Eastern Hemlock, *Tsuga canadensis*)

6b. Needles ¾ in. to 1.25 in. long, no petiole; bubbles in bark (Family Pinaceae; Balsam Fir, *Abies balsamea*)

Deciduous Tree Keys (Broadleaf Trees and Shrubs)

OPPOSITE Branching

1a. Leaves simple (2)

1b. Leaves compound (3)

2a. Leaves palmate, 3–5 lobes (Family Sapindaceae—Maples, *Acer* spp.)

2b. Leaves entire (4)

3a. Palmate, 5–7 leaflets (Family Sapindaceae—Horsechestnut, *Aesculus hipposcastanum*)

3b. Pinnately arranged leaflets (5)

4a. Leaves large, heart-shaped (Family Bignoniaceaea—Northern Catalpa, *Catalpa speciose*)

4b. Shrub-sized tree (under 16 ft., 5 m; multiple trunks) (6)

5a. Leaflets 7–11, single-winged samaras (Family Oleaceae—Ashes, *Fraxinus* spp.)

5b. Leaflets 3–9 (7)

6a. Leaf veins parallel to margin (Family Cornaceae—Dogwoods, *Cornus* spp.)

6b. Leaves simple, entire (mostly), berry-like fruits
Family Caprifoliaceae—Honeysuckles, *Lonicera* spp.; Elders, *Sambucus* spp.; Viburnums, *Viburnum* spp.;
Family Rubiaceae—Buttonbush, *Cephalanthus occidentalis*

7a. Leaflets 3–5 (sometimes 7–9), double-winged samaras (Family Sapindaceae—Boxelder, *Acer negundo*)

7b. Leaflets 3, lantern-shaped fruit (Family Staphylaceae—American Bladdernut, *Staphylea trifolia*)

ALTERNATE Branching

1a. Leaves simple (2)

1b. Leaves compound (5)

2a. Leaves lobed (3)

2b. Leaves entire (4)

3a. Leaves pinnately lobed
Family Fagaceae—Oaks, *Quercus* spp.;
Family Magnoliaceae—Tulip-tree, *Liriodendron tulipifera*

3b. Leaves palmately lobed (Family Platanaceae; American Sycamore, *Platanus occidentalis*)

4a. Tree size (single trunk typical; >16 ft., 5 m)
Family Fagaceae—Beech, *Fagus* sp.; Alder, *Alnus* spp.; Birch, *Betula* spp.; Hornbeam, *Carpinus* sp.; Hophornbeam, *Ostriya* sp.
Family Rosaceae—Cherry, *Prunus* spp.; Hawthorn, *Crataegus* spp.
Family Salicaceae—Willow, *Salix* spp.; Poplar, *Populus* spp.
Family Tiliaceae—American Basswood, *Tilia americana*
Family Ulmaceae—Elm, *Ulmus* spp.

4b. Shrub size (multiple trunks, branching; <16 ft., 5 m)
Family Elaeagnaceae—Autumn Olive, *Elaeagnus umbellata*
Family Fabaceae—Redbud, *Cercis canadensis*
Family Hamamelidaceae—American Witch Hazel, *Hamamelis virginiana*
Family Lauraceae—Sassafras, *Sassafras albidum*; Spicebush, *Lindera benzoin*
Family Rosaceae—Serviceberry, *Amelanchier* spp.; Crabapple, *Malus* spp.; Multiflora rose, *Rosa multiflora*
Family Rhamnaceae—Glossy Buckthorn, *Frangula alnus*

5a. Leaflets rounded
Family Fabaceae—Black Locust, *Robinia pseudoacacia*; Honey-Locust, *Gleditsia triacanthos*

5b. Leaflets pointed
Family Anacardiaceae—Staghorn Sumac, *Rhus typhina*; Poison Sumac, *Toxicodendron vernix*
Family Juglandaceae—Black Walnut, *Juglans nigra*; Shagbark Hickory, *Carya ovata*
Family Rosaceae—American Mountain-Ash, *Sorbus americana*
Family Simaroubaceae—Tree-of-Heaven, *Ailanthus altissima*

Deciduous Trees in WINTER

Some might say it is easier to identify trees in winter because the leaves are not obstructing the view of the branches. Examining the twigs and bark will lead to an accurate identification of your tree. Refer to the *Questions to Ask Yourself* section on *Deciduous Trees* prior to the *Tree Identification Keys.*

Although an explicit key for winter tree identification is beyond what this book shares, here is a sampling of some common deciduous tree twigs found in the region that might help you narrow down your identification. Search for online resources to assist you further.

Twigs in Winter. Compare and contrast the buds and other features on the twigs. Note the position, color, and shape. The first three deciduous twigs are opposite branching (white ash, red maple, and sugar maple), whereas the last six exhibit an alternate arrangement (aspen, yellow birch, paper birch, American beech, black cherry, and red oak). Knowing the branching pattern will narrow your search for the tree type.

Tree Bark. Another way to botanize in the winter (or anytime, really), is to know common bark types. You can find real satisfaction from recognizing trees from a distance. Realize that young trees do not always display the textbook bark features as they do in maturity. Refer to these textures to refine your tree identification searches in the field guide.

Smooth: American beech, musclewood, quaking aspen
Papery: paper birch, yellow birch, river birch; scotch pine (peeling toward top)
Vertical Strips: cedars, sugar maple, red maple, silver maple, shagbark hickory, ironwood
Flakes: black cherry, American sycamore
Blockey: red pine, white pine (mature)
Lenticels: chokecherry, locust
Furrows: ash, oak

Source: C. J. Fitzwilliams-Heck
American Beech

Source: C. J. Fitzwilliams-Heck
Musclewood

Source: C. J. Fitzwilliams-Heck
Quaking Aspen

Source: C. J. Fitzwilliams-Heck
Paper Birch

Yellow Birch

River Birch

Scotch Pine

Northern White Cedar

Sugar Maple

Shagbark Hickory

Ironwood

Black Cherry

Source: C. J. Fitzwilliams-Heck

American Sycamore

Source: C. J. Fitzwilliams-Heck

Red Pine

Source: C. J. Fitzwilliams-Heck

White Pine

Source: Jeanie Williams

Chokecherry

Source: Jeanie Williams

White Ash

6.9 Field Guide: Trees and Shrubs Species Accounts

The trees found in this section are common, or of special interest or concern, within the Great Lakes basin and Upper Midwest region. Take notes on key characteristics such as leaf or needle shape, branching pattern (deciduous trees), bark, and other identifying features—as you did with the Meeting a Tree chart activity. Some further research online or in other books may be necessary. This is not a comprehensive collection of trees you may encounter. However, studying this chapter is the beginning of a better understanding of trees, forests, and natural resources.

Using this section. Groups of trees and shrubs are classified into conifers and deciduous, and the broadleaf group is further divided into opposite then alternate branching patterns. The species are organized by alphabetized family name, and identifying characteristics have been provided. If your species does not appear in this guide, then you may have a hybrid or cultivated species, or it simply did not make it to the book. By determining its family or genus with this field guide, you will have an easier time searching its identification in another resource. Nonnative species have been identified where applicable.

Conifers (Evergreen; Needleleaf)

Pine Family (Pinaceae)

© Jeff Holcombe/Shutterstock.com

Larch Genus (*Larix*)

Tamarack (*Larix laricina*)
Height: up to 80 ft. (24 m)
Needles: tufts of 12 or more 1 in. (2.5 cm) near end of branches; bright blue-green; deciduous
Key Features: newly exposed bark reddish-purple; cones yellow-brown, papery
Habitat: swamps, bogs, fens; moist, well-drained uplands
Notes: eaten by many wildlife; porcupines often kill it for the sweet inner bark

© Aleksander Bolbot/Shutterstock.com

© Marie C Fields/Shutterstock.com

Pine Genus (*Pinus*)

Eastern White Pine (*Pinus strobus*)

Height: up to 100 ft. (30 m)
Needles: bundles of 5; 2–5 in. long; soft, flexible
Key Features: bark deeply cracked, brown when mature; female cones 8 in. (20 cm), cylindrical, resinous, and male mid-crown, small, yellow, clustered at branch tip
Habitat: well-drained soil; sandy loam; rock ridges; bogs
Notes: saplings can persist in understory for 20 years

© Craig Hinton/Shutterstock.com

Red Pine (*Pinus resinosa*)

Height: up to 70 ft. (21 m)
Needles: bundles of 2; 6 in. (15 cm) long; brittle (snaps easily)
Key Features: bark red-brown large, flat plates or blocks (mature), flaky (young); female cones woody, open 2 in. (5 cm) and male small, granular, clustered at branch tip
Habitat: sandy soil; rocky slopes; dry to moist areas with sun; pine plantations
Notes: seeds favored by songbirds, red squirrels harvest tree cones, other small mammals gather ground seeds

© LFRabanedo/Shutterstock.com

© Craig Hinton/Shutterstock.com

Scotch Pine (*Pinus sylvestris*)

Height: up to 60 ft. (18 m)
Needles: bundles of 2, twisted; 2 in. (5 cm) dark blue-green that fades in winter
Key Features: bark bright orange, peeling in strips; female cones 2–3 and male small, yellow clustered at tip of new branches
Habitat: upland sites; sandy loam; tree plantations
Notes: nonnative; can be aggressive in sandy, acidic soils

© Erik Agar/Shutterstock.com

Jack Pine (*Pinus banksiana*)

Height: up to 70 ft. (21 m)
Needles: bundles of 2; 1 in. (2.5 cm) yellowish-green; slightly twisted widely forked
Key Features: curved female cones grow laterally along branches pointing toward branch tip, scales serotinous (sealed with resin)
Habitat: dry, sandy or rocky glacial outwash plains to moist soil
Notes: mature cones remain closed on tree for many until heat from fire or sunlight melts the scale's resin; the endangered Kirtland's warbler requires large, pure stands

Spruce Genus (*Picea*)

© Lukas Gojda/Shutterstock.com

© Stefan Schug/Shutterstock.com

White Spruce (*Picea glauca*)

Height: up to 75 ft. (23 m)
Needles: single, 4-sided, short, stiff blue-green; white lines all sides; curves upward along branches
Key Features: twigs hairless; bark thin scales; female cones 2 in. (5 cm) cylindrical, thin, smooth, close-fitting scales
Habitat: along streams and lakes; rich, moist soil
Notes: provides food and shelter for a variety of animals

© Nick Pecker/Shutterstock.com

Black Spruce (*Picea mariana*)

Height: up to 75 ft. (23 m)
Needles: single, 4-sided, short, stiff grayish-green spirally arranged
Key Features: female cones 1 in. (2.5 cm) purplish-brown on short stalks; tree tops often narrow or spindly
Habitat: bogs, swamps, but occasionally well-drained soils
Notes: shallow-rooted tree susceptible to weather damage

© Kseniya Bogdanova/Shutterstock.com

Blue Spruce (*Picea pungens*)

Height: up to 50 ft. (30 m)
Needles: single, 4-sided, short, stiff, sharply incurved blue-green silver
Key Features: cones cylindrical 4 in. (10 cm) with irregular scales
Habitat: landscaping; tree plantations; native to central and western Rocky Mountains
Notes: nonnative; popular ornamental tree; Colorado spruce

© Sarycheva Olesia/Shutterstock.com

Norway Spruce (*Picea abies*)

Height: up to 60 ft. (18 m)
Needles: single, 4-sided but slightly flattened, dark yellowish-green
Key Features: numerous drooping branches; cones up to 7 in. (17 cm) cylindrical
Habitat: landscaping; tree plantations
Notes: nonnative; popular ornamental tree; used as a windbreak

Hemlock Genus (*Tsuga*)

© Melinda Fawver/Shutterstock.com

Eastern Hemlock (*Tsuga canadensis*)

Height: up to 70 ft. (21 m)
Needles: flat, dark green top with whitish banded underside; thread-like stalk; grows in two rows
Key Features: ½ in. cones; inner bark bright reddish-purple
Habitat: cool, moist, shady sites
Notes: tip of tree usually droops; provides dense cover and food for wildlife; nonnative invasive, hemlock woolly adelgid (*Adelges tsugae*) feeds on sap of hemlock branches and can kill needles and branches, resulting in tree death

Fir Genus (*Abies*)

© Mikolaj Kepa/Shutterstock.com

Balsam Fir (*Abies balsamea*)

Height: up to 60 ft. (18 m)
Needles: flattened needles, dark shiny green, growing around branch in two rows, stalkless
Key Features: dark purple, erect, barrel-shaped resinous cones; bark dull green, smooth with resin blisters
Habitat: swamps to well-drained soils
Notes: shallow-rooted easily toppled by weather; easily killed by fire because of thin bark

Cypress Family (Cupressaceae)

 © R_Johnson/Shutterstock.com	 © Marinodenisenko/Shutterstock.com
Eastern Red-cedar or Juniper (***Juniperus virginiana***) **Height:** up to 50 ft. (15 m) **Needles:** scale-like (mature) overlapping, dark green **Key Features:** cones berry-like, dark blue waxy; crown narrow or broadly pyramidal **Habitat:** poor, dry soils; abandoned fields **Notes:** male and female cones separate trees; reddish wood resists decay; used by many birds for food and cover	**Northern White Cedar** (***Thuja occidentalis***) **Height:** up to 70 ft. (21 m) **Needles:** scale-like, overlapping, dull green, flat fanlike display; raised resin glands on underside **Key Features:** cones small, tan; 4–6 overlapping pairs **Habitat:** limestone bluffs; fields; swamps, bogs **Notes:** aromatic wood; food and shelter for many wildlife

Deciduous (Broadleaf)

Opposite Branching

Tree species arranged alphabetically by family name

© Vitechek/Shutterstock.com

Family Bignoniaceae

© Christian Musat/Shutterstock.com

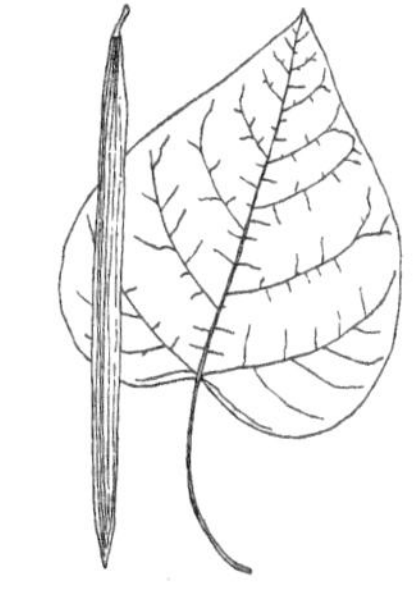

© Morphart Creation/Shutterstock.com

Northern Catalpa (*Catalpa speciose*)

Height: up to 50 ft. (15 m)
Leaves: simple, heart-shaped with slender point; dark green; hairy underside
Key Features: showy flowers, white with purple and yellow specks
Habitat: bottomlands; introduced in edge habitats
Notes: fast-growing; weak branches often break

Family Caprifoliaceae

© SunGrownNomad/Shutterstock.com

Elder Genus (*Sambucus*)
Common Elderberry (*Sambucus canadensis*)

Height: up to 10 ft. (3 m)
Leaves: compound, usually 7 elliptical, sharply toothed leaflets
Key Features: creamy white flowers in broad clusters; purple-black berrylike fruits
Habitat: moist sites; disturbed areas
Notes: twigs pithy, white; fruits and flowers used in foods but leaves, bark, roots toxic

Honeysuckle (*Lonicera*)

© nurism/Shutterstock.com

Amur Honeysuckle (*Lonicera maackii*)
Height: up to 15 ft. (4 m)
Leaves: simple, entire, dark green, egg-shaped with slender tip and tapered base
Key Features: arched branches; white, two-lipped fragrant flowers; dark red berries on short stalks persist into winter
Habitat: swamps to upland; disturbed sites
Notes: nonnative; aggressive; fruits toxic

Viburnum (Viburnum)

Nannyberry (*Vibernum lentago*)
Height: up to 30 ft. (9 m)
Leaves: simple, oval, curved in at tip, sharply toothed
Key Features: white, fragrant, broad-clustered flowers; fruits long-stalked dark blue, berrylike
Habitat: swamp and forest edges; roadsides; thickets
Notes: hardy, fast growing; root suckers; fruits provide food for many wildlife

© Gerald A. DeBoer/Shutterstock.com

© Yurich/Shutterstock.com

American Highbush Cranberry (*Viburnum trilobum*)
Height: up to 20 ft. (6 m)
Leaves: simple, maple-like leaf with deeply cut 3 spreading pointed lobes; sparsely coarse-toothed
Key Features: flowers white, 5 petals; fruit red to orange, juicy berry-like drupes in branched clusters
Habitat: near water, cool woodlands or ravines in moist, rich sites
Notes: berries persist in winter and serve as food for many wildlife

Family Cornaceae

© Zocchi Roberto/Shutterstock.com

Dogwood Genus (*Cornus*)

Flowering Dogwood (*Cornus florida*)

Height: up to 30 ft. (9 m)
Leaves: simple, entire, oval, veins parallel to margin and curve toward pointed tip
Key Features: showy white 4-petal flowers; red, berrylike fruit clusters
Habitat: well-drained soil; deciduous forests
Notes: many birds and small mammals eat the bitter fruits; attractive ornamental plant

© julie deshaies/Shutterstock.com

© Jennifer Bosvert/Shutterstock.com

Red-Osier Dogwood (*Cornus sericea*)

Height: up to 10 ft. (3 m)
Leaves: simple, entire, parallel veins arched toward tip
Key Features: bark red; multitrunked; white flowers succeeded by waxy white berries
Habitat: rich, moist sites in woodlands; floodplains
Notes: thicket-forming shrub

Family Oleaceae

Ash Genus (*Fraxinus*)—*Fraxinus* sp. populations threatened by emerald ash borer (*Agrilus planipennus*).

© Andrew Sabai/Shutterstock.com

© vergasova/Shutterstock.com

(in the autumn)

White Ash (*Fraxinus americana*)

Height: up to 80 ft. (24 m)
Leaves: compound, pinnate, 7 ovate leaflets, smooth margin toothed sparsely; light underside with bumps
Key Features: single-winged samaras clustered on stalk; bark diamond-shaped furrows (mature trees)
Habitat: rich upland, well-drained soils
Notes: leaf scars u-shaped; wood used in many products

© Maxal Tamor/Shutterstock.com

Green Ash (*Fraxinus pennsylvanica*)

Height: up to 70 ft. (21 m)
Leaves: compound, pinnate, 5–9 (usually 7) leaflets, same color both **sides**; shallow teeth along margin above mid-leaf
Key Features: single-winged samaras clustered on stalk; bark **diamond**-shaped furrows
Habitat: wet soils on floodplains with other hardwoods
Notes: leaf scars semicircular; can be invasive; genus has **similar** species; all threatened by emerald ash borer (*Agrilus planipennus*)

Black Ash (*Fraxinus nigra*)

Height: up to 50 ft. (15 m)
Leaves: compound, 7–11 stemless leaflets lance-shaped, finely toothed; reddish-brown fuzz at base
Key Features: bark scaly, with shallow, straight fissures; blue-black buds
Habitat: streambanks, floodplains, wetland borders
Notes: shade intolerant yet can remain in water for weeks; grows with black spruce, eastern white-cedar, silver maple

Family Rubiaceae

© Arty Alison/Shutterstock.com

Buttonbush (*Cephalanthus occidentalis*)

Height: up to 15 ft. (4 m)
Leaves: simple, oval, shiny, dark green; sometimes in whorls of 3
Key Features: fragrant white globe-shaped flowers; fruits round, bumpy reddish-brown
Habitat: wetland borders; thickets
Notes: used in landscaping; tolerates a range of habitats; attracts numerous butterflies; used as food and cover by wildlife

Family Sapindaceae

Maple Genus (*Acer*)—the nonnative, invasive Asian longhorned beetle (*Anoplophora glabripennis*) favors and kills many *Acer* sp., among other tree types, by tunneling in the wood in some parts of the region

© AWesleyFloyd/Shutterstock.com

© Bodor Tivadar/Shutterstock.com

Sugar Maple (*Acer saccharum*)

Height: up to 100 ft. (30 m)
Leaves: simple, palmate, smooth margin, 5 lobes (sometimes 3); few irregular, blunt-pointed teeth; round notches
Key Features: dark gray bark, long winged strips (mature); u-shaped, parallel samaras (winged seeds)
Habitat: dry to moist soil, sandy or rich; uplands to valleys
Notes: wood and sap desirable

© Holly Guerrio/
Shutterstock.com

Red Maple (*Acer rubrum*)
Height: up to 90 ft. (27 m)
Leaves: simple, palmate, 3–5 lobes, irregularly double-toothed; light underside; shallow, sharp notches; turns red in autumn
Key Features: red-winged samaras following flowers
Habitat: forest swamps; sometimes dry uplands
Notes: flowers early spring; grows quickly; sap half as sweet as sugar maple

© annalisa e marina durante/
Shutterstock.com

Silver Maple (*Acer saccharinum*)
Height: up to 100 ft. (30 m)
Leaves: simple, palmate, 5–7 lobes, deep concave sinuses, irregularly coarse-toothed; silvery-white underside
Key Features: yellow-red flowers in dense, almost stalkless clusters; samaras spread at 90° angle
Habitat: near streams, swamps, lakes
Notes: fast-growing; often an ornamental but needs space and drops many seeds and leaves; many birds and small mammals eat seeds or live in hollowed trunks

Mountain Maple (*Acer spicatum*)
Height: up to 30 ft. (9 m)
Leaves: simple, palmate, 3 prominent lobes above midl-eaf; irregularly saw-toothed; wedge-shaped notches
Key Features: clumped, crooked trunks; twigs velvety, yellow-purple gray; flowers erect, branched clusters
Habitat: mixed woods; thickets; swamps and ravines
Notes: important for erosion prevention; deer and moose seek new growth

© Maxal Tamor/Shutterstock.com

Norway Maple (*Acer platanoides*)

Height: up to 70 ft. (21 m)
Leaves: simple, palmate, 5 (rarely 7) lobes, few large bristle-tipped teeth, dark green; turns yellow in autumn
Key Features: twigs have milky "juice"; samaras almost 180°, bark has low ridges, uniformly textured
Habitat: wide range of soils; disturbed areas; thickets
Notes: nonnative; planted in urban sites; resistant to many diseases; can be aggressive

© Julia Sudnitskaya/Shutterstock.com

Boxelder Maple or Ashleaf Maple **(*Acer negundo*)**

Height: up to 70 ft. (21 m)
Leaves: compound, pinnate, 3–5 leaflets (sometimes 7–9)
Key Features: twigs shiny, waxy, greenish-purple; often has multiple trunks
Habitat: near water; often in disturbed areas
Notes: many clustered, drooping samaras; male and female flowers on separate trees

© spline_x/Shutterstock.com

© Anastasiia Malinich/Shutterstock.com

Horsechestnut (*Aesculus hippocastanum*)—the nonnative, invasive Asian long-horned beetle (*Anoplophora glabripennis*) can kill these species by tunneling in the wood in some parts of the region

Height: up to 60 ft. (18 m)
Leaves: compound, palmate, 5–9 wedge-shaped stalkless leaflets attached to long petiole
Key Features: buds shiny, dark brown sticky at twig tips; showy erect long cone-shaped flower clusters
Habitat: landscaping; disturbed sites
Notes: nonnative; often planted as landscape tree; branches unpleasant smelling when bruised; seeds, leaves, bark toxic; similar appearance to native Ohio buckeye (*Aesculus glabra*) found in southern part of region

Family Staphylaceae

American Bladdernut (*Staphylea trifolia*)

Height: up to 20 ft. (6 m)

Leaves: compound, 3 finely toothed leaflets

Key Features: drooping bell-shaped flowers; thin, veiny lantern-shaped fruits with brown seeds that rattle inside

Habitat: forest edges; moist, rich floodplains or hillsides

Notes: shade-tolerant; spreading roots send up suckers to form thickets

Alternate Branching

Family names arranged alphabetically below.

Family Anacardiaceae

© Nick Pecker/Shutterstock.com

Staghorn Sumac (*Rhus typhina*)
Height: up to 20 ft. (6 m)
Leaves: compound, pinnate, long, narrow 11–31 lance-shaped leaflets with reddish central stalk; dark green top and whitish underside; leaf scar horseshoe-shaped
Key Features: flowers yellowish-green, dense, erect; fruits scarlet, fuzzy, erect, dense, cone-shaped clusters; branches exude milky juice; pith hollow
Habitat: woodland borders; old fields; dry to mesic sites
Notes: sometimes an ornamental; spreads via root suckers; stabilizes slopes; food for many birds and mammals

© Sabrina Nichols/Shutterstock.com

Poison Sumac (*Toxicodendron vernix*)
Height: up to 25 ft. (8 m)
Leaves: compound, pinnate, 7–13 ovate leaflets smooth-edged (sometimes wavy); dark top and white beneath; essentially hairless; wingless central stalk; leaf scar shield-shaped
Key Features: buds purplish-brown, hairy at branch tips; flowers tiny yellow-green, nodding 5-petals, clustered; fruits white, pearl-like drupes, dry, arching clusters
Habitat: swampy woodlands, open; often with tamarack; spotty distribution within region
Notes: severe skin rash can result from contact (wash with soap and water) and inhalation if burned; game birds and songbirds eat the "berries"

Family Betulaceae
Alder Genus (*Alnus*)

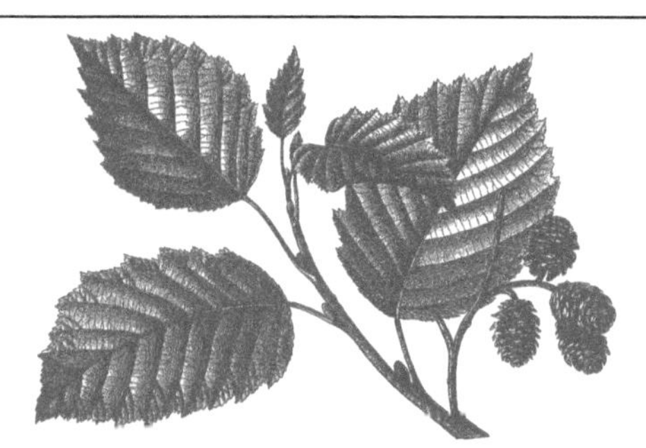

© Hein Nouwens/Shutterstock.com

© Vladyslav Lehir/Shutterstock.com

Tag Alder (*Alnus incana*)

Height: up to 25 ft. (8 m)

Leaves: simple, dull dark green above, pale green beneath, thick; 6–12 straight veins per side, broadly ovate with edges undulating and sharp double teeth

Key Features: buds reddish-brown; cone-like catkins; narrow, 2-winged nutlets hanging through year; trunk pale orange, horizontal pores

Habitat: wet, open sites

Notes: important streambank stabilizer, provides shade, enriches soil; early flowers sought by bees; wildlife food

Birch Genus (*Betula*)

© J. L. Levy/Shutterstock.com

© Morphart Creation/ Shutterstock.com

Paper Birch (*Betula papyrifera*)

Height: up to 70 ft. (21 m)

Leaves: simple, entire, coarsely double-toothed

Key Features: white (to red-brown) bark peels off trunk is thin strips; catkins (flowers) yellow-green

Habitat: moist, sandy soil; grows in burned sites

Notes: white birch; tough, pliable bark; when several bark layers removed soon blackens and dies

© simona flamigni/Shutterstock.com

Yellow Birch (*Betula alleghaniensis*)
Height: up to 100 ft. (30 m)
Leaves: simple, 8–12 straight veins per side ending in large tooth with 2–3 smaller intervening teeth
Key Features: bark shiny reddish-brown (young), bronze, tightly curled papery shreds (mature); cone-like oval fruit erect on twigs
Habitat: shady, rich, moist sites; hardwood forests, swamps, streamside
Notes: important lumber source; aromatic, wintergreen-scented twigs and leaves; torn off bark can scar or kill tree

© Ginn Tinn Photography/Shutterstock.com

River Birch (*Betula nigra*)
Height: up to 80 ft. (24 m)
Leaves: simple, oval, coarsely double-toothed edges, wedge-shaped base
Key Features: bark pale reddish-brown, papery, scales and horizontal lines (young) or thick scales (mature)
Habitat: streambanks, wet woodlands
Notes: often planted in landscapes

Hornbeam Genus (*Carpinus*)

© FoxEyePhotography/Shutterstock.com

Musclewood or American hornbeam or Blue-beech **(*Carpinus caroliniana*)**
Height: up to 30 ft. (9 m)
Leaves: simple, oval, double-toothed, bluish-green above and yellowish underside; straight veins (occasionally slightly forked)
Key Features: bark smooth, fluted (sinewy muscle-like) blue-gray; ribbed nutlets; leaf-like bract hanging in clusters
Habitat: rich, moist, shady understory; bottomlands
Notes: often grows with ironwood (*Ostrya virginiana*); dense, strong wood

Hophornbeam Genus (*Ostriya*)

© Erika J Mitchell/
Shutterstock.com

Ironwood or Eastern hophornbeam **(*Ostrya virginiana*)**

Height: up to 50 ft. (15 m)

Leaves: simple, entire, 2 rows, dark yellowish-green and soft, double-toothed tapered to a point; straight veins

Key Features: bark narrow, shaggy, vertical, peeling strips; trunk maximum diameter 12 in. (36 cm); unusual fruit clusters light greenish, flattened, papery inflated sacs

Habitat: hillsides, ridges; well-drained forests

Notes: often grows with musclewood (*Carpinus caroliniana*); extremely dense, hard wood

Family Elaeagnaceae

© weha/Shutterstock.com

© Marinodenisenko/Shutterstock.com

Autumn Olive (*Elaeagnus umbellata*)

Height: up to 12 ft. (4 m)

Leaves: simple, lance-shaped, pointed, silvery and brown scales on leaves in spring, turning darker in summer; underside obvious silvery scales

Key Features: fragrant, small, white to yellow, 4-lobed flowers; fruits yellow-orange (unripe) to red with brown dots (ripened), round, drupe-like berries

Habitat: sandy soils, disturbed sites

Notes: nonnative and very aggressive

Family Ericaceae

© Jeannie Oh/ Shutterstock.com

© Anastasiia Mavrina/ Shutterstock.com

© Sophia Granchinho/ Shutterstock.com

Labrador Tea or Bog Labrador Tea or Swamp Tea (***Rhododendron groenlandicum***)
Height: 1–3 ft. (30–90 cm)
Leaves: simple; leathery texture; curls as edges; wrinkly top and rusty-orange hairy undersides; evergreen; 1–2.5 in. (20–60 mm) long and 0.12–0.6 in. (3–15 mm) wide
Key Features: leaves; tiny white flowers in semispherical clusters (very fragrant; sticky); fruit capsules fuzzy, oval
Habitat: bogs; fens; wet shorelines
Notes: shrub; may be toxic to humans; important to some bird species for nesting material; not preferred browse of deer and moose; similar to leatherleaf (*Chamaedaphne calyculata*) but has dull green above with silvery scales beneath leaves and has white, bell-like flowers in terminal racemes

Family Fabaceae

© iablik/Shutterstock.com

Black Locust (*Robinia pseudoacacia*)
Height: up to 60 ft. (18 m)
Leaves: compound, pinnate, 7–21 oval leaflets notched at tip; spine pair at base
Key Features: pods long, brown, flat and orange seeds
Habitat: rich, moist soils; disturbed sites
Notes: nonnative to region; aggressive and outcompetes native vegetation (especially in sandy soils); invasive Asian long-horned beetle (*Anoplophora glabripennis*) may kill some *R. pseudoacacia* by tunneling in the wood in some parts of the region

© Mariola Anna S/ Shutterstock.com

Honey-Locust (*Gleditsia triacanthos*)
Height: up to 80 ft. (24 m)
Leaves: compound, once or twice pinnately divided, 18–30 oblong leaflets with no terminal leaflet; dark green top with pale underside
Key Features: pods long, brown (later in season), leathery, flat, spirally twisted and drop over winter without opening; bark has long, branched thorns
Habitat: moist, rich lowlands; disturbed sites and spreads
Notes: sweet flowers attract pollinators; mammals eat seeds and sweet pulp of the pods; native to southern portion of region; invasive

© Morphart Creation/ Shutterstock.com

© Melinda Fawver/ Shutterstock.com

© Nadia Levinskaya/Shutterstock.com

Redbud (*Cercis canadensis*)
Height: up to 20 ft. (6 m)
Leaves: simple, heart-shaped, thick, glossy or leathery; dark green top and pale below; 5–9 prominent veins radiating from stalk; stalk swollen near blade
Key Features: showy, pea-like pink flowers abundant in clusters along branches; fruits red-brown, flat pods
Habitat: moist soil; streamsides
Notes: often planted as ornamental; native in southern part of region; first flowers around 5 years; reduces leaf surface area on hot days by curling broad leaves; subfamily Cercidoideae

Family Fagaceae
Beech Genus (*Fagus*)

© hans engbers/Shutterstock.com

© kristof lauwers/Shutterstock.com

© Kyle Selcer/Shutterstock.com

American Beech (*Fagus grandifolia*)
Height: 80 ft. (24 m)
Leaves: simple, elliptical; sharp, widely spaced teeth
Key Features: bark smooth, light gray; buds unique long, slender, brown; fruits spiny husk with 2–3 triangular nuts; brown leaves remain on tree in winter
Habitat: hardwood forests, gentle slopes; bottomlands
Notes: nuts sought by many birds and mammals that disperse seeds; porcupines kill by girdling; fungus and invasive insect kills trees; beech leaf disease

Oak Genus (*Quercus*)—acorns; clustered terminal buds

© D. Kucharski K. Kucharska/Shutterstock.com © Ihor Hvozdetskyi/Shutterstock.com

Northern Red Oak (*Quercus rubra*)

Height: up to 90 ft. (27 m)
Leaves: simple, pinnate, 7–11 bristle-tipped lobes; notches rounded
Key Features: acorns oval, shallow hairy cup; bark vertical fissures with flat ridge tops
Habitat: uplands; well-drained loam
Notes: many mammals eat the acorns; *Quercus* sp. threatened by oak wilt, especially found with *Q. rubra*. Fungus *Bretziella fagacearum* generally begins as a bark infestation ultimately causing leaf discoloration, wilt, defoliation, and death; spreads by insect, wind, or underground via tree roots.

© Bildagentur Zoonar GmbH/Shutterstock.com

Pin Oak (*Quercus palustris*)

Height: up to 70 ft. (21 m)
Leaves: simple, pinnate, deeply cut sinuses, 5-lobed (occasionally 7–9); bristle-tipped; hair tufts on underside
Key Features: acorns rounded, shallow cup; downward-hanging lower limbs (often dead)
Habitat: wet bottomlands
Notes: urban and natural settings; fast-growing; early bloomer; mammals and waterfowl eat the acorns

Black Oak (*Quercus velutina*)

Height: up to 80 ft. (24 m)
Leaves: simple, pinnate, shiny dark green, 5–7 bristled lobes, parallel-sided, u-shaped
Key Features: Acorns ragged-edged cup; dark grayish-black cracked into rectangles
Habitat: dry, well-drained sandy or gravel soils; mixed with pine or hickory
Notes: inner bark yellow to orange; mammals and birds eat the acorns

© Bildagentur Zoonar GmbH/Shutterstock.com

Eastern White Oak (*Quercus alba*)
Height: up to 100 ft. (30 m)
Leaves: simple, pinnate, bright green topside and pale below, 7–9 rounded major lobes
Key Features: acorn cup shallow, knobby scales
Habitat: riverbanks; sandy plains or hillsides; moderately well-drained soil
Notes: important hardwood for building; seems to have least impact from oak wilt

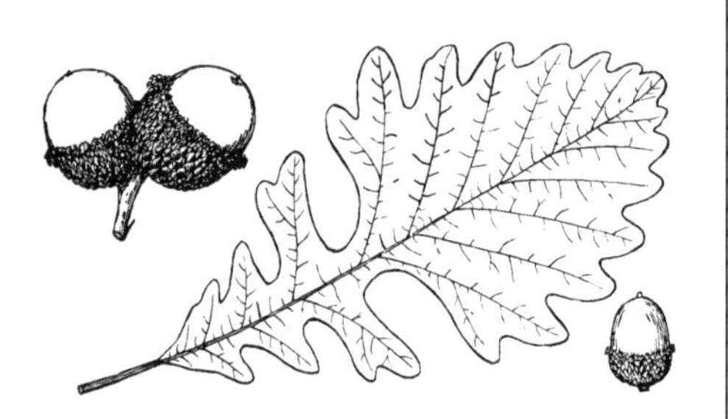

© Morphart Creation/Shutterstock.com

Bur Oak (*Quercus macrocarpa*)
Height: up to 100 ft. (30 m)
Leaves: simple, pinnate, 2–4 rounded lobes (or 6–8) below a coarse-toothed upper wide wedge-shaped lobe
Key Features: acorn cup deep, fringed
Habitat: dry grasslands or moist bottomlands
Notes: like other oaks, leaves may persist in winter

Family Hamamelidaceae

© cristo95/Shutterstock.com

© lightrain/Shutterstock.com

American Witch Hazel (*Hamamelis virginiana*)
Height: up to 30 ft. (9 m)
Leaves: simple, entire, dark green top and paler beneath; asymmetrical base; wavy-scalloped margins; 5–7 straight parallel ascending veins per side
Key Features: flowers bright yellow, twisted ribbon-like; fruits short, thick 2-beaked woody capsules
Habitat: shady sites; deep, rich soil; open woodlands with dry, sandy soil
Notes: slow-growing; sometimes in landscaping for showy, fragrant flowers and persistent fruits

Family Juglandaceae

© Nikolay Kurzenko/Shutterstock.com

© K Quinn Ferris/Shutterstock.com

Black Walnut (*Juglans nigra*)

Height: up to 90 ft. (27 m)
Leaves: compound, 15–23 lance-shaped leaflets; no terminal leaflet (or very small)
Key Features: fruits fleshy, yellowish-green, hard inner black nut; twigs heart-shaped leaf scar and bud in notch (winter); dark twig pith
Habitat: well-drained soil; bottomlands and slopes
Notes: highly-prized hardwoods; walnut husks rich with tannins; juglone toxin released from roots and prevents other deciduous trees from rooting (including other walnuts)

© rsev97/Shutterstock.com

© guentermanaus/Shutterstock.com

Shagbark Hickory (*Carya ovata*)

Height: up to 80 ft. (24 m)
Leaves: compound, 5 elliptical, finely toothed leaflets (3 at tip much larger)
Key Features: bark with long, loose strips curled up from trunk; fruits round, green, thick, woody husk enclosing thin-shelled nut
Habitat: dry to moist sites; upland slopes; moist bottomlands
Notes: high-quality wood; important food source for squirrels

Family Lauraceae

© Kathy Clark/Shutterstock.com

Sassafras (*Sassafras albidum*)

Height: up to 50 ft. (15 m)
Leaves: simple, ovate or 2–3 lobed; bright green; fragrant; leaves yellow to red in autumn
Key Features: bark zigzagged, fragrant, corky ridges; branches crooked and wide-spreading; fruits dark blue, berry-like drupes
Habitat: dry to moist; disturbed sites; sandy, acidic soils
Notes: colonizes quickly via root suckers; game birds and mammals occasionally feed on fruits; dried bark banned for use in foods because of carcinogenic nature

© mizy/Shutterstock.com

© rsev97/Shutterstock.com

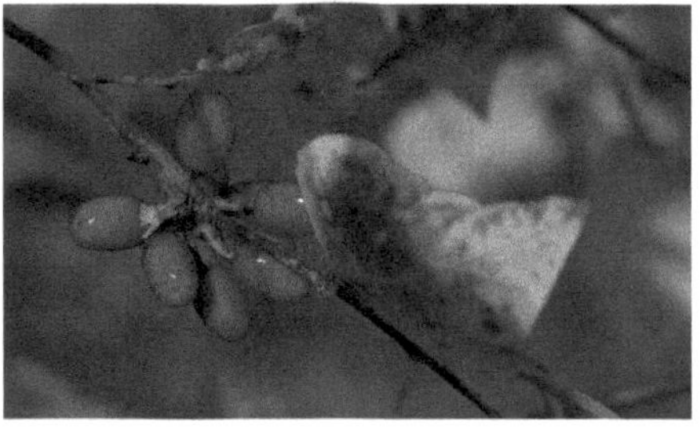

© rsev97/Shutterstock.com

Spicebush (*Lindera benzoin*)

Height: up to 20 ft. (6 m)
Leaves: simple, ovate, usually larger toward tip, smooth edged
Key Features: aromatic; flowers yellow, dense showy clusters at joints of previous year's twig; fruit bright red, spicy-smelling, berry-like drupes football-shaped
Habitat: moist deciduous forests and swamps
Notes: agreeable spicy fragrance; planted as ornamental

Family Magnoliaceae

Tulip-tree (*Liriodendron tulipifera*)

Height: up to 100 ft. (30 m)
Leaves: simple, long with squared, notched tips with 2–3 lobed sides
Key Features: flowers resemble tulip head, orange at base with light greenish-yellow petals; fruits cone-like clusters green to straw-colored
Habitat: deep, rich moist shaded sites; well-drained soils
Notes: valued timber tree; fast-growing; used in landscaping; bees gather large amounts of nectar from flowers; birds and mammals depend on it for food

Family Myricaceae

Sweet-fern (*Comptonia peregrina*)

Height: up to 4 ft. (123 cm)
Leaves: dark green (summer), fern-like (coarse, tooth-like lobes); narrow; 2–4 in. long (5–10 cm); very fragrant; light green in spring and brown in fall
Key Features: foliage fragrance when crushed; flowers are small yellowish catkins (April-May); shiny stems with resin dots; old stems copper or purple color; small nutlet clusters
Habitat: sandy, gravelly open areas; poor soil
Notes: used for erosion control; can fix nitrogen; spreading, colonizing plant (spreads twice the height of shrub); not a fern!

Family Platanaceae

© Beck Polder Photography/Shutterstock.com © Melinda Fawver/Shutterstock.com

American Sycamore (*Platanus occidentalis*)

Height: up to 100 ft. (30 m)

Leaves: simple, palmate, 3–5 shallow lobes, large-toothed; bright green topside, pale underside

Key Features: bark mottled, light brownish-gray, peeling irregular flakes; fruits light brown, round

Habitat: moist soil; bottomlands

Notes: fast-growing; planted as ornamental shade tree; leaves mistaken for maple; the nonnative, invasive Asian long-horned beetle (*Anoplophora glabripennis*) may kill some *P. occidentalis* by tunneling in the wood in some parts of the region

Family Rhamnaceae

© spline_x/Shutterstock.com

Glossy Buckthorn (*Frangula alnus*)

Height: up to 20 ft. (6 m)

Leaves: simple, elliptical, fine-toothed, up to 3 in. (7.5 cm), 5–10 parallel veins

Key Features: flowers yellow on short stalks; fruits red to black drupes

Habitat: moist soil; disturbed sites

Notes: nonnative; aggressive and displaces natives rapidly; birds spread fruit

Family Rosaceae

© Bragapictures/Shutterstock.com

© Breck P. Kent/Shutterstock.com

American Mountain-Ash (*Sorbus americana*)

Height: up to 30 ft. (9 m)
Leaves: compound, 13–17 lance-shaped, sharp-toothed leaflets; taper pointed, short-stalked
Key Features: white, broad-clustered flowers (5 petals each); fruits dense, red-orange fleshy
Habitat: swamp borders; drier uplands; thickets; coniferous forests
Notes: slow-growing, short-lived; fruits acidic and high in tannins; birds and mammals feed on fruits; similar to invasive, nonnative tree-of-heaven

Cherry, Plum, Peach Genus (*Prunus*)

© Ariene Studio/Shutterstock.com

© Joseph Jacobs/Shutterstock.com

Wild Black Cherry (*Prunus serotina*)

Height: up to 90 ft. (27 m)
Leaves: simple, entire, thick, waxy, dark green top with pale underside with white or rusty hairs on lower midvein; fine, incurved teeth
Key Features: mature bark rough with squared, out-curved scales (often compared to burnt potato chips); young bark dark, smooth with conspicuous horizontal lenticels
Habitat: disturbed sites; well-drained soils
Notes: easily worked and desired wood; all parts of tree toxic except cherry flesh; food source for many birds and small mammals; deer eat fresh green leaves in spring, but leaves inedible in autumn

© Cynthia Shirk/Shutterstock.com

© Nick Pecker/Shutterstock.com

© Nick Pecker/Shutterstock.com

Chokecherry (*Prunus virginiana*)

Height: up to 20 ft. (6 m)

Leaves: simple, elliptical, widest above mid-leaf, finely toothed

Key Features: bark dark grayish-brown, smooth or fine scales and prominent lenticels; fruits dark reddish-purple

Habitat: open woodlands or exposed area; rich, wet soil

Notes: all parts of tree toxic but the fruit; grows prolifically from root suckers or stumps; fast-growing, short-lived; pioneer species; stabilizes soil; birds eat berries and disperse

Hawthorn Genus (*Crataegus*)

© Dina Rogatnykh/Shutterstock.com

Downy Hawthorn (*Crataegus mollis*)

Height: up to 35 ft. (13 m)

Leaves: simple, oblong-elliptical, dull green topside pale beneath and slightly hairy; impressed veins above; base tapered; toothed; parallel veins

Key Features: numerous horizontal branches; pale gray branchlets; thorny branches and trunk; white, 5-petaled clustered flowers (20 stamens) at end of branches in early spring; large brightly colored fruit late August into September (apple-like fruits) reddish to yellow

Habitat: rich, moist woodlands and streamsides; disturbed sites

Notes: many hawthorn species; target of gypsy moths and leaf rusts

Serviceberry Genus (*Amelanchier*)

© Ronald Wilfred Jansen/Shutterstock.com

Smooth Serviceberry (*Amelanchier laevis*)

Height: up to 30 ft. (9 m)

Leaves: simple, entire, oval-elliptical and abruptly pointed, dark green top and pale beneath (mature); red and folded when young; 25 sharp teeth; 10 or fewer veins per side; hairless stalks

Key Features: bark thin, smooth gray; buds narrow, pointed egg-shaped; drooping clusters of showy white 5-petaled flowers; fruits dark red-purple berry-like

Habitat: mixed woodlands; clearings; well-drained sites

Notes: twigs have faint almond scent; flowers and fruits attract many birds; similar to downy serviceberry, and often grown together

Crabapple (*Malus*)

© By John A. Anderson/Shutterstock.com

© Morphart Creation/Shutterstock.com

Wild or **Sweet Crabapple (*Malus coronaria*)**
Height: up to 25 ft. (8 m)
Leaves: simple, ovate to rounded, bright green above and paler below; coarsely toothed at tip, somewhat lobed at base; hairless when mature
Key Features: showy, fragrant, pink to white flowers; waxy small apples (pomes) and can persist through winter; many thorn-like spurs on branches; contorted branches
Habitat: deciduous forest understory; disturbed areas
Notes: planted as ornamental (varieties); interesting patterns found in wood; food source

© Tamotsu Ito/Shutterstock.com

Multiflora Rose (*Rosa multiflora*)
Height: 10–17 ft. (3–5 m)
Leaves: compound 2–4 in. (5–10 cm), 5–9 leaflets with feathery stipules
Key Features: showy white or pink flowers in clusters, blossoms 1.5 in. (4 cm) across in early summer; hips reddish-purple; climbing shrub
Habitat: disturbed areas, wetland edges, forest understory
Notes: nonnative, aggressive, invasive, difficult to manage; cutting promotes growth

Family Salicaceae

Willow Genus (*Salix*)—the nonnative, invasive Asian long-horned beetle (*Anoplophora glabripennis*) may kill some *Salix* sp. by tunneling in the wood in some parts of the region

Black Willow (*Salix nigra*)
Height: up to 80 ft. (24 m)
Leaves: simple, lance-shaped, curved tips, uniformly green top and beneath; finely toothed
Key Features: deeply furrowed dark bark; twigs surrounded by short stipules (scalelike leaves)
Habitat: along waterways; wet soils
Notes: only willow with uniform leaf color with obvious stipules; branches easily broken and can take root

© magnetix/Shutterstock.com

© Krisztian Tefner/Shutterstock.com

Weeping Willow (*Salix babylonica*)
Height: up to 60 ft. (18 m)
Leaves: simple, lance-shaped green topside, gray-green below
Key Features: long, drooping branches; twigs bright yellow; catkins
Habitat: along waterways; wet lawns; disturbed sites
Notes: often hybridizes; branches fall often; typically last to lose leaves and first to bud out

© venars.original/Shutterstock.com

© Lana Shulga/Shutterstock.com

Pussy Willow (*Salix discolor*)
Height: up to 20 ft. (6 m)
Leaves: simple, narrow, widely spaced rounded teeth; dark green topside with whitish underside
Key Features: furry white catkins on twigs (spring)
Habitat: streamside, wet meadows, swamps
Notes: first catkins to appear in spring (before leaves); only collect branches from female shrubs because they do not shed pollen; catkins important food for bees and insects in spring

Poplar Genus (*Populus*)—the nonnative, invasive Asian long-horned beetle (*Anoplophora glabripennis*) may kill some *Populus* sp. by tunneling in the wood in some parts of the region

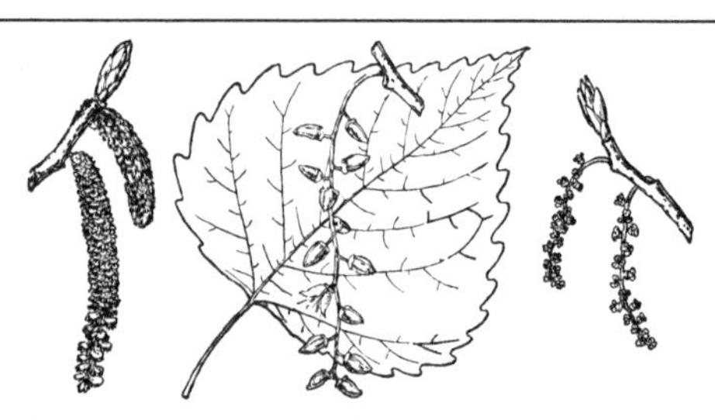

© Morphart Creation/Shutterstock.com

Eastern Cottonwood (*Populus deltoides*)

Height: up to 100 ft. (30 m)

Leaves: simple, triangular, coarse-rounded teeth; petioles flat

Key Features: bark gray with deep furrows (mature; smoother when young); buds sticky (not fragrant); fruit capsules green with cottony seeds

Habitat: bottomlands; floodplains or sand dunes

Notes: fast-growing but if in cityscape raises cement by moisture-seeking roots; hybrids exist

© rsev97/Shutterstock.com

© Morphart Creation/Shutterstock.com

Quaking Aspen (*Populus tremuloides*)

Height: up to 70 ft. (21 m)

Leaves: simple, entire nearly round, shiny green above and lighter beneath; fine, rounded teeth; long, flat petioles

Key Features: smooth bark pale gray-green with dark scars, does not peel

Habitat: variable; moist to dry; uplands, rocky, sandy, loam, clay soils

Notes: trembling aspen; leaves shake or tremble in the slightest breeze; clones created by root suckers than can cover entire forests; hundreds of species rely on tree for food and shelter; often mistaken for birches (this bark not papery!)

© rsev97/Shutterstock.com

(leaf in autumn)

Bigtooth Aspen (*Populus grandidentata*)
Height: up to 60 ft. (18 m)
Leaves: simple, entire, large blunt teeth; flat petioles
Key Features: smooth yellow to darkening bark, furrowed near base; hanging catkins; silky, white hairy fruit tufts
Habitat: prefers moist, fertile soil; well-drained uplands or on disturbed sites
Notes: fast-growing, shade-tolerant pioneers; holds soil in place after disturbance; twigs and buds whitish-down in spring; favorite of wildlife

© ManeeshUpadhyay/Shutterstock.com

Balsam Poplar (*Populus balsamifera*)
Height: up to 80 ft. (24 m)
Leaves: simple, entire, oval tapered to a tip, glossy dark green topside and silvery beneath; round petiole
Key Features: buds long, pointed, sticky, fragrant; drooping flower clusters with cottony seeds; mature bark dark gray furrowed
Habitat: along watercourses; bottomlands
Notes: fast growing; sends up sprouts from roots and stumps

Family Simaroubaceae

© simona pavan/Shutterstock.com

Tree-of-Heaven (*Ailanthus altissima*)
Height: up to 80 ft. (24 m)
Leaves: compound, pinnate, 11–41 leaflets, up to 30 in. (70 cm)
Key Features: small, green, erect flowers in long clusters; dense winged, hanging nutlet clusters
Habitat: dry soil; ubiquitous
Notes: nonnative, aggressive; often planted as ornamental

Family Tiliaceae

© Tatyana Azarova/Shutterstock.com

© miroha141/Shutterstock.com

©ClubhouseArts/Shutterstock.com

American Basswood (*Tilia americana*)

Height: up to 100 ft. (30 m)
Leaves: simple, entire, heart-shaped, coarsely-toothed
Key Features: flower and nutlets hang from strap-like bract
Habitat: cool, moist woods often near water; bottomlands in damp loam
Notes: one of softest, lightest hardwoods; fast-growing and long-lived; an attractive ornamental with fragrant flowers; easily damaged; heavily browsed by deer and rabbits

Family Ulmaceae

© Streltsova Anna/Shutterstock.com 

American Elm (*Ulmus americana*)

Height: up to 100 ft. (30 m)
Leaves: simple, oval, abruptly pointed, asymmetrical base, double teeth, prominent straight veins; rough above
Key Features: overall vase-shaped tree; trunks have v-shaped crotches; fruits flat with nutlet centered within membranous wing (samara), hairy at edges only, hanging clusters
Habitat: protected slopes; open sites of old fields or pastures
Notes: most species lost to Dutch elm disease (a sac-fungus, Ascomycota); some disease-resistant species bred; hosts many species of moths and butterflies

Herbaceous Plant Species—Natural History Journal Pages

*Choose an **herbaceous (vascular, nonwoody)** plant found in or near your Nature Area to complete the following details. Use your field guide and any other reputable resource. Remember, when making sketches: use pencil, draw simple shapes for the general outline, add labels to identify parts of what you are drawing and to provide more detail (i.e., size), and erase unnecessary lines upon completion.*

1. Full Common Name: ______________________________

   ____

   Scientific Name (*Genus species*): ______________________________

   ____

   Family Name (-acea): ______________________________

2. Sketch the overall plant shape with identifying characteristics labeled.

3. Shade the map of North America to indicate the plant's range.

4. Description of plant:

   Height: ______________________________

   Flower (if present): ______________________________

   Leaf: ______________________________

   Seasonality (when does it bloom?): ______________________________

5. Draw the mature flower, fruit, and seed (include size, color, type of flower/fruit/seed arrangement, and other important features).

6. Draw the leaves (note leaf size, type, and arrangement).

7. Where would you find this plant? Identify its preferred or required habitats (especially regarding shade and water).

8. Discuss the potential ecological interactions this plant could have with other organisms in its natural environment (as a food source, home, threats, concerns, etc.).

9. Describe the edibility, medicinal, or poisonous qualities this plant has to humans (include notes on how to gather and preserve it, or discuss precautions and remedies).

10. Provide other notes of interest about your herbaceous plant not previously mentioned.

11. List all resources used (**books:** author's last name, first initial, year, and book title; **websites:** name of site, date retrieved, and web address):

Coniferous Tree—Natural History Journal Pages

Choose a **CONIFEROUS** *(evergreen; needleleaf) tree that is found in or close to your nearby Nature Area/Sit Spot to complete the following details. Use your field guide and other reputable resources. Remember, when making sketches: use pencil, draw simple shapes for the general outline, add labels to identify parts of what you are drawing and to provide more detail (i.e., size), and erase unnecessary lines upon completion.*

1. Full Common Name: White Spruce

   Scientific Name (*Genus species*): Picea glauca

   Family Name (-aceae): Pinaceae

2. Draw the overall picture (silhouette) of the tree with identifying characteristics labeled (i.e., height, bark type, key features).

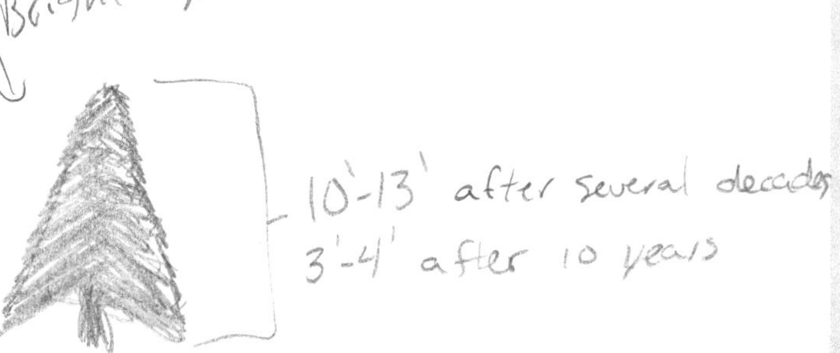

3. Provide a description of tree (size and shape, trunk, any other important identifying characteristics).

   Cone Shaped, foilage all the way to the ground. Bright green.

4. Shade the map of North America to show the conifer's range.

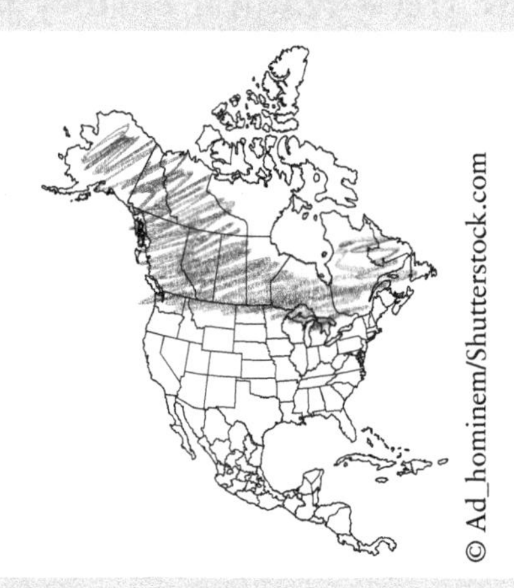

© Ad_hominem/Shutterstock.com

5. Draw mature male and female cones (label: size, color, and important features).

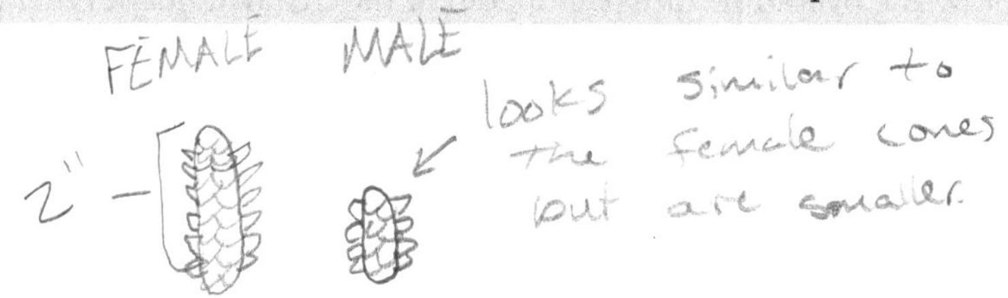

6. Draw a tree branch with its leaves (needles). Label needle size, needle characteristic (i.e., flat, square, needles per cluster, scaly), and twig characteristics.

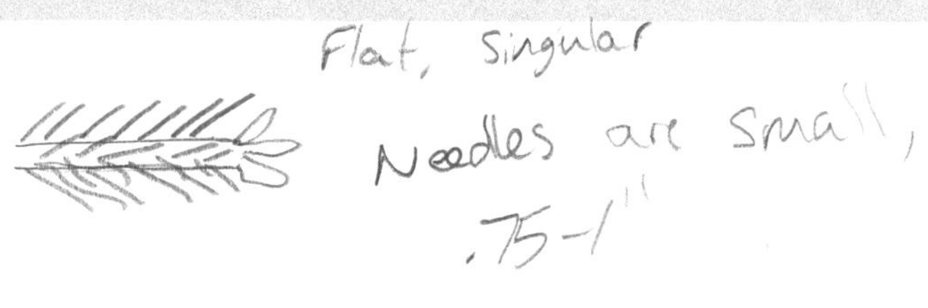

7. What is the conifer's preferred or required habitats (especially regarding sun exposure, water, and soil).

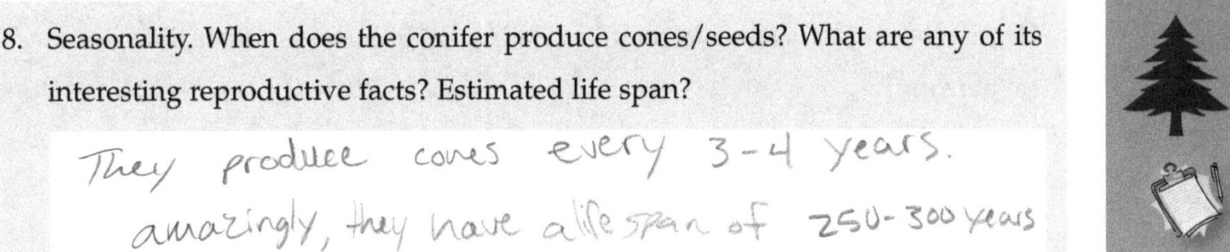

8. Seasonality. When does the conifer produce cones/seeds? What are any of its interesting reproductive facts? Estimated life span?

They produce cones every 3-4 years. amazingly, they have a life span of 250-300 years

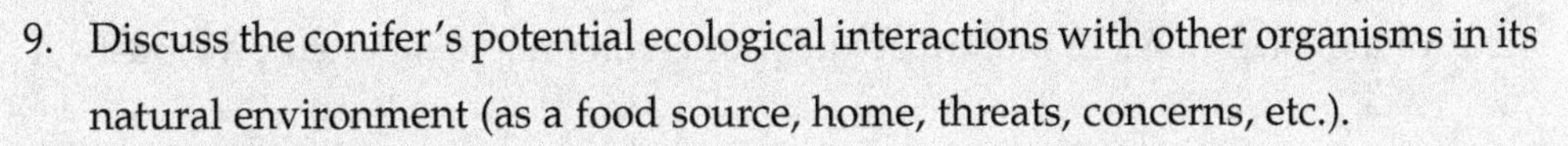

9. Discuss the conifer's potential ecological interactions with other organisms in its natural environment (as a food source, home, threats, concerns, etc.).

Provides food and shelter to many animals. fires can actually help with the production of cones.

10. a. What are the conifer's edibility, medicinal qualities, or poisonous aspects to *humans* (tip: use those words to help with an internet search along with tree name)?

Needles and pitch can be used to treat lung congestion along with Joint pain relief.

10. b. If edible/medicinal: How is the tree [parts] prepared/gathered/preserved and how is it used?

Needles are stripped from branches

If poisonous: Describe precautions with this tree.

Not poisonous to humans.

11. List the conifer's survival, craft, or daily uses to humans (other than discussed above).

pitch can be made into glue.

12. Provide other notes of interest about your conifer not mentioned previously.

looks like it could be a christmas tree.

13. List all resources used (**books:** author's last name, first initial, year, and book title; **websites:** name of site, date retrieved, and web address):

bplant.org

Fitzwilliams-Heck C.J. 2020,2021, A practical guide to Nature Study.

Deciduous Tree/Shrub— Natural History Journal Pages

Choose a ***DECIDUOUS*** *(broadleaf) tree that is found in or close to your nearby Nature Area/Sit Spot to complete the following details. Use your field guide and other reputable resources. Remember, when making sketches: use pencil, draw simple shapes for the general outline, add labels to identify parts of what you are drawing and to provide more detail (i.e., size), and erase unnecessary lines upon completion*

1. Full Common Name: ____

   ____

   Scientific Name (*Genus species*): ____

   ____

   Family Name (-aceae): ____

2. Draw the overall picture (silhouette) of the deciduous tree with identifying characteristics labeled (i.e., height, bark type, key feature).

3. Provide a description of the deciduous tree/shrub (size and shape, trunk, any other important identifying characteristics).

4. Shade the map of North America to show the deciduous tree's range.

5. Draw the mature flower, fruit, and seed (label: size, color, and important features—i.e., number of petals).

6. Draw the tree's branch with leaves (label: branching pattern, simple or compound, leaf size, margin, veins, twig characteristics, etc.).

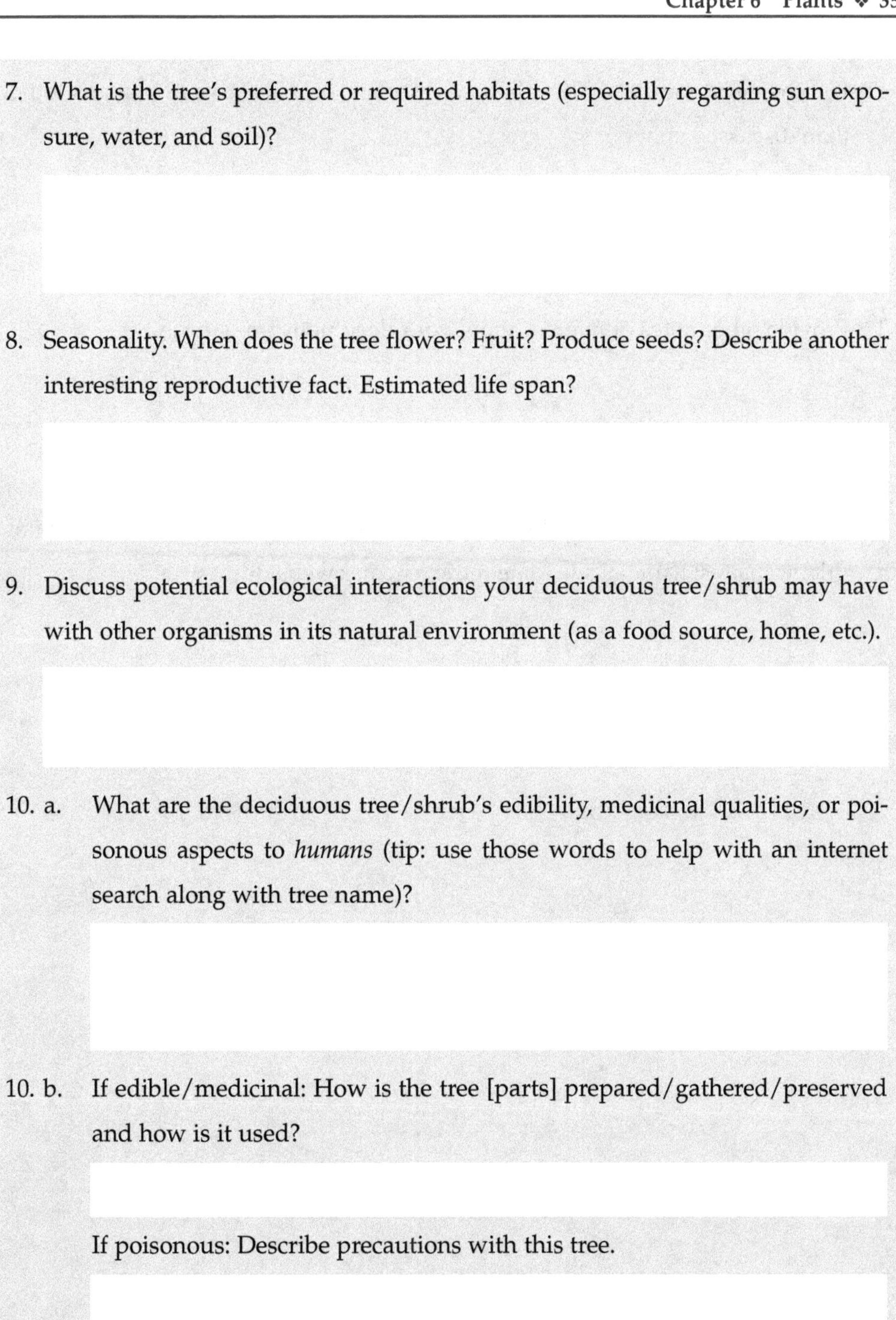

7. What is the tree's preferred or required habitats (especially regarding sun exposure, water, and soil)?

8. Seasonality. When does the tree flower? Fruit? Produce seeds? Describe another interesting reproductive fact. Estimated life span?

9. Discuss potential ecological interactions your deciduous tree/shrub may have with other organisms in its natural environment (as a food source, home, etc.).

10. a. What are the deciduous tree/shrub's edibility, medicinal qualities, or poisonous aspects to *humans* (tip: use those words to help with an internet search along with tree name)?

10. b. If edible/medicinal: How is the tree [parts] prepared/gathered/preserved and how is it used?

If poisonous: Describe precautions with this tree.

11. List the deciduous tree/shrub's survival, craft, or daily uses to humans (other than discussed above).

12. Provide other notes of interest about your deciduous tree/shrub.

13. List all resources used (**books:** author's last name, first initial, year, and book title; **websites:** name of site, date retrieved, and web address):

7 Invertebrate Animals

*Although you may not see or hear animals near where you live, more than likely, they exist and probably to a greater extent than you may realize. The wildlife community interacts within the environment outside your door in complex relationships as introduced in Chapter 4. In these last two chapters, we explore basic **zoology**, or the study of the animal kingdom. **Animals** are multicellular, eukaryotic organisms that consume organic matter, respire oxygen, grow, have the capability of movement, and can reproduce sexually. In zoology, you study the structure, distribution, behaviors, evolution, and classification of animals.*

*This chapter covers the **invertebrates** (in'ver-deb-rits), or animals without a backbone. The organisms will be presented in a somewhat evolutionary progression, going from the most primitive to the most complex. The tour starts with the protozoans of Kingdom Protista, then Kingdom Animalia's freshwater invertebrates, and the most time spent on arthropods (which includes insects and spiders). General characteristics, identifying features of each animal group, conservation, and naturalist tips for observation are discussed in many of the animal sections. Each group will have images of common representatives found in the region—many of which are found elsewhere. Understanding and recognizing what lives near you can be applied to learning new species when visiting other places. Your newly developed, or honed naturalist skills will prove effective and efficient when discovering and interpreting new environments.*

7.1 Animal-Like Organisms: Kingdom Protista

Before we delve into the study of animals, consider the "animal-like" organisms that existed prior to modern animals, and still exist today. Kingdom Protista, as discussed in Chapter 1, contain not only the slime molds and phytoplankton but also encompasses the **protozoans**—or "first animals." These mostly microscopic, single-celled organisms are abundant in freshwater ecosystems, especially if those ecosystems are organically enriched. Despite their simple structure and organization, protozoans carry out all the processes of life including digestion, respiration, excretion, osmoregulation, irritability and response, and reproduction. In a pond, protozoans are perhaps the most abundant organisms and as primary consumers form an important part of the food chain.

Protozoans are classified based upon their style of locomotion or type of organelle used for locomotion. The three main modes for movement are pseudopodia, flagella, and cilia. Pseudopodia, or "false feet," are found with *Amoeba* spp. These organisms do not exhibit free swimming yet they are considered a common freshwater species. You will often find these species on the underside of lily pads and other vegetation in shallow water. The flagella, or "whiplike processes," are found in *Euglena* spp., a common genus in freshwater ponds that often gives a greenish tinge to water if present in sufficient numbers. Lastly, some organisms possess cilia, or "hairs," that cover its body. *Paramecium* spp. are an example that are common in ponds with considerable vegetation.

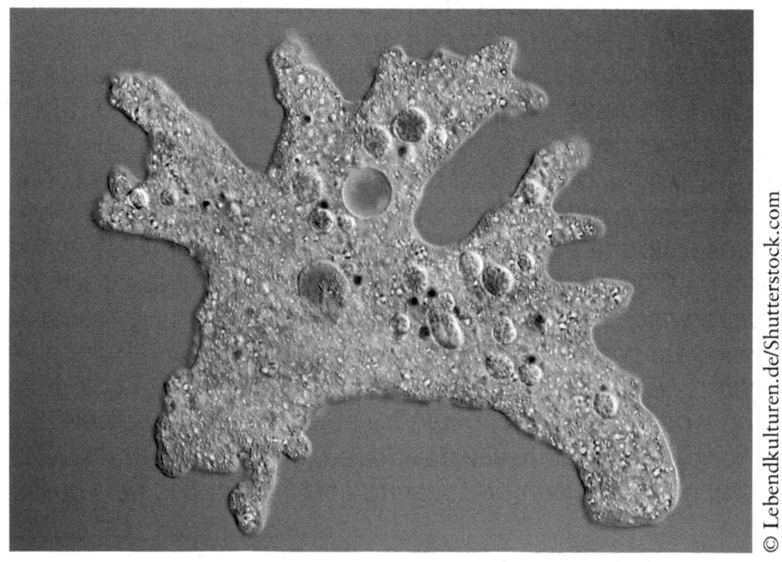

Amoeba proteus within Phylum Sarcomastigophora and Subphylum Sarcodina. Habitats include slow streams, clear water ponds, often living on the bottom of lily pads or on shallow vegetation.

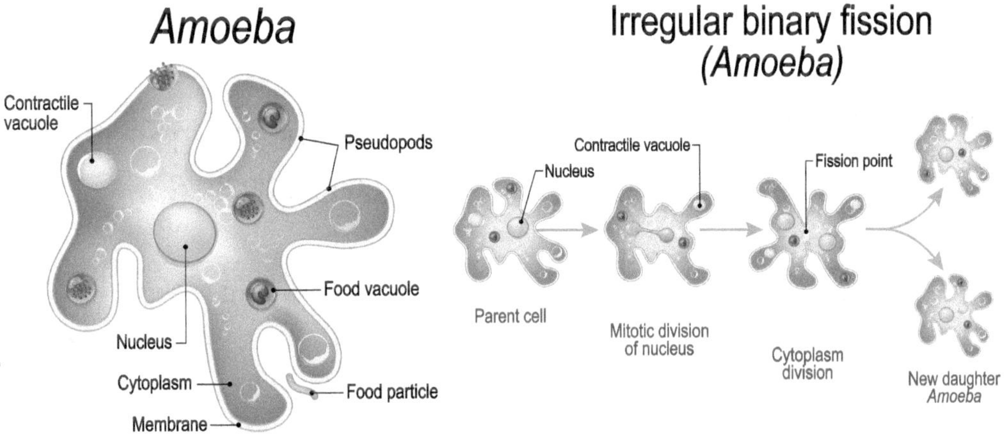

Amoeba

Contractile vacuole
Pseudopods
Food vacuole
Nucleus
Cytoplasm
Food particle
Membrane

Irregular binary fission (Amoeba)

Contractile vacuole
Nucleus
Fission point
Parent cell
Mitotic division of nucleus
Cytoplasm division
New daughter Amoeba

Food-getting and digestion of *Amoeba* involves moving toward food via its pseudopodia to engulf prey like phyto- and zooplankton. It uses its food vacuole for digestion. Respiration occurs through the membrane, or plasmalemma, which allows gas exchange. Reproduction, or binary fission, happens when it reaches its full size. These protists will respond to stimuli like all animals.

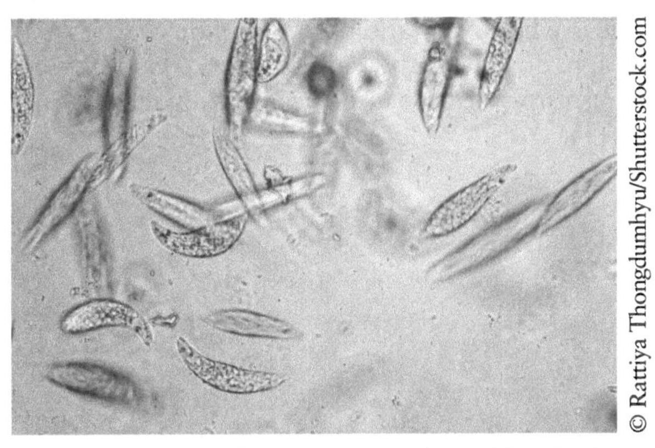

You will find *Euglena* spp. propels itself in freshwater streams and ponds with its flagella, and will give a green color to the water if present in sufficient numbers.

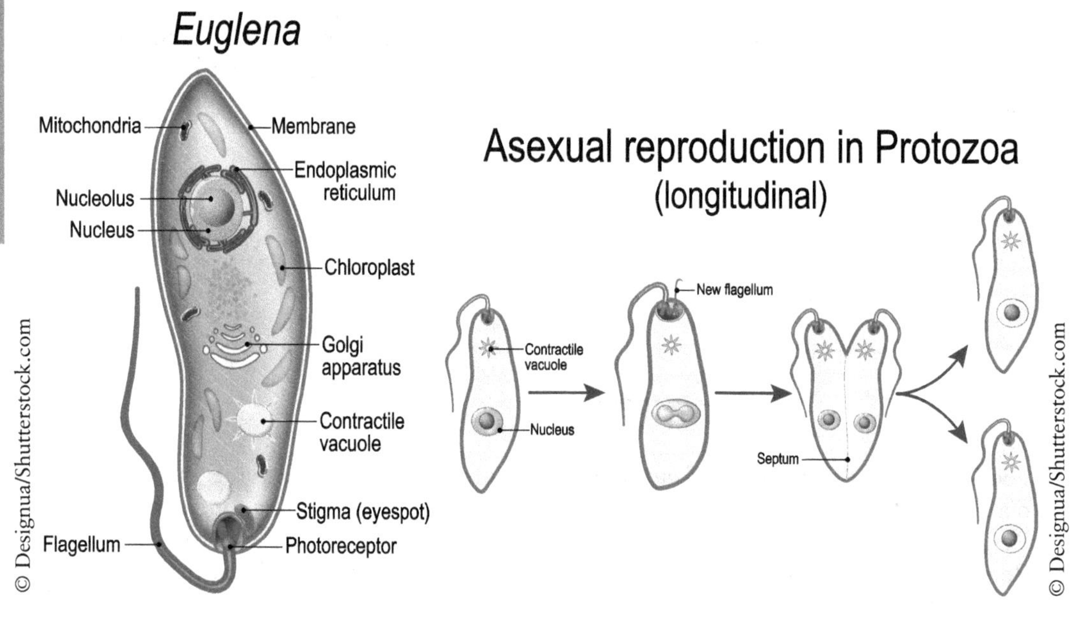

Euglena spp.'s classification may be debatable due to its abilities to obtain nutrients. These organisms can photosynthesize (autotrophs) and can absorb dissolved nutrients through their pellicle. This also makes them heterotrophic, or more specifically, **saprozoic.** Additionally, the pellicle is where respiration and excretion of gases and wastes occur. For reproduction, *Euglena* spp. divide lengthwise. Lastly, in unfavorable conditions like the lack of water, the organisms may **encyst** to prevent drying out.

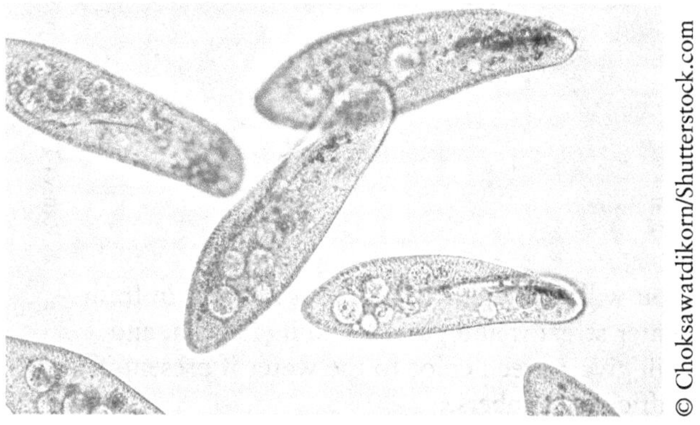

Paramecium caudatum from Phylum Cilophora inhabits freshwater pools, ponds, and sluggish streams.

Locomotion is a result of beating its numerous cilia covering its body, while the organism rotates along its longitudinal axis. The body is flexible and will bend accordingly. Its behavior includes avoiding reaction. While observing this species, you will see it bumping into an object, backing up, and "trying" again.

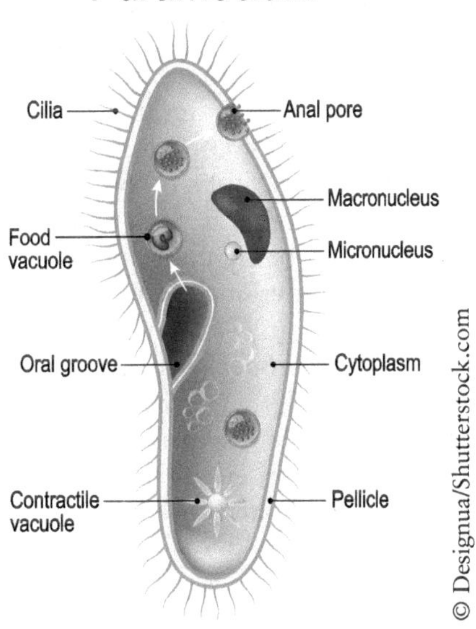

Paramecium spp. eat bacteria, algae, and other small organisms by using cilia in the oral groove to sweep food into its body where food is collected in a food vacuole. The food vacuoles circulate through the cytoplasm until food completely digests. Wastes are released through the anal pore. Respiration happens through the body surface.

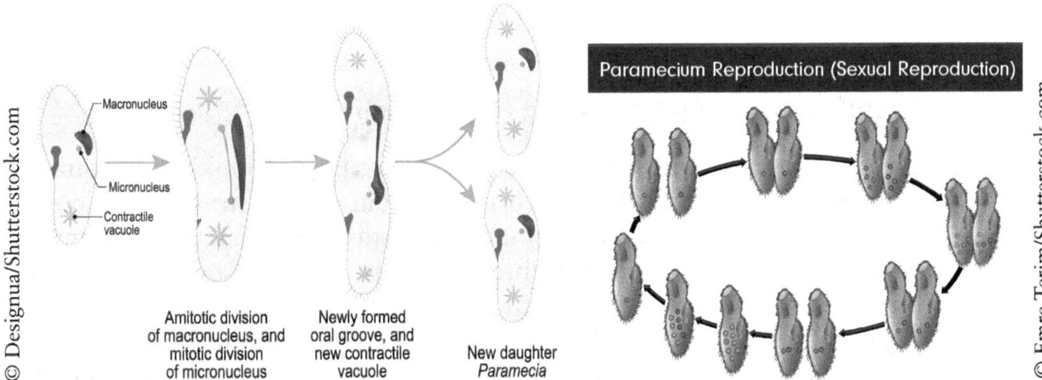

Paramecium spp. reproduces asexually by division, or sexually by attaching its oral groove with another individual and swapping their genetic material, followed by separation and and division.

Nature Journal: As discussed in previous chapters, pond and stream studies can be very exciting with potentially unfamiliar habitats and organisms to discover. Practice good safety when near water to observe or make temporary collections of pond water. Let somebody know where you will be, wear gloves as a precaution, and if you must go in the water, stay in areas with water below your knees and wear waders if you have them.

To observe organisms in their natural environment, use a magnifier to examine the undersides of lily pads and other vegetation. You can also make an underwater viewer by submerging the bottom of a clear glass jar. Lastly, try taking a sample of the pond water in a jar to use a magnifier through the glass, or put a small sample under a microscope for further studies. Remember to leave no trace when you finish making observations—take all your equipment and belongings! Record your findings. Complete a Current Conditions report (as seen in Chapter 1) for where you made the observations or collections, draw the specimens observed at the greatest magnification used, and research more about your experiences.

Review Questions

1. What are protozoans? Where are they found? What are the possible means for locomotion? Provide an example of a species type for each mode of locomotion.
2. Explain the characteristics, habitat, niche, and basic biology of *Amoeba*, *Euglena*, and *Paramecium* spp.

7.2 Freshwater Invertebrates: Kingdom Animalia

If you have spent time making observations of water samples, looking on aquatic plants, examining submerged woody debris or within the substrate of the pond or stream, you may have encountered small organisms that are actually animals. The major groups of freshwater invertebrate animals witnessed may fit into one these categories: sponges, cnidarians, rotifers, worms, mollusks, or arthropods.

We begin our journey in Kingdom Animalia with the simplest or most primitive of organisms, the freshwater invertebrates. In other words, these are animals found in lakes or rivers and do not have a backbone within their body. They are also often visible with the unaided eye (although a magnifier definitely helps in some cases).

Phylum Porifera. **Phylum Porifera,** or the sponges, have a body entirely of pores, which allow for respiration with the water flow. Although found mostly in marine environments, you will also discover them growing upright or encrusted on submerged rocks or logs in shallow freshwater as colonies of 1 in. (2.5 cm).

Sponges have a body of silica that provides support to their structure. They feed on floating or swimming microscopic plants and animals that get trapped in their pore-riddled body as water circulates through. Due to their sedentary lifestyle and dietary requirements, they offer an indication of relatively good water quality. They can reproduce by various means. One is by gemmule formation or

Spongilla lacustris

encasing a fragment of the sponge in winter for protection and allowing it to grow when conditions become favorable again. Another means is asexually by budding, or simply buds form in the spring and drift away to begin a new colony. Lastly, they can reproduce sexually in the summer by releasing sperm to swim into another sponge for fertilization, which results in the birth of free-swimming larvae that had developed within the sponge's inner cavity.

Phylum Cnidaria. With the **Phylum Cnidaria,** all representatives possess stinging cells (nematocysts or cnidocytes) on its tentacles. These are the "jellyfish," and they exist in the region as a nonnative, invasive species. These freshwater animals only grow to about 1 in. (25 mm) in diameter as the adult, free-swimming medusa stage. They will have about 400 small tentacles packed around the margin of its umbrellalike "head." Food and waste are both taken in and expelled out of the same underside opening, respectively. They drift in the water with tentacles extended awaiting prey such as zooplankton to inject the paralyzing poison from the nematocysts. However they do not have the capacity to harm humans. Freshwater cnidarians can be found in slow-moving rivers or ponds and lakes.

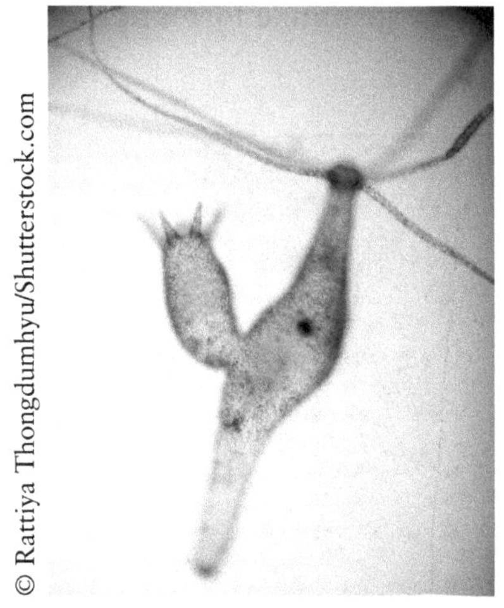

Cnidarian in its juvenile or polyp form. Note the budding or means of asexual reproduction taking place on the left side of the image. Its tentacles are at the top with its mouth at its center.

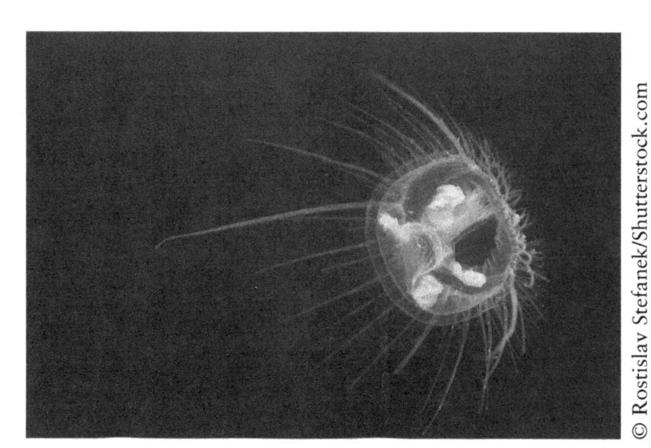

The adult cnidarian, *Craspedacusta sowerbii* in a Midwest lake.

The cnidarians, juvenile stage begins as a small polyp attached to underwater substrates as colonies. They can move slowly across the substrate or "somersault," turning end over end. Food that floats by can become entrapped in their tentacles. This stage can also potentially encyst over the winter allowing it to reemerge in better conditions.

Phylum Rotifera. In your pond samples, have you seen any microscopic organisms under your magnifying lens or microscope that seems to have a rotating movement on one end that propels it through the water? You may likely have **Phylum Rotifera** in your sample. These widely distributed "wheeled" animals inhabit the littoral zones, soils, mosses, and lichens. They feed on algae or prey upon other **zooplankton**, or animals that drift in the water column. Although, zooplankton have mechanisms for locomotion they simply go with the flow of the water (a wide array of this sized animal can apply to this definition). Like other zooplankton, rotifers serve as an important food source for worms, crustaceans, and fish larvae. Rotifers can also encyst to remain dormant if a pond dries up to awaken in favorable conditions—water.

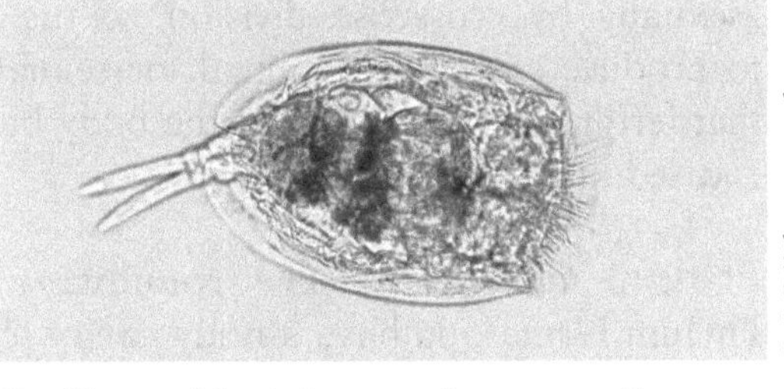

Rotifer euchlanis is one of many rotifer types. Note its cilia on the front end (right side of image). This will "wheel" it through the water. The opposite end, its base or foot, can secrete a "glue" to attach to objects.

Phylum Platyhelminthes. The "worms" in the animal kingdom can be classified into one of three phyla, all of which are unrelated groups except for sharing a similar general shape. The most primitive group being **Phylum Platyhelminthes.** Most species in our region are from Class Turbellaria, and we can classify many of them as planarians. You will find these animals in quiet bodies of water. Some say that you can catch planarians by baiting a hook with a small piece of raw liver near the bottom in the littoral zone. The light sensitive animals remain hidden among and under submerged vegetation and rocks during the day. Planarians are fun to watch under a magnifying glass or dissecting microscope.

Planarians are scavengers with their mouth midway on the underside of its less than 1 in. (2.5 cm) body. They move along surfaces by means of hairlike cilia covering their underside and eat smaller animals, living or dead. These free-living flatworms do not possess a body cavity (**acoelomate**), they lack an anus, have no circulatory system, and respire by diffusion. Flatworms reproduce

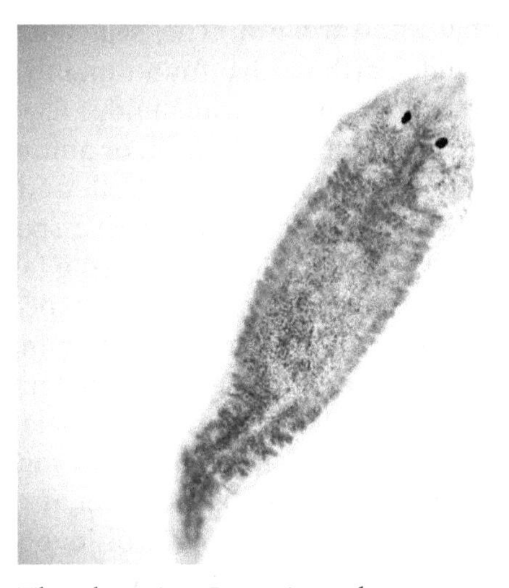

The planarian *Dugesia* sp. has two prominent eye spots that detect light. This is the largest free-swimming flatworm in North America and common in marshes or littoral zones of ponds.

asexually by crosswise division of the body to produce a complete animal, or sexually so that fertilized eggs outside of the body become encased in a hard cocoon.

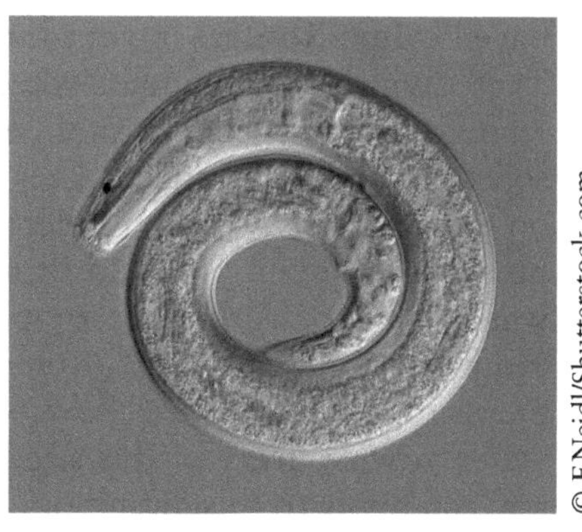

Phylum Nematoda. The roundworms of **Phylum Nematoda** have a body cavity (coelomates), therefore have a round cross section. These unsegmented roundworms have a size range from very small 0.1 mm to very long 1 m (those parasitizing small intestines of mammals). Nematodes, free in nature, will be found in soil and found in the bottom mud or debris in standing water. In the water column, you can identify them by their constant whiplike thrashing motion. Species' niches may be predatory, herbivorous, or parasitic. They also help cycle nutrients but can either promote or reduce decomposition and plant productivity. Nematodes provide a good food source for small animals and serve as a host for some bacteria and fungi. Some scientists proposed that a diversity of nematodes may indicate good soil quality.

Phylum Annelida. The segmented worms in our region, **Phylum Annelida,** include earthworms, found mainly on land; and leeches, found mostly in freshwater. The earthworms, under Class Oligochaeta, feed on organic matter while burrowing in soil or muck, or among decaying vegetation or floating algal masses. They offer physical, chemical, and biological benefits to soil fertility. However, several nonnative and invasive species of earthworms exist in our regional forests. The concern exists with the fast rate they can consume the organic layers of the soil. The result causes an expedited shift in plant community assemblages—causing die offs or stunting of young plants or saplings. Native earthworm species also exist in our region from families Megascolecidae (*Diplocardia* spp.) and Sparganophilidae, which are semiaquatic. The size ranges, dependent on species, from 10 mm (0.40 in.) to 360 mm (14 in.) long. The earthworm has complete physiological systems; however, it respires through its skin. They have a digestive system running the length of its body, a closed circulatory system, and have hermaphroditic capabilities by possessing both male and female reproductive parts. Research further information and videos on earthworms to provide insight on these common animals discovered in nature.

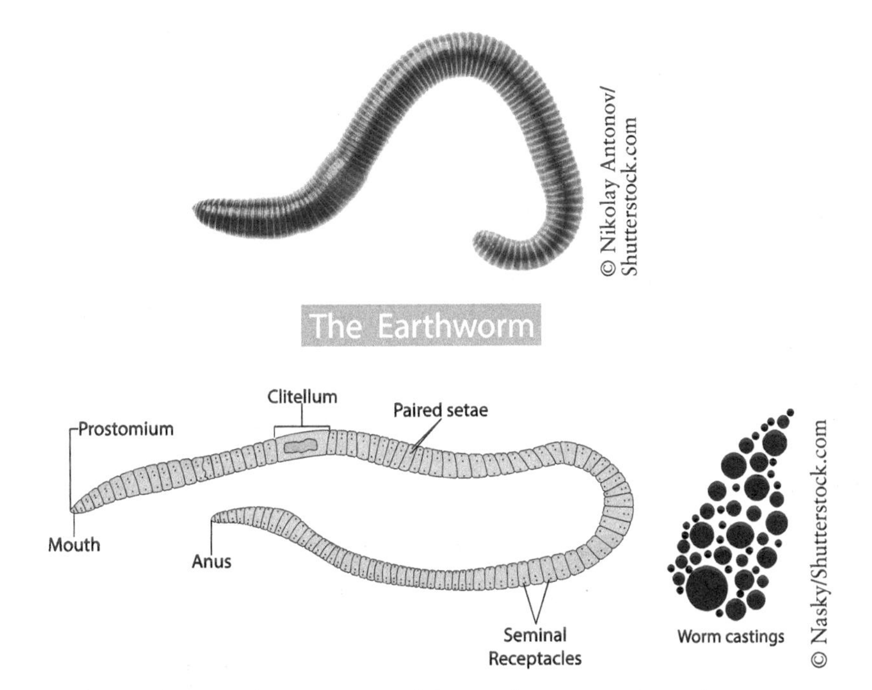

The diagram shows the exterior anatomy of the earthworm with castings, or its waste, which serves as a rich, organic fertilizer in compost and the soil.

Leeches, flat and segmented, inhabit calm, dark waters of the littoral zone among vegetation. They move along substrates by going end over end alternating mouth and tail suckers, whereas others swim freely. Parasitic species have specialized jaws compared to the minority of scavenger and carnivorous species.

A leech, within Phylum Annelida, is closely related to the earthworm. Their sizes range from 1 cm (0.4 in.) to 7.5 cm (3 in.; or longer).

Phylum Mollusca. The animals within **Phylum Mollusca** include those with a soft body typically surrounded by a shell. In freshwater, you will find snails and clams of which have secreted their shell. On land, you will also discover snails in moist environments and slugs, who do not have a shell. Over 80,000 species of mollusks have been recognized. Their population structure is used as bioindicators for monitoring the quality of aquatic environments. In some areas, native freshwater clams have declined significantly. They exist as an important food source for some animals and have been utilized by humans as a resource. Concerns of terrestrial mollusks arise in agricultural fields where they have caused crop damage. In many of the lakes in the Great Lakes and Upper Midwest regions, two invasive mussels, the zebra (*Dreissena polymorpha*) and its larger and more concerning cousin quagga (*D. bugensis*), have caused significant alterations to aquatic ecosystems due to the intense filtering abilities it possesses and lack of a true predator. Further studies of this phylum would offer a fun and informative aquatic and terrestrial nature study.

Zebra mussels (*Dreissena polymorpha*), an invasive species, on a boat prop and a close-up view.

Phylum Arthropoda. Lastly, the largest phyla of invertebrates belong to **Phylum Arthropoda**—a very diverse group. These animals have externally jointed legs with a segmented outer skeleton (**exoskeleton**). They have well-developed circulatory, digestive, nervous, and reproductive systems. Arthropods exhibit an array of complex behaviors relative to other invertebrates. Three classes are quite abundant in our freshwater systems—Classes Insecta (i.e., dragonflies), Arachnida (i.e., spiders), and Crustacea (i.e., crayfish). These, among other arthropods, are discussed more in depth in the next section.

Review Questions

1. In what kingdom are sponges, hydra, flatworms, roundworms, leeches, clams, and crayfish found? Name the phylum for each of these examples.
2. Provide a characteristic for each phylum discussed in this chapter that distinguishes it from other phyla.

7.3 Arthropods

> ***Nature Journal:*** Before you begin learning the basics about this amazing phylum, reflect on the following: (1) Think about the last time you encountered a "bug"—an insect, a spider, or a tick—something without a backbone. (2) How did it make you feel? (3) Describe what it looked like. Remember to include colors and size (relate it to an actual measurement or in relation to a familiar object like a coin size, finger, fist, etc.). (4) Where did you find it? (5) What was it doing? (6) How would you describe or interpret its behavior?

Within Kingdom Animalia, we have the largest and most diverse Phylum—Arthropoda (meaning "jointed feet," or in other words—externally jointed). These animals do not have a skeleton; therefore, no vertebrae or backbone exist—they are invertebrates. However, they have an external skeleton – an **exoskeleton.** This exoskeleton protects the muscles and organs within the animal's body. The muscles pull from a distance on the inside of the skeleton to obtain leverage and aid in locomotion. For their size, arthropods are the strongest animals. An exoskeleton cannot grow, so it periodically must shed. You may find these shed carapaces in nature. At first, you might think you see an insect, but it has actually grown out of its covering. That insect has moved on with its newly grown exoskeleton. This segmented exoskeleton is made of chitin, which is a nitrogenous material secreted by the animal. The chitin hardens the exoskeleton, waterproofs it, and protects the animal from potential dangers and abiotic factors.

In the head of arthropods, you will find the highest degree of **cephalization**, or the greatest concentration of sense organs. There are two different types of eyes that could exist. Depending on the arthropod, it will have either compound or simple eyes. **Compound eyes** have multiple lenses, are image-forming, and covers

A fly's compound eye (25x magnification).

most of the head's surface area. Due to the eye's huge size and vantage points the lenses can have on its surrounding environment, it can be difficult for predators to catch them. The simple eyes (**ocelli**) detect light and dark, and do not form images. Other types of sensory perception involves the arthropods **antennae** with which it senses its environment as a touch or taste assessment.

Class Insecta

The largest arthropod class is Insecta (Kingdom Animalia, Phylum Arthropoda, Subphylum Hexapoda, Class Insecta). Those scientists who study insects are called **entomologists.** Their research could focus on a species' population or communities in which they live. Perhaps the focus is on behavior, breeding, evolution, or relationships to other animals or plants or to their environment. The topic possibilities are virtually endless and can help us better understand these organisms and their roles in nature.

The characteristics of insects include the presence of six legs (or three pairs), three body parts (head, thorax, and abdomen), and a compound or simple eye structure. Observe the features of the bee in the diagram. You will notice the eyes, antennae, and mouthparts attached to the head. Another noteworthy feature on the head (not seen on the diagram) includes the organ called the tympanic membrane that detects sound vibrations (yes, they can hear!). Posterior to the head, you have the **thorax** where you will find the wings (if present on the species) and all six legs. Often, insects have two pairs of wings, but the presence and number of wing pairs is species dependent. The wings are typically membranelike, and transparent with veins. In Order Coleoptera (the beetles, like "ladybugs"- keep reading to find

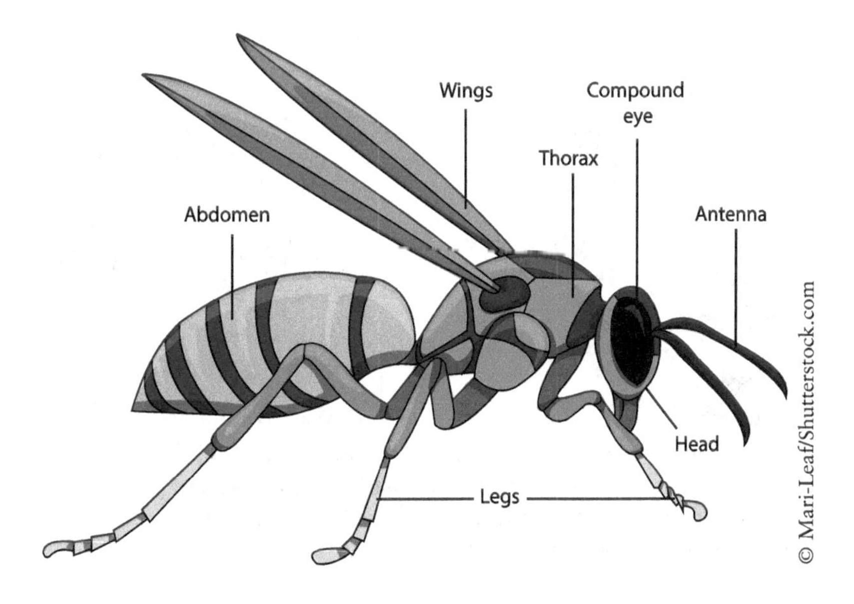

out why these are not truly "bugs"), the first pair of wings are modified into hard covers that protect the membranous flight wings beneath it. Order Diptera's (the flies) have very small secondary wings that are used for balance. In the last body segment, the **abdomen** contains the reproductive parts, such as the female's ovipositor for depositing eggs and if the insect is in the Order Hymenoptera (i.e., bees and wasps), they may have a stinger attached to the tip of the abdomen.

Human Connections

In terms of value to humans, we need insects! Regardless of any person's feelings about them, insects provide benefits to humans and the ecosystem. In fact, 99.6% of all insects are considered beneficial. One-third of human food relies upon pollination from insects, and insects prey upon other insects we may consider pests or potentially carry disease. For example, the dragonfly preys upon mosquitoes. They help keep mosquito populations from growing too large and dominating an area. Additionally, insects also nourish many fish species as the insects deposit eggs on the water's surface or while many exist in the aquatic juvenile stage.

The other 0.4% of insects considered "pests" are classified as such because of the actual harm or destruction they can cause to humans or human possessions. These insects can destroy food crops, forests, clothing, or serve as vectors for human transmissible disease. Another great aspect of insects is that they can be aesthetically pleasing. The colors or displays can be considered pretty to observe. Once you start noticing and watching insects, and perhaps examining them up close you can develop an appreciation and fascination of these animals.

A bee pollinates a flower. As the bee lands on a petal, the pollen from the flower attaches to the animal's brushy parts of the body. The bee visits another flower and those pollen grains can pollinate the female part of that flower, the pistil.

Insect Success

Insects have existed on this planet for over 300 million years. How have they survived all these years? Think about it. Their success can be attributed to many of their characteristics or behaviors. Their attributes as an r-selected species provides most of the answers (see Chapter 4). Most insects have a small size. How is this helpful? It can help make it easier for them to fit into places out of view of potential predators. They also have a fast reproduction rate to enable them to adapt relatively quickly to a changing environment. Most have wings to escape dangers. Their exoskeleton is hard and flexible to provide protection and agility. Many exhibit crypsis or resembling something in its habitat that helps them escape notice (e.g., looking like a stick or dead leaf). Or, perhaps they are mimics and look like something dangerous (e.g., bee fly looks like a bee but cannot sting). Some are not palatable or bad tasting to predators. Some insects can utilize different food sources and habitats by going through metamorphosis. You will also find varying diets of insect orders, so this helps ensure there is something for everyone (their mouthpart characteristics can help you identify their niche). As you learn more about insects, you will also discover their resiliency to adverse conditions. Many can resume life functions after freezing or drying out. Lastly (?), have you noticed more insects seem to emerge at night? Most insects are nocturnal. With the limited light, it makes it difficult for predators to detect them. No wonder these animals have been successful for so long!

Insect Physiology

Breathing exists within insects but is done differently than in humans. In the **respiratory system** of insects (see the grasshopper diagram), spiracles, air sacs, and trachea are used. Spiracles are openings on the surface of the insect's body. Gases pass through these openings (inhalation and exhalation) and travel through tracheoles (tubelike structures) and then moves to air sacs. The air cycles through these features and into each cell continuously. This type of respiratory system limits the animal's size.

The **digestive system** of insects is comprised of the following organs: crop, gastric caecum, stomach, ileum or small intestine, and colon. These organs are listed in the order in which food passes. The crop is used for food storage, the

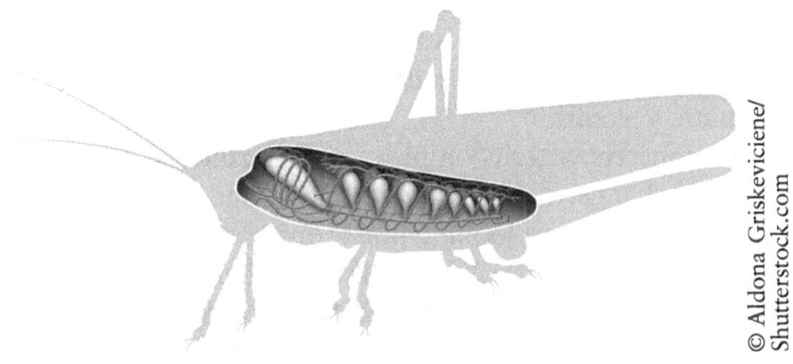

Respiratory system of a grasshopper.

gastric caecum are pouches used for additional digestive surface area, the stomach is where digestion of food begins, food absorption occurs in the ileum (or small intestine), and the function of the colon is for food and waste product elimination. Insects also have excretory, reproductive, and cardiac systems.

Observing the mouth of an insect can tell you more about its niche. It is adapted for acquiring its favored food item. The four common mouthparts include sponge, sucking, siphon, and chewing. Flies have a mouthpart like a sponge so it can soak up nutrients on almost anything it lands upon. Insects that draw blood or "bite" have sucking mouthparts (i.e., deer fly and mosquito). Some insects have a mouth like a straw or a siphon. It is long and tubular to allow it to access flower heads with nectar held deep inside (i.e., butterflies and moths). Insects like grasshoppers have chewing mouthparts to eat the plants on which they depend.

Metamorphosis

The word metamorphosis translates as something having "many body shapes." In other words, the organism will undergo significant change depending on the species. Depending on the species, the types of changes that can occur include a change with the body shape, the niche (job or role in its ecosystem; typically a change with what the animal eats or what eats it), how the animal's physiology operates (internal functioning), or where the animal lives. Many insects are born in the water from eggs, live in a larval form or nymph, possibly pupate, then emerge from the water as a flying adult.

Different types of metamorphosis exist. The most common type of metamorphosis with insects is "complete"—approximately 88% of all insects demonstrate this life cycle. It has four distinctive stages: egg, larva, pupa, and adult. Each stage is very different from the other. Every stage has its own niche. Beetles and butterflies are examples of insects that move through this process.

With incomplete metamorphosis (about 12% of insects), three stages exist-egg, nymph, adult. The biggest change is that the habitat and niche differ for the nymph and adult. A great example is the dragonfly. They live in freshwater for the first two stages, then emerge as an adult, which lives in terrestrial environments. They can fly, and will have a new diet yet still are carnivorous. Insects with this type of life history will typically live near water as adults since they need to lay their eggs in the water. Dragonflies will gently deposit their eggs on the surface of the water, often near aquatic vegetation.

For terrestrial insects with incomplete metamorphosis, they will exhibit gradual, or simple life cycle change. Grasshoppers crickets, and mantids are a common example. From the egg, a miniature looking grasshopper emerges as the nymph stage then matures with multiple shedding of its exoskeleton before transforming into the adult stage. The habitat and niche for this type of life cycle remain the same for the nymph and adult stages. A few insects do not exhibit metamorphosis. The silverfish insect is an example that simply goes from egg to adult without any changes except for size.

Holometabolism
(complete metamorphosis)
Adult
Egg
Larva
Pupa
(Grub)
Rhino Beetle
Holometabolism
(complete metamorphosis)
Egg
(Caterpillar)
Larva
Adult
Pupa
Swallowtail Butterfly
(Chrysalis)
Hemimetabolism
(incomplete metamorphosis)
Dragonfly
Adult
Egg
Nymph
Paurometabolism
(incomplete metamorphosis)
Egg
Nymph
Adult
Cricket
Paurometabolism
(incomplete metamorphosis)
Eggs
(in Ootheca)
Adult
Nymph
Praying Mantis
Ametabolism
(no metamorphosis)
Egg
Adult
Juvenile
Silverfish

At each stage of metamorphosis, a new niche is filled. Remember, not all insects go through each of these stages. Now when you encounter insects in these various stages, you will have an understanding of the role they serve in the ecosystem in that point in time.

With complete metamorphosis, the following descriptions are the roles each life cycle stage serves the ecosystem.

Egg—A female insect lays eggs which contain the developing insect. They can be laid singly, in masses, or encased. Eggs can serve as nutrients for other animals.

Larva—Larvae hatch from eggs and do not look like adults (usually have a wormlike shape). Caterpillars, maggots, and grubs are all just the larval stages of insects. Larvae molt their skin several times and they grow slightly larger. The niche is to eat a vast amount of food during this stage. Larvae also serve as a source of food for other animals (i.e., birds, fish, rodents).

Pupa—Larvae make cocoons around themselves. Larvae do not eat while inside the cocoon. Their bodies develop into an adult shape with wings, legs, internal organs, and so on. This change takes anywhere from four days to many months. The niche is the transformation and as a potential food source.

Adult—Inside the cocoon, the larvae change into adults. After a period of time, the adult breaks out of the cocoon. The adult may have a new diet and will also be prey for other animals.

For incomplete metamorphosis, stages differ as well as their niches.

Egg—Often encased to protect and hold eggs together.

Nymph—Eggs hatch into nymphs that look like small adults but usually do not have wings. They eat the same type of food as an adult (a carnivore remaining a carnivore) but lives in a different habitat. They shed or molt exoskeletons and replace them with larger ones several times as they grow. Most nymphs molt 4–8 times. Their niche is to eat, grow, and be food for other animals.

Adult—They stop molting when they reach an adult size and have also grown wings. The niche is a new diet and habitat, and a food source for other animals.

Nature Journal: Become familiar with the field guide in this section so you can make some efficient deductions about what insects you find. Recognizing the insect orders will give you a basic understanding of that group and help make for quick identification.

Spend time looking for insects. Bring along your magnifier. Check out the ideas below for increasing your chances of finding a diversity of insects. Record when your specimens were found, what they looked like, behavior, and habitat. Complete Natural History Journal Pages to help you research the insects further (found at the end of the arthropod section).

Now it's time to branch out. At your Sit Spot, if you have a nearby tree or shrub with reachable branches, simply put a light-colored cloth beneath the branch. Give the branch a shake or tap it with a stick. Discover what has fallen!

Try getting swept up with more insect investigations. If you have a fine-meshed net, use it to sweep through tall grasses (in a pinch, a nylon stocking over a pulled wire hanger can work). With the net facing down, move its opening across the vegetation in a pendulum-like motion. Check out what you collected!

Digging in with further investigations. Create a pitfall trap in your Nature Area or in your garden. Don't worry, you will not injure the organisms – just be sure to check your pitfall often. You may likely want to eventually create several of them. This could also be a fun field investigation (see Chapter 2). All you need is a trowel (small, hand-digging shovel), two plastic cups, four rocks, and a piece of wood. Think about where you might discover the most diversity of insects (or whatever your question may be)—under trees, among your herb garden, next to a wetland, bare soil, and so on. Dig a hole for the cup so that the top is at ground level. Stack the cups and insert them in the hole. You can remove the inner cup later on for investigating your catch while leaving the other in place to keep collecting. Arrange the four rocks around the hole and place the wood on top of them to shade the hole. Inspect the hole regularly and record what you find. Fill your holes when you have finished with your study.

Lastly, leave a light on. At night, turn on a porch light to attract insects for better viewing. To really get a look at what is drawn to the light, hang a white sheet near the light (without letting the light touch the fabric). Your discoveries may amaze you!

Remember to keep a detailed record of your observations and questions in your Nature Journal. A nature study in entomology can be done anywhere!

Class Arachnida—Spiders, Ticks, and Mites

Another type of common arthropod are the arachnids—these are not insects! Common examples of Class Arachnida in our region include spiders, harvestmen, chiggers, ticks, and mites. Just like other things in nature, arachnids have a niche. They may eat insects or other arachnids considered as pests to humans. You can recognize an arachnid if it possesses the basic characteristics of eight walking legs (four pair), one to two body parts, no antennae, ocelli (simple eyes only), no wings, exoskeleton with external joints, and mostly carnivorous. How are these characteristics different than insects?

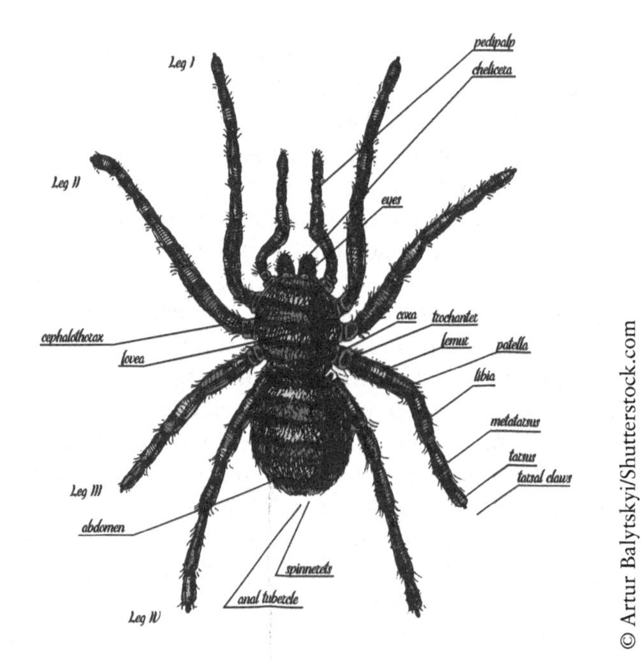

Spiders

Most people will think of spiders as an example of arachnids. For practical purposes, this will be the emphasis for this section and the field guide. Order Araneae are the most obvious, common, and diverse of the class.

> ***Nature Journal:*** How do you feel about spiders? When was your last encounter with one? Where was it? View the section of this book where some spider species are pictured and described. Do you recognize any of them? List the familiar spiders and those you would like to find in nature. Then, try to find them! Keep detailed records of your observations.

The basic features of spiders include two body parts (the cephalothorax, which combines the head and thoracic regions; and the abdomen), eight pairs of simple eyes (16 total), and poison fangs. Of the four pairs of legs, the first pair will often be used for feeding and defense with the second pair used in feeding, moving, and as a reproductive function. The last pairs of legs are used for walking. All spiders have venomous fangs but most are too small to harm us, or do not have enough venom to cause health issues. Two venomous spiders in our region are the northern or black widow and the brown recluse, but are uncommon to rarely found.

Other physiological features of spiders include a respiratory system that functions slightly different than insects with the presence of book lungs rather than air sacs. They have a simple digestive system due to only ingesting liquids, yet all spiders are carnivores. Spiders will either be hunters or will spin webs to capture their prey. The spider secretes anticoagulants and digestive enzymes externally over its captured prey, and digests the prey outside of the spider's body. In other words – it will catch the prey, inject it with venom to subdue it, then inject enzymes to liquefy the organs to suck it out (with its sucking stomach organ) and into the pharynx then the stomach. The caeca contain pouches for additional liquid food storage, and the intestine will absorb the food's nutrients.

Although not all spiders spin webs to capture prey, they may still produce the silk for other purposes. The spinnerets, fingerlike glands located in the spiders' abdomen, produces the spider's silk. The silk is a liquid protein that is stronger than human muscle per weight, and so elastic it can stretch more than 300% its length before breaking. It is used for one or more of the following things depending on the species—safety lines, swathing prey, egg sac formation, ballooning, or webs.

The silk released as a safety line (dragline), trails behind the spider to lower it or help land as it becomes attached with what it touches. Some spiders will use their silk to swath their prey for a later meal, and others create a silken case for their cluster of numerous eggs they lay. Spiderlings, the juveniles a few days after hatching (which have grown into a miniature version of the adult after being born tiny and blind), will "balloon" or parachute via their silk as the autumn winds disperse them to a new area to inhabit. Some adults will do this as well. Due to hundreds of spiderlings potentially emerging in one place, the area cannot support all of them and cannibalism often occurs after the young feed on their stored yolk sac. Ballooning allows for a quick dispersal. After leaf drop in October, look for these threads in the sky or hung up on trees. How many can you count?

Spiderlings of an orb weaver grouped together after molting and starting to disperse.

Those spiders who spin webs can also be considered some of the animal world's best architects with the web intricacies they create and the immense tensile strength they possess. A web's function is primarily a hunting snare, so it needs to be durable. Think about whether you have seen spider webs in nature. What did it look like? Where did you see the webs? Spider species who spin webs will only spin one type of web. This is part of the natural history of the spiders you would want to know. When you find a web, you can have an idea of who spun it, but you will not know for certain the species unless you see the animal. The best time to find webs is on a dewy morning or after a rain when the sun is out. The webs glisten in the sun making it a fun thing to focus on while on a walk. If you accidentally walk into one, the spider may consume the remaining silk threads and recycle it into a new web construction in about a half hour.

There are four common spider web types you may encounter. The *cobweb* has an irregular pattern and is usually just sticky on edges while the spider sits inverted in the center awaiting potential prey to become entangled in its construction. We may mostly think of these webs indoors, but they also exist in abundance in nature. The families that create cobwebs include Theridiidae, Pholcidae, and Dictynidae (see the field guide at the end of the section for examples of these spiders).

The *sheet web* will be found among the grass or bushes, and is flat, convex or concave; mostly nonsticky with it only tacky along trapping threads above the sheet. The spider sits inverted beneath the sheet, pulling victims through the bottom of the flat web. You will mostly discover sheetweb weavers in our region that are from Family Linyphiidae.

For the next two webs, you may discover the shape and intricacy very fascinating. The *funnel web* is flatter than the sheet web and has a funnel opening in the center where the spider sits in retreat waiting to dash quickly out and grasp the entangled prey. The web threads are not sticky. You may find them indoors in corners, windowsills, basements, and outside tucked in corners around buildings or among the grasses. Family Agelenidae are the funnel weavers.

The *orb web* is the one with large, circular, intricate patterns. You will usually find them vertical and high above the ground suspended between small trees and shrubs. It has threads that anchor it that are not sticky,

Funnel web with a spider.

but those spiraling to the center are sticky and will hold the prey. The spider sits in the hub or off to the side or rolled up in a leaf as a retreat awaiting its prey. Orb weaving spiders are not typically dangerous to humans. The three orb weaving families are Uloboridae, Tetragnathidae, and Araneidae.

An orb web with a tiny spider at the center.

Nature Journal: On a dewy morning, head out to your Nature Area and search for webs. Photograph and draw what you see. Try to find the spider who constructed the web and see if it is in the field guide. Complete the Natural History Journal Pages at the end of this chapter for the spider species you found.

On a summer's night, take a flashlight or headlamp to look for spiders. Wolf spiders can easily be observed traveling on the ground, and fishing spiders on water surfaces that have a large body with eyes that will reflect the light as green! In addition, you might discover orb weavers building webs at night and may be more prevalent near the water. Make reflections in your journal of what you discover in nature and from your research.

Subphylum Crustacea—Zooplankton and Crayfish

In this subphylum, we have the crustaceans. The regional examples include zooplankton (i.e., copepods and daphnia) and crayfish (see this chapter's field guide for an image and description). Other examples in the group, but not in the region include lobsters, crabs, and shrimp. Have you seen any of these in the wild? How about in an aquarium?

Many microcrustaceans, the zooplankton, are important primary consumers in aquatic ecosystems. As discussed in Chapter 4, the primary consumers level in a food chain indicates it eats "plants" (herbivorous zooplankton would eat mainly

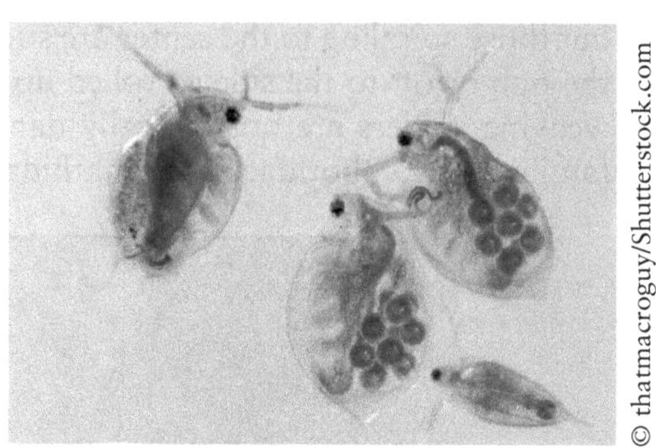

Upon close examination of pond or lake water, you may discover the zooplankton, copepods and *Daphnia* sp. (0.01–0.20 in.; 0.2–5 mm)

the phytoplankton, algae that are actually protists). These organisms are preyed upon by secondary consumers (i.e., larger zooplankton, and fish).

Larger crustaceans like the crayfish, is a food source for larger animals such as birds and aquatic mammals. Humans also use many crustaceans as food. How can you tell if an animal is a crustacean? You will likely first recognize its basic arthropod characteristic of having an externally jointed carapace. A crustacean's exoskeleton is relatively hard and composed of calcium carbonate. They have two pairs of antennae, paired appendages, and two major body parts (the cephalothorax and abdomen). Males will often have large, specialized "arm/s" to clasp the female. They also have a digestive system with a two-part stomach with internal "teeth" and a digestive or "green" gland, which serves the function of a liver (i.e., purifying blood). Crustaceans respire using gills.

Class Diplopoda—Millipedes

Another arthropod are the multisegmented diplopods or millipedes. We have these animals in the region but only reach up to 1–2 in. (2.5–5 cm). They have two pairs of legs per segment (segments are fused with most species having under 100 legs). These herbivores have odor glands used for defense. To see an example of a millipede, see the field guide at the end of this section.

Class Chilopoda—Centipedes

In a separate class, the chilopods or centipedes, have a different morphology or appearance than millipedes. They have one pair of legs per segment that typically extend outward from the body (see the field guide for an example). Centipedes will inject venom to paralyze their small prey. You will often find these animals in mesic conditions, or near your shower drain.

Nature Journal: Track your encounters with arthropods—insects, arachnids, crustaceans, diplopods, and chilopods. Consider conducting a field investigation for insects and implement some of the methods for collecting or attracting them. At the least, create a chart to make quick comparisons among different groups. Important headings to include are abiotic conditions like time, date, temperature, and weather conditions. Provide the habitat and location of where you observed the animal. Include a description and a small sketch and discuss what it was doing. Lastly, try to identify it to the most specific taxonomic level as possible. Include the information of class, order, family, genus, and species. Feel free to personalize the chart for your needs—if you want to focus on a certain class or habitat, then you may want to separate charts accordingly.

Review Questions

1. Discuss the characteristics that all organisms within the Phylum Arthropoda share with each other.
2. What is an entomologist?
3. Discuss possible reasons why insects have been so successful.
4. What percentage of insects are actually pests? Why are they considered pests? Describe four ways insects are beneficial to nature and humans.
5. What does a typical insect look like? Name the bodily characteristics and where they are located on the body.
6. What is metamorphosis? Describe the different types of metamorphosis. What is an insect example for each type of metamorphosis? How does metamorphosis contribute to an animal's survival? Discuss the niche of each stage.
7. What are the four basic types of mandibles of insects? How are they used? Name an example of an insect for each mouth type. What can the mouth type tell you about the animal?
8. From the field guide section that follows, discuss the distinguishing characteristics between the major insect orders (Coleoptera, Lepidoptera, Hymenoptera, Orthoptera, Odonata, Hemiptera, and Diptera).
9. Give examples of members from the Class Arachnida. What are the common features of arachnids. How are spiders beneficial?
10. What are spinnerets? How can they be used? Describe the four web types.
11. Discuss characteristics of Crustacea, Diplopoda, and Chilopoda.

7.4 Field Guide: Insects, Spiders, and Other Arthropods

When you see an arthropod, at the very least, try to group it into one of the categories you see listed in this section. Is it from Class Insecta, Arachnida, Crustacea, Diplopoda, or Chilopoda? See the previous section as a reference and review features of each group in this field guide.

Arthropod Identification Key

Use the icon silhouettes in the gray shaded box to match features of your specimen in question. This will lead you to the corresponding arthropod order and affiliated numbered section.

Class Insecta

The orders are arranged by abundance, with the most abundant or most common insects listed first.

1. **Beetles—Order Coleoptera**
2. **Butterflies and Moths—Order Lepidoptera**
3. **Bees, Wasps, and Ants—Order Hymenoptera**
4. **Grasshoppers, Crickets, and Katydids—Order Orthoptera**
5. **Dragonflies and Damselflies—Order Odonata**

6. **True Bugs—Order Hemiptera**
7. **Flies—Order Diptera**
8. **Miscellaneous Insect Orders—Walking Stick, Mantis, Earwig, Large-winged Insects, Cockroach, Antlion, Snow Flea**

Class Arachnida

9. **Spiders**
10. **Ticks**

Subphylum Crustacea

11. **Crayfish**
12. **Isopods—Pillbugs and Sowbugs**

Class Diplopoda

13. **Millipede**

Class Chiriopoda

14. **Centipede**

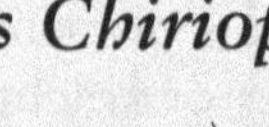

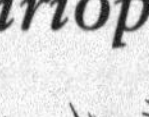

Arthropod Species Accounts

Class Insecta - The Insects

In this section, you will read a very brief overview given for the insect orders you may encounter. Know the basic characteristics discussed for each order so that you can make easy deductions with your identifications. The next step would be to learn common members of each of the orders. Focus on **morphology** of the animals to help with identification—meaning, what they look like.

When making identifications, pay attention to size, color, markings or patterns on the body, and the presence and orientation of the wings. Take note of the animal's head and the length and features of the antennae, characteristics of the wings, leg length, and if any reproductive structures or stingers are attached to the abdomen. Habitat would also provide an indication of what your organism in question might be. Learning characteristics of the animals within the major insect orders will not only help you make quick identifications, but it can also help you recognize potential niches they may fill.

1. Beetles—Order Coleoptera

("koleos" = sheath; "-ptera" = wings)

Most of the insects you encounter will likely be a beetle. Order Coleoptera is the most diverse and numerous of all organisms. Members of the order can easily be identified by the straight line down the center of its back. The forewings, or **elytra**, will join at the center of the back while at rest. This thick, outer cover of the abdomen lifts to reveal the folded membranous hind wings underneath used in flight. Their body shapes may appear elongated, cylindrical, or hemispherical. Beetle species' antennae vary considerably. Beetles can be found in nearly all habitats, and are leaf chewers, predators, scavengers, or wood borers. They also exhibit complete metamorphosis with most beetles only taking one year to go through it, whereas those living in wood or soil can take longer. Many beetles can even make sounds by rubbing parts of their body together (**stridulation**).

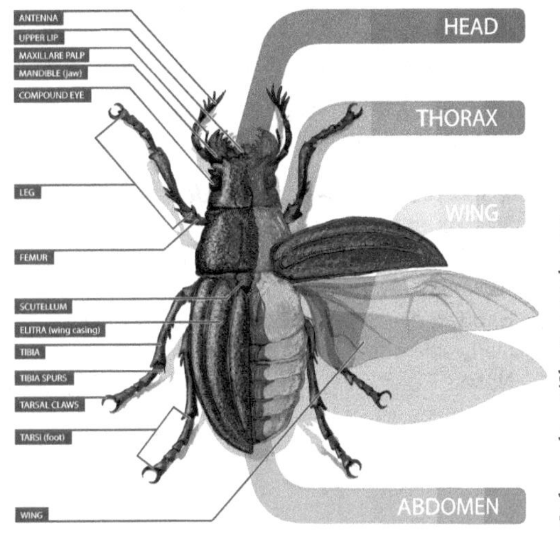

Diagram of a beetle.

Ground Beetles and Tiger Beetles—Family Carabidae

Small to medium sized between ⅛ and ½ in. (up to 1.25 cm) long; flat bodies; head narrowest body segment; antennae length moderate and threadlike; long legs; generally black or brown with some brightly colored; important predators in gardens

© Brett Hondow/Shutterstock.com

Fiery Hunter (*Calosoma calidum*)
Size: up to 1 in. (2.5 cm)
Key Features: black with iridescent rows of red and yellow dots or punctures along wing covers
Habitat: fields; beaches; deciduous woods
Notes: hunts on the ground

Vivid Metallic Ground Beetle (*Chlaenius sericeus*)
Size: up to ½ in. (2.5 cm)
Key Features: iridescent green to red to blue; dull in low light
Habitat: near water
Notes: emits pungent odor when disturbed; eggs laid in mud balls attached to plants; eats larger injured or dead insects

© Malachi Jacobs/Shutterstock.com

Six-Spotted Tiger Beetle (*Cicindela sexguttata*)
Size: up to ½ in. (1.25 cm)
Key Features: green iridescent elytra; white spots vary 0–8
Habitat: deciduous forest; open areas; gravel roads; sunny trails; logs; rocks
Notes: move with quick bursts; active during day; adults active May–July

© Paul Sparks/Shutterstock.com

Festive Tiger Beetle (*Cicindela scutellaris lecontei*)
Size: up to ½ in. (2.5 cm)
Key Features: iridescent green to purple; edged with white band
Habitat: dry, sandy areas; among pines
Notes: active April–September

Leaf Beetles—Family Chrysomelidae

Typically under ½ in. (12 mm) with variable shapes; many brightly colored with spots, stripes or other markings (sometimes confused with ladybird beetle family); found on leaves and flowers on all types of plants, but species particular to specific plants

© Brett Hondow/Shutterstock.com

Spotted Cucumber Beetle (*Diabrotica undecimpunctata*)
Size: ¼ in. (0.5 cm)
Key Features: pale greenish-yellow with black spots on elytra; three distinct body parts; long antennae
Habitat: observed on over 200 plant species; cucumber and corn plants
Notes: serious pest on corn

Scarab Beetles—Family Scarabaeidae

Varied group; mostly convex; short antennae

© HildeAnna/Shutterstock.com

Japanese Beetle (*Popillia japonica*)
Size: 0.6 in. (16 mm)
Key Features: iridescent green with brownish elytra and five white side brushes of hair
Habitat: landscapes, gardens; feeds on over 300 species (including rose family, Rosaceae)
Notes: nonnative, invasive

© Ernie Cooper/Shutterstock.com

June Beetle or June Bug or May Beetle **(*Phyllophaga* sp.)**
Size: ½—1.5 in. (12–35 mm)
Key Features: elongated; brown; domed
Habitat: deciduous forests (feeds on leaves of oak, willow, birch, ash)
Notes: attracted to light at night; create a buzz when flying

Family Gyrinidae

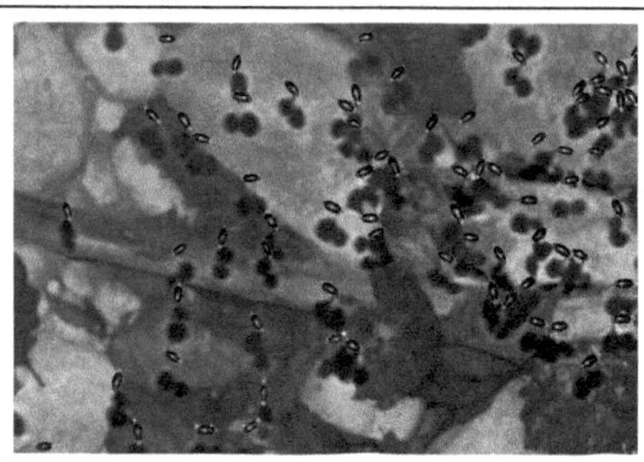

© Vitolga/Shutterstock.com

© Pan Xunbin/Shutterstock.com

Whirligig Beetle (*Gyrinus natator*)

Size: up to ½ in. (1.25 cm)
Key Features: black or bronzed; nine faint grooves on wing covers
Habitat: swirl (gyrate) across water surface
Notes: eyes divided to see above and below surface simultaneously; predator

Carrion Beetles—Family Silphidae

© Rusty Dodson/Shutterstock.com

American Carrion Beetle (*Necrophila americana*)

Size: up to ¾ in. (3 cm)
Key Features: broadly oval; black head, yellow pronotum with black splotch in center; wrinkly wing covers black with yellow posterior tip
Habitat: found on dead animals
Notes: mistaken for bumblebee in flight; feed on drying carcasses; June–August

Family Geotrupidae

© Erik Agar/Shutterstock.com

Earth-Boring Dung Beetle (*Geotrupes splendidus*)
Size: up to ½ in. (2.5 cm)
Key Features: metallic bronze, purple, greenish-black; rounded oval body
Habitat: cow manure
Notes: constructs shallow tunnel beneath dung for female to lay eggs

Metallic Wood-Boring Beetles—Family Buprestidae

Up to 1 in. (2.5 cm); elongated body and flattened with tapered posterior; head somewhat retracted

© Herman Wong HM/Shutterstock.com

Emerald Ash Borer (*Agrilus planipennis*)
Size: up to ½ in. (2.5 cm)
Key Features: iridescent or metallic mainly green with reddish-bronze highlights
Habitat: favors young, native ash (*Fraxinus* sp.) trees
Notes: nonnative invasive decimating ash trees in region; larvae tunnel under bark eventually killing tree; late May–August

Click Beetles—Family Elateridae

Elongated, oval, flat body; dark colors; prothorax seems "loose" with prolonged sharp points; moderate length antennae

© William Cushman/Shutterstock.com

Eyed Click Beetle or Eyed Elater (***Alanus oculatus***)
Size: 1.25 in. (4 cm)
Key Features: large, conspicuous black eye spots with whitish rings surrounding each on pronotum; body black with small white specks
Habitat: decaying logs; open wooded areas; camouflaged on tree trunks during day
Notes: May-June; common at lights at night; makes "click" as it snap a spine into chest groove and propels it into air

Fireflies or Lightening Bugs or Lightening Beetles—Family Lampyridae

© khlungcenter/Shutterstock.com

Firefly or Lightening Bug (***Photuris* sp.**)
Size: ½ in. (2.5 cm)
Key Features: flat, narrow, elongated body soft, dark wing covers outlined in yellow and along the elytra; orange pronotum with dark spot; head generally hidden
Habitat: fields; open woods; wetlands
Notes: on dark, warm summer nights, larvae and adults glow greenish-yellow flashes; males flash while flying and females respond while on the ground

Ladybird Beetles—Family Coccinellidae

Oval, convex bodies orange to red with black spots; heads mostly hidden; short antennae. Feeds on aphids and other pests. Nonnative **species, seven-spotted lady beetle** (*Coccinella septempunctata*) threatens our native ladybird beetles, and **multicolored Asian lady beetle** (*Harmonia axyridis*) eat pests but also problematic when overwintering in people's homes.

© Paul Reeves Photography/Shutterstock.com

Spotted Lady Beetle (*Coleomegilla maculata*)
Size: ¼ in. (5 mm)
Key Features: reddish-pink with 10 spots on wing covers and two on pronotum
Habitat: fields; gardens; wetlands (usually)
Notes: feeds mostly on pollen, some aphids; spring-summer

Stag Beetles—Family Lucanidae

Large beetles up to 1 in. (25 mm) with elongated dark body; males have large mandibles

© Jay Ondreicka/Shutterstock.com

Pinching Beetle or Reddish-brown Stag Beetle **(*Lucanus capreolus*)**
Size: 1 in. (22–35 mm)
Key Features: dark reddish-brown; males with long, curved upper jaw (sicklelike)
Habitat: deciduous forests; logs and stumps
Notes: attracted to lights at night

Long-Horned Beetles—Family Cerambycidae

Elongate, cylindrical bodies; long antennae, half or greater than body length; colors vary

© Paul Kulinich/ Shutterstock.com

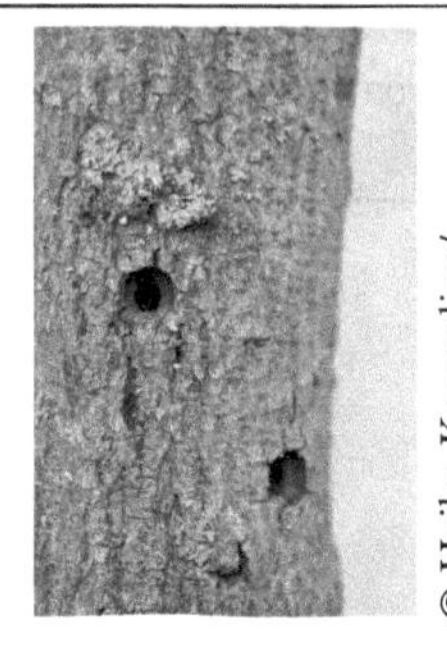

© Heiko Kueverling/ Shutterstock.com

Asian Long-Horned Beetle or ALB (***Anoplophora glabripennis***)

Size: 1.25–1.5 in. (4–5 cm)

Key Features: shiny black with white spots; alternating white and black bands on very long antennae typically extended posteriorly

Habitat: mainly maple trees (*Acer* spp.) but also confirmed on many other tree types (i.e., birch, poplar, and willow)

Notes: nonnative, very invasive (not found in all parts of region); branches attacked first with larvae feeding on outer sapwood and larger larvae chewing tunnels into heartwood; large exit holes left on trunk

© Mircea Costina/Shutterstock.com

White-Spotted Pine Sawyer (*Monochamus scutellatus*)

Size: ¾ in. (5 mm)

Key Features: dark with single white spot on male and mottled brown and white on females

Habitat: living, dead, or dying conifer trees (pines, firs, spruce)

Notes: smaller than nonnative invasive, ALB

© IrinaK/Shutterstock.com

Cottonwood Borer (*Plectrodera scalator*)

Size: 1.25–1.5 in. (4–5 cm)

Key Features: large, black and white; antennae as long as body and often extending horizontally (perpendicular to body)

Habitat: prefers eastern cottonwood trees, but also poplars and willows; larvae feed in roots and root collar

Notes: does not leave an exit hole like ALB; young trees killed or weakened by larvae but mature trees often not seriously injured

2. Butterflies and Moths—Order Lepidoptera

("scaled-wings")

Order Lepidoptera is the second largest order—the butterflies and moths. They have two pairs of scaly wings and mouthparts that are either suckers or siphons that coil beneath their head when not used. Lepidopterans go through complete metamorphosis.

A quick way to determine the difference between butterflies and moths is to observe the following:

Color: butterflies typically have brighter wing colors than moths

Time of day: butterflies are active during the day, and most moths are active at night

Body: butterflies tend to have a smooth body, whereas moths usually have a bristled or "bushy" body (or "hairy")

Antennae: the sensory organs on the head tend to differ between the two types; butterflies tend to have straight antennae with "clubs" on the end, and moths' antennae are typically "comblike."

At rest: the wings of butterflies at rest will be displayed and will sometimes close upward toward the sky; most moth species' wings will rest on the back of their body after they land.

Brush-Footed Butterflies—Family Nymphalidae

Largest family of butterflies. Many with orange and black wings; front bushy legs reduced with second and third pairs used to walk

© Anton Kozyrev/Shutterstock.com

Mourning Cloak (*Nymphalis antiopa*)
Size: wingspan up to 4 in. (10 cm)
Key Features: topside dark maroonish-brown with pale yellow edges and bright blue spots line the black line between the maroon and yellow; underside with gray markings and same yellow edges
Habitat: deciduous forests; all habitats
Notes: usually first butterfly observed in spring; one of the longest life spans among butterflies

© TheLazyPineapple/Shutterstock.com

Great Spangled Fritillary (*Speyeria cybele*)
Size: wingspan 2.5–3.5 in. (6–8 mm)
Key Features: orange above with five black dashes near base of forewing with many black dashes at base of hind wing; two rows black crescents along edges; lighter underside
Habitat: fields and woodland edges
Notes: various species of native violet plants host larvae

© Frode Jacobsen/Shutterstock.com

Eastern Comma (*Polygonia comma*)
Size: wingspan 2–2.5 in. (4.5–6.5 cm)
Key Features: orange with upper side forewings with dark markings and hind wings all black; underside with striped light and dark brown
Habitat: woodlands near water sources
Notes: hibernates winter as adult in sheltered spot, glycerol prevents freezing solid (lighter coloration); feeds on sap

© Richard Walpole/Shutterstock.com

Red Admiral (*Vanessa atalanta*)
Size: wingspan up to 2.5 in. (6 cm)
Key Features: topside mostly dark with reddish-orange slashes on forewing and hind wing; underside cryptic with blues and pinks
Habitat: open fields
Notes: hibernates or migrates; males territorial and perch in daytime until sunset; stinging nettle host plant for larva development

© Shutterschock/Shutterstock.com

Painted Lady (*Vanessa cardui*)
Size: wingspan 2.5 in. (6 cm)
Key Features: topside brown with reddish-ochre, olive basal areas alternating with black and white near tips of forewings; paler underside
Habitat: open fields
Notes: migrates to Mexico and California in fall

© Sari ONeal/Shutterstock.com

Monarch (*Danaus plexippus*)
Size: wingspan up to 4 in. (10 cm)
Key Features: topside orange with veins and margins black and small white spots in margins; underside paler; males have a black spot on hind wings
Habitat: fields
Notes: migrate in fall to mountains of Mexico; bitter tasting to birds due to milkweed; resembles viceroy

© Paul Reeves Photography/Shutterstock.com

Viceroy (*Limenitis archippus*)
Size: wingspan up to 3 in. (88 cm)
Key Features: resembles monarch; has a bar across hind wing
Habitat: near streams; willows
Notes: bitter tasting to predators; demonstrates Muellerian mimicry; migrates

© Sari Oneal/Shutterstock.com

© Sari Oneal/Shutterstock.com

Red-Spotted Purple (*Limenitis arthemis astyanax*)

Size: wingspan 3–3.5 in. (7 cm)
Key Features: iridescent blue with whitish-blue spots along topside margin; similar underside but with more pronounced orangish-red spots near tips of wings
Habitat: deciduous forests; edges; roadsides
Notes: often in flight for short durations close to ground; rests in sun high on top of trees; when at rest antennae straight forward; hybridizes

© Dawn photos/Shutterstock.com

Common Wood-Nymph (*Cercyonis pegala*)

Size: wingspan 2–3 in. (5–7 cm)
Key Features: light to medium brown with two large eyespots on forewing; underside with fine wavy lines especially lower part of hind wing
Habitat: meadows; woodland edges
Notes: low, bouncy flight

Swallowtails—Family Papilionidae

Includes our largest sized butterflies with wingspans varying from close to 3 to 5 in. (9–13 cm); long conspicuous "tails" on hind wings; typically dark wings with yellow markings or yellow wings with dark markings

© Maria T Hoffman/Shutterstock.com

Eastern Tiger Swallowtail (*Papilio glaucus*)
Size: wingspan up to 3.5 in. (9 cm)
Key Features: yellow with black stripes
Habitat: woodlands, open areas; flowers; puddles (>12 swallowtails in June)
Notes: Similar to the Canadian tiger swallowtail

© Paul Roedding/Shutterstock.com

Black Swallowtail (*Papilio polyxenes*)
Size: wingspan up to 3.5 in. (9 cm)
Key Features: black with yellow spots on wings and blue spots with small red eyespot on hind wings
Habitat: open fields; gardens; less common in heavily forested areas
Notes: lays eggs on plants of parsley family (Apiaceae)

Whites and Sulphurs—Family Pieridae

Common in open areas with body mostly white or yellow; wings folded above back while at rest

© Christian Musat/Shutterstock.com

Cabbage White (*Pieris brassicae*)
Size: wingspan 1.5 in. (4 cm)
Key Features: white wings with black-tipped forewings; male with one black dot, females with two
Habitat: fields; gardens; common at mud puddles drinking
Notes: lays eggs on plants of mustard family (Brassicaceae); May–September

© gardenlife/Shutterstock.com

Clouded Sulphur or Common Sulphur **(*Colias philodice*)**
Size: 1.75 in. (4.5 cm)
Key Features: yellow wings with very narrow black borders on upper side (some females whitish-green)
Habitat: open fields
Notes: lays eggs on plants of legumes (Fabaceae); common at mud puddles; May–October

Blues, Copper, and Hairstreaks—Family Lycaenidae

Small, fragile with slender body up to ½ in. (1 cm); wing colors vary blue, brown, black; antennae often with white rings; eyes typically encircled with white

© IanRedding/Shutterstock.com

Spring Azure or Holly Blue (***Celastrina argiolus***)
Size: ½ in. (1 cm)
Key Features: pale silvery-blue wings with pale white dots; underside pale whitish with row of black dots
Habitat: moist woodlands
Notes: lays eggs on dogwood trees and viburnums

© David James Chatterton/Shutterstock.com

American Copper (*Lycaena phlaeas*)
Size: ½ in. (1 cm)
Key Features: topside forewings bright orange with dark border and 8–9 black spots; hind wings dark with orange border; undersides similar but paler
Habitat: woodlands and open fields
Notes: common; very active in bright sun

Skippers—Family Hesperiidae

Small, thick bodied; hooked antennae tips

© Betty Shelton/Shutterstock.com

Silver-Spotted Skipper (*Epargyreus cla0rus*)
Size: wingspan 1.5 in. (4 cm)
Key Features: dark brown with gold bands on forewings and silver spot on underside hind wing
Habitat: open areas; forest edges; riparian area
Notes: feed on nectar of many flower types (almost never yellow varieties), mud, feces

Giant Silkmoths—Family Saturniidae

© thelittleflower/Shutterstock.com

Luna Moth (*Actias luna*)
Size: wingspan 4 in. (11.5 cm)
Key Features: lime green wings with white body; long "tails"; eyespot on each hind wing
Habitat: deciduous forests; edges; attracted to light at night
Notes: larvae feed on birch (*Betula* spp.)

© David Havel/Shutterstock.com

Cecropia Moth (*Hyalophora cecropia*)
Size: wingspan up to 7 in. (16 cm)
Key Features: brownish with red near base of forewing; crescent red spots with white centers on all wings (larger on hind wings); white with reddish band running across all four wings; hairy reddish body fading to reddish-white posteriorly and alternating red and white bands on abdomen
Habitat: deciduous forests (prefers maples, *Acer* spp.)
Notes: threats to populations include caterpillars hosting parasitic wasps and flies, squirrels consuming pupae, pruning of trees, leaving outdoor lights on

Sphinx Moths—Family Sphingidae

Fast fliers that hover at flowers like hummingbirds

© Steve Schlaeger/Shutterstock.com

White-Lined Sphinx (*Hyles lineata*)
Size: wingspan 2–3 in. (5–8 cm)
Key Features: topside dark brown with tan stripe from base to tip; white lines cover veins; hind wing dark with broad pink middle band; underside tan with darker markings; stout furry body with underside with six distinct white stripes
Habitat: various; gardens; meadows
Notes: larvae voracious eaters; can damage gardens or crops

Gypsy Moth (*Lymantria dispar*)—Family Erebidae

© Ihor Hvozdetskyi/ Shutterstock.com

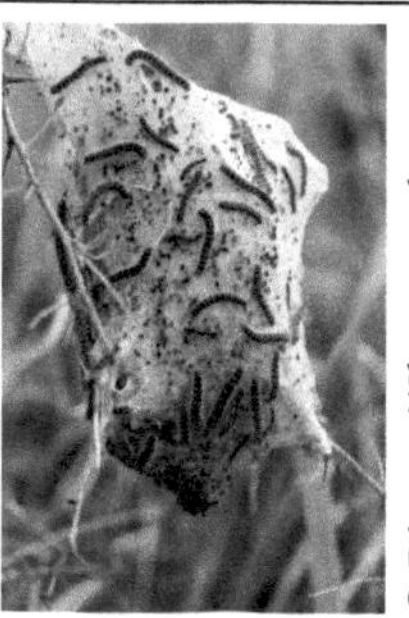

© Digoarpi/Shutterstock.com

Size: 2 in. (5 cm)

Key Features: light brown with darker markings going across wings when at rest (female larger and mostly white with some brown markings); feathery antennae

Habitat: deciduous, coniferous, or mixed forests

Notes: Nonnative, invasive; major forest pest that defoliates trees; control efforts difficult

3. Bees, Wasps, and Ants—Order Hymenoptera

("membranous-wings")

The third largest order are the hymenopterans—bees, wasps, and ants. Although sometimes referred to as "social insects," most species live alone. The common characteristic is the narrow or pinched "waist." The body segments are very pronounced, especially between the thorax and abdomen. They have two pairs of thin, transparent wings with the hind wings being smaller, and in some species completely absent. They will either be chewers or suckers and go through complete metamorphosis. This order has the only insects with stingers; however, not all have stingers, and those that do are females only.

Bees—Family Apidae

Vary in size; black and yellow; back legs adapted to carry pollen; long tongues for nectar; honeybee and bumblebees create colonies (single fertile female or "queen"; often 10s of 1000s of nonreproductive females or "workers"; small fraction of fertile males or "drones") organized by complex communication through pheromones and movement.

© Daniel Prudek/Shutterstock.com

Honeybee (*Apis mellifera*)

Size: ½ in. (1 cm) (queen slightly larger)
Key Features: reddish-brown with black and orangish-yellow rings on mostly hairless abdomen; fuzzy thorax; pollen basket on hind legs; blackish-brown legs
Habitat: meadows; gardens; and edges of deciduous forests
Notes: social; brought to North America in 1600s; perennial colonies in human-made beekeeping hives or garden bee condos, or in hollowed trees; depended upon for agriculture pollination; threatened by pests and diseases from *Varroa* mite and colony collapse disorder

© Elliotte Rusty Harold/Shutterstock.com

Bumblebee (*Bombus impatiens*)

Size: ¾–1 in. (17–23 mm) (size varies among roles)
Key Features: entire body fuzzy; queens and workers are black with a yellow thorax and first abdominal segment; males yellow head; pollen basket on hind legs
Habitat: deciduous forest; fields
Notes: social; colonies are annual up to 500 workers; very important pollinator

© Elliotte Rusty Harold/ Shutterstock.com

Carpenter Bee (*Xylocopa virginica*)

Size: ¾ in. (2 cm)

Key Features: mostly black, metallic (shiny) body with purple tint; yellow fuzzy thorax with black spot on back

Habitat: nests in dead or solid wood; sometimes home exteriors

Notes: solitary; may find holes in exposed wood in outdoor structures

Yellowjackets, Hornets, and Wasps—Family Vespidae

Mostly social insects with an annual nest; smooth bodies (few hairs) with narrow wings folded over back when at rest

© Paul Reeves Photography/ Shutterstock. com

Aerial Yellowjacket or Yellow Hornet (***Dolichovespula armenia***)

Size: ½ in. (1 cm)

Key Features: Black thorax with yellow markings and striped black and yellow abdomen

Habitat: open areas or edges near forests; from ground to tall trees

Notes: prey on variety of insects; stingers used repeatedly; similar to eastern yellowjacket (*Vespula maculifrons*)

© Ernie Cooper/ Shutterstock.com

© Manu M Nair/ Shutterstock.com

Bald-Faced Hornet (*Dolichovespula maculata*)

Size: 0.7 in. (14 mm)

Key Features: mostly black with white markings on face and on abdomen tip

Habitat: forest margins

Notes: constructs paper nest high off ground attached to branches; not a true hornet (actually a yellowjacket)

© Ramona Edwards/ Shutterstock.com

© Jay Ondreicka/ Shutterstock.com

Potter Wasp (*Eumenes fraternus*)
Size: 0.6–0.8 in (15–20 mm)
Key Features: black body with whitish-yellow markings on head, thorax, and ringed around abdomen; very skinny extended "waist"
Habitat: shrubby areas; edge zones
Notes: construct clay pot structures as nest, lays egg and places a moth or beetle larva; adults feed on nectar; solitary

© David Brimm/ Shutterstock.coms

Northern Paper Wasp (*Polistes fuscatus*)
Size: 1–1.75 in. (15–21 mm)
Key Features: dark body with yellow stripes; two orange spots on abdomen; color pattern varies greatly
Habitat: woodlands and edges
Notes: needs wood for making nests

Thread-Waisted Wasps—Family Sphecidae

© Maciej Olszewski/ Shutterstock.com

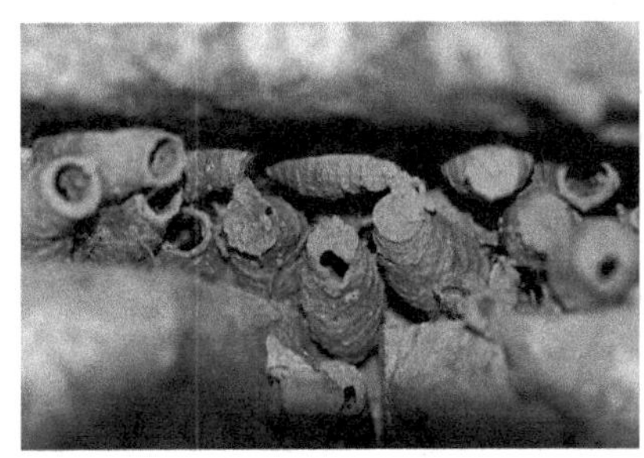

© Ruth Swan/Shutterstock. com

Black and Yellow Mud Dauber (*Sceliphron caementarium*)
Size: 0.94–1.1 in. (24–28 mm)
Key Features: conspicuous pedicel ("connector" from thorax to abdomen); black with yellow legs and markings on thorax
Habitat: water edges; man-made structures; rock ledges
Notes: legs dangle in flight; builds larva home of mud balls (often connected or lengthened) attached to hard substrate; larvae are fed spiders

Miscellaceous Families

© NERYXCOM/Shutterstock.com

**Common Sawfly (*Tenthredo* sp.)—
Family Tenthredinidae**

Size: 0.1–0.8 in. (2.5–20 mm)
Key Features: lack the narrow, threadlike waist of other hymenopterans; often mistaken for bees or wasps because of noticeable ovipositor (females only; not a stinger)
Habitat: forests and edges
Notes: sawlike ovipositor for all species used to cut into plants to lay eggs

© Paul Reeves Photography/Shutterstock.com

**Pigeon Horntail (*Tremex columba*)—
Family Siricidae**

Size: 20–30 mm (1–3 in.)
Key Features: ovipositor obvious (not a stinger); orange-brown head and black with yellow bands on mid-section; yellow legs
Habitat: deciduous forests; larvae feed on dead or dying trees
Notes: stingless wasp

© koohlman/Shutterstock.com

**Giant Ichneumon (*Megarhyssa* spp.)—
Superfamily Ichneumonoidea**

Size: 2–3 in. (5–6 cm)
Key Features: ovipositor very long and obvious (longer than body; cannot be withdrawn into body); genus with diverse size and color (black or black with thin yellow banding)
Habitat: woodlands
Notes: some sting; some parasitize horntails; control many pest insects by larvae as parasites in beetles and flies

© Camyoshi/Shutterstock.com

**Eastern Cicada Killer (*Sphecius speciosus*)—
Family Crabronidae**

Size: 0.6–2 in. (1.5–5 cm)
Key Features: long, robust with hairy, black and reddish thorax and black with light yellow stripes on abdomen; wings brownish
Habitat: open fields; lawns
Notes: solitary wasp; preys on cicadas; females larger than males and carry killed prey to burrow for nesting

Ants—Family Formicidae

Small to medium size (1/20–½ in.; 0.5–5 mm); dark and some reddish; complex social structure; workers wingless whereas reproductive females and males are winged after emerging from pupae

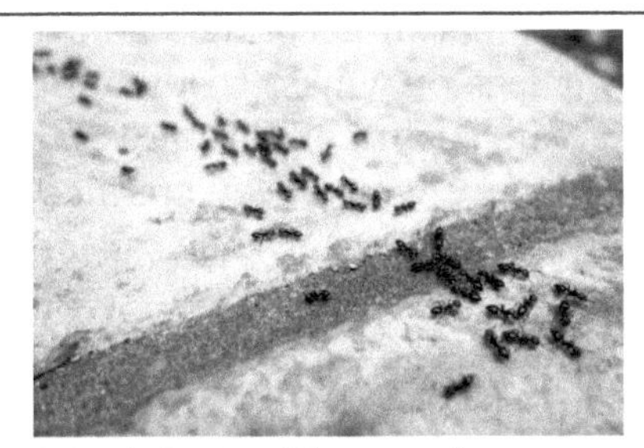

© Dario Lo Presti/Shutterstock.com

Pavement Ant (*Tetramorium caespitum*)
Size: 1/8 in. (5 mm)
Key Features: reddish-brown black
Habitat: nest in soil, under objects, in logs, under pavement
Notes: common

© Amelia Martin/Shutterstock.com

Carpenter Ant (*Camponotus novaeboracensis*)
Size: ¼ in (6 mm)
Key Features: black head and shiny abdomen with reddish thorax
Habitat: forests; nests in dead trees and logs
Notes: does not eat wood but chews it to make nest cavities; eats living and dead insects

© Karin Jaehne/Shutterstock.com

Mound-Building Ant (*Formica sp.*)
Size: 1/8 in. (5 mm)
Key Features: blackish-brownish red
Habitat: fields
Notes: may construct large mounds

4. Grasshoppers, Crickets, and Katydids—Order Orthoptera

("straight-wings")

The "musical" and cryptic insects of grasshoppers, crickets, and katydids belong to Order Orthoptera. Most orthopterans can generate sound by rubbing specialized organs together on their legs or wings (stridulation). With enlarged hind legs, they can jump great heights or distances, and many with two pairs of wings can fly. If they have wings, they are inconspicuous at rest. Crickets do not have wings. Their forewings have a leathery texture, and the hind wings remain folded (or absent). Orthopterans are chewers, and some may be considered a pest to agricultural areas. Most species are omnivores, and some will prey upon living or scavenge dead insects. They go through incomplete metamorphosis.

Short-Horned Grasshoppers—Family Acrididae

Named for their short antennae; diurnal; active summer to fall.

© Randy R/Shutterstock.com

Carolina Grasshopper (*Dissosteira carolina*)
Size: 1.25–2.25 in. (3–6 cm)
Key Features: cryptic at rest with brownish-tan gray and mottled, but showy wings when it flies over dirt roads or bare ground; open wings have brownish-black with yellow margins
Habitat: open fields
Notes: often mistaken for a butterfly; one of region's largest grasshoppers; most common in disturbed areas

© Gerald A. DeBoer/Shutterstock.com

Two-Striped Grasshopper or Yellow-Striped Grasshopper **(*Melanoplus bivittatus*)**
Size: 1 ¼--2 ⅛ in. (3.0–5.5 cm)
Key Features: relatively large; yellowish-green body with pale yellow stripes along top of body
Habitat: farms and gardens; grassy areas; moist areas
Notes: *Melanoplus* spp. known pests to agricultural fields; common

Katydids—Family Tettigoniidae

Green insects with very long antennae (usually longer than body); wings resemble a leaf

© Jason Patrick Ross/ Shutterstock.com

Broad-winged Bush Katydid (*Scudderia pistillata*)

Size: up to 2 in. (5 cm)

Key Features: pale green with paler underside body; wings resemble green, horizontal, angular leaves; large hind legs; long, forward-reaching antennae

Habitat: meadows with goldenrod and asters; damp areas

Notes: great camouflage during day; calls at night; heard more than seen; song of a lispy buzzing with rapid sequences of buzzing notes with increasing utterances for five sets with pauses lasting several minutes; daytime song sound is a raspy, rattle series.

Crickets—Family Gryllidae

© Michael Siluk/ Shutterstock.com

Field Cricket (*Gryllus pennsylvanicus*)

Size: 0.6–1.0 in. (15–25 mm)

Key Features: dark black to dark brown (some reddish); black antennae longer than body; wings held flat over abdomen

Habitat: fields and forest edges where it burrows into soil; around areas of human habitation

Notes: omnivorous feeding on insects and seeds; chirps slow and methodical; mostly nocturnal; summer to fall; they have wings for producing sound (males) not for flight

5. Dragonflies and Damselflies—Order Odonata

A common, large, colorful flying insect order observed near water; eggs laid in water near aquatic vegetation and nymphs have multiple molts in water before emerging as adults (you may find its shed exoskeleton!); predators in both environments. The nymphs can act as a bioindicator of good water quality. Dragonflies have fat-bodied nymphs and adults hold their wings out to the sides when resting. Damselflies have skinny nymphs, and adults hold their wings together over their backs when resting. They go through incomplete metamorphosis.

Darners—Family Aeshnidae

Large, slender body with green or blue; feeds on insects in flight often in masses during late summer afternoons

© Jeanne Raises/Shutterstock.com

Common Green Darner (*Anax junius*)

Size: up to 3 ⅛ in. (8.5 cm); wingspan 3 in. (8 cm)
Key Features: large dragonfly; green thorax with blue or purplish abdomen
Habitat: fields; wetlands
Notes: nymphs feed on other aquatic insects, tadpoles, larval fish; adults catch insects on the wing such as mosquitoes, flies, and moths; migrates into Texas and Mexico

Baskettails—Family Corduliidae

Baskettails of this family hold egg ball at tip of abdomen then quickly dips the tip into water to deposit a 6 in. (15 cm) long rope of eggs

© cliff collings/Shutterstock.com

Common Baskettail (*Epitheca cynosura*)

Size:

Key Features: brown, hairy thorax; hind wings have dark patches at the base; end of abdomen curves out like a dog wagging its tail

Habitat: near shallow aquatic environments

Notes: May to mid-July; swarms over meadows in late afternoon

Skimmers—Family Libellulidae

Conspicuous; perch frequently on emergent aquatic vegetation; observed away from water in open areas

© Jeanne Raises/Shutterstock.com

Twelve-Spotted Skimmer or Ten-Spot Skimmer (***Libellula pulchella***)

Size: 2 in. (5 cm) long

Key Features: alternating white and dark brown-black spot on outstretched wings; abdomen brown with yellow spots (male) or stripes (female) along its sides.

Habitat: mostly aquatic habitats; open fields; forest margins

Notes: late May through September

© Lflorot/Shutterstock.com

Common Whitetail (*Plathemis lydia*)

Size: 2 in. (5 cm) long

Key Features: males have white chunky abdomen with brown-black bands on otherwise translucent wings ("checkered"); females' body have zigzag abdominal stripes with same wing pattern

Habitat: ponds, marshes, slow streams

Notes: mosquito predator

Broad-Winged Damselflies—Family Calopterygidae

Large damselflies; metallic colored; along rivers

© Paul Reeves Photography/Shutterstock.com

Ebony Jewelwing (*Calopteryx maculata*)

Size: 1.5–2.2 in (39–57 mm)

Key Features: emerald green, shiny body with all black wings (male); female duller wings with white tips

Habitat: forest edges; streamside vegetation

Notes: preyed upon by many insects, birds, and fish; uses many plant types for shelter; similar to river jewelwing (*Calopteryx aequabilis*) that has only black-tipped wings

Narrow-Winged Damselflies—Family Coenagrionidae

© Paul Reeves Photography/Shutterstock.com

Familiar Bluet (*Enallagma civile*)

Size: 1–1.5 in. (28-39 mm)

Key Features: delicate, slim; mostly blue body with narrow black bands (females less colorful); clear wings with some visible dark veins held together against body while perched

Habitat: wetlands; ponds

Notes: resembles blue-fronted dancer, *Argia apicalis* but that species has wings together above the body while at rest

6. True Bugs—Order Hemiptera

"half-wings"

People sometimes mistake members of Order Hemiptera for beetles. Notice the carapace. The wings do not meet at the center in a straight line but will overlap. As those forewings rest on the carapace of hemipterans, they form an "x" on their back. These are considered true bugs. To call any insect a "bug" is not accurate. Some common names have caused confusion. For example, the "ladybug" is actually a beetle not a bug—research what it looks like. How do the forewings rest on the carapace? The ladybug is now referred commonly as the lady beetle or ladybird beetle. Hemipterans have uniformly thickened wings roofed over their abdomens, whereas some members have no wings. Insects in this order have piercing-sucking mouthparts, and a sawlike ovipositor for depositing eggs in plant tissue or bark. They go through incomplete metamorphosis.

Large Milkweed Bug (*Oncopeltus fasciatus*)—Family Lygaeidae

Size: 0.5–0.75 in. (10–18 mm)

Key Features: reddish-orange and black "x" pattern on back; black band across center (male) or two black dots (female)

Habitat: open fields; milkweed plants

Notes: mostly eats milkweed seeds; often used as a model organism for laboratory experiments

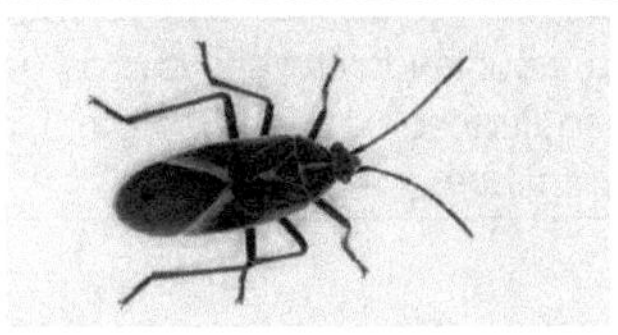

Eastern Boxelder Bug (*Boisea trivittata*)—Family Rhopalidae (scentless bug group, but this species stinks!)

Size: 0.4 in. (12.5 mm)

Key Features: dark brownish-black with red wing veins and markings

Habitat: mainly boxelder trees (also other *Acer* sp. and ash, *Fraxinus* sp.)

Notes: feed on young seeds of trees; strong smelling; deters predation with aposematic colors (bad-tasting compounds); eat, lay eggs, matures on boxelders (*Acer negundo*); clusters together in large groups to sun themselves

Stink Bugs or Shield Bugs—Superfamily Pentatomoidea (of the Suborder Heteroptera)
Scutellum (hardened part of thorax that covers top part of abdomen) triangular; five-segmented antennae; sucking mouthparts; threatens many crops; strong smelling

Green Stink Bug (*Chinavia hilaris*)
Size: 0.5–0.75 in. (13–18 mm)
Key Features: bright green with narrow yellowish edges; large shieldlike shape (triangular)
Habitat: woodlands; crops; orchards; gardens
Notes: adults feed on seeds if present, otherwise eats plant tissue

Spined Soldier Bug (*Podisus maculiventris*)
Size: 0.3–0.5 in. (8.5–13 mm)
Key Features: pale brown to tan; shield shaped with prominent "shoulders" or spurs behind head
Habitat: crops
Notes: considered beneficial; preys on gypsy moth caterpillars and larvae of leaf beetle pests

Brown Marmorated Stink Bug (*Halyomorpha halys*)
Size: 0.7 in. (1.7 cm)
Key Features: dark brown topside; creamy-brown underside; coloration varies (gray, brown, black, copper); marmorated pattern (variegated or veined like marble); alternating dark bands on thin outer edge of abdomen; light banded antennae
Habitat: crops; fields; building structures
Notes: nonnative, invasive; groups together; nymphs and adults feed on over 100 plant species (many crops); very stinky when provoked

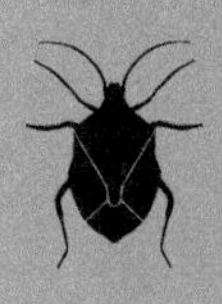

Leaf-Footed Bugs—Family Coreidae

Name refers to leaflike extensions on some species; oval shaped body with four segmented antennae; many veins in forewing membranes; stink glands

Western Conifer Seed Bug or WCSB ***(Leptoglossus occidentalis)***

Size: 0.6–0.8 in. (16–20 mm)

Key Features: variable colors or bands of brown and black; alternating bands along outer wing edges on flaring abdomen sides

Habitat: coniferous forest

Notes: nonnative, invasive; buzzing noise in flight; sprays a bitter, foul-smelling chemical substance; adapted to suck plant sap of young cones causing underdevelopment (pest of conifer plantations)

Giant Water Bug (*Lethocerus americanus*)—Family Belostomatidae

Size: 1–2 in. (2.5–5 cm)

Key Features: medium-dark brown with legs extended and forearms forward; large "x" on back where wings cross

Habitat: ponds, swimming on bottom among vegetation

Notes: seen by people on warm summer evenings (in search of mates or drawn to lights); feeds on insects, tadpoles, fish larvae, snails, young salamanders, small snakes; "toe biter"; overwinters in muck

Common Water Strider (*Aquarius remigis*)—Family Gerridae

Size: 0.5 in. (1 cm)

Key Features: dark brown to black; middle and hind legs obvious and extended

Habitat: surface of lakes and slow rivers

Notes: uses shorter front legs for seizing prey; sharp mouthpart (rostrum) used for piercing and sucking nutrients from prey; feeds on mosquito larvae and other insects living just under or at water's surface; fine "hairs" on leg tips allows them to skate on top of water

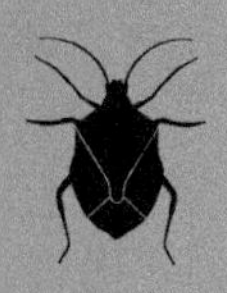

© Sue Robinson/Shutterstock.com

Water Boatman (*Sigara* sp.)—Family Corixidae

Size: 0.5 in. (12 mm)

Key Features: dark brown with small light specks; four long, rear, yellow-brown legs extended and shaped like oars; two shorter front legs; jerky movements

Habitat: surface of ponds; lakes

Notes: can be numerous in lakes with synchronized swimming movements; similar to backswimmer (*Notonecta* sp.), but that species swims on its back with hind legs swimming in unison for smoother movement than water boatman; feeds on algae, nematodes, zooplankton

Leafhoppers—Family Cicadellidae

Plant feeders that suck sap from grasses, trees, shrubs; hind legs modified for jumping; wings peaked over abdomen and angular head slanting inward like those of other hoppers and similar families.

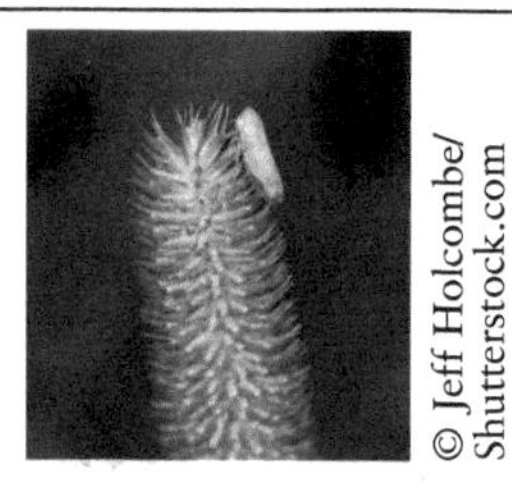

© Jeff Holcombe/Shutterstock.com

Potato Leafhopper (*Empoasca fabae*)

Size: ⅛ in. (3 mm)

Key Features: pale green, iridescent body; distinctive white "H" between head and wing base; hop

Habitat: croplands (about 30% of regional population); fields; woodlands; parks

Notes: serious crop pest (potatoes, beans, apples, clover, alfalfa); migrates to mixed hardwoods of along Gulf of Mexico

Spittlebugs or Froghoppers—Family Cercopidae

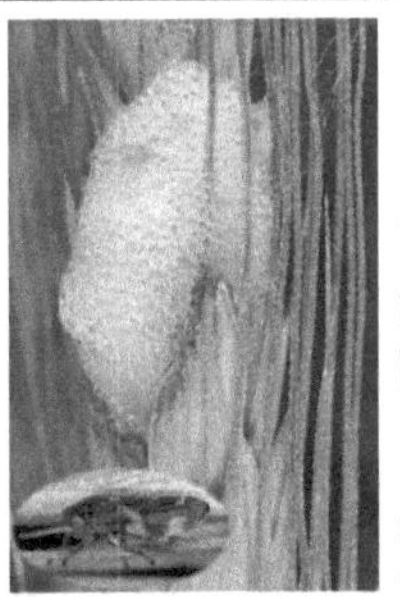

© Tomasz Klejdysz/ Shutterstock.com

Meadow Spittlebug (*Philaenus spumarius*)

Size: 0.2–0.3 in. (5–7 mm)

Key Features: coloration variable (up to 20 different colors known)—usually brownish-yellow with black and brighter patches; jumps

Habitat: fields; gardens

Notes: jumps to escape predation; runs and flies; larvae self-generate foam nests observed on flowers in spring and summer and adults leave when foam dries completely (about 10 days)

© Mary Terriberry/ Shutterstock.com

© Fotoz by David G/ Shutterstock.com

Cicada (*Magicicada* spp.)—Family Cicadidae

Size: 0.9–1.5 in. (2.4–3.5 cm)

Key Features: dark green with brown and black markings on large, stout body (all black for some species); wings membranous longer than abdomen and hold tentlike over abdomen; short antennae; bulging eyes

Habitat: deciduous forest

Notes: do not sting; do not bite; buzzing and ticking sounds (species dependent); buzz produced by males vibrating sound organ inside thorax and abdomen to attract female; 13 or 17-year cycles for synchronous emergence of mature nymphs from ground ("periodical cicadas"—research when the next emergence will be near you!), but other species not on a synchronized schedule ("annual cicadas" because some are seen each summer but still spend four to eight years underground); feeds on roots of deciduous trees surviving on xylem fluids while nymphs; adults active four to six weeks; mated females lay eggs in woody plants; sometimes called "locusts" but not related

Aphids—Family Aphididae

Very destructive pest of cultivated plants; weakens plant from sucking sap, carrying plant viruses, and disfiguring plants with "honeydew" deposits that eventually grows a sooty mold; capable of asexual reproduction; does not survive soapy water sprays

Peach-Potato Aphid (*Myzus persicae*)

Size: 1/16–⅛ in. (1.8–2.1 mm)

Key Features: black head and thorax with yellowish-green abdomen and dark patch on back; body color varies due to diet and temperature

Habitat: many garden and crop plants; lays eggs in fruit (*Prunus* sp.) trees

Notes: some females develop wings, other adults do not

7. Flies—Order Diptera

"two-wings"

This group has species most everyone is familiar—the flies, mosquitoes, and gnats. You may consider them pesky, but realize many have beneficial qualities or may not have an economical or human health consequence. Members of Order Diptera have one pair of obvious wings. The other pair is reduced in size and have a typically shriveled appearance. This second pair of wings is used for balancing. They have sucking or sponging mouthparts and go through complete metamorphosis.

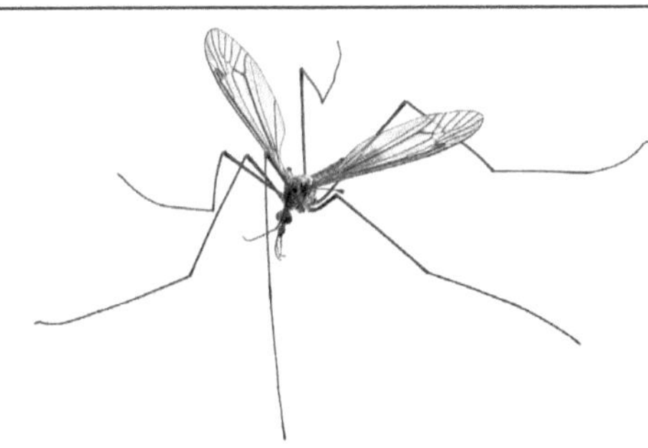

© Cristina Romero Palma/Shutterstock.com

Crane Fly—Families Tipulidae and Pediciinae

Size: 0.25–1.5 in. (7–35 mm); wingspan 0.25–2.5 in. (1–6.5 cm)

Key Features: resembles oversized mosquito; slender body with stiltlike legs (easily come off); V-shaped groove on thorax

Habitat: prefers moist and aquatic habitats

Notes: do not sting or bite; aquatic (semiaquatic) larvae ("leatherjackets") elongated, cylindrical and taper at front end, head retracted; valuable prey animal for many; larvae eat insects and plant matter but adults do not eat (incapable of killing or eating other insects) and have short life spans; phantom crane flies mostly black with white "feet" and two other bands up legs (Ptychopteridae, *Bittacomorpha clavipes*)

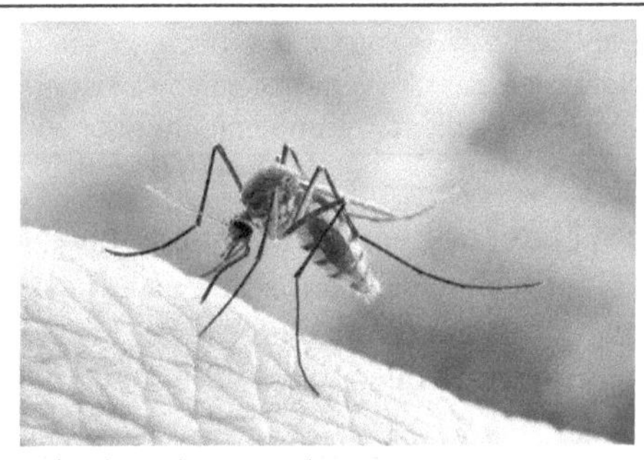

© frank60/Shutterstock.com

Mosquito—Family Culicidae

Size: 0.1–0.2 (3–6 mm)

Key Features: slender, dainty body; dark gray or black

Habitat: near standing water (wetlands, lake, puddles)

Notes: several flowers pollinated by mosquitoes (males); females' mouthpart pierces skin of vertebrates and some other arthropods (blood contains protein and iron needed to produce eggs); peak activity dusk and dawn; saliva transferred to host causing rash or disease; larvae live in epilimnion (at surface) of water

© xpixel/Shutterstock.com

Black Fly (*Simulium* spp.)—Family Simulidae
Size: 1/16–¼ in. (1.6–6.4 mm)
Key Features: black, dark brown, or gray; short legs; broad wings; humpbacked
Habitat: near moving water especially near wooded areas
Notes: females bite vertebrate animals while males feed on nectar; not known to carry disease; common spring-early summer (or throughout summer in some areas); attracted to color blue

Horse Flies and Deer Flies—Family Tabanidae

Medium to large sized insects with big, bright eyes; colors vary by species from black, brown, to yellow; females feed on blood of mammals (vector for diseases); males feed on nectar and pollen

© skynetphoto/Shutterstock.com

Black Horse Fly (*Tabanus atratus*)
Size: 1–2 in. (2.5–5 cm)
Key Features: all black (or smaller species gray or black); often with bright green eyes
Habitat: varied woodlots and edges; tall grasses; margins of wetlands or ponds
Notes: often observed on paths and roads near wooded areas; decrease in flies when winds increase or temperature drops; attracted to light

© Bruce MacQueen/Shutterstock.coms

Deer Fly (*Chrysops* spp.)
Size: up to 0.4 in. (up to 10 mm)
Key Features: triangular shape while at rest; tannish body; wings transparent with distinctive dark brown to black patches at the center and tops
Habitat: varied woodlots and edges; tall grasses; shaded wetlands
Notes: often fly around a person's head; active May-September; strong fliers; attracted to dark moving objects and carbon dioxide

© Worraket/Shutterstock.com

Common Green Bottle Fly (*Lucilia sericata*)—Family Calliphoridae (Blow Flies)
Size: 0.4–0.6 in. (10–14 mm)
Key Features: bright, metallic, blue-green or golden body with black markings
Habitat: flowers; dead or decaying organic matter; feces
Notes: used in forensic entomology for determining time of death (usually first one at the scene after death); May–September

8. Miscellaneous Insect Orders—Walking Stick, Mantis, Earwig, Large-winged Insects, Cockroach, Snow Flea

© Melinda Fawver/Shutterstock.com

Northern Walkingstick (*Diapheromera femorata*)—Order Phasmida, Family Heteronemiidae
Size: 2.5–3 in. (6.4–7.7 cm)
Key Features: resembles a small, slender twig; colors vary (shades of brown or green)
Habitat: high in forest canopy
Notes: no wings; slow moving; feed on deciduous tree leaves (oak, cherry, basswood preferred) and can defoliate; occasionally seen on tree trunks or on walls of buildings (usually near tall trees); incomplete metamorphosis

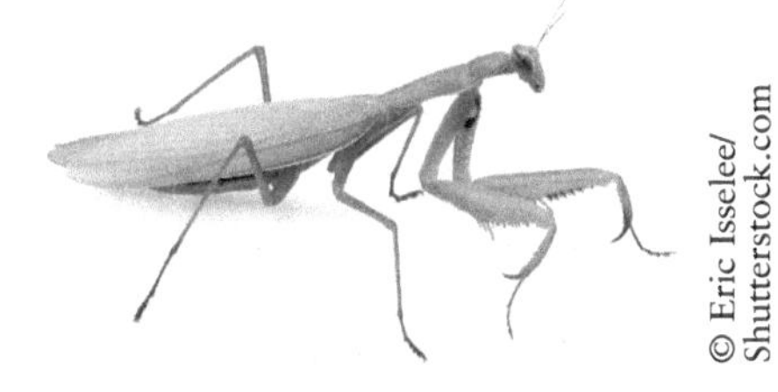

© Eric Isselee/Shutterstock.com

Chinese Mantis (*Tenodera sinensis*)—Order Mantodea (Mantids)
Size: 3–5 in. (7–10 cm)
Key Features: slender, bright green (some with brown stripes on front wings); obvious legs; raptorial forelegs
Habitat: meadows; gardens
Notes: introduced in late 1800s; feed on other insects pests and beneficial (females sometimes catch small vertebrates without impacting populations like reptiles, amphibians, and hummingbirds); food for other mantids and birds; cannibalistic (half of all matings females cannibalize males); egg masses covered with hardened foamy substance on trees to prevent desiccation (nymphs emerge and quickly disperse to not become prey to siblings); another introduced species European mantis (*Mantis religiosa*), and native Carolina mantis (*Stagmomantis carolina*) found in southern part of region; incomplete metamorphosis

© Sarah2/Shutterstock.com

European Earwig (*Forficula auricularia*)—Order Dermaptera

Size: 0.5–2 in. (12–15 mm)

Key Features: obvious, strong pinchers on abdomen tip (males widely separated; females slender straight pinchers); brownish-blackish; flattened body; membranous wings hidden underneath first pair that are short and leathery

Habitat: nocturnal; cracks and crevices during day and under organic debris (leaves, rocks, logs); found on flower blossoms

Notes: do not infest ears (or wigs); cannot inflict pain to humans; secretes foul-smelling liquid when provoked; feed on decaying matter and some living plant material; gradual metamorphosis

© Alex Puddephatt/ Shutterstock.com

© Fredlyfish4/Shutterstock.com

Mayflies—Order Ephemeroptera

Olive-Winged Drake Mayfly or Giant Michigan Mayfly (***Hexagenia limbata***)

Size: up to 2 in. (5 cm) including "tails" (cerci)

Key Features: variable colorations of pale brown to yellow; elongated, soft-bodied; two to three very long cerci; four (or two), clear membranous wings held upright while at rest; conspicuous eyes; aquatic larvae flattened on rocks or burrowing with three tails

Habitat: near lakes and slow rivers; nymphs burrow in mud underwater

Notes: early July large emergence occurs in masses; lives about one day to reproduce as adults then dies; often called fish flies; thrive in clean water; complete metamorphosis

© Thomas Porter/Shutterstock.com

© K Steve Cope/Shutterstock.com

Stoneflies—Order Plecoptera
Giant Stonefly (*Pteronarcys* spp.)—Family Perlidae

Size: up to 1.5 in. (3.8 cm)
Key Features: dark color; yellowish or green in early spring or summer; wings held flat on body past abdomen; long antennae; aquatic larvae flat, brown with two tails
Habitat: near rivers; nymphs under rocks in clean, highly oxygenated water
Notes: poor fliers; winter stoneflies (Family Capniidae) on snowy banks of rivers on warm winter (walk, feeding on algae); larvae indicator of good water quality; incomplete metamorphosis

© Elliotte Rusty Harold/Shutterstock.com

© FJAH/Shutterstock.com

Northern Caddisfly (*Pycnopsyche* spp.)—Order Trichoptera

Size: up to 1 in. (2.5 cm)
Key Features: resembles moths (not related); brown often mottled pattern on forewings; transparent hind wings
Habitat: aquatic environments; nymphs in clean, highly oxygenated water attached to rocks
Notes: larvae builds and lives in shelter or case made of objects nearby (minerals, rocks, twigs, plant matter)—other species are net-spinners or may be free swimmers with no case or nets; clean water indicator; adults poor fliers; complete metamorphosis

© Henrik Larsson/Shutterstock.com

© MP cz/Shutterstock.com

Alderfly or Lacewings **(*Sialis velata*)—Order Neuroptera**

Size: 3/16–¼ in. (5–6 mm)

Key Features: long, black body; smoky wing color with a broad base

Habitat: riparian areas; aquatic larvae

Notes: attracted to lights; May-June; females guard eggs by perching on nearby emergent aquatic plant stems; larvae brown, 1 in. (2.5 cm) with multiple horizontal legs and pointed, bristly abdomen tip; pupate on shore in leaf litter; complete metamorphosis

© SIMON SHIM/Shutterstock.com

© Jason Patrick Ross/Shutterstock.com

Dobsonfly (*Corydalus cornutus*)—Order Megaloptera

Size: up to 2 in. (5 cm)

Key Features: large, brown; tiny white spots on forewings and lay against body at rest in a V-pattern; males have large mandibles (¾ in.; 2 cm) used for fighting other males

Habitat: riparian areas

Notes: males cannot bite people (females may if provoked); July to August emergence; eggs laid on branches over streams or lakes and larvae drop into water; large larvae (hellgrammite) in water up to 1–2 in. (2.5–5 cm), dark brown with spiky legs horizontal along length, mandibles visible; pupate on land in leaf litter; complete metamorphosis

© Melinda Fawver/Shutterstock.com

Wood Cockroach (*Parcoblatta* spp.)—
Order Blattaria

Size: 1 in. (2.5 cm)
Key Features: flattened, oval body; light to medium brown; head hidden from view; wings longer than abdomen (first pair leathery; second more membranous)
Habitat: live outdoors in wooded areas; rotting logs
Notes: not the species that invades homes; feeds on decaying organic matter; incomplete metamorphosis

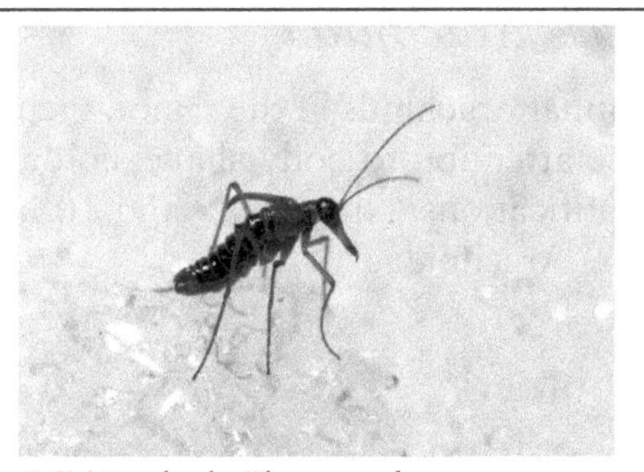

© Jiri Prochazka/Shutterstock.com

Snow Flea (*Hypogastrura nivicola*)—
Order Collembola (a type of springtail)

Size: 0.2 in. (<6 mm)
Key Features: seen as many black specks on snow; jumping
Habitat: moist areas on vegetation or logs; on top of snow on warm winter days
Notes: not true fleas; not a threat to humans or pets; a springtail (no wings; launches via hinged organ, furcula, at rear of abdomen); found all year; alive in winter due to proteins acting as antifreeze; feed on decaying plant matter; simple metamorphosis

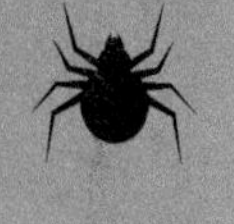

Class Arachnida

Familiar arachnids in the region include spiders, harvestmen, ticks, and mites. Pay close attention to body shape, color, markings, and habitat when trying to make identifications. Numbers 9 and 10 in the following part of the field guide covers Class Arachnida.

9. Spiders

Species determination of Order Araneae, or spiders can be difficult due to needing to examine mouthparts, eyes, or leg features at high magnification. If you were able to magnify the mouthparts of the spiders in our region, you would see that they all belong to the suborder Araneomorph of which their fangs clasp prey from the sides in a horizontal motion. The eyes of the larger spiders are most noticeable and will help with identification. However, these features and others are beyond the scope of this field guide. In this section, you will become familiar with a practical collection of examples with details of their characteristics to help you make identifications to varying taxonomic levels.

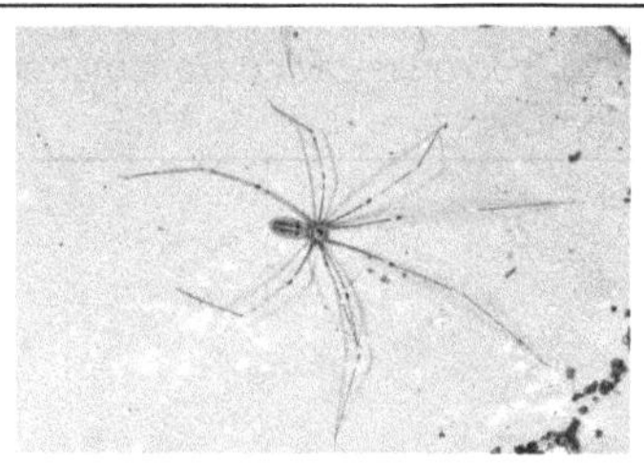

© Sergio Schnitzler/Shutterstock.com

Cellar Spiders—Family Pholcidae
Long-Bodied Cellar Spider (*Pholcus phalangioides*)
Size: body up to 0.3 in (8 mm); legs 5–6 times its body length (spans up to 2.8 in; 7 cm)
Key Features: pale grayish to yellowish; elongated or circular body; pale legs; very long, thin legs
Habitat: indoors; cobweb
Notes: observed in homes on ceilings or in basements; shakes body when disturbed; beneficial in eating spiders; not "daddy longlegs" or harvestman of forests

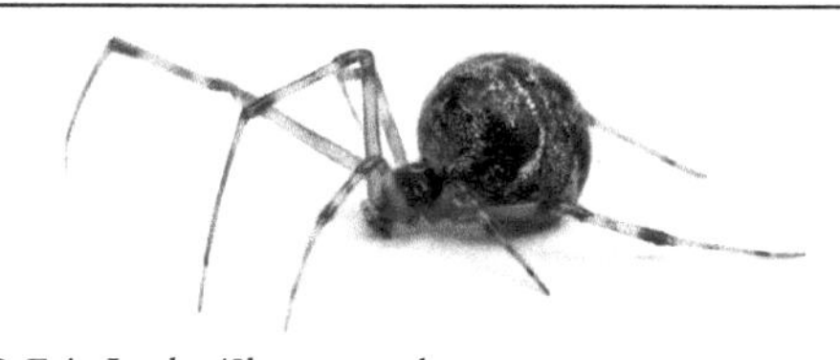

© Eric Isselee/Shutterstock.com

Cobweb Weavers (Comb-Footed Spiders)—Family Theridiidae
Common House Spider (*Achaearanea tepidariorum*)
Size: body up to 0.3 in. (8 mm); legspan up to 0.8 in. (20 mm)
Key Features: brownish-gray with white specks; bulbous abdomen; legs yellowish with darker rings
Habitat: under rocks, boards; corners of buildings
Notes: males and females on same irregular cobweb

Sheetweb Weavers—Family Linyphiidae
Hammock Spider (*Pityohyphantes costatus*)

Size: up to 0.25 in. (7 mm)
Key Features: abdomen light with darker herringbone pattern with light spots; dark, forked stripe from eyes down carapace; legs yellow with darker rings or spots
Habitat: shrubs or lower limbs of trees; outside of buildings
Notes: weaves flat sheets of random intricate mesh

© Anya Douglas/Shutterstock.com

Orb Weavers—Family Araneidae
Yellow Garden Argiope (*Argiope aurantia*)

Size: up to 1.1 in. (28 mm); legspan up to 2.8 in.; 70 mm)
Key Features: abdomen black with yellow bands; front legs all black sometimes with narrow yellowish-orange band, other legs with yellowish-red femur
Habitat: sunny fields or edges protected from wind; tall vegetation to make web; eaves of outbuildings
Notes: orb web up to 2 ft. (6 m); not aggressive; may bite if disturbed but venom harmless to humans

© Jason Patrick Ross/Shutterstock.com

Nursery Web Spiders—Family Pisauridae
Dark Fishing Spider (*Dolomedes tenebrosus*)

Size: up to 1 in. body (26 mm); legspan up to 3.5 in (90 mm)
Key Features: pale to dark with many light-dark chevron markings on abdomen; legs dark banded
Habitat: wooded areas; on trees; near water; outbuildings
Notes: stalks prey on land and water; eats prey larger than itself; no web; nursery web made from silk and folded leaves for young

© Clayton Nichols/Shutterstock.com

Wolf Spiders—Family Lycosidae
Striped Wolf Spider sp. (*Gladicosa* sp.)

Size: body up to 0.5 in. (14 mm); legspan up to 1.4 in. (35 mm)
Key Features: grayish-brown with two wide bands along edge of abdomen; legs long, rust colored with light rings with spines
Habitat: woodlands (oaks preferred); leaf litter
Notes: run after prey; no web; hunts at night (occasionally seen during day)

© scubaluna/Shutterstock.com

Funnel Weavers—Family Agelenidae
Barn Funnel Weaver (*Tegenaria domestica*)

Size: up to 0.5 in. (12 mm); legspan up to 1.1 in. (30 mm)
Key Features: dark orangish-brown to gray; elongated, somewhat flat and straight abdomen; striped legs; two black stripes on cephalothorax; chevrons running across abdomen; short bursts of movement
Habitat: under rocks; in buildings; woodlands; dark corners
Notes: active and agile hunters; flees from light

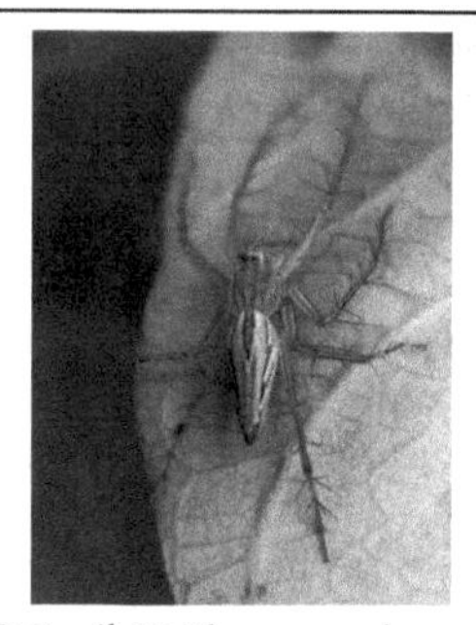

© Faqih03/Shutterstock.com

Lynx Spiders—Family Oxyopidae
Striped Lynx Spider (*Oxyopes salticus*)

Size: body up to 0.3 in. (7 mm); legspan up to 0.6 in. (15 mm)
Key Features: tan to cream; stripes; pointed abdomen
Habitat: grassy, herbaceous areas; low trees; crops
Notes: hunts in daytime (actively pursues prey then ambushes); no web

© Manfred Ruckszio/Shutterstock.com

Crab Spiders—Family Thomisidae
Goldenrod Crab Spiders (*Misumena vatia*)

Size: up to 0.4 in. (10 mm); legspan up to 0.8 in. (20 mm)
Key Features: yellow or white (flower dependent; takes nearly a month to change body color); reddish wavy band along sides of abdomen; bulbous abdomen; outstretched yellow legs (first and second pairs crablike)
Habitat: open fields; common on goldenrod plants (*Solidago* spp.)
Notes: ambushes prey during day; grabs prey with front legs but holds it with jaws; no web

© Alen thien/Shutterstock.com

Jumping Spiders—Family Salticidae
Bronze Jumper (*Eris militaris*)

Size: body up to 0.3 in. (8 mm); legspan up to 0.4 in. (10 mm)
Key Features: brownish-bronze gray with black patches on elongated abdomen; some banding on legs
Habitat: sunny places; low woodland vegetation; edges
Notes: hunts in daytime; hop around; no web

Venomous Spiders

Most spiders have venom glands. For the most part, they do not have enough poison to harm humans. Allergic reactions to the spiders in our region are likely due to the body chemistry of the person and not the venom. The two venomous spiders we have are the brown recluse and the northern widow. Neither of them are native to the region, and seldom seen.

© Sari Oneal/ Shutterstock.com

Family Sicariidae
Brown Recluse or Fiddleback Spider **(*Loxosceles reclusa*)**
Size: body up to 0.5 in. (12 mm); legspan up to 1.4 in. (35 mm)
Key Features: grayish-brown; distinctive dark violin-shaped marking on top of cephalothorax; neck of violin points toward rear; three pairs of eyes (only feature used for reliable identification)
Habitat: among debris; undisturbed wood piles; indoors
Notes: rare in our region; not aggressive; bites when disturbed (wound develops crust and redness surrounding it; when crust falls off it leaves deep crater in skin; difficult to heal for months); webs loose and irregular under bark and rocks

© Jay Ondreicka/Shutterstock.com

Family Theridiidae
Northern Widow (*Latrodectus variolus*)
Size: up to ½ in. long (1.25 cm)
Key Features: Female northern widow spiders have round, shiny black abdomens, with two touching red triangles (the hour-glass marking) on the underside of abdomen
Habitat: undisturbed woodlands; grasslands; under rocks, logs; entrances to animal burrows; shrubs
Notes: rare in our region; timid, nonaggressive; bite often unnoticed (leading to, an hour later, strong abdominal cramps that can affect breathing); web horizontal sheet with irregular vertical threads

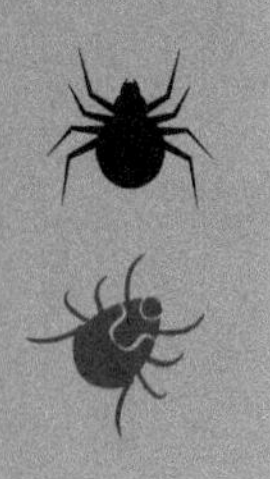

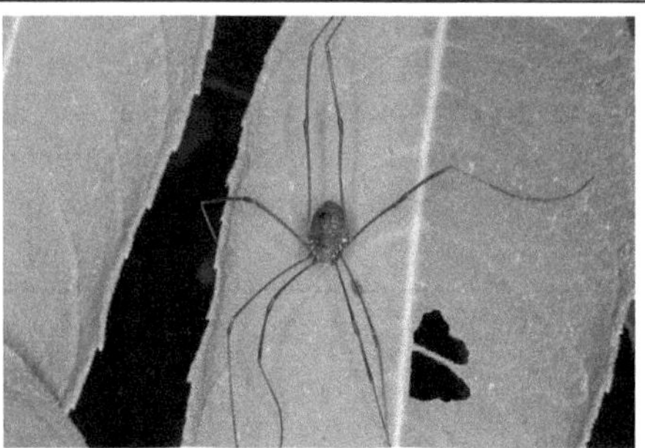

© Elliotte Rusty Harold/ Shutterstock.com

Order Opiliones, Family Sclerosomatidae
Daddy Longlegs or Harvestman ***(Leiobunum vittatum)***—this is not a spider!
Size: body ½ in. (1.25 cm); leg span up to 2 in. (5 cm)
Key Features: oval body; brown; one body part; long legs bent above body; one pair of eyes
Habitat: organic matter
Notes: shed a leg when attacked; no venom; no web; kill weak invertebrates

10. Ticks—Order Ixodida, Family Ixodidae

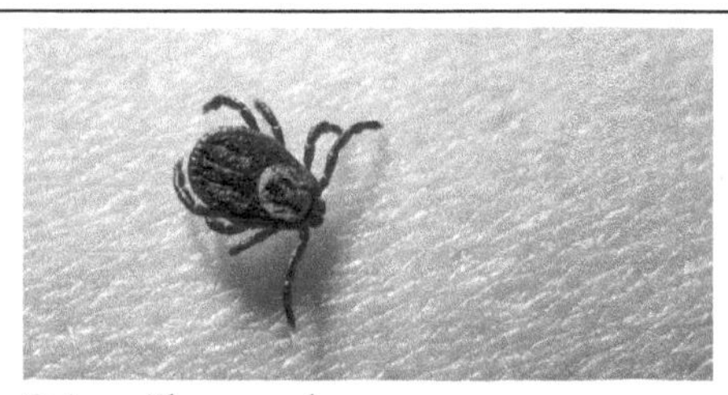

© Anest/Shutterstock.com

American Dog Tick (*Dermacentor variabilis*)
Size: up to 0.2 in. (5 mm)—females will become engorged with blood after feeding
Key Features: brownish-red with lighter gray-silver markings on the "upper shield" (**scutum**); eight segmented outward legs with forelegs directed forward
Habitat: foliage; tips of plants
Notes: attach to pet fur and humans; responsible for Rocky Mountain spotted fever, tularemia, tick paralysis; lurk on foliage waiting for hosts

© KPixMining/Shutterstock.com

Black-Legged Tick or Deer Tick **(*Ixodes scapularis*)**
Size: 0.1 in. (3 mm)
Key Features: females orangish-red abdomen and black "upper shield" (scutum); black legs; if abdomen grayish-blue color, female engorged with blood or may be dark with cream edge
Habitat: among tall vegetation and shrubs; questing at about knee-high
Notes: vector for Lyme disease; nymph stage difficult to detect and can attach to humans

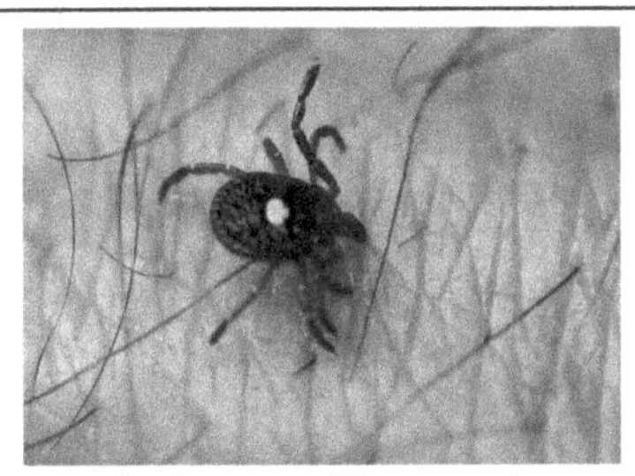

© Melinda Fawver/Shutterstock.com

Lone Star Tick (*Amblyomma americanum*)

Size: up to 0.2 in. (5 mm)—females will become engorged with blood after feeding
Key Features: brown with whitish spot near center of back (female) or white streaks or spots along margins (male)
Habitat: wooded areas with thick understory (where white-tailed deer are present, host); edges; high grass
Notes: nonnative; starting to come into range; can carry ehrlichiosis affecting humans and dogs; can cause person to develop a meat allergy manifesting in anaphylaxis

Mites (*Arrenurus* spp.; *Leptus* spp.)

Size: 0.04 in. (<1 mm)
Key Features: unsegmented body
Habitat: soil, plants, parasite, or epibiont (living on another in a mutualistic or commensalistic relationship)
Notes: majority beneficial in soil and aquatic environments, but some carry disease or cause allergic or skin reactions; pest species of honey bees

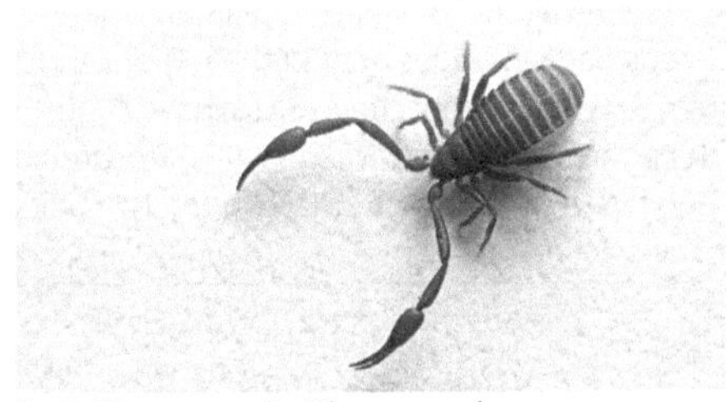

© Gallinago_media/Shutterstock.com

Order Pseudoscorpiones

House Pseudoscorpion (*Chelifer cancroides*)

Size: 0.08–0.31 in. (2–8 mm) long
Key Features: pear-shaped body; pincers resemble scorpions
Habitat: leaf or pine litter; soil; under tree bark; under rocks; in homes near dusty books
Notes: not a scorpion; not dangerous; preys upon clothes moth and beetle larvae, any, mites, small lice, booklice; often commensalistic relationships for purpose of transport

Crustaceans, Millipedes, and Centipedes

The invertebrate animals for groups 11–14 have distinctive body differences. Look closely and read further for details of representative species.

11. Crayfish—
Subphylum Crustacea, Order Decapoda, Family Cambaridae

© Geza Farkas/ Shutterstock.com

Size: 2–5 in. (5–13 cm) long

Key Features: looks like a miniature lobster; look for a "chimney" of mud pellets near water's edge—an indicator of their extensive burrowed tunnels

Habitat: ponds; streams; under rocks in the muck

Notes: Feed on living or dead plant material or small animals. Swim backward if disturbed. Active at night, hide during day. Numerous native species exist in the region, and two species threaten ecosystems of some regions. The rusty crayfish, *Orconectes rusticus*, have robust claws and often have spots on the side of its carapace. The aggressive red swamp crayfish, *Procambarus clarkii*, have dark red raised red spots covering its body and claws with a black stripe on the top of its abdomen. Dispose of bait properly. Many nonnative species are released this way.

12. Isopods—Subphylum Crustacea, Class Isopoda

© Cristina Romero Palma/ Shutterstock.com

© Cristina Romero Palma/ Shutterstock.com

Pill Bug or Roly-Poly or Woodlouse (***Armadillidium vulgare***)
Subphylum Crustacea, Class Isopoda
Size: 0.7 in. (18 mm)
Key Features: dark gray-black; eight segments; curls up when disturbed
Habitat: moist habitat with decaying plant matter
Notes: studied extensively; lives up to three years; harmless

© PHOTO FUN/Shutterstock.com

Sow Bug (*Oniscus asellus*)
Size: up to 1 in. (2.5 cm)
Key Features: brownish-gray; oval shape with overlapping plates; flattened; 14 legs
Habitat: damp places
Notes: needs decaying organic matter to survive; does not curl up when disturbed

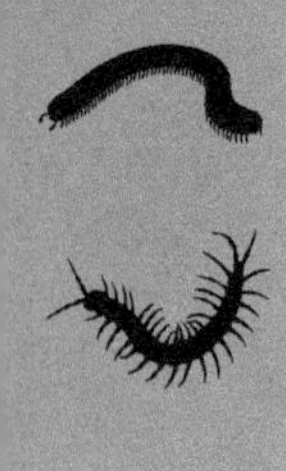

13. Millipede (*Narceus americanus-annularis*)—Class Diplopoda

© Kaleb Kroetsch/Shutterstock.com

Size: 1–1.5 in. (2.5–3.5 cm) long
Key Features: dark brown; up to 400 very short legs (2 pairs per body segment); moves slowly; curls tightly if threatened
Habitat: beneath rocks, logs, leaf litter; active at night
Notes: moves slowly; feeds on decaying organic matter

14. Centipede or House Centipede (*Scutigera coleoptrata*)—Class Chilopoda

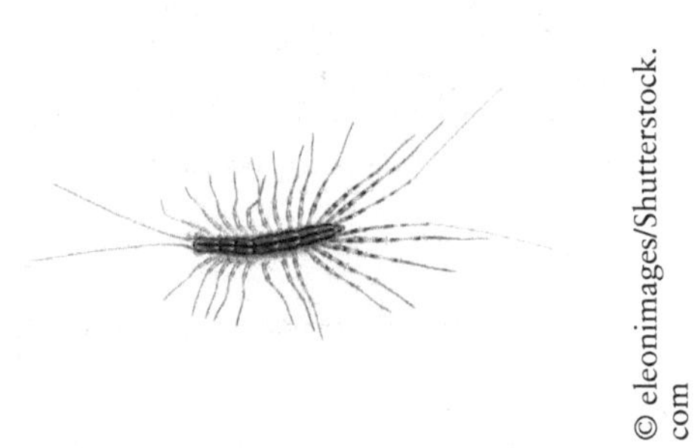

© eleonimages/Shutterstock.com

Size: 1–1.5 in. (2.5–3.5 cm) long (seemingly bigger with legs and antennae, up to 5 in.; 12.7 cm)
Key Features: up to 15 pairs of legs; yellow-gray body; moves quickly
Habitat: in the house; outdoors under rocks, boards, and moist organic matter
Notes: eats insects and arachnids; strong venom injected into prey from separate fang (more of a sting than a bite)—seldom affect humans; if so, minor swelling, minimal pain; mostly nocturnal; nonnative

Arthropods—Natural History Journal Pages

Choose an **insect, spider, or crustacean from your field guide that you suspect lives nearby.** *If you cannot find the information you need in your book to complete the following exercise, then research another reputable resource. Consider completing another report for other arthropods you find or have interest in.*

1. Full Common Name: ____________________

   Scientific/Latin Name (*Genus species*): ____________________

   Family Name: ____________________

   Order: ____________________

   Class ____________________

   Phylum: ____________________

   Kingdom: ____________________

2. Draw the overall picture of the **adult arthropod** with identifying characteristics labeled (size, color, mouthparts, body segments, etc.).

3. Discuss and draw the arthropod's life cycle (complete, incomplete, simple/gradual, or no metamorphosis).

4. Shade the map of North America to show the arthropod's range in which it is found.

5. What is the animal's preferred or required habitats. Include information regarding home, breeding place, and life cycle habitat change.

6. What are the potential ecological interactions your arthropod could have with other organisms in its natural environment (as a food source, what it eats, etc.)?

7. Discuss a behavior of the organism that interests you.

8. Other than actually seeing it, describe how you would know this arthropod was in the area in terms of the evidence it would leave (sound, eaten leaves or bark, etc.).

9. Is this organism considered a beneficial or pest insect to humans? Explain your answer.

10. Discuss the edibility, medicinal, or poisonous or dangerous qualities your arthropod may possess for humans. Include notes on how to gather/preserve the arthropod, or precautions that may need to be taken.

11. Discuss something of interest about your arthropod not previously mentioned in this report.

12. List all resources used (**books**: author's last name, first initial, year, and book title; **websites**: name of site, date retrieved, and web address):

8 Vertebrate Animals

We will now spend the rest of our zoological studies thinking about chordates or animals with a backbone, also referred to as **vertebrates** *(ver'-ta-brits). Just as with the invertebrate animals described in Chapter 7, these organisms are multicellular eukaryotes that consume organic material, breathe oxygen, are mobile, and reproduce sexually. They also serve an important role in food webs. Remaining in Domain Eukarya and Kingdom Animalia, we shift our focus to Phylum Chordata. Starting with the first vertebrate appearing in the fossil record, fish, we will move through characteristics of this animal, make connections to their ecological niche and our lives, then wrap up the section with a field guide to some common fish in our region. This format and details will also be shared for the other vertebrate animals – amphibians, reptiles, birds, and mammals. The chapter culminates with a chart where you can compare all these chordate characteristics.*

8.1 Fish

Nature Journal: Let's reflect on fish. Describe the last time you experienced a fish (the occasion, how you felt about it, your observations). Perhaps the last time you experienced a fish was when you caught one in a nearby lake or river, or perhaps it was in an aquarium or on television or a book—or it was your meal last night! What characteristics do you know about fish?

Discuss what you know about their body, habitats, and so on? What is the name of a fish found in your closest river or lake? Discover if the fish is in this book's field guide by looking in the index or flipping through the pages. What is one thing you learned about that fish? What would you like to know about the topic of fish?

Overview

When considering biological and habitat diversity, no animal group can "outdo" fish. It is the most diverse class of vertebrates. There are more fish species than all vertebrates combined. They live in almost every aquatic environment—deep and shallow water, fresh and salt water, desert pools, the Antarctic. Some of the species have incredible capabilities such as hibernating out of water, gliding out of water, producing electricity, illumination, traveling at speeds up to 60 mph (97 km/hr), they could be immobile, or parasitic—the list could go on. How do we know these things? A scientist who studies fish is an **ichthyologist**. They will study various aspects of fish like their habitat, distribution, diversity, behavior, trophic relationships, and physiology.

The basic characteristics of many fish would include certain features such as fins for locomotion, scales and mucus for protection and streamlining, a swim bladder to maintain position in the water, a lateral line used as a hearing and touch organ, and gills to breathe underwater.

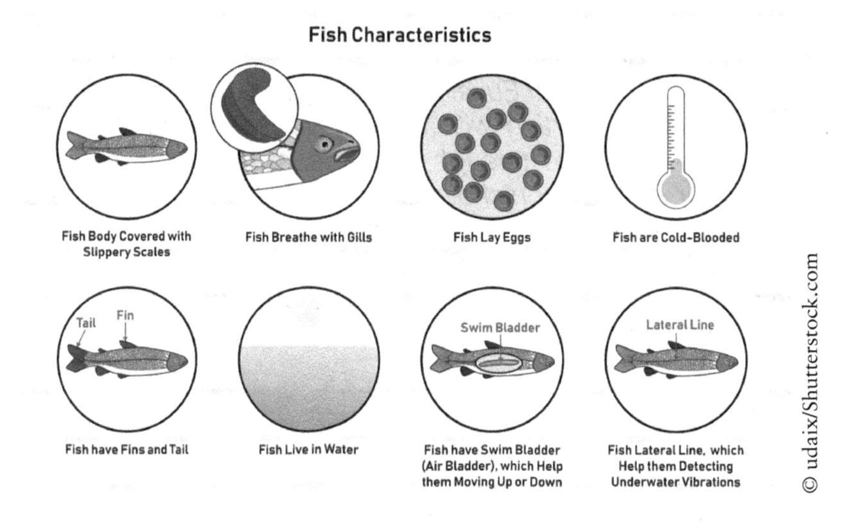

© udaix/Shutterstock.com

Fish Classes

Lamprey—Class Agnatha

The most primitive of all fish belong to the Class Agnatha ("without jaws"). These were the first chordates to appear in the fossil record 280 million years ago (mya) and are considered the oldest type of fish alive. Examples of this fish in the Great Lakes include native and nonnative lamprey species.

Using the diagram of the nonnative, invasive sea lamprey, *Petromyzon marinus*, become familiar with

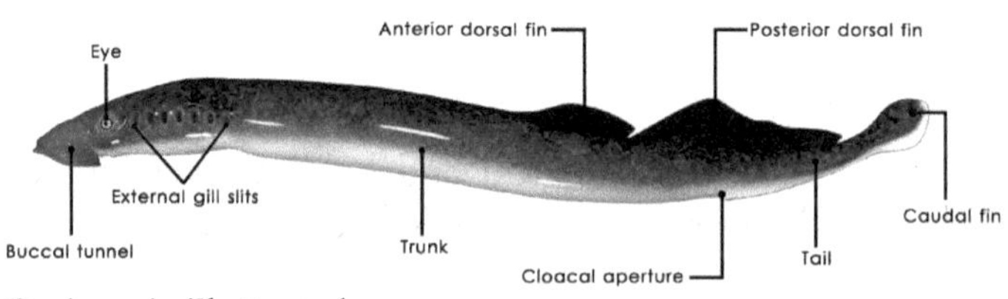

© sciencepics/Shutterstock.com

its morphology or physical features. Take note of the body plan features—notice the nostril, the "dots" for external gill slits (they are not covered like the bony fish discussed later), and you should notice the unique fin arrangement as well.

Lamprey do not possess a lateral line nor a swim bladder. They have no scales. Notice no paired fins, but has dorsal and caudal fins. The dorsal fins help with determining if you have a native or nonnative lamprey. The nonnative sea lamprey have two separate dorsal fins. All lamprey have well-developed eyes and olfactory system. Please note they are not eels! Eels belong to a different fish order and have actual bone, jaws, and pectoral fins—lamprey do not.

Not all lamprey are parasitic. In our region, we have both native nonparasitic and parasitic, and only one nonnative and it is parasitic.

The native lamprey types found in the Great Lakes:

Nonparasitic: American Brook and the Northern Brook Lamprey

Parasitic: Chestnut Lamprey and Silver Lamprey

Nonnative Invasive (parasitic): Sea Lamprey

The mouth of the parasitic lamprey is a large, sucking disk with sharp, horned teeth. It will adhere to surfaces and once it finds and latches to its prey (cold-blooded animals, mainly bigger fish), it will start to bore its filelike tongue into the animal and sucks its blood for nutrients. Once the lamprey is satiated, it will detach. Nonparasitic lamprey do not have a mouthpart like the one featured here; in fact, they do not eat as adults.

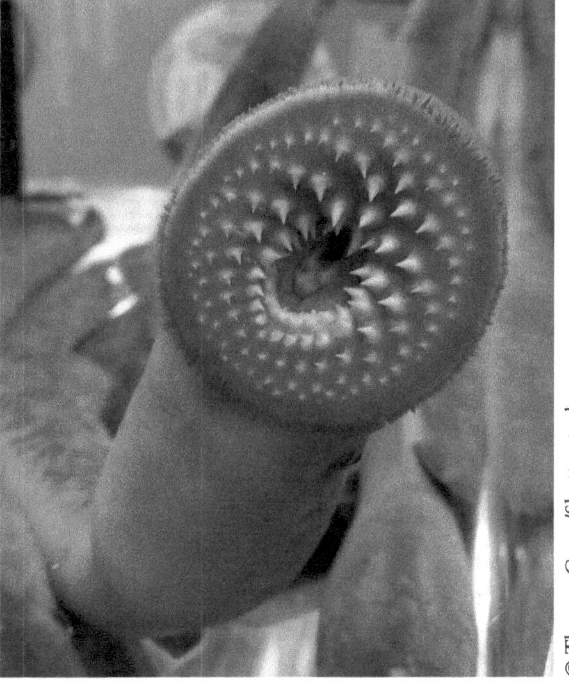

© Theresa Grace/Shutterstock.com

Adult sea lampreys swim up streams to spawn and then die. Fertilized eggs hatch into small, wormlike larvae that burrow into stream bottoms. The larvae feed on debris and small plant life (algae) for an average of

3–6 years before they transform into the parasitic adult (upward of 17 years). The adults migrate into the Great Lakes (in home range, they would migrate into the ocean) where they spend 12–20 months feeding on fish. The life cycle span, from egg to adult, can take an average of five to eight years to complete. For nonparasitic forms, the process is much shorter because as adults they do not feed but focus on reproduction.

Lampreys have been enormously destructive since they invaded the Great Lakes. Sea lamprey attach to fish with a sucking disk and sharp teeth. They feed on body fluids, often scarring and killing host fish. During its life as a parasite, each sea lamprey can kill 40 or more pounds of fish. Sea lamprey are so destructive that under some conditions, only one of seven fish attacked by a sea lamprey will survive. The parasite preys on all species of large Great Lakes fish such as lake trout, salmon, rainbow trout (steelhead), whitefish, chubs, burbot, walleye, catfish, and even sturgeon. Sea lamprey have had a serious negative impact on the Great Lakes fishery. Because sea lampreys did not evolve with naturally occurring Great Lakes fish species, their aggressive, predaceous behavior gave them a strong advantage over their native fish prey.

A sea lamprey parasitizing a salmon.

To date, eradicating sea lamprey has been 90% successful with the monitored application of a lampricide (3-trifluoromethyl-4-nitrophenol, or TFM) that targets only the larvae of lamprey. It takes millions of dollars per year to monitor and apply. However, the control of lamprey protects the Great Lakes fisheries industry estimated at nearly $5 billion per year. Other control methods are continually being explored.

One control option includes synthesized pheromones that have shown independent influences on the sea lamprey behavior. A **pheromone** is a naturally produced chemical substance that some animals can release into the environment. One pheromone may serve as a migratory function, and the odor emitted from larvae are thought to lure maturing adults into streams with suitable spawning habitat. The other option used involves a sex pheromone emitted from males that is capable of luring females long distances to very specific locations. This is rather powerful given the fact it occurs in complete darkness and many lampreys at this stage in their life have strongly degraded eyesight.

The pheromones released could be used in a targeted effort at environmentally friendly lamprey control by drawing in mass lamprey to capture. The research into sea lamprey genetics as well as pheromones may result into a successful, effective management technique that could one day drastically reduce the need for TFM treatments of spawning grounds.

Activity: Research the latest approaches to sea lamprey control near you, and the types of studies conducted to help mitigate the issue.

Sturgeon—Class Actinopterygii

The only member of this class in our Great Lakes is the lake sturgeon (*Acipenser fulvescens*). Their partly cartilaginous skeletal features separate it into this class. Evolutionarily, this is an ancient fish with skin of bony plates on its back (adults) and a bottom feeder. They have a very long life span with females living upward of 150 years. Its population numbers have declined over the years by pollution, fisheries' indiscriminate killing to prevent accidental damage to fishing gear, and overharvesting of sturgeon for its value in caviar and isinglass. However, great conservation and restoration efforts have taken place since 2001 and now very limited sturgeon fishing seasons are permitted.

Bony-Fish—Class Osteichthyes

Most fish in our region belong to this class, the "bony fish." They have a bony skeleton and have more bones than any of the vertebrates. The fish also have a bony flap called the operculum that covers the gills, swim bladders to maintain position in the water, and most have scales covering the surface of their body.

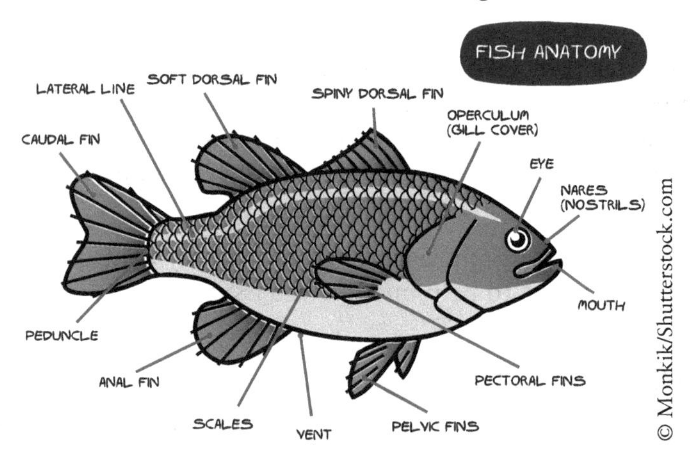

Activity: The next time you catch a fish, buy a whole one at the grocery store (with the scales and head still intact), or while watching them at an aquarium spend time identifying the external features of the animal. Use the fish anatomy diagram to help you. In many of our fish, you can see the lateral line running the length of the body. You will discover that the arrangement and shape of features and the body plan of different species vary.

Body Shapes. Fish vary in size and forms. From the small goby at just half an inch (1.25 cm) to a whale shark (not in our regional waters) of up to 70 ft. (21 m) and 25 tons. Their form or shape depends on their life habits. Fast swimmers will appear streamlined and slow, bottom-dwelling fish will often have a flattened structural appearance. Consider the shape of the fish. Its shape can tell you something about its niche—where it lives and what it does in its environment.

When observing the general shape of fish, think of it as if you were nose-to-nose with the fish and looking down the length of its body. The common shapes you might encounter include a narrow body where it is taller than they are wide like the sunfish and largemouth bass. This allows for high maneuverability among aquatic plants. Flatfish have a ventrally flattened body where the body's height is shorter than its width, which is common among bottom dwellers. A torpedo or bullet-shaped body indicates a fast fish and often found in open or strong currents like various trout or salmon. Lastly, a tubular or cylindrical shape can enable the fish to fit into small spaces like eels.

Another perspective of the fish's body plan is to consider its mouth orientation. This can also tell you something about its niche or behavior. Many mouth types exist, but some are more common than others. If you see the mouth oriented downward or on the ventral surface, you can deduce it is primarily a bottom feeder or swims above where it finds food. These fish often have barbels to help sense its surroundings and locate food sources. Upward orientation indicates they feed at the surface and likely wait for prey to appear above them before striking from below. Their diet will mostly be insects or smaller fish that swim near the surface. Most fish have a terminal mouth located in the middle of the head and point forward with both jaws the same length. These fish are mainly opportunistic omnivores and can feed at any location.

Coloration. Did you know fish can change color? This is called **metachrosis.** Knowing this ability exists, helps the naturalist realize that fish colorization varies within the same species. This needs to be taken into consideration when trying to make an identification. The color will change based on the habitat and depth at which the fish spent most of its time. **Dermal chromatophores** are a feature within the pigment of the animal that is responsible for the color change (you will see this feature in other animal types too!). Other types of coloration in fish are in reference to camouflage and spawning. Regardless of how you look at it, this is evolution at work. Think about how natural selection plays a role between fish coloration and population dynamics.

Dermal chromatophores enable the fish to have a dark coloration in a dark habitat or at deep depths, and have a light coloration to help camouflage it in a light habitat or in shallow depths. Fish living in turbid (cloudy) and muddy-bottomed water tend to be darker than fish that live in clear water with light sand bottoms. Another type of camouflage includes countershading. You will notice this on many fish. The dorsal, or upper surface of the body is dark, and the ventral, or underside is lighter and often silvery. Pretend you are potential prey of this animal. If you are swimming above this fish, you may not notice him because all you would see is dark below you. The same situation would occur if you swam beneath him and you would just see light colors above.

Spawning coloration is yet another variation you might see in a species. During certain times of the year, many fish species (typically the males) have beautiful colors that advertise their spawning condition or willingness to spawn. After this time of reproduction, the bright coloration will fade. Regardless of color variations in fish for whatever reason, there are some markings that will remain like the presence of the lateral line (if usually visible) and any dark lines, bars, or spots. Focusing on those features will help with making an accurate identification during any conditions. Another notable fish feature to help identify the species is its fins.

Types of Fins. Fin type, arrangement, and size vary among fish families. Fins can be used for a variety of things like helping to stabilize and prevent fish from rotating while swimming; they permit stops, fast starts, maneuvering, and help maintain the fish's position in the water. Use the fish anatomy diagram pictured previously to help you become familiar with the location of the following fin types. Understand that the position and types of fins vary between fish families. Body shape, darkened markings, and fins will help you identify fish species.

Pectoral: attached to pectoral girdle; for balance, maneuvering, propulsion.

Pelvic: attached to pelvic girdle; for balance.

Dorsal: some are spiny-rayed, others are soft-rayed (bendable); found on back; stabilize the animal against rolling and assist in sudden turns.

Adipose: found on trout and catfish; fatty fin without fin rays; located on back; smaller than dorsal fin; stabilizes fish

Anal: near anus; stabilizes fish in water

Caudal: tail fin; shape depends upon lifestyle of the fish; most important—moves fish forward.

> ***Nature Journal:*** Select a fish species from the field guide section of the book. Sketch it. Write its common and scientific names. Label each fin type (refer to the diagram found earlier in this section). How would you describe its body shape? Based on its mouth orientation, what do you suppose it feeds upon or where? What can the color of the fish observed in the picture tell you about where the fish lives? Compare your deductions to what you research about that species. Clarify in your descriptions how the fish's fins, shape, mouth, and coloration can potentially benefit the animal's lifestyle or niche.

Scales. Scales offer protection to the fish throughout its life. They offer the main protection for the body as an overlapping "armor." Scales are connective tissue covered with calcium. As the fish grows, scales enlarge. The scales can be used to determine the age of a fish. Without harming the fish, fisheries biologists can gently flake off a scale to determine its age. A similar concept to aging trees, annuli or the annual growth rings are counted on the scale while examining it under a microscope (a pocket microscope works too). However, rather than counting every ring, groupings of rings or bands are counted. In temperate water, when annuli are close together it can be interpreted as one winter; whereas, a large gap between rings indicates a long growing season or fast growth.

Several types of scales exist, but the most common are ctenoid and cycloid scales. More primitive scales have less protection offered to the fish than advanced or more durable scale forms. Ctenoid scales are rough with a comblike edge as with the perch and bass. Cycloid scales are smooth as observed in carp and salmon. Sturgeon and gar have very hard scales termed, ganoid; and catfish and lamprey have no scales.

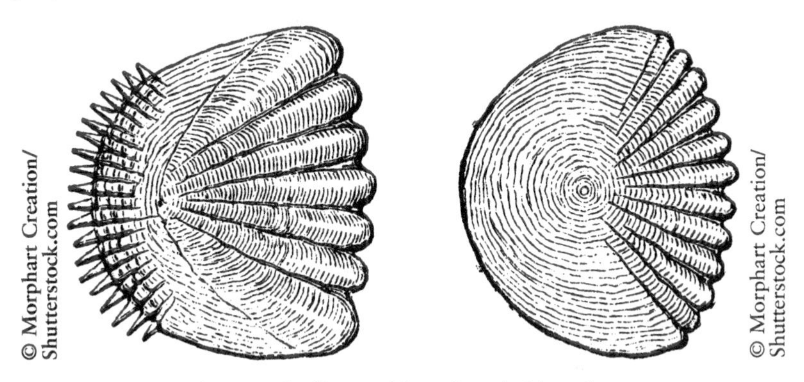

Drawings of magnified ctenoid and cycloid scales. Although difficult to determine from these images, realize that the fish's age is read by examining groupings of annuli. The cycloid scale on the right would be approximately 5–7 years old.

Age and Growth. Fish grow until death. They do not possess internal genetic mechanisms to limit growth. Growth rates are dependent on heredity, length of the growing season based on water temperature and food supply. Optimum conditions for the species result in rapid and continued growth. **Stunting** is when fish do not grow to their average species' size. This may occur when too many small fish exist for the available food supply. Not enough predators exist in that food chain to regulate population numbers of the primary and secondary consumers. Another reason for smaller than expected fish could be the result of intense competition between young for the same food supply. Managing our fisheries by limiting the number, size, and season of the fish you catch will help with the quality of fish in your lake or stream.

Mucus. Have you ever felt a live fish? How did it feel? Slimy? That slime, or mucus actually helps the fish. It can make it easier for the fish to swim, and it has antimicrobial properties that can prevent infection. So, the next time you catch-and-release fish, please take the benefits of mucus into consideration. Be sure you do not have lotion or sunscreen on your hands, and wet your hands before handling the fish so that you minimize the amount of mucus removed and be sure you do not have lotion or sunscreen on your hands.

FISH

Swim Bladder. Another neat feature of many fish is the presence of the swim bladder, also called the air or gas bladder. Its purpose is to help hold the fish's position in the water. This is called **hydrostasis.** In some fish, it can also be used as a lung to breathe (gars), to hear or detect vibrations with special bones called Weberian ossicles, or to make sound when held taut.

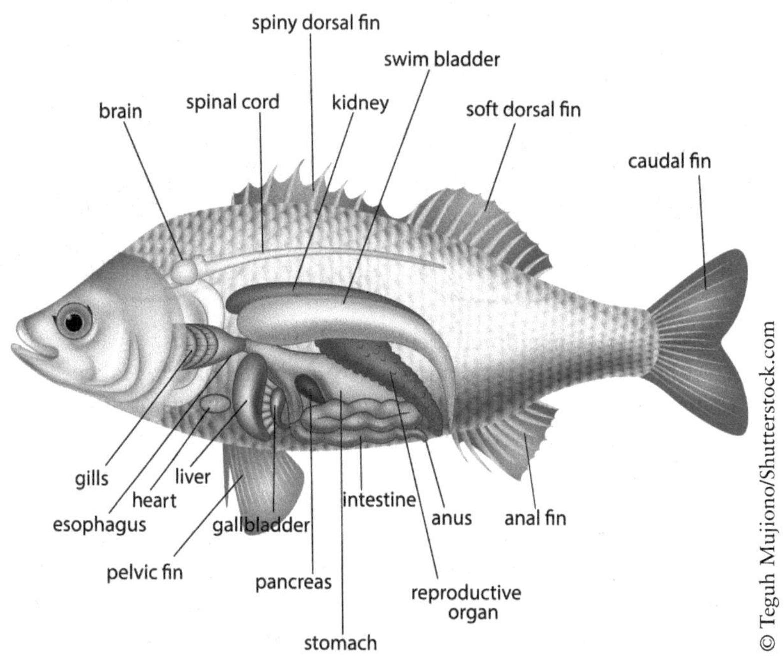

If you fish to help with your personal food supply (abide by local regulations, of course), consider examining the internal structures of the fish as you clean it for consumption. Can you find all the features and organs in the diagram?

Physiology

The physiology of an organism refers to its organs and organ functions. For fish physiology, read about their respiration, circulation, nervous system, digestion, and respiration. ***Activity:*** Think about how fish relate to other vertebrates. Start filling in the Chordate Characteristics Comparison Chart found at the end of Chapter 8.

Respiratory System

Have you ever seen fish in an aquarium? They constantly open and close their mouths, right? This is part of their respiration process. They take in gulps of water to help them extract the required oxygen. For this to occur, fish also need water to pass over their gills for respiration. Seventy-five percent of their oxygen is acquired from one gulp of water. Some fish, like salmon, need high levels of oxygen so they will be found in cool, fast moving water. Catfish can tolerate low oxygen levels and will be found in warmer, slower moving waters than salmon.

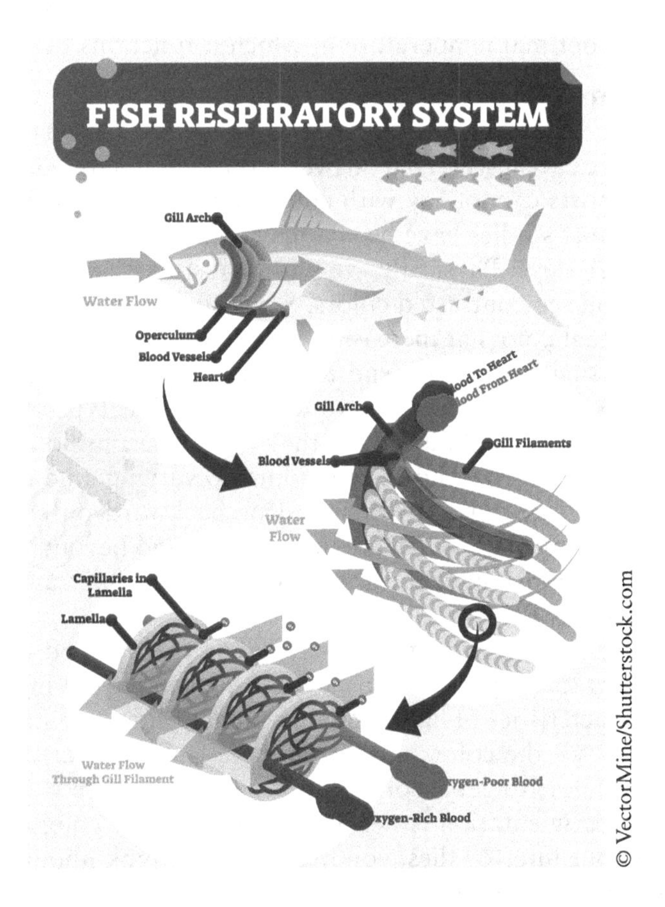

© VectorMine/Shutterstock.com

Circulatory System

Fish blood circulates from the heart to the gills to allow for purification before traveling to other parts of the body. The blood pumps in a predictable manner. Find the heart in the diagrams of the fish anatomy and on the respiratory system image. It is on the ventral side of the fish close behind the mouth. From there, it goes to the gills. Its flow is **anterior** (toward the head) on the **ventral** (belly) side then **posterior** (toward the tail) on the **dorsal** (back) side. The circulatory system is comprised of the heart, blood, and blood vessels. It is a two-chambered heart. The atrium is the receiving chamber for the blood and the ventricle is the chamber that pumps the blood through the body. The blood vessels are largest closest to the heart and get smaller as it networks through the tissue. As with other physiological and anatomical features, you can compare and contrast vertebrate animals.

In terms of the fish's capabilities for thermal regulation, it is considered ectothermic, often referred to as cold-blooded. This means that its internal body temperature varies with environmental temperature. Fish cannot regulate its internal body temperature as we can and other **endotherms**, or warm-blooded animals. Animals have an optimal temperature in which it functions best without stress.

Nervous System

The way a fish detects stress is mainly through their nervous system and their senses. How well can fish see? That answer is difficult to answer. It depends on the species, and scientists cannot say with confidence what fish see or what registers in their brains. Most studies have been done on either physical or chemical examinations of the fish's eye. Physically, we know most fish can obtain a clear image, detect motion, and see contrasted colors. Some studies have also shown that fish can recognize actual colors at increasing light intensities.

Most fish in shallow waters tend to detect color more readily likely due to increased light penetration compared to deep water. The type of fish and its habitat could limit seeing capabilities, but they also have physiological constraints. They have a fixed pupil size and cannot adjust to varying amounts of light. Their lens is nonflexible and can move forward or backward only slightly. This can create an unclear image. Most fish are nearsighted, and beyond 3 ft. (1 m) objects become blurry. Many fish have the ability to see almost 360° around them except the areas directly posterior.

If you have ever gone fishing, you may have experimented with different colored lures or flies (perhaps you also factored light intensity into your selection; water alters characteristics of light thereby changing colors of submerged objects). Fish can possibly see the colors red, green, blue, and yellow. Know that red and blue are the last colors to get absorbed, so if you fish in deep water use something blue. Further experimentation is needed to predict what colors may attract fish. With selecting your lures or flies, you may want to think about what creates the greatest contrasting color in the water, or what you are trying to mimic looks like in the water. Black is likely the most visible color under most conditions, and tackle that has two very contrasting colors may prove effective.

How effective are the other fish senses? Notice that fish do not have external ear openings. They have an ear bone, called the **otolith** that detects sound vibrations. The sense of smell is also important to fish and many species can detect food, danger, and its spawning area. The sense most poorly developed is taste. However, if you observe **barbels** (hairlike structures around the mouth) on the fish, then that species can "taste" or sense its environment.

Arguably, the most important sensory organ of fish is the **lateral line**. This feature can sometimes be seen along the length of the fish as seen in diagrams in this section and pictures of fish species in the field guide. The lateral line is made up of a series of openings. Those openings receive sound or other vibrations in the water, which will travel through the attached canals. Those paths lead to sensory hairs along the muscle that signal nerve impulses sent to the brain. The brain interprets those signals as what caused the vibration. The lateral line has auditory and tactile sensory perceptions or can be considered its hearing and touch organ. It may determine current direction, object proximity, water temperature, enables them to detect potential predators, locate possible prey, and to help stay grouped while traveling together in schools.

A magnified photo of the lateral line, a sensory organ that runs the length of a fish's body.

Also part of its nervous system is some species' ability for **bioluminescence.** Primarily deep-water fish may have the light-producing chemicals that is used for communication, territory, or attracting mates. Biofluorescence, not due to a chemical reaction, is demonstrated from the absorption of blue light and emitting a different color like red, orange, or green. The pigment becomes oxidized without giving off heat. More research is needed to determine and understand these phenomena in our regional waters.

Digestive System

To help us further understand fish, let's consider its basic digestive process. Digestion begins with the mouth. In order to digest something, it has to be ingested first. Fish in the temperate zone feed primarily during warm periods. Their metabolism and feeding frequency slows down during cold periods.

As you thought about previously, consider the mouth orientation of a fish species to help you learn about its behavior. Now, consider the teeth present in a fish species. These features will provide further clues to their diet (niche). All types of fish diets exist, so corresponding adaptations for these diets also exist. Planktivorous fish eat primarily plankton (i.e., phytoplankton and zooplankton) and aquatic insects. The larger game fish anglers covet are carnivores and eat other vertebrates (i.e., amphibians, reptiles, birds, and mammals). Those that eat primarily other fish are termed **piscivorous.**

Feeding behaviors can be classified as predators (kills prey for food), grazers or browsers (feeds on plant life), parasites (attaches to and feeds on other fish), or scavengers (feeds on dead fish or fish parts). The shape of the tooth will indicate its abilities. A predatory fish would have sharp teeth that could cut, pierce, or hold onto prey. Fish with a preference for crustaceans would have teeth adapted for crushing. Plankton-eating fish will have teeth modified as strainers.

The digestive system is comprised of a variety of organs. The organ structure or function may differ depending upon the species' diet. Fish may have a relatively wide esophagus that allows it to swallow food whole. The stomach in predators is elongated and straight back from the esophagus to also make swallowing entire prey easier. Why would a herbivore have more surface area in the intestine than a carnivore? Plant matter requires more breakdown of tissue to obtain the required nutrients. Remember, plants have a cell wall of cellulose. Parasites, like some lamprey, will have a simple and straight digestive system because they do not need to digest tissue—just blood. The small intestine uses enzymes secreted from the pancreas to help it break down food into digestible parts and absorbs the nutrients. The pyloric caeca are pouches of the intestine at the point where it meets the end of the stomach. It provides a greater surface area to help with digestion. This is mainly seen in herbivores. Lastly, the large intestine's role is to eliminate indigestible wastes.

Reproductive System

The reproductive system also helps us interpret behaviors of the fish we observe or seek. Fish reproduce in different ways. Most lay eggs, while some bear live young. Most eggs are fertilized after the release from the female's body, and some are fertilized inside the female. Almost all game fish, those fish that are sought out by anglers are egg layers.

Mating (**spawning**) in egg-laying fish, usually occurs once a year at a particular time. Spawning season is when eggs of the female and sperm (**milt**) of male are ripe. Most fish will exhibit external fertilization where eggs are released by the female and fertilized externally by the male's milt (**oviparous**). Internal fertilization is when the eggs are fertilized inside the female reproductive tract by the male's sex organ. Two types of strategies exist for internal fertilization, ovoviviparous or viviparous. **Ovoviviparous** is when eggs are fertilized internally and kept inside the female reproductive tract nourished by a yolk until hatched. Whereas, **viviparous** is when eggs are fertilized internally but nourished directly by the female's reproductive tract.

The incubation period of the eggs is temperature dependent in external fertilization, since fish are ectotherms. All fertilized eggs require special conditions for successful development. Sunlight, oxygen, water agitation, salt, chemicals, water temperature, and other factors have an influence on egg development and need to be just right for successful development and hatching.

Migration. Usually once or twice per year, depending on species, a mass movement of fish will occur. This is called migration and is induced by a need to spawn. Salmon exhibit this classical migration habit (in our region, the lake trout (*Salvelinus namaycush*) and brook trout (*Salvelinus fontinalis*) are the only native salmonids!). Other mass movements are not considered migration and would include fish movements because of sudden adverse conditions arising or to take advantage of a daily feeding habit.

There are different types of migration that exist dependent on the species. **Anadromous** fish live in the ocean during adulthood and spawn in freshwater streams (salmon and sea lamprey). Those who spend adulthood in freshwater and spawn in the ocean are termed **catadromous** (American eel). **Potadromous** fish spend adulthood in large freshwater and spawn in streams (lake sturgeon). Seasonally, some just migrate to shallower water to spawn (bass). Something unique to consider is how our nonnative fish the salmon, rainbow trout, and sea lamprey are anadromous in their native waters but they have adapted to their new environment and are now potadromous!

Fish Management

If we want to maintain or create a healthy, diverse fish population, then fish management is necessary. Management is based on the scientific data collected by fisheries biologists. Habitat protection and improvements often results from the information analyzed. They also set size and number limits of catches for certain times of the year for each species. These regulations allow fish populations to reach potentially optimal growth and reproduction. A sustainable recreational (and commercial) fishing industry is also a goal. Natural resource agencies may sometimes supplement fish to lakes and rivers based on need or demand. The **stocking** of fish comes from the rearing of certain fish in **hatcheries** that are raised until they can survive in the wild. Salmon, trout, and walleye often cannot reproduce enough for demand and will be stocked in their appropriate habitats.

What Can I Do?

There are at least a couple things you can do to help make a positive difference in our fisheries. Buy a fishing license—even if you do not fish. Consider gifting licenses. The small fee charged may help support the management efforts within your state or province. Remember, the entire ecosystem is connected. You are not just helping the fish; you are helping protect and preserve the waters in which they rely.

Participate in creel surveys. The name "creel" originates from when anglers typically used a creel to collect their catch, traditionally in a wicker basket. You can volunteer to collect and offer your creel data. A volunteer or government employee may do **creel surveys**. At access sites, a person talks with anglers to learn about the fishing conditions of that lake or river. Realize that it is very difficult to monitor all our waterways, so this is the next best thing. Research the types of volunteer and community science opportunities near you that focuses on fisheries.

People who do not participate in recreational fishing can still discover and enjoy these amazing animals. Simply wear polarized sunglasses to discover what lives in the water. You can purchase them for a low cost at almost any place that sells sunglasses. These types of glasses allow you to view underwater by blocking the glare you see on the surface of the water. The surface is horizontal; therefore, the light is reflected horizontally. These glasses block those horizontal light waves. They will also help block harmful UV rays. Discover fish migrations in nearby rivers (i.e., salmon or trout) or learn when and where fish spawn in lakes—this can be an exciting time for fish viewing and anglers.

If conditions are favorable for swimming, wear a dive mask or goggles to observe what swims around you. A snorkel would allow for continued underwater viewing without emerging for air as frequently. You could also wade in the shallow area and use a minnow net to catch small animals or use your bare hands. Consider learning to fish. Watch online videos, research it, participate in a fishing clinic, or ask somebody to teach you. Realize you do not have to keep the fish you catch but learn to be smart with your catch-and-release practices, so you do not harm the fish. Be sure to read the fishing regulations for where you live, buy a license, and practice good water safety.

Review Questions

1. What is an ichthyologist?
2. Discuss the basic features of a fish's external anatomy and body shape. Identify the six main fin types. Where are they located and what is their function?
3. Explain coloration in fish. What is the purpose of fish scales? Most have one of what two types? Discuss the purpose of fish having mucus on their bodies.
4. Describe the potential functions of the swim bladder.
5. Discuss the physiological systems of the fish (respiratory, circulatory, nervous, digestive, and reproductive).
6. Discuss the growth, growth rate variables, stunting, and aging techniques of fish.
7. Identify and explain four types of fish migration.
8. Review the fish families in the field guide section. What are the distinguishing characteristics between families? Draw the general outline or silhouette of each family to help with quick identification.
9. Why are fish managed? How are they managed?
10. List ways you can help protect our fisheries. Name some ways you can enjoy fish.

Field Guide: Fish

In this field guide, the fish are grouped according to family. This method of classification indicates similar basic body plans, physiology, and behaviors. Use the fish silhouettes or icons to determine which family to investigate for your fish in question. Focus on body shape and fin position when making your initial observations.

© FARBAI/
Shutterstock.com

Fish Family Identification Key

At the very least, know your fish by the family in which it belongs. In the gray shaded box, choose the best match for your fish's body shape and fins, then turn to the corresponding numbered section and icon to start perusing some options within that family.

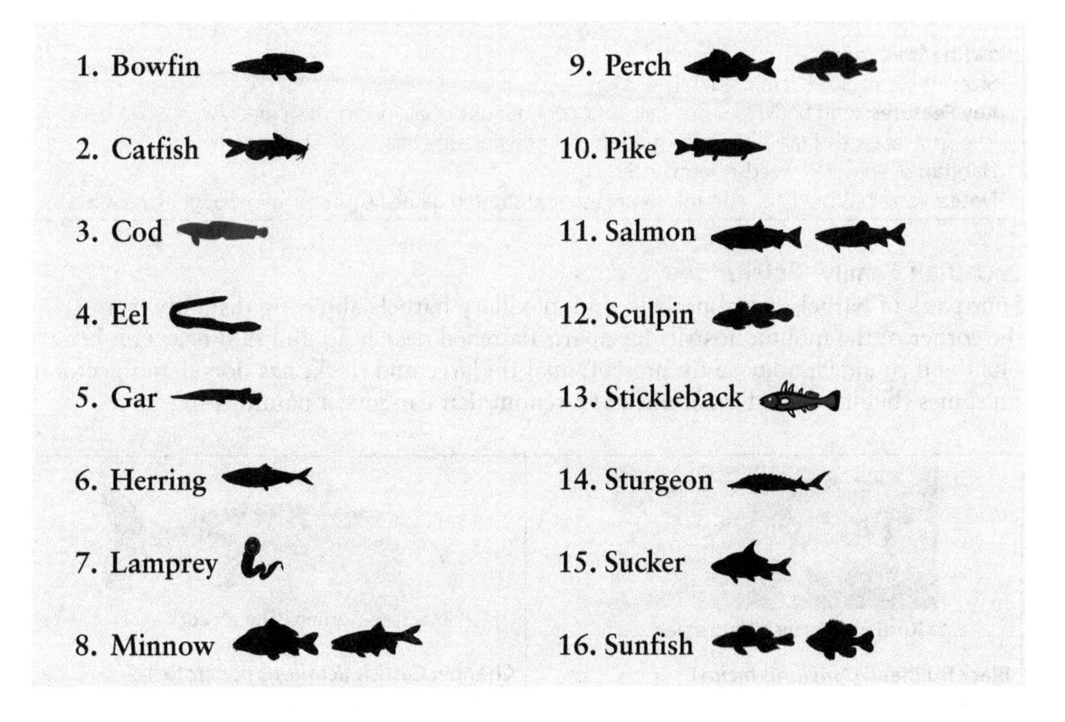

Note: *Some features in the field guide images may vary from what you may observe in an actual specimen (i.e., color, markings, and fin shape). This can be due to the habitat in which the fish was collected, its age, time of year (i.e., breeding season often intensifies some species' coloration) or if the fish has been out of the water for any length of time (some fins will lie flatter when out of the water, as in some of these pictures).*

Fish Species Accounts

1. Bowfin Family—Amiidae

A "living fossil"; uses gills to breathe, but can use swim bladder as lung in stagnant water; can survive as long as bottom stays moist

© Miroslav Halama/Shutterstock.com

Bowfin (*Amia calva*)

Size: 12–24 in. (30–61 cm); 2–5 lb (1–2.3 kg)
Key Features: long body; rounded tail; long, continuous dorsal fin; brownish green with white belly; "eye spot" at base of tail; serpentine motion of dorsal fin; large head with sharp teeth
Habitat: deep water; weeds; warm water
Notes: voracious predator; controls rough fish and stunted game fish; not usually sought by anglers

2. Catfish Family—Ictaluridae

Four pairs of barbels: nasal barbels, and maxillary barbels above lip distantly from the corner of the mouth; nostrils far apart; flattened near head and thin near tail; head blunt and rounded; adipose fin present; anal fin large and thick; has dorsal and pectoral fin spines (bullheads and madtoms have venom that can give a painful sting)

© Rostislav Stefanek/Shutterstock.com

Black Bullhead (*Ameiurus melas*)

Size: 8–10 in. (20–25 cm); 4–6 oz (113–170 g)
Key Features: dark-based barbels around mouth; black to olive; dark greenish to yellow sides; ventral side creamy yellow; large anal fin; bar at base of caudal fin sometimes obvious; rounded tail; scaleless skin
Habitat: shallow, slow streams or ponds; tolerates turbid waters
Notes: very abundant; often stunted because of overpopulation; other similar species: yellow and brown bullheads, stonecat, madtom

© Sergey Goruppa/Shutterstock.com

Channel Catfish (*Ictalurus punctatus*)

Size: 12–20 in. (30.5–51 cm); 3–4 lb (1.4–1.8 kg)
Key Features: long barbels around mouth; grayish back and sides with black spots; white belly; forked tail; large adipose fin
Habitat: clean, fast-moving streams with deep pools; tolerates turbid water
Notes: sport fish among anglers; males build nests in sheltered areas like undercut banks or logs

3. Cod Family—Gadidae

[Burbot]: long, cylindrical bodies, flattened heads, large mouths, unique fin structure; single barbel (whiskerlike projection) extends below tip of lower jaw; body coloration is largely mottled green, except for a cream-colored underside; relatively poor swimmers as a result of their unique body shape

Burbot (*Lota lota*)

Size: 20 in. (0.5 m); 2–8 lb (1–4 kg)

Key Features: tubular body; brown and mottled scales; cream-colored chin and belly; small barbel at each nostril; longer barbel on chin; rear dorsal fin long and above the anal fin (similar size and shape) and near the caudal fin; first dorsal fin short

Habitat: deep, cold, clear rivers and lakes in northern parts of region; benthic

Notes: voracious predator of small fish (and other things); digs trench where it waits to ambush prey; low swimming endurance; breeds in shallow water under ice

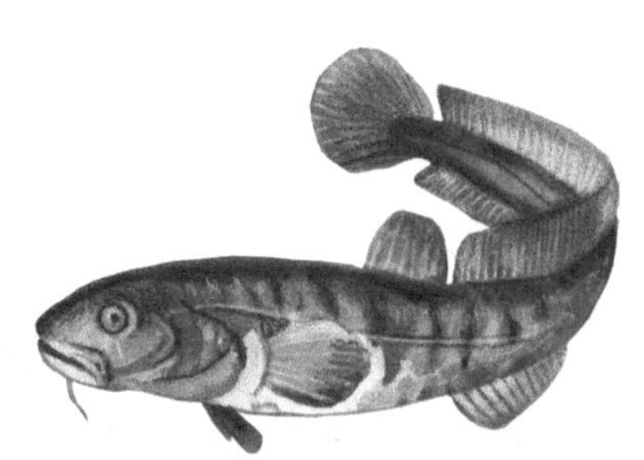

© Les_Frolova/Shutterstock.com

4. Eel Family—Anguillidae

Snakelike body covered with tiny embedded scales (and a thick slime layer); paired eyes and nostrils; mouth has movable jaws with many sharp teeth; dorsal, caudal and anal fins are fused into one long fin extending around posterior of body; upper half of body is olive green to brown, grading from pale yellow to white on ventral side

© Ruben Martinez Barricarte/Shutterstock.com

American Eel (*Anguilla rostrata*)

Size: 24–36 in. (0.6–1 m); 1–3 lb (0.4–1.4 kg)

Key Features: long, tubular body; large pectoral fins; dark brown with yellowish sides and white belly; large mouth; continuous dorsal, tail, and anal fins

Habitat: bottoms of soft sediment streams

Notes: most active at night; not a popular food fish

5. Gar Family—Lepisosteidae

Long, cylindrical fish with distinctive long jaws containing needlelike teeth; fins arranged to generate quick bursts of speed useful for ambushing prey; hard, diamond-shaped ganoid scales that are covered with ganoin, an enamel-like substance that takes a high polish; a unique characteristic of gar is their ability to breathe atmospheric air

© Reimar/Shutterstock.com

Longnose Gar (*Lepisosteus osseus*)

Size: 24–36 in. (0.6–1 m); 2–5 lb (1–2.3 kg)

Key Features: long, cylindrical body with snout twice as long as head; needlelike teeth; olive brown usually with dark spots on sides; single dorsal fin near tail fin just above anal fin; platelike scales

Habitat: backwaters of large rivers and floodplain lakes

Notes: eats small fish; warm, deep water; tolerates turbid water; effective with controlling rough fish populations by stalking then uses a fast, sideways slash for capturing; known to grow to 5 ft.

6. Herring Family—Clupeidae

Streamlined for swimming, body is relatively deep and flattened laterally (side-to-side), with a distinctly forked tail (caudal fin); compressed body and silvery scales serve as camouflage in the open waters of the ocean, scattering light and helping to conceal herring from predators attacking from the deep

© Reimar/Shutterstock.com

Alewife (*Alosa pseudoharengus*)

Size: 4–8 in. (10–20 cm)

Key Features: silvery bluish shiny with white belly; dark spot behind gills above pectoral fin; large mouth with protruding lower jaw

Habitat: open water of Great Lakes and some inland lakes

Notes: this nonnative (Atlantic) herring is subject to large summerkills; not well adapted to freshwater; sea lamprey depletes its predators causing large increases in alewife populations; forage food for salmon and Lake Trout; used commercially as animal food

7. Lamprey Family—Petromyzonidae

Native and nonnative species; parasitic or nonparasitic; slender body without pectoral or pelvic fins; circular mouth (no jaws); large eyes

Native Lamprey (*Ichthyomyzon* spp., *Lampetra* spp.)
Size: 6–12 in. (15–30 cm)
Key Features: tubular shaped body with no paired fins; long dorsal fin continuous to the tail; round mouth; seven paired gill openings
Habitat: quiet pools of streams as juveniles and lakes possible as adults
Notes: native lamprey common names: silver and chestnut lampreys are parasitic; northern brook and American brook lamprey are nonparasitic—all have little impact on other fish populations

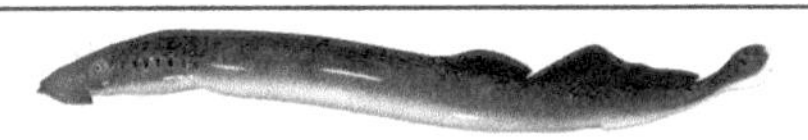

© sciencepics/Shutterstock.com

Sea Lamprey (*Petromyzon marinus*)
Size: 12–24 in. (30–60 cm)
Key Features: tubular body; dorsal fin deeply divided or separate and extends to the tail; no paired fins; round mouth; seven paired gill openings
Habitat: pools of freshwater streams as juveniles; free-swimming adults in Great Lakes (ocean in coastal regions)
Notes: dies after spawning; more information found in chapter text

8. Minnow Family—Cyprinidae

No scales on the head; no teeth in the jaws; no spines in the fins; dorsal fin with less than 10 soft rays

© Sergey Yeromenko/ Shutterstock.com

© Nosyrevy/Shutterstock.com

Asian Carp (*Cenopharyngodon* spp., *Hypophthalmichthys* spp.)
Size: 16–22 in. (41–56 cm); 5–50 lb (2.3–22.7 kg)
Key Features: mostly gray; large body; upturned mouth; low-set eyes
Habitat: large, warm rivers or lakes
Notes: invasive species introduced in the United States to control algae in aquaculture, water treatment, and farm ponds (1970s) but due to flooding have found open water in region; voracious plankton eaters that could disrupt entire food web of nonnative waters; boat motors cause some species to leap from water with potential of injuring boaters; bighead carp pictured (*Hypophthalmichthys nobilis*)

© Edvard Ellric/Shutterstock.com

Common Carp (*Cyprinus carpio*)
Size: 16–18 in. (41–46 cm); 5–20 lb (2.3–9.0 kg)
Key Features: brassy brown to dark olive with white belly; some red on tail and anal fins; two pairs of barbels; round, extendable mouth; scales have a dark spot at base with dark margin
Habitat: warm, shallow water; among vegetation
Notes: nonnative; can uproot aquatic vegetation disrupting habitat for other fish or waterfowl; food source for humans worldwide

© IrinaK/Shutterstock.com

Hornyhead Chub (*Nocomis biguttatus*)
Size: 4–12 in. (10–30 cm)
Key Features: grayish-olive brown sometimes with dark stripe on side and black spot at base of tail; red spot behind eye; breeding males with multiple nubby "horns" on head
Habitat: streams or in lakes near stream mouths
Notes: used as bait for walleye

Fathead Minnow (*Pimephales promelas*)
Size: 3–4 in. (8–10 cm)
Key Features: olive with gold sides and white belly; black lateral line with a widened area at base of tail; dark blotch on dorsal fin; rounded snout
Habitat: shallow, weedy, turbid waters with no predators
Notes: common; used as bait

Matthew Patterson, USFWS

Common Shiner (*Notropis cornutus*)
Size: 4–12 in. (10–30 cm)
Key Features: silvery body, dark green back; usually with dark stripe along length; breeding males with bluish head and pinkish body and fins
Habitat: slow or standing waters
Notes: common bait; stubborn when hooked

9. Perch Family—Percidae

Elongated; two dorsal fins separate or barely joined; spines front dorsal fin; rough scales

© Sean McVey/Shutterstock.com

Johnny Darter (*Etheostoma nigrum*)

Size: 2–4 in. (5–10 cm)
Key Features: tannish-olive with dark blotches and tan sides; *x*, *y*, and *w* patterns on sides; breeding males darker with black bars
Habitat: rivers and lakes; among rocks
Notes: small swim bladders allow for sinking rapidly and making a quick "dart"; rests on bottom; eats insect larvae and zooplankton

© Dan Thornberg/Shutterstock.com

Walleye (*Sander vitreus*)

Size: 14–17 in. (35–43 cm); 1–3 lb (0.5–1.4 kg)
Key Features: dark grayish-olive brown with golden; spines in first dorsal and anal fins; sharp teeth; white spot on bottom of tail
Habitat: lakes and streams
Notes: favorite of anglers; sees well in low light

© Keith Publicover/Shutterstock.com

Yellow Perch (*Perca flavescens*)

Size: 8–11 in. (20–28 cm); 6–10 oz. (3–5 kg)
Key Features: yellowish-green with potentially some orange; six to nine dark vertical bars; lower fins yellowish-orange
Habitat: clear open water of lakes and streams
Notes: common; sport fish throughout year; populations may be tied to alewife populations

10. Pike Family—Esocidae

Long, slender; predators with "duck bills" and strong, sharp teeth; greenish; one dorsal fin farther down back; known to hybridize with other pike

 © Krasowit/Shutterstock.com **Grass Pickerel (*Esox americanus*)** **Size:** 12 in. (30 cm) **Key Features:** olive green to brownish-yellow; wavy bars on sides (usually more pronounced than in the picture); dark "teardrop" below eyes; rounded forked tail **Habitat:** shallow, weedy areas of lakes and rivers **Notes:** often mistaken for young northern pike	 © miha de/Shutterstock.com **Northern Pike (*Esox lucius*)** **Size:** 18–24 in. (46–61 cm); 2–5 lb (1–2.3 kg) **Key Features:** long body; dark green with light bean-shaped spots **Habitat:** often in weedy areas of lakes and rivers **Notes:** fast predator of small fish; an angler favorite

11. Salmon Family—Salmonidae

Predatory, coldwater game fish (includes trout); prominent, single dorsal fin, and greatly reduced fleshy adipose fin; moves upstream to spawn

 © slowmotiongli/Shutterstock.com **Brook Trout (*Salvelinus fontinalis*)** **Size:** 8–10 in. (20–25 cm); 8 oz. (0.2 kg) **Key Features:** dark olive, grayish-black with wormlike markings on back and sides have red-dish-tan spots; lower fins reddish-orange with white front edge; squared tail (sometimes slightly forked) **Habitat:** cool, clear streams or small lakes; sandy to gravelly bottom and vegetation **Notes:** cannot tolerate water above 50°F (10°C); common near headwaters of spring-fed streams	 © Vinne/Shutterstock.com **Brown Trout (*Salmo trutta*)** **Size:** 11–20 in. (28–51 cm); 2–6 lb (1–3 kg) **Key Features:** olive to golden-brown; large dark spots on sides (sometimes also on dorsal and tail), also red spots with lighter rings along sides **Habitat:** open water and clear, cold streams with gravel bottoms **Notes:** introduced late 19th century; secretive fish; feeds at night

© Pi-Lens/Shutterstock.com

Lake Trout (*Salvelinus namaycush*)

Size: 7–10 lb (3–5 kg)
Key Features: dark grayish-green with white spots on sides; unpaired fins; deeply forked tail; inside mouth white
Habitat: cold, deep, clear water; northern species
Notes: overfished in northern lakes of the 1950s and affected by the parasitizing sea lamprey, now part of vigorous stocking program (Lake Michigan)

© Edvard Ellric/Shutterstock.com

Rainbow Trout (*Oncorhynchus mykiss*)

Size: 20 in. (51 cm); 3–8 lb (1.4–3.6 kg)
Key Features: brown to blueish-green with silver on lower sides often with pinkish stripe; dorsal and tail fins covered with small black spots
Habitat: cool streams, preferably fast moving; tolerates cool, clear lakes and coasts of large lakes
Notes: introduced from Pacific Northwest in late 1900s; a favorite sport fish; steelhead trout and rainbow trout are the same species genetically, but lead different lifestyles—steelhead migrate and rainbows remain in rivers; steelhead are all silver with a white mouth and spotted tail, and in rivers will have an all spotted body with radiating rows of spots on tail with all white mouth and gums; migrate from Great Lakes (ocean in its native range) into rivers where they were born to spawn (then returns to lake, unlike other salmon who die after spawning)

© Keith Publicover/Shutterstock.com

Chinook Salmon (*Oncorhynchus tshawytscha*)

Size: 24–30 in. (60–76 cm); 15–20 lb (7–9 kg)
Key Features: iridescent greenish-blue with silver below lateral line; small spots on back and tail; dark mouth inside; breeding males are olive brown to purple with a hooked snout
Habitat: open water of Great Lakes and associated spawning streams
Notes: an important sport fishery; mature in four years; fall spawning runs; introduced from Pacific Ocean in 1960s (other salmon introduced include coho, Atlantic, and pink)

Cisco (*Coregonus artedi*)

Size: 10–12 in. (25–30 cm); 12 oz. (340 g)
Key Features: silvery with faint pinkish tone and dark back; light-colored tail; small mouth
Habitat: shoals of Great Lakes; deep, cool, clear inland lakes; northern species
Notes: many "smoked whitefish" sold are actually ciscoes; productive commercial fish; caught year-round (winter through the ice; summer by flies)

Lake Whitefish (*Coregonus clupeaformis*)
Size: 18 in. (46 cm); 3–5 lb (85–142 g)
Key Features: silvery with dark brown on back and tail; tail deeply forked; snout longer than lower jaw; small mouth; two flaps between each nostril
Habitat: large, deep, cool inland lakes; shallows of Great Lakes
Notes: important commercial fish; often swims in schools; swims deep in summer heat (up to 200 ft.; 60 m); populations greatly reduced in 1950s by sawmill debris ruining spawning beds, overfishing, and sea lamprey

12. Sculpin Family—Cottidae

Bottom dwellers; can crawl on leglike fins; hides among rocks and vegetation during day; stout body with large, somewhat flattened head; eyes near top of head; two separate dorsal fins

© 23frogger/Shutterstock.com

Mottled Sculpin (*Cottus bairdii*)
Size: 4–5 in. (10–13 cm)
Key Features: winglike pectoral fins; all fins seem large for its body; blotchy brown; broad head with eyes almost on top; no scales
Habitat: cool streams; clear lakes; rocky, vegetated bottom
Notes: common; often found in trout streams

© Rostislav Stefanek/Shutterstock.com

Round Goby (*Apollonia melanostomus*)
Size: up to 12 in. (30.5 cm)
Key Features: gray (young); black and brown splotches (mature); black spot on rear of first dorsal fin; fused pelvic fin
Habitat: brackish water; stays on rocks and other substrates in shallow water; open sandy areas
Notes: nonnative, invasive and aggressive with voracious appetites; eats aquatic insects, snails, zebra mussels

13. Stickleback Family—Gasterosteidae

Small body; elongated; stout dorsal spikes; narrow base on caudal fin; bony plates on sides

© Alex Coan/Shutterstock.com

Brook Stickleback (*Culaea inconstans*)

Size: 2–4 in. (5–10 cm)
Key Features: torpedo-shaped; very narrow area just before tail (caudal peduncle); front dorsal fin has short, separate spines; small, sharp teeth
Habitat: cool, shallow areas of streams and lakes
Notes: aggressive predators of small aquatic animals and sometimes algae; males build and defend nest for females to deposit eggs (even in captivity); ninespine stickleback found only in Great Lakes.

14. Sturgeon Family—Acipenseridae

Heavy, torpedo-shaped body with very tough skin and prominent rows of bony plates or shields; large, ventral, suctorial mouth and four barbels in front of it; back and sides olive-brown to gray; underside white; fins dark brown or gray with the single, dorsal fin far back near the caudal fin

© James.Pintar/Shutterstock.com

Lake Sturgeon (*Acipenser fulvescens*)

Size: 20–55 in. (0.5–1.4 m); 5–40 lb (0.2–1.13 kg)
Key Features: medium gray to brown with white belly; bony plates; tail has longer upper lobe than bottom; blunt snout; four barbels; openings between eye and gill corner (spiracles)
Habitat: large quiet rivers
Notes: do not reproduce until 20 years old; can live >75 years

15. Sucker Family—Catostomidae

Bottom feeders with underslung mouth, suckerlike with "lips"; no adipose fin; spineless fins; smooth scales; teeth in throat

© RLS Photo/Shutterstock.com

White Sucker (*Catostomus commersonii*)
Size: 12–18 in.; 1–3 lb (0.5–1.4 kg)
Key Features: tan with gray sides and faint dark patches; lighter fin colors tinged with orange; snout extends beyond upper lip slightly
Habitat: variable
Notes: very common; tolerant of most water conditions; used for human and animal consumption; used as bait; may compete with trout larvae for food soon after hatching

16. Sunfish Family—Centrarchidae

Body is deeply compressed laterally; attachment of pelvic fins is far forward, nearly beneath the pectoral fins; three or more spines at front of anal fin; scales are ctenoid (rough edges); large eyes

© Rostislav Stefanek/Shutterstock.com

Largemouth Bass (*Micropterus salmoides*)
Size: 12–20 in. (30.5–51 cm); 1–5 lb (0.5–2.3 kg)
Key Features: dark green with dark lateral band and light-colored belly; lower jaw extends beyond rear margin of eye; forward facing mouth
Habitat: shallow, vegetated lakes and rivers
Notes: popular game fish; voracious predator and eats almost anything; strong fighters and often jump when hooked; often only kept for food under two pounds and from clear water

© dcwcreations/Shutterstock.com

Smallmouth Bass (*Micropterus dolomieui*)
Size: 12–20 in. (30.5–51 cm); 1–4 lb (0.5–2 kg)
Key Features: dark greenish-bronze, mottled with usually dark vertical bands; light belly; red eye; large mouth, forward-facing; stout body
Habitat: clear, fast rivers; clear lakes with gravel or rocky shorelines
Notes: game fish noted for strong fight and jumps

© Steve Oehlenschlager/Shutterstock.com

Black Crappie (*Pomoxis nigromaculatus*)
Size: 7–12 in. (18–30 cm); 10–16 oz. (0.3–0.5 kg)
Key Features: black and green blotches, darker on top; back arched with depression above eye
Habitat: calm, clear rivers and medium-sized lakes with vegetation and deep basins
Notes: a year-round favorite of anglers; aggressive carnivore; feeds at night; nests in shallow water

© Matt Howard/Shutterstock.com

Bluegill (*Lepomis macrochirus*)
Size: 6–9 in. (15–23 cm); 5–10 oz (0.14–0.3 kg)
Key Features: taller than wide; dark olive-green blending to grayish-copper brown on sides with vertical dark bars and yellow underside; dark gill spot; darkened spot at rear margin of dorsal fin
Habitat: rivers and lakes with vegetated shorelines or bays; transition between plant zone and deep water among the vegetation
Notes: popular among anglers; feeds on insects and small fish; feed at surface; hybridizes with other panfish

© Rostislav Stefanek/Shutterstock.com

Pumpkinseed (*Lepomis gibbosus*)
Size: 6–8 in. (15–20 cm); 6–10 oz. (0.17–0.3 kg)
Key Features: brownish-olive with speckled sides of orange, blue, yellow, green spots; black gill spot with light margin and red-orange crescent
Habitat: vegetated lakes and slow rivers; cooler water than bluegill
Notes: common; gathers in schools around docks and submerged logs; hybridizes with other panfish

Fish—Natural History Journal Pages

Choose a fish found in the Field Guide that interests you. Use the book and other resources to research the following information.

1. Full Common Name: Smallmouth Bass

   Scientific/Latin Name (*Genus species*): Micropterus dolomieui

   Family Name: Centrarchidae

   Order: Perciformes

   Class: Actinopterygii

   Phylum: Chordata

   Kingdom: Animalia

2. Draw the overall picture of the **adult fish with identifying characteristics labeled** (include: size, all colors, dark markings, and other special features).

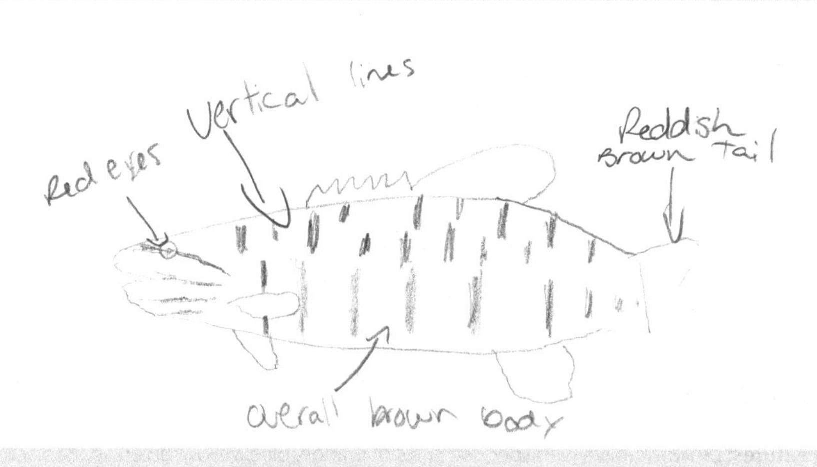

3. In your own words, provide a description of what this fish looks like.

   gold/brown, vertical lines, Red eyes, Brighter in clearer waters, darker in darker waters, 2-5 lbs 14-18 inches

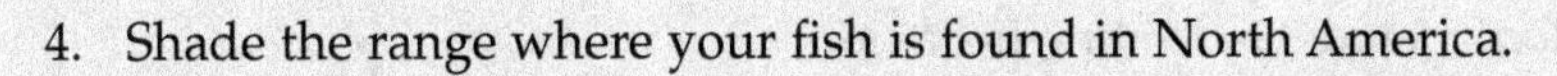

4. Shade the range where your fish is found in North America.

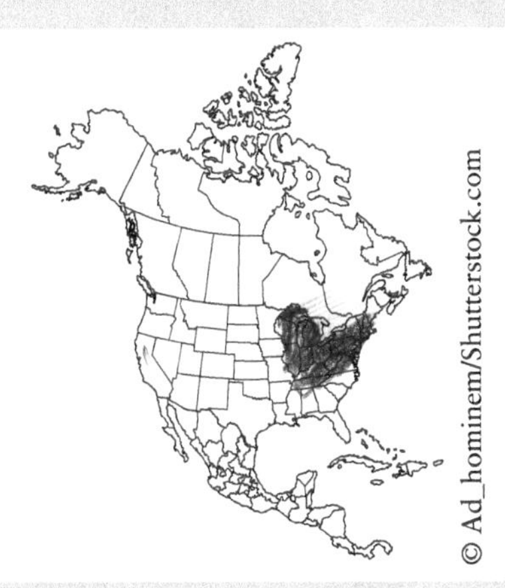

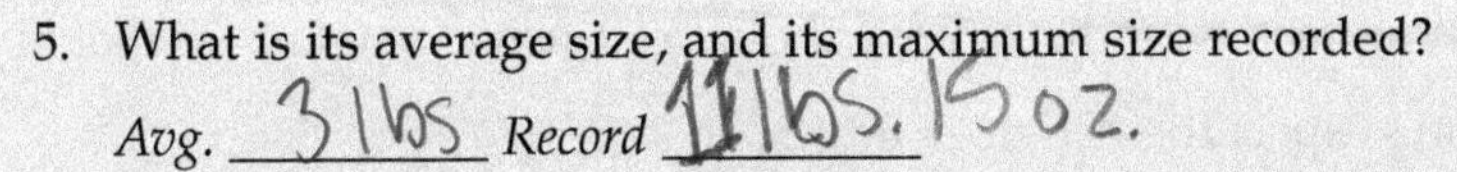

5. What is its average size, and its maximum size recorded?

   Avg. 3 lbs *Record* 11 lbs. 15 oz.

6. a. Where is your fish's preferred habitat (include differences in habitat for juvenile and adult if applicable).

   Clear, fast rivers; clear lakes with rocky Shorelines or gravel

   b. Describe where you would find this fish in relation to where you live or from your Sit Spot (include an actual river or lake name and distance to reach it).

   My Sit spot is on the muskegon river, So I could find it pretty close to me.

7. a. What does its food consist of?

   Small fish and crayfish, young of other Smallmouth bass.

b. In addition, include its feeding behavior.

Feed mostly during daylight hours
most active at dawn and dusk.
feed at night during warm months.

8. Discuss its reproductive behavior.

Males make a bed on rock and
gravel and a female will lay her eggs when
the water is around 55-60°F.

9. What is another "fun fact" you discovered about your fish's behavior—what is it and why do you suspect the fish does it?

The lines on the side of the fish will
fade with age.

10. Discuss a method of conservation or management for your fish—to protect/restore/control its population.

Smallmouth bass have a season when
the can and cannot be kept, they
also have size restrictions for keeping.

11. List all resources used (**books:** author's last name, first initial, year, and book title; **websites:** name of site, date retrieved, and web address):

8.2 Amphibians

Nature Journal: A quick primer—without researching it, list the characteristics of what you think defines an amphibian. Name as many amphibians as you can in 30 seconds. Describe the last time you experienced an amphibian.

Have you heard frogs call in the warm months of the year? A sign of spring to many is hearing the early season calls of chorus frogs and spring peepers. Yet another area of focus for the naturalist is on amphibians. Discovering and studying these nearby animals can be a thrilling nature study activity. Someone who studies and researches amphibians *or* reptiles is called a **herpetologist**.

A short-hand reference for these animals—amphibians and reptiles—is "herps." It can be a nature study goal to discover as many herps as you can in one day or in a year! Herps are not closely related but share some common traits. The low numbers of amphibians and reptile species combined is relatively small compared to other organisms. For example, there are approximately 4500 amphibian species and about 7000 reptile species, globally. That totals 11,500 species compared to over 9000 bird species and countless plant species. With these relatively few "herp" species and enough similar traits between the groups, they are often referenced together. However, these animal species are different enough to put them into separate taxonomic classes, Class Amphibia and Class Reptilia.

To find amphibians, you will need to know some basic things about their natural history. It would also help to research the preferred habitat and behaviors for these species. Amphibians and some reptiles, like turtles, will often be found near water bodies or in mesic or damp forests. They are also common in moist, humid areas.

Activity: Visit a pond's edge to quietly scout for visible frogs. Walk through a woodlot and gently roll a log to find salamanders. If you do not have logs in the woodlot, consider placing a log or wood board in the shade (where you have permission to do so), and look under that log or board daily to discover what is living there—you might find some salamanders eventually!

Although not advised, if you decide to handle an amphibian, make sure your hands are wet and all bug spray or lotion has been washed off. Support the animal's entire body. If observing small animals who were found in the water, use a water-filled jar for making observations and to prevent their skin drying out and to reduce stress. After a short time of gentle handling and observation, return the animal to the same place you found it. Wild animals are best left uncaught and observed from a respectable distance.

Overview

The definition of the term, *amphibious* is something that can exist or live in at least two different environments (i.e., land and water). Amphibians are animals that live essentially two lives at some point in their life cycle. Their lives usually begin in the water and metamorphoses into a stage that can live on land. Examples of amphibians in this region include frogs, toads, salamanders, newts, and mudpuppy species.

Thinking specifically about the taxonomy of the amphibians, we are considering eukaryotic organisms with membrane-bound organelles (cell structures) and animals with a backbone. Therefore, they are in Domain Eukarya, Kingdom Animalia, Phylum Chordata, and due to their amphibious nature (along with some other attributes soon to be discussed), they are in Class Amphibia—"living two lives."

We can observe complete metamorphosis, or a change in body form and ecological niche from the larval to adult stage in frogs and toads. Metamorphosis also occurs with salamanders and newts, but the only significant change is habitat preference from the larval to adult stage. The adults of most of the species live on land yet need to breed in or near water. Larva often resemble adults, and all have a carnivorous diet throughout its life cycle. This process may be better understood with a brief evolutionary background.

Amphibian Evolution

Amphibians first appeared in the fossil record about 370 million years ago (mya) in what was considered the Middle Devonian. The earliest amphibians were the first land-dwelling vertebrates. Meaning, these were the first animals with a backbone (vertebrate) adapted to live on land and able to breathe atmospheric air. Amphibians appeared in the fossil record after fish (a water-dwelling vertebrate).

The first types of amphibians were considered "tetrapods" or having "four feet." They had a fused pelvic and pectoral girdle. Each structure had bones to support lower and upper limbs, respectively. Fish do not have these skeletal structures. Early amphibians had finlike limbs resembling fish fins. The two pairs of lateral limbs (pectoral and pelvic fins) were lobed, and unlike fish, had an internal skeletal structure enabling the animals to crawl from pond to pond. The bones of the lateral fins were the basis for five-fingered appendages (pentadactylous), a single upper limb bone, two lower limb bones, multiple wrist/ankle bones, and hand/foot bones.

An additional feature that amphibians possess contrary to fish, is having developed primitive lungs to breathe atmospheric air. Land-dwelling amphibians also do not have gill slits for breathing and they can hear airborne sounds. To help us make further distinctions about amphibians, we will compare the class's orders.

Frogs and Toads—Order Anura

Order Anura (an-ura or "lacking tails") includes approximately 4000 species of frogs and toads globally. Reference the field guide after this section to see what species are found in the Great Lakes basin and Upper Midwest region. Typically, a group classified as true frogs will have smooth and moist skin whereas toads have bumpy, dry skin. However, the treefrogs also tend to have somewhat rough skin as they spend most of their lives in the trees.

Anurans are mainly adapted for jumping or maneuvering across land. The adults are carnivorous and consume insects as a main source of their diet. The adults have a long, sticky protrusible tongue attached toward the front of their mouth that they flick out to capture its prey. Often, adult frogs or toads will sit motionless waiting for its prey to move so it can sense its location, then use its tongue to "grab" it. Another important sense anurans possess is hearing. Frog and toad adults have an external eardrum to allow for excellent hearing capabilities. In some species, like the bullfrog and green frog, you can see this eardrum (**tympanum** or **tympanic membrane**) on the sides of its head.

Have you ever touched a frog? They may feel slimy to the touch. Most anurans have mucus they secrete through the skin. In some species the mucus may have poisonous qualities if predators consume it. Another skin consideration is its color. Frogs and toads may have bright coloration or exhibit great camouflage. In our region, species will typically blend in with the background they are found.

The prominent tympanic membrane on the bullfrog (*Lithobates catesbeianus*).

Look at the hind limbs on the frog skeleton. Notice how the leg bones are fused—the tibia and fibula are one bone. This provides power and stability to the jumping frog. Moreover, the very long hind toes will help them jump. The backbone, or vertebral column is very short and rigid to provide a smooth launch into the jump. You will find a similar structure in the forearms. Here, they also have fused bones—the radius and ulna are one. This helps provide strength and support when the frog lands on them first after a jump.

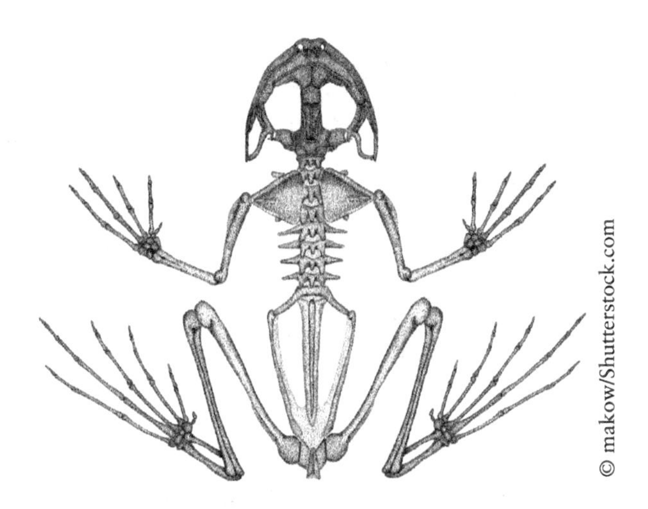

Metamorphosis

In terms of metamorphosis, anurans exhibit the most change of other species in this class. It takes approximately 80 days or more to go from the egg to larva to adult stages (rate is dependent on species and habitat conditions). Eggs are laid in water by the adult female. A larva, or tadpole, emerges with a long-finned tail, lateral line, breathes through gills and eats plants. These features resemble those of a fish. As time progresses, the tadpole develops hind limbs then forelimbs, and respiration advances from external to internal gills then to lungs. Once these features mature, the frog can tolerate a terrestrial environment. The adult will retain a small tail feature that serves as a fat reserve. In this **metamorph** stage, the tail eventually becomes absorbed as a source of energy. As an adult, frogs and toads have legs, breathe via lungs, have a second pair of external eardrums, a carnivorous digestive system, and many can live in terrestrial and aquatic (or mesic/moist) environments. You may mostly observe toads away from water sources and have more of a crawl than a hop like the frogs. Other terrestrial amphibians can be found in the next order.

Complete metamorphosis of the wood frog (*Lithobates sylvaticus* or *Rana sylvatica*). Follow the life cycle from egg; embryo; tadpole (larva); tadpole with two, then four legs; metamorph; to finally adult.

Nature Journal: Keep record of the frogs and toads you see and hear. Note the time, day, weather conditions, location, what it looked like, its behavior, and identify it. Make it a goal to find as many of the species in our field guide as you can. This could be a great foundation for a descriptive field investigation, and a survey as part of a community science project for a local organization.

Activity: Consider creating a toad house. During hot, dry weather, toads prefer to take refuge in cool, damp places. You can create one of these houses near a water source, among your landscaping, or in your garden. Use a chipped flowerpot and turn it upside down with a rock on top to hold it in place. Enlarge the chip, if necessary, to make an opening for the toad to enter. Keep record of if, when, and where the toads prefer their new house. This could also be a descriptive, comparative, and correlative field investigation (see Chapter 2 for more details on setting up a study).

Salamanders—Order Caudata

For the next group of herps, we consider the amphibians belonging to Order Caudata. This translates to "tailed ones." Examples include salamanders, newts, and mudpuppies. If unfamiliar with this group, refer to the field guide section. What features do you observe with these animals? How do they differ than the anurans? You should notice the elongated body plan and presence of a long tail. The **costal grooves** will be visible on many salamanders, which are simply distinctive indentations between ribs.

In Order Caudata, you will find aquatic or terrestrial forms. At first glance, they may seem lizardlike. However, you must consider the habitat in which you find them. Moreover, these animals have moist skin unlike their reptilian cousins, the lizards, which are dry and scaly. Comparing the way salamanders walk to how lizards walk differs greatly. Salamanders can walk with a side-to-side body bending movement. Whereas, lizards mainly move just their limbs as they propel themselves forward. Salamanders also do not have claws or scales like lizards. The red-backed salamander is a very common amphibian found under logs in nearby mesic forests.

Activity: As you did for the Anurans, record your observations of species from Order Caudata. Seek them out! During the hot, dry part of the summer salamanders will mostly be active at night. Grab a flashlight or headlamp and look for them along the water's edge, under logs, or gently peel back moss patches to find them. Record your discoveries to help look for trends and improve your nature study.

Amphibian Anatomy and Physiology

Some basic amphibian anatomy includes skeletal features mentioned previously for frogs and toads who have no tail and have girdles for jumping. Salamanders have many tail vertebrae and differing toe numbers between front and hind limbs. Salamanders also can regenerate certain body parts if they get removed (typically from predators like birds). Another defense strategy is metachrosis. Amphibians can change skin color to match their habitat surroundings due to their dermal chromatophores (just like fish). It can take up to an hour for many species to change color.

Circulatory System

Amphibians are ectothermic with body temperatures that vary with the environment. Due to their inadequate lung capacity within their simple sac lungs (or altogether lacking in some salamanders), they depend upon the permeability of gases through their moist skin to assist in respiration. In this process, oxygen and carbon dioxide diffuse through the skin. Amphibians also have a 3-chambered heart that pumps oxygen to the rest of the body.

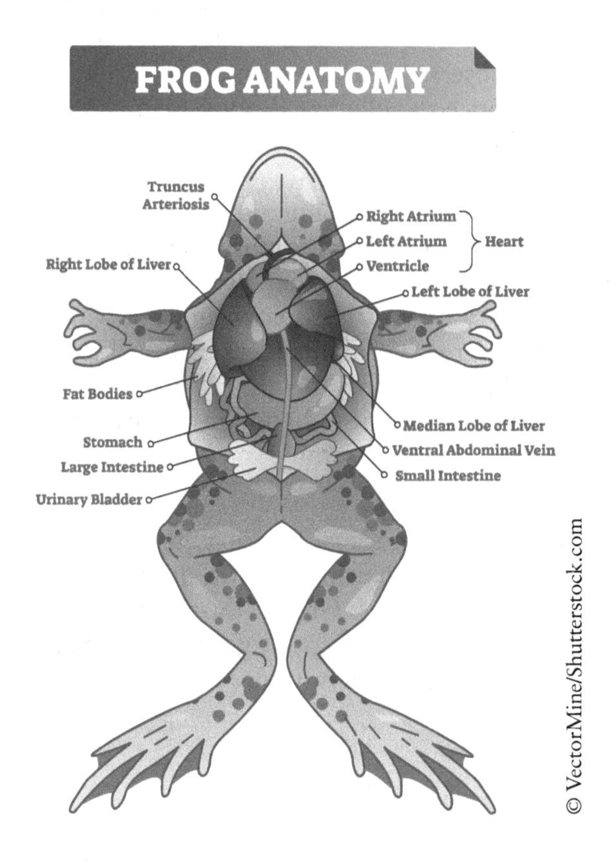

Digestive System

The digestive system of amphibians will seem familiar. They have the same organs as humans with similar functions. The esophagus is the food passageway to the stomach where it holds food and begins enzymatic digestion. The small intestine finishes the digestion of food and starts to absorb it, and the pancreas secretes digestive enzymes so food is broken down to molecular size. The liver secretes bile for fat digestion, then the large intestine absorbs water and compacts feces for elimination.

Nervous System

Regarding the nervous system, the brain is simple. Amphibians are not capable of extensive learning.

Hearing is well developed in anurans with a tympanic membrane (ear drum) and columella (ear bone). Vibrations from sound waves hit these structures and produce the hearing sense. Amphibian eyes are primitive. Lenses can only move forward and backward giving the animals limited focusing ability. This means that many frogs are unable to locate prey unless it moves. As far as taste, they have well developed taste buds on their tongue and the surface of the mouth. This allows them to taste their environment. Lastly, the lateral line enables the aquatic stage of amphibian species to sense movement in the water, and to detect proximity of objects and other animals.

Reproductive System and Egg Laying

Reproduction for most frogs and toads will involve external fertilization in the process of **amplexus.** With his front legs, the male grasps the female from behind. This will eventually initiate egg laying by the female, which can vary in time due to species type. Next, the male fertilizes the eggs. Members of Order Caudata will exhibit internal fertilization. This is when a male releases a packet of sperm (spermatophore) in the water that the female picks up internally which fertilizes the eggs.

Amphibian eggs require water, or at least a moist environment. The female deposits eggs on aquatic vegetation in the water, will be free floating, or in the case of the red-backed salamander will be under rocks or logs on land. The eggs will develop a gelatinous texture once they encounter the water. Frog eggs will form a mass, yet you will still see the contour of each egg. Individual toad eggs will typically form a continuous, gelatinous stringlike configuration in the water. Most salamanders will also lay eggs in the water, but the egg mass will have a layer of gel surrounding it. If you look closely you will still see the outline of the eggs.

All amphibian eggs lack a shell. This makes them semipermeable to their environment, therefore dependent and susceptible to the water's conditions. Water, oxygen, and carbon dioxide will flow in and out of egg membranes, which will sustain the embryo development. Water temperature also influences hatch time with warm water resulting in relatively fast development from egg to adult, and colder water slows development. In most amphibian species of the world, many eggs are laid at one time with little to no parental care. With some species in other regions, few eggs are laid with more parental care provided; or perhaps the eggs are transported on their back, mouth, or stomach!

Activity: Spend time looking for egg masses among aquatic vegetation in the spring and then the tadpoles swimming en masse and adults sitting still along the water's edge in the summer. This is a fascinating nature study to witness. Find a good resource online for identifying amphibian egg masses. In some regions, you may find community science programs that have you monitor vernal pools for salamander or frog eggs!

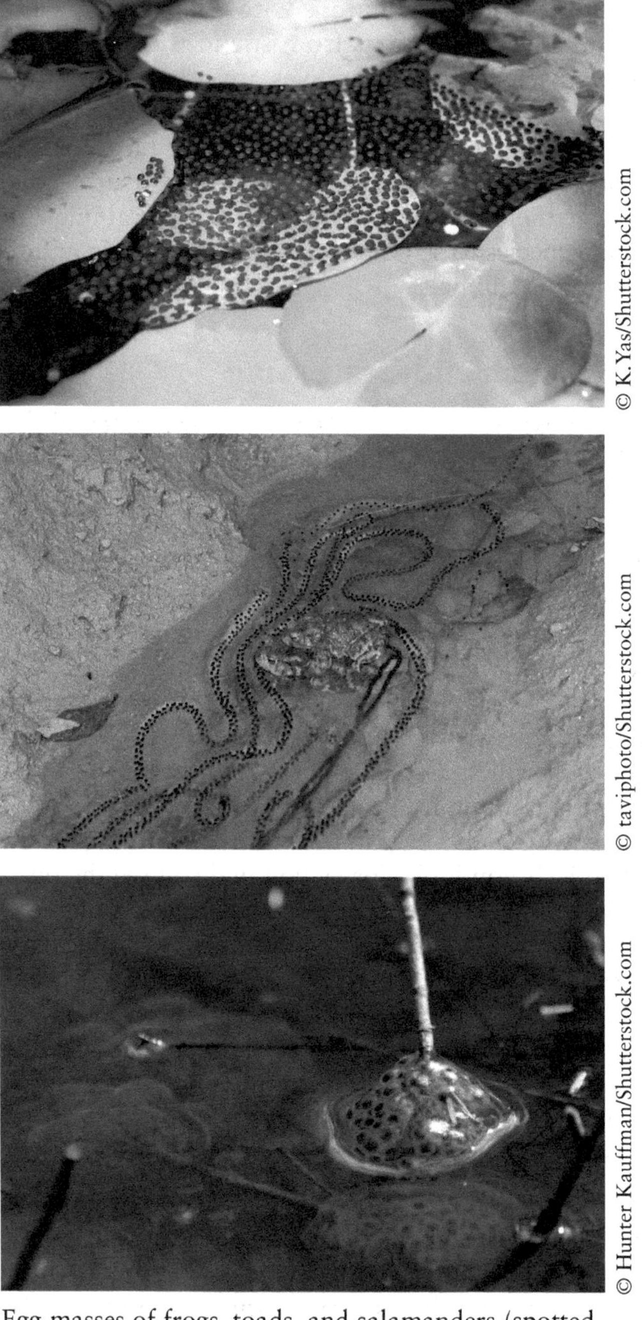

Egg masses of frogs, toads, and salamanders (spotted salamander, *Ambystoma maculatum*)

Vocalizations

Frogs and toads are calling in the spring and summer. Research the various vocalizations online for your region. You can easily start to identify their calls and know which species are nearby. This is a skill that can be applied to collect data for community science programs. Monitoring frogs and toads in wetlands tells a story. The relative species richness and diversity serves as an indicator for habitat quality. This data proves especially meaningful if you monitor the area through time to determine if the calls have changed. The data can indicate an increase or decrease in species number or variety. This can also signify a habitat has changed.

The animals mainly call to mark territory and attract mates. Frogs and toads have vocal cords in their larynx similar to mammals. Male frogs have a vocal sac that balloons out during vocalization. Vocal communication initiates the breeding season with the territory call. The male frog or toad will mark its territory with a particular call. This signals to other males of the same species that this is the space they want to lure in a female for mating. Next is the advertisement call. Once the territory is established, the male will vocalize a different call to attract a female of the same species. This call has a specific pattern and frequency so the female knows where to go. Studies have shown females have an eardrum and inner ear structure responsive to male species' call frequency. Prior to and during the advertisement calling, males will release a molecule (pheromone) into the water where they breed. The female's olfactory (smelling ability) can detect the hormone making her more receptive to mating. Once amplexus and reproduction ends, the

The male spring peeper (*Pseudacris crucifer*) with his vocal sac ballooning during vocalization.

male will give a release call. This call also sounds similar to his vocalization when captured by a predator.

Migration

Amphibians will migrate during breeding season. They move to a nearby freshwater habitat like a pond, lake, river, or wetland. They have the breeding location memorized and will follow the same path annually—regardless of whether the water source still exists. The animal will enter and exit the same spot each year! This mass movement will happen at night, and during humid or rainy conditions. Traveling during these times helps them maintain moist skin to respire more efficiently thereby enabling them to travel farther. Amphibians will use various means for navigation such as the sun and moon positions, smell, and other landmarks.

Other than the warmer temperatures in the spring, light serves as a major trigger or cue for breeding season to begin. The parietal eye (sometimes referred to as the **pineal gland**) is on the back or top of their head. It can detect light—the sunrise, sunset, and moonlight. Increased or decreased light duration triggers a response cueing hormonal, physiological, and behavioral changes in many (or most) organisms. Migrating to the breeding pond will involve the amphibians using sense of smell, landmarks, and celestial navigation or star positions to help them find their breeding pond.

Dormancy

Have you ever found a frogsicle or salamandapop? You might find a "frozen" amphibian during the early fall when temperatures tend to fluctuate greatly. The parietal eye's detection of less daylight will signal to the animal the approaching winter months. Prior to dormancy, amphibians produce extremely high levels of sugar that serves as a type of antifreeze. When temperatures go below freezing, the animals turn mostly to ice and will not function in any way. Before they get to this state, they will usually start to find their overwintering location. When the cool weather arrives, many amphibians bury themselves in the ground under tree stumps and leaf litter below the frost line. The land hibernators include the spring peeper, chorus frog, gray tree frog, wood frog, toads, and most salamanders. Mink frogs, green frogs, leopard frogs, and bullfrogs remain at the bottom of ponds until the thaw.

Dormant amphibians live off stored glycogen and fat during those days of "slumber." Their blood sugar levels are 100 times higher than normal. During this "frozen state" (actually, more of a slushy inside with about 65% of the body's water converted to ice), the animal has no vital functions. There is no heartbeat, no blood circulation, their breathing has ceased, and no detectable brain activity exists. However, once the surrounding temperature warms to their preferred conditions, all vital functions return 1–2 hours after they thaw.

Amphibian Decline

For the past few decades, a rapid decline in amphibian populations has occurred worldwide—even in pristine habitats. Due to their semipermeable skin, amphibians are susceptible to water chemistry changes. Although you might observe amphibians in an area seemingly untrammeled by humans, it does not mean that the area is unaffected by our habits. The issue of runoff or acid rain carried by wind currents from an industrial area can easily have a negative effect on these animals. They may die off in an area, or deformities result, which in some cases can also appear in their offspring. If you ever discover amphibians with unusual morphological characteristics—or simply look abnormal with perhaps additional limbs, please take photographs if possible and report it to your local natural resource agency.

Other considerations for these animals' decline include the concern of habitat fragmentation. The reduction of trees and vegetation will increase runoff concerns (see Chapters 3 and 4). The elimination of wetlands is of special concern to these aquatic-dependent species. Contaminants that the runoff picks up and what the air currents bring is at the detriment of these sensitive populations. Nonnative, invasive species also pose a threat with the added competition for food and space, and the introduction of potential diseases. Some people also will harvest amphibian species unsustainably and use them in the illegal pet trade operations. Other concerns revolve around climate change, decreased rainfall, and increased UV radiation. Their sensitive skin and egg masses are intolerant of any of these issues.

Why Should We Care? What Can We Do?

Amphibians offer many reasons why we should care about them. They provide us with insight to the relative environmental quality of where they live. Since they live on land and in water, and have sensitivities to change, they are good bioindicators of the environment. Amphibians also help control what we might consider pests, by eating mosquitoes and flies. Some parts of frogs and toads have been used in modern medicine, such as for the study of pain killers and the high level of morphine found in the skin of some amphibians. Overall, these incredible animals can help advance environmental and medical studies.

As you have read, many factors can act against the amphibian populations. How many of those concerns involve humans? These consequences to our actions serve as an indicator that our habits need to change in regard to our impact on the environment. If you find a declining population of amphibians near you, take an inventory of how the landscape is being used. Try to determine where the areas of concern exist and make an action plan for remediation. Other ways to help these animals and thereby our environment, is to participate in community science programs to inventory frogs and toads by listening to their vocalizations at key points in the season, assess mass migratory movements, and report any habitat concerns you witness.

Nature Journal: Like you have for other species, keep track of the amphibians you observe. Use the Natural History Journal Pages at the end of this section as a template to guide you through learning about a selected frog, toad, and salamander within or near your Nature Area.

Review Questions

1. Name the kingdom, phylum, and class for amphibians. Discuss the two orders discussed in the section.
2. When did the first land-dwelling vertebrates appear? Discuss amphibian anatomy and physiology that enables these animals to live on land.
3. What is meant by amphibians having "two lives"? Describe the characteristics of the metamorphic stages of frogs and toads. How is metamorphosis different for salamanders?
4. Discuss the anatomy and physiology of amphibians.
5. Describe the three types of vocalizations.
6. Discuss amphibian dormancy in terms of the basic physiological changes that occur.
7. Discuss the possible causes for amphibian population declines.
8. What can citizens do to help amphibian populations?

Field Guide: Amphibian Species Accounts

© Bahruz Rzayev/Shutterstock.com

© RedlineVector/Shutterstock.com

The frogs, toads, and salamanders chosen for this field guide were based on the most widespread species across the Great Lakes basin and Upper Midwest. Try to make connections about key features within families and to other families of seemingly similar species.

When using this field guide to identify a frog species, consider its relative size in comparison to other species. This book has frogs organized from small to large, and by group—treefrogs, true toads, and true frogs. For salamander species, they will appear from small to large woodland varieties then the aquatic species listed at the end of the section.

Frogs

Treefrogs—Family Hylidae

Most species have large suction pads at the tips of their toes; capable of climbing vertical surfaces; small species under 2 in. (5 cm); clear and melodious vocalizations in spring

© RealityImages/Shutterstock.com

Northern Cricket Frog (*Acris crepitans*)
Size: 0.75–1.50 in. (19–38 mm)
Key Features: bumpy skin with variable colors and vertical body striping; dark triangle between eyes; complete dark, wavy banding on legs; pronounced webbing on hind feet
Habitat: edges of ponds, streams; floodplains and mudflats; more of southern range of region
Notes: not arboreal like other treefrog family members; known to hibernate far from breeding pond in upland areas; continual ticking sound for its vocalization

© Matt Jeppson/Shutterstock.com

Western Chorus Frog or Striped Chorus Frog **(*Pseudacris triseriata*)**
Size: 1.6 in. (40 mm)
Key Features: smooth skin, grayish-green brown; three dark stripes on back sometimes not continuous; dark eye stripe; upper lip with light stripe
Habitat: woodland swamps; grasslands
Notes: one of first to vocalize in region; most obvious during breeding season vocalizing loudly around ponds; rising trill sound that sounds like running a thumb down the teeth of a comb; small, thin-legged frogs; nearly webless toes small and round with no suction pads like their treefrog relatives

 © Jason Patrick Ross/ Shutterstock.com	 © Natalia Kuzmina/ Shutterstock.com
Spring Peeper (*Pseudacris crucifer*) **Size:** 0.98–1.5 in. (25–38 mm) **Key Features:** smooth skin; dark "x" on back; dark bar between eyes; colors vary with usually tan-nish-gray olive **Habitat:** low plants; vernal pools among thicket; woodlands **Notes:** one of earliest frogs to vocalize; call day and night; mostly heard and not seen because remain hidden among dense plants; peeps every 1–2 seconds	**Gray Treefrog (*Hyla versicolor*)** **Size:** 1.5–2.0 in. (3.8–5.1 cm) **Key Features:** bumpy skin; great variability in color from gray to green depending on substrate; legs with dark banding pattern and bright yellow inner thigh **Habitat:** trees and shrubs near watersource **Notes:** less prone to drying out compared to its relatives; call sounds like a loud trill sometimes in response to loud noises; close relative of Cope's gray treefrog

True Toads—Family Bufonidae

Most species of true toads will have pronounced cranial crests (bony ridges on top of the head); conspicuous parotoid glands behind the eyes (swellings where a toxic fluid secretes to deter predators); bumps also secrete a toxin to predators if ingested; toads do not cause warts and are not harmful to the touch

 © Steve Byland/ Shutterstock.com	 © Fotoz by David G/ Shutterstock.com
Fowler's Toad (*Anaxyrus fowleri*, formerly *Bufo fowleri*) **Size:** 2.0–3.7 in. (5.0–9.5 cm) **Key Features:** brownish-gray to olive green with rusty red and darkened bumpy area; three or more bumps per dark spot on back; pale stripe on back; belly whitish with a dark spot **Habitat:** sandy areas with freshwater nearby **Notes:** less tolerant than American toad; puffs up, urinates, and secretes a foul-smelling liquid when disturbed; may also feign death to deter predators; a short, 4-second screechy shriek	**American Toad (*Anaxyrus americanus*, formerly *Bufo americanus*)** **Size:** 2–4 in. (5–10 cm) **Key Features:** bumpy skin and variable in color; stocky body; one to two brown bumps in each dark spot on back; parotoid gland elongated, separate from cranial crest; may have a light line on middle of back **Habitat:** gardens; lawns; wetlands; forests **Notes:** eats many insects; nestles into damp soil during day; if captured, it will inflate, urinate, and release a foul smell; a long, soft, screechy trill last-ing about 30 seconds

True Frogs—Family Ranidae

Found mainly in water; most vulnerable to pollution; dorsolateral ridge usually present with exception of the bullfrog; tympanic membrane more visible in these species

© Jay Ondreicka/Shutterstock.com

Wood Frog (*Lithobates sylvaticus*, formerly *Rana sylvatica*)

Size: 2.0–2.8 in. (51–70 mm)
Key Features: tannish to dark brown; dark mask; upper jaw with light line; ridges prominent
Habitat: wetlands; wet, shady woodlands
Notes: one of the first frogs to call in spring; makes a weak quacking call

© Matt Jeppson/Shutterstock.com

Pickerel Frog (*Lithobates palustris*, formerly *Rana palustris*)

Size: 1.5–3.0 in. (4.0–7.5 cm)
Key Features: dark, square blotches running in two parallel rows at center of back; yellow ridges; jaws with light stripe; inner thigh bright yellowish-orange
Habitat: wetlands; cold streams; woodlands
Notes: low-pitched snore or croak; secretes toxic fluid to deter predators

© Kris Holland/Shutterstock.com

Northern Leopard Frog (*Lithobates pipiens*, formerly *Rana pipiens*)

Size: 2–3.5 in. (5–9 cm)
Key Features: greenish-brown with light colored belly; large, roundish spots bordered with lighter color; ridges light colored; upper jaw with light stripe
Habitat: wetlands; damp meadows
Notes: travels up to a mile to breeding pond; vocalization is a short, guttural, low snore followed by clucking sound

© Adam Tremel/Shutterstock.com

Mink Frog (*Lithobates septentrionalis*, formerly *Rana septentrionalis*)

Size: 1.9–3.0 in. (4.8–7.6 cm)
Key Features: mostly green with brown blotches and cream-colored belly and bright green lips; light (female) or yellow throat (male); ridges may be absent; splotchy hind legs
Habitat: rivers; cold lakes
Notes: musky onion odor released if mishandled; call sounds like sticks being hit together; requires cold water; populations declining

© Tom Reichner/Shutterstock.com

Green Frog (*Lithobates clamitans*, formerly *Rana clamitans*)
Size: 2.0–3.9 in. (5.0–10 cm)
Key Features: mostly green or brown with variable dark splotches; prominent ridges; green upper lip; tympanum membrane same size as eye (female) or twice its diameter (male); bright yellow throat (male); partially webbed hind feet
Habitat: wetlands, rivers, lakes
Notes: spends most of its time along water's edge and dives deep to hide from potential danger; vocalization is a single "plunk" sound

© Ilias Strachinis/Shutterstock.com

Bullfrog (*Lithobates catesbeianus*, formerly *Rana catesbeiana*)
Size: 3.6–6.0 in. (9.0–15 cm)
Key Features: olive green or brown with brownish blotches and whitish-yellow belly spotted with grayish-yellow; upper lip bright green; no ridges; yellow throat (male); brown irises with horizontal, almond-shaped pupils
Habitat: all permanent freshwater
Notes: our region's largest frog; almost always near water; does not leave its 40 square foot (3.7 square meter) area; vocalization is a deep "ddrruumm" heard up to a quarter mile away; screams when grabbed and can "run" across water surface

Salamanders

Lungless Salamanders—Family Plethodontidae

These animals have no lungs; slow movers; inhabits moist woodlands

© Jason Patrick Ross/Shutterstock.com

Four-Toed Salamander (*Hemidactylium scutatum*)
Size: 2.0–3.0 in. (5.0–7.5 cm)
Key Features: reddish-brown with gray sides and white belly with black spots; v-shaped grooves on back; constricted at base of tail; feet with four toes
Habitat: forested bogs
Notes: our region's smallest salamander; difficult to find; females lay eggs on sphagnum moss hanging over water and guards site until larvae emerge and fall into water

© Steve Bower/Shutterstock.com

Red-Backed Salamander (*Plethodon cinereus*)
Size: 2.0–4.0 in. (5.0–10 cm)
Key Features: thin body; red back with dark sides often flecked with white; large bulging eyes
Habitat: deciduous and coniferous forests
Notes: one of smallest salamanders; does not lay eggs in water; females lays eggs under logs or rocks and guards young after hatching for up to three weeks; can jump when startled by slapping tail against ground

Mole Salamanders—Family Ambystomatidae

Most burrow in moist ground or leaf litter; only seen at night or during breeding season at ponds; generally has very prominent costal grooves (indents along sides)

© Jay Ondreicka/Shutterstock.com

Blue-Spotted Salamander (*Ambystoma laterale*)
Size: 3.0–5.0 in. (7.5–13 cm)
Key Features: dark brown to blue-black with bluish-white patches and spots
Habitat: moist deciduous woods; mixed coniferous forests with ponds
Notes: very cold-tolerant; thrashes around when attacked; mating occurs in water at night; found mostly under logs in woods and mostly only seen on land during its breeding migration

© Jay Ondreicka/Shutterstock.com

Spotted Salamander (*Ambystoma maculatum*)
Size: 4.0–9.0 in. (10–23 cm)
Key Features: stout body; dark brown or black with two rows of irregularly spaced yellow or orange spots down back; unspotted grayish belly; wide head
Habitat: moist forests; woodland ponds
Notes: lowers head and waves tail when threatened; poison released to deter predators; lives in underground burrow during day

© Joe Farah/Shutterstock.com

Tiger Salamander (*Ambystoma tigrinum*)
Size: 7.0–13 in. (18–33 cm)
Key Features: dark brown to black with light, irregular blotchy spots; broad head; small eyes; short legs
Habitat: diverse; prairies; lakes; wetlands; woodlands
Notes: aggressive predator of insects and earthworms; often discovered on roads during migration; adults tend to burrow

© Jason Patrick Ross/Shutterstock.com

Eastern Newt (*Notophthalmus viridescens*)—Family Salamandridae (Newts)
Size: 2.0–5.0 in. (5.0–13 cm)
Key Features: terrestrial adults are dark brown to olive green (maybe with red markings); no costal grooves; yellow belly with black dots; line through eyes; tail vertically flattened; eft (land-dwelling juvenile for one to three years after water emergence) has a bright orange to reddish-brown coloring
Habitat: wetlands; moist woodlands; rivers; lakes
Notes: unique life cycle (see key features); some skip eft stage and transform into aquatic adult

© Morphart Creation/ Shutterstock.com

Mudpuppy (*Necturus maculosus*)—Family Proteidae (Mudpuppies)

Size: 10–16 in. (25–40 cm)

Key Features: long, thick body; gray to rusty brown and irregular spots; dark red external gills at base of head; short legs; tail vertically flattened; square head; little eyes

Habitat: in the water of lakes and rivers

Notes: completely aquatic; gills shorter in water with high oxygen levels; feeds at night; active all year

Amphibians—Natural History Journal Pages

Choose an amphibian from the Field Guide section of the book. Consider completing a natural history for at least one frog or toad (Order Anura), and one species from Order Caudata (salamander, newt, or mudpuppy).

1. Full Common Name: Northern leopard frog

   Scientific/Latin Name (*Genus species*): Lithobates pipens

   Family Name: Ranidae

   Order: Anura

   Class: Amphibia

   Phylum: Chordata

   Kingdom: Animalia

2. Draw the overall picture of the adult amphibian with identifying characteristics diagramed (label: size, colors, and markings).

Draw and diagram the juvenile stage.

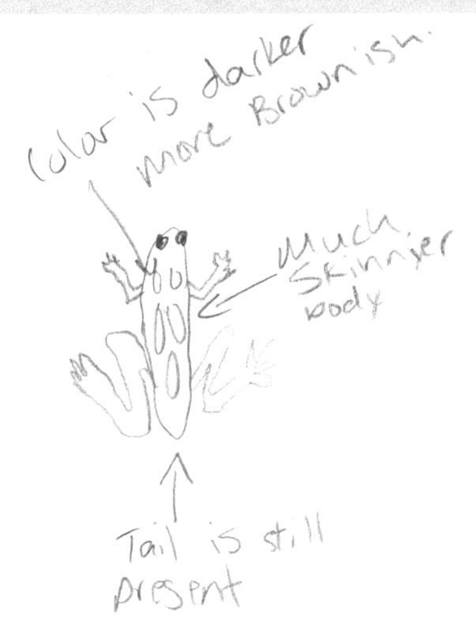

3. Provide a detailed description of your amphibian.

Colors range from green to brown
5-9 cm in length, dark spots on body
with yellowish stripes on either side of Back

4. On the map of North America, shade the range or area where your amphibian is found.

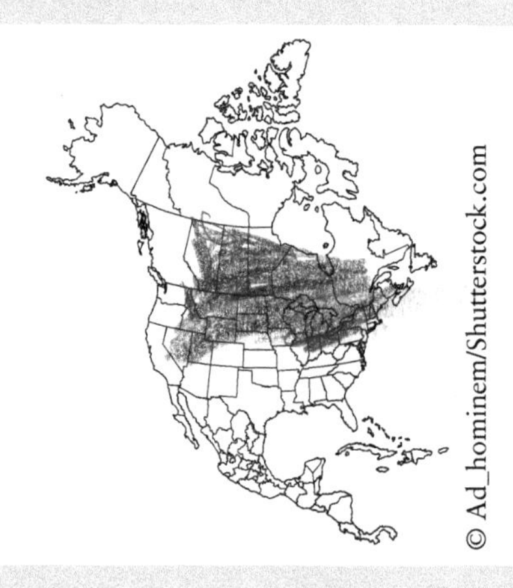

5. a. Where is its preferred habitat?

wetlands, damp meadows

b. Explain why you think you have a good or poor habitat at your nature spot for this amphibian.

I think I have a decent habitat
because my nature spot is on a
River.

6. What does the amphibian eat in its natural habitat?

Terrestrial invertibrates, spiders, larvae, slugs, earthworms

7. Describe an aspect of its behavior that you find interesting—what is it and why do you suspect it does it?

They will wait for prey to come to them. I think they do this because their movement isn't very efficent for the meal they get.

8. Discuss a method of conservation or management for your amphibian—to protect/restore/control its population:

we need to keep our waters clean and reduce pollution. These frogs are very volnuralable when it comes to pollution

9. List all resources used (**books:** author's last name, first initial, year, and book title; **websites:** name of site, date retrieved, and web address):

ontario wildlife foundation

8.3 Reptiles

> ***Nature Journal:*** Now that you have learned the basic characteristics of amphibians, reflect on how you think reptiles are the same and how they differ. Name as many reptiles as you can in 30 seconds. Describe the last time you experienced a reptile. What do you want to know about reptiles?

Reptiles belong to the same higher level taxonomic groups as amphibians except for its class—Domain Eukarya, Kingdom Animalia, Phylum Chordata, but Class Reptilia. There are approximately 6,000 reptile species worldwide.

Overview

Members of Class Reptilia in the region include turtles, snakes, and lizards. You will find these reptiles on land or in freshwater. Typically, reptiles will have cryptic coloration and can stay motionless for a long time. This helps them have great camouflage and go undetected to the unobservant passerby. Reptiles also have other incredible adaptations making them fun to study.

Distribution

Where can you find reptiles? The greatest species richness of reptiles occurs where it is warmest. The numbers get less as you move away from the equator. Globally, the United States and Canada ranks relatively low in species richness. You will find more suitable habitats for a variety of reptiles in warm climates.

Reptile Evolution

Reptiles appeared after amphibians in the evolutionary history, about 300 million years ago. They dominated the earth for 100 million years occupying all niches-on land, in the air, and in the sea, until a mass extinction occurred. A popular theory is that a giant asteroid or meteorite, as wide as 6 miles across, struck the Earth. This collision would have produced an explosive force and damage similar to that of a nuclear war.

Vast amounts of dust, debris, and ash would have spread quickly around the planet after the impact, blocking out the sunlight reaching Earth's surface. This led to a relatively rapid climate change with increased radiation levels in the biosphere. Lack of sunlight is believed to have lasted for several months and to have caused temperatures to drop worldwide from 66°F (19°C) to about 14°F (–10°C). Food chains were destroyed by a series of events that followed. Without sunlight to perform photosynthesis, many plants would have died, causing a domino-like effect on other organisms in the food chain, including the dinosaurs.

A feature or adaptation of reptiles that has improved the survivorship of the class within its preferred habitat is the "land egg." The egg is laid on land compared to the eggs we have learned about previously in the animal kingdom. This allows the animal to not depend on the presence of water for its existence or for generations of populations to continue. The land egg exists in all reptilian orders (read more about the egg under the Reproductive System section).

Types of Reptiles

We will focus on two different groups within Class Reptilia. First, the Order Testudines, which include turtles and tortoises. In the Great Lakes and Upper Midwest region, we only have turtles present, not tortoises. Most turtles in the area spend their time in aquatic habitats. People often mistake them for Class Amphibia because of the amount of time spent in and around water. Some turtles are found farther away from water and spend more of their time on land—like the Eastern box turtle, *Terrapene carolina*. Tortoises will be found in dry, hot climates and spend their time on land. Therefore, they will have adaptations that reflect the habitat in which they are found. A major reason turtles are classified in Class Reptilia is that they have scaly, nonglandular skin, which indicates they do not breathe through their skin, nor do they produce mucus like amphibians.

The next reptile order is Order Squamata. Snakes and lizards are members of this order due to their shared characteristics. The major difference between the two is one does not have legs and the other does. An order not found in the Great Lakes region are those members of Order Crocodilia. Crocodiles and alligators are found in areas where the climate is warm year-round and have reliable water sources. They can be found in southeastern North America. Regardless of where the reptiles inhabit, all will have shared characteristics as we will discover.

Reptilian Skin and Skeletal Features

As you read about reptiles, think about the similarities and differences between them and other animal classes. Reptiles have scaly, dry, nonglandular skin to prevent water loss. They shed the skin periodically (**ecdysis**) either all at once (snakes) or in patches (lizards). Just like amphibians and fish, reptiles are capable of metachrosis and possess chromatophores to allow for color change. You will find scales made of chitin on the ventral (belly) side of snakes. Chitinous material also comprises the bony external plates of turtle shells.

The shed skin of a snake. What you likely find is the skin shed in its entirety but will be inside out. The snake moves out of its skin after it has already grown its new layer. Imagine you peeling off a sock from your foot—inside out. That is how the snake's skin is shed.

Turtles have limited breathing room. Their respiration is restrained because of being bound to their shells. The top shell (**carapace**) has the backbone (vertebrae)

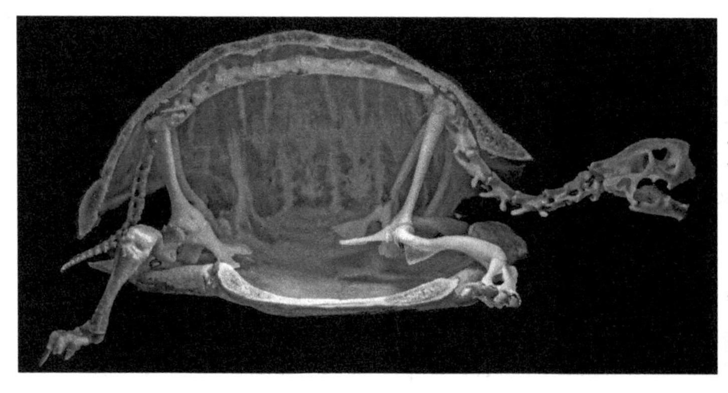

attached to it. Attached to the bottom shell (**plastron**), you would find their breastbone (sternum). The **bridge** is the section of shell on the sides that connects the carapace and plastron.

Why do snakes lack pectoral or pelvic girdles? They have no arms or legs! The absence of these and characteristics of other skeletal features of snakes are adaptated for swallowing prey whole. For example, each one of their bones is attached by elastic-type ligaments to allow for maximum stretching and bending. Additionally, snakes do not have a breastbone, their jawbones unhinge, and the ribs expand to allow for ingesting and digesting entire prey items.

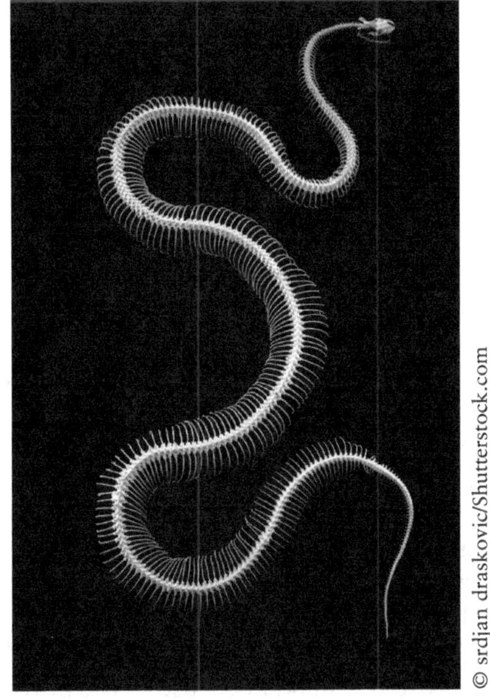

Snake skeleton. Note the vertebrae, ribs, and tail.

Reptile Physiology

As in other representatives discussed so far in Kingdom Animalia, reptiles are ectothermic or "cold-blooded." This means that their body temperature is regulated by the external environment. You will not see a turtle, snake, or lizard in the cold winter months in our region! In the summer, reptiles can be seen basking on logs, rocks, or pavement to warm their body and get some necessary vitamin D. Some may also seek shade when the days are too hot and dry but emerge on the warmed surfaces at night.

Circulatory System

The reptile circulatory system is more advanced than amphibians. They have more heart chambers and function at a greater efficiency. Reptiles have a 3 ½–4 chambered heart with two atria and a ventricle with an incomplete ventricular septum. It can pump oxygenated and deoxygenated blood throughout the body.

Respiratory System

Unlike amphibians, reptiles rely solely on lungs for respiration. They also have **alveoli,** balloonlike sacs within the lungs that provide a greater surface area for gas exchange. Snakes have one lung. This saves space in their elongated body plan. Turtles cannot expand their ribs to breathe because their lungs are essentially attached to the carapace and the viscera attached to the plastron.

Digestive System

Snakes require adaptations to assist in digesting their prey of which they often ingest whole. As their fangs (sharp teeth) pierce and hold their prey, digestion begins in the mouth with their salivary glands. The strong digestive enzymes secreted will help start to break down the prey's tissues. The ingestion process can sometimes take a long time and prey can become lodged in the throat. The trachea, or breathing tube, is set forward to allow the snake to breathe in that scenario.

Venomous snakes, which only comprise about one-third of 2,000 snakes (and a few lizards), will have venom glands to produce a poison through their fangs to inject within the prey. It usually will paralyze it initially, and ultimately kill it. Two types of venom exist, and any snake will only have one or the other. Hemolytic poison affects oxygen-carrying ability of the blood within the prey. Neurotoxic poison affects the nervous system and prevents nerve impulse transmission leading to a rapid death. Be mindful of where you step and where you sit in areas with snakes. If ever bitten, which is a rare occurrence, seek medical attention immediately. Do not try to capture the snake, because that can result in further injury. If possible, take a picture of the snake or take notice of color and markings. Remain calm, cleanse the area, and keep the bite lower than your heart.

Order Testudines do not have fangs, they do not even have any teeth (not even the snapping turtles!), and their tongue is unable to extend beyond its mouth. Order Squamata is the only reptile group with a tongue that can stick out. In fact, snakes and some lizards rely on this feature to sense their environment.

Nervous System

The cerebellum of the reptile is well developed. This region of the brain is responsible for coordination of the animal. Reptiles also have unique sensory perceptions, especially snakes. Many reptiles have lateral vision in which they move their head from side to side to locate and fix prey. Snakes have no moveable eyelids but have a transparent "cap" called the **brille**. Their eye movement is limited because of this, but objects are magnified. Snakes lack external ear openings and are deaf to airborne sounds but will respond to vibrations detected in the ground using their **columella** (ear bone). Snakes rely primarily on the senses of touch and smell.

Snakes also have a forked tongue that fits into a pocket (**Jacobson's organ**) in the roof of its mouth. This organ contains sensory epithelium for taste and sensing of chemicals in the environment. The forked tongue has an increased surface area to gather molecules in the air. After the snake "samples" that air, it will "read" its tongue in the Jacobson's organ. The tongue fits perfectly on this organ where the molecules are processed by the nerves and the brain. The snake can determine if potential food or danger are near based on the taste of its surroundings.

The venomous snakes in our region all belong to the Family Viperidae. These vipers have a **loreal pit** for thermoreception (hence the family's common name, pit vipers) that picks up heat radiated from the prey. This receptor is found between and just below the nostril and eye in most viper species. Other key features of these snakes include almost all with a triangular head, vertical pupil of the eye, and some have a rattle at the tip of their tail used to communicate stress. They also are ovoviviparous, where the fertilized eggs are held within the female and give birth to live young.

Reproductive System

As mentioned previously, the first "land egg" is seen with reptiles. Most reptiles are oviparous, or egg layers. Some are ovoviviparous (eggs remain inside female), while a few are viviparous (live births). With the egg layers, dozens of eggs are typically laid in the ground with no parental care. The land egg, or **amniotic egg**, will have a leathery outer texture to protect the growing embryo, and four membranes within the egg's shell to nurture it rather than depending on water as with amphibian eggs.

There are four membranes inside the egg. The chorion encloses the embryo and other membranes. The amnion is a fluid-filled sac with the embryo contained within it. The allantois allows for the passing of embryonic wastes and is highly vascular for respiration to occur. Lastly, the yolk sac contains the necessary nutrients for optimum embryo growth.

Environmental factors have been shown to influence the sex of reptiles while they are in the eggs. This is called temperature-dependent sex determination (TDSD). Turtle development is especially sensitive to temperature and moisture levels. They can achieve bigger hatchlings with wetter conditions than average, and develop into female turtles during high temperatures (mild temperatures may result in females with the snapping turtle). The critical time of sex determination is the middle third of the incubation time. Due to microclimates, more variation in sex ratios exists between nests rather than within nests.

Excretory System

A feature that is worth mentioning about the excretory system of most reptiles is the reabsorption of liquid urine. Lacking a urinary bladder, most urine is reabsorbed, and uric acid waste is released as a pastelike form—much like that of a bird.

Growth and Longevity

Reptiles grow rapidly the first few years of life. However, ectotherms cannot function in our winters, so energy is not typically expended on growth during the cold temperatures. Growth and environmental conditions will also influence breeding. Their growth slows upon reaching full maturity and most energy goes into reproduction. Some reptiles, especially turtles do not breed for the first time until later in life, sometimes not until their 15th year.

Reptiles in this region also need to conserve energy because of the climate and limited food supply. They will **brumate,** or go through a semihibernation state in cool to cold temperatures. This is when the animal sleeps most of the time and eats very little. They tend to go below the frost line during extended periods of inactivity. Some turtles may brumate underwater and snakes tend to brumate together in large numbers in a protected place—sometimes with other snake species. Turtles experience many brumations in life, because they are probably the longest living vertebrates, some living in our region to over 75 years. In warm climates, reptiles will tend to grow larger and live 50-200 years!

Reptile Threats

As with all organisms, habitat loss and degradation threaten populations. Wetland draining causes drastic decreases of turtles, and shoreline disturbances result in nesting and egg loss. Introduced and nonnative invasive species competes with native species for food and space to live. New diseases are also introduced by nonnative species and can be linked to pollution. Environmental pollution—in the water, air, or within the ground degrades habitats. An advantage of reptiles' nonporous skin is that reptile populations may not be as vulnerable to pollution as amphibians. However, global climate change will also alter reptilian habitats due to differences in the local temperatures and precipitation levels. Lastly, reptiles face decline because of unsustainable collecting for pets, food uses, decoration, and personal status.

What Is Being Done?

Protective laws and regulations exist for reptiles. Through inventories, wildlife regulatory agencies maintain lists of rare, threatened, and endangered species. Those that you find on these lists usually have special protection and management strategies in place dependent on the state or province. If you have concerns or questions regarding any species, please contact your government wildlife agency. Look for community science opportunities involving inventories.

Activities: Hopefully, your curiosity is piqued to head out and find reptiles. Go to a nearby pond or river in the summer that has logs emergent from the water. Bring your binoculars to discover turtles basking on those logs. If you have an open field, consider placing a wood board flat on the ground. This creates habitat for snakes. Snakes can use those boards to retreat beneath or bask upon.

Nature Journal: Complete Natural History Journal Pages for what you discover (the template is found after the field guide section). Start your species inventory list. Perhaps you will want to conduct a field investigation of the reptiles nearby. Add a reptile species to your Chordate Characteristic Comparison Chart found at the end of Chapter 8.

Review Questions

1. What class do reptiles belong? Discuss each order. Give examples.
2. How are reptiles distributed in North America?
3. When did reptiles first appear in the fossil record? What were possible reasons for the dinosaur extinction?
4. Discuss the epidermis or skin of reptiles. How is it adapted to living on land? What are the shells of turtles called? Describe the skeletal modifications in snakes.
5. Discuss the circulatory system. Explain the basic structure and functioning of the reptile heart. Are reptiles cold- or warm-blooded?
6. Explain the digestive system characteristics of reptiles.
7. Why do snakes have just one lung? What are the various ways turtles can respire? Why is it important for turtles to have alternate ways to breathe?
8. Discuss the nervous and sensory systems in reptiles.
9. Do reptiles lay eggs or have live births? Which happens most often? Is fertilization internal or external? How is the reptilian egg adapted to land? Describe the characteristics of the reptilian egg.
10. How do snakes and lizards function without a urinary bladder?
11. What is the typical growth pattern for reptiles? What reptile lives the longest? Describe reptile dormancy.
12. Discuss the various threats to reptiles and what can be done to protect these animals or habitats in which they live.

Field Guide: Reptile Species Accounts

The animals in this section represent the reptiles that occur across the largest area of the Great Lakes basin and Upper Midwest region. The turtles are listed first, then snakes, and finally a lizard example. The species in each section are organized by relative size from smallest to largest. Make connections between animals of the same family.

Turtles

Aquatic or terrestrial; slow-moving reptiles enclosed in hard, scaly, or leathery shells; head retractable (except snapping turtle); thick legs

© Ryan M. Bolton/Shutterstock.com

Common Musk Turtle or Stinkpot (***Sternotherus odoratus***)**—Family Kinosternidae**

Size: 3.0–5.0 in. (7.5–13 cm)
Key Features: smooth, high-domed carapace olive-gray to dark; two pale stripes on sides of head
Habitat: streams; ponds
Notes: most always near water; secretes musky fluid to deter predators; has a conspicuous hinge on plastron to help "close up shell" when body is retreated; most of its nests are preyed upon; not found in northern reaches of region

© fivespots/Shutterstock.com

Spotted Turtle (*Clemmys guttata*)—Family Emydidae

Size: 3.0–5.0 in. (7.5–13 cm)
Key Features: black carapace with yellowish spots; spots on skin (or not); light colored plastron with dark blotchy edges
Habitat: lakes and rivers with soft, muddy bottom
Notes: observed in water or on land; omnivore (eats algae, duckweed, water lily, crayfish, tadpoles, aquatic insects); populations negatively impacted by habitat loss and collectors; more centrally found in region

Common Map Turtle (*Graptemys geographica*)—Family Emydidae

Size: 4.0–13 in. (10–32 cm)
Key Features: low central ridge on olive brown carapace that has thin, yellowish rings that resemble contour lines of a map; yellow angular spot behind eye
Habitat: rivers and lakes with soft sediment
Notes: very aquatic; prefers mollusks and crayfish (strong jaws!)

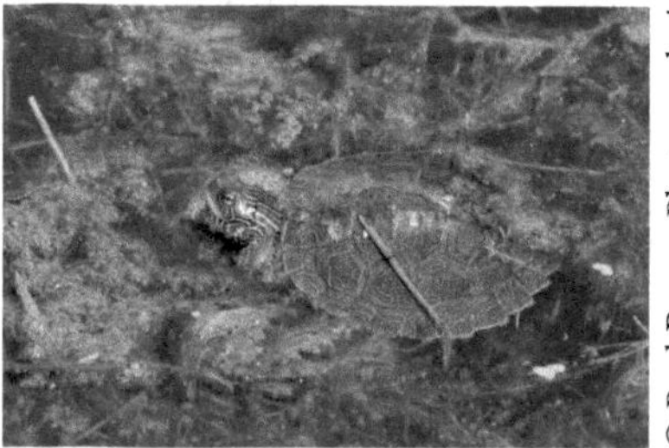

© Paul Reeves Photography/Shutterstock.com

© Gregory Johnston/Shutterstock.com

Painted Turtle (*Chrysemys picta*)—
Family Emydidae
Size: 4.0–9.0 in. (10–23 cm)
Key Features: dark olive to black carapace with red edges; skin dark with yellow striping; plastron yellowish (Midland subspecies has red stripes)
Habitat: lakes; rivers; permanent water sources
Notes: very common (tolerant) turtle; observed swimming under ice in lake; Red-eared sliders found in southern range have a prominent red stripe behind eye (sometimes orangish-yellow)

© Seth LaGrange/Shutterstock.com

Eastern Box Turtle (*Terrapene carolina*)—
Family Emydidae
Size: 4.0–8.0 in. (10–20 cm)
Key Features: high-domed carapace with black and yellow lines, spots, or blotches; skin also resembles carapace pattern; dark plastron with yellow blotches
Habitat: mesic forests; wet fields
Notes: hinged plastron to enable a more complete closure during retreat; seed and fruit eater; males travel up to half-mile; long-lived terrestrial turtle; found in more southern range of region

© fivespots/Shutterstock.com

Wood Turtle (*Glyptemys insculpta*)—
Family Emydidae
Size: 5.0–9.0 in. (13–23 cm)
Key Features: brown pyramidal plates on carapace; orangish skin; yellow plastron with dark markings
Habitat: woodland streams; wetlands
Notes: not as common as other turtles (found in central part of region)

© Ryan M. Bolton/Shutterstock.com

Blanding's Turtle (*Emydoidea blandingii*)—
Family Emydidae
Size: 5.0–11 in. (13–28 cm)
Key Features: high dark carapace with small, yellowish spots or blotches; chin and throat yellow
Habitat: wetlands; lakes; quiet streams
Notes: hinged plastron to allow full retreat into shell; gentle turtle but hisses when disturbed; seen under ice in winter; can live up to 50 years; eats plants, small fish, tadpoles; crustaceans

© Natalia Kuzmina/Shutterstock.com

Spiny Softshell Turtle (*Apalone spinifera*)—Family Trionychidae

Size: 5.0–18 in. (13–45 cm)
Key Features: thin, flat, olive green, soft, leathery carapace with dark markings; skin matches carapace; spiny projections along front edge of carapace; long, tubular snout; webbed feet
Habitat: fast rivers; lakes
Notes: stays under water for long time; fast swimmer; shy

© Jeff Holcombe/Shutterstock.com

Snapping Turtle (*Chelydra serpentina*)—Family Chelydridae

Size: 10–19 in. (25–48 cm)
Key Features: big head with thick neck; beaked snout; powerful legs; dark greenish-brown or black carapace with toothed rear margin; small plastron; long tail; webbed feet; long, sharp claws
Habitat: quiet rivers; lakes; permanent water sources
Notes: can weigh up to 50 lb (23 kg); beware of bite—it can extend its neck and turn quickly to strike; no teeth, but very strong jaws with sharp beak; often has algae or mud on its back; omnivore; travels great distance to lay eggs; can live up to 30 years

Snakes

Limbless, scaly reptile; long tapered body

Red-Bellied Snake (*Storeria occipitomaculata*)—Family Colubridae

Size: 8.0–10 in. (20–25 cm)
Key Features: small, thin snake (pencil-sized) with brownish-black back with four thin stripes or one wide pale stripe ; neck with three light spots or collar; red-orange belly
Habitat: mesic forests; wet fields; bogs
Notes: eats slugs; prey of many; not an egg layer, female gives birth

© Gerald A. DeBoer/Shutterstock.com

© Jason Patrick Ross/Shutterstock.com

Brown Snake or DeKay's Brown Snake (***Storeria dekayi***)**—Family Colubridae**

Size: 9.0–12 in. (23–30 cm)

Key Features: brownish-gray with light center stripe and small black, parallel rows of dark spots; ventral side light brown or pink with small dark spots

Habitat: moist deciduous forest near water; wetlands; parks; fields

Notes: secretive snake; small home range; eats worms and snails (teeth specialized for removing shells); common prey animal; dens in same place each winter

© Matt Jeppson/Shutterstock.com

Ring-necked Snake (*Diadophis punctatus*)—Family Colubridae

Size: 10–15 in. (25–38 cm)

Key Features: black to dark gray with smooth scales; obvious yellow neck ring (usually present); belly reddish-orange to yellow with black spots

Habitat: damp forests; grasslands; rock outcrops with vegetation (south facing)

Notes: small and secretive; nocturnal; tends to coil and hide its head while exposing bright underside; tail often coiled in a tight spiral; emits a foul liquid when handled

© Jason Patrick Ross/Shutterstock.com

Smooth Green Snake or Grass Snake (***Opheodrys vernalis***)**—Family Colubridae**

Size: 12–24 in. (30–61 cm)

Key Features: bright green with pale belly; smooth scales; white lips and chin; tongue red with black tip

Habitat: fields; forest edges; open woodlands

Notes: small; defenses include camouflage, fleeing, coil and strike (too small to inflict painful bite to humans); insectivore; habitat loss and insecticides threaten populations

© Jason Patrick Ross/Shutterstock.com

Queen Snake (*Regina septemvittata*)—Family Colubridae

Size: 12–30 in. (30–76 cm)

Key Features: dark to tan (sometimes olive) with light yellowish lateral stripe and lower row of scales on sides; pale yellow bellow with stripes; yellowish-white chin and throat; small head

Habitat: clear, shallow rivers with rocky bottoms; lakes; ditches

Notes: slender water snake (excellent swimmer); sometimes climbs branches; eats mostly crayfish (populations declining in areas where habitat and food altered)

© Ryan M. Bolton/Shutterstock.com

© Jay Ondreicka/Shutterstock.com

© IHX/Shutterstock.com

© Matt Jeppson/Shutterstock.com

Eastern Hognose Snake or Puff Adder (***Heterodon platirhinos***)**—Family Colubridae**

Size: 15–30 in. (38–76 cm)

Key Features: snout upturned; wide neck; thick body; tan to brown or reddish with dark, squarish blotches on back and round blotches on sides; dull yellow or gray on chin and belly with tail lighter shade (snake can appear as one color and can be dark in color)

Habitat: sandy soils; woodland edges

Notes: digging snake (uses its uniquely shaped snout); quite the performer—raises head and flattens neck like a cobra before hissing and striking; feigns death; sometimes vomits and turns over with belly up, mouth open, tongue out but "comes alive" when threat is over

Massasauga (*Sistrurus catenatus*)—
Family Viperidae

Size: 15–48 in. (38–76 cm)

Key Features: thick body; brown to gray with dark blotches on back with smaller alternating spots on sides outlined with white line; two stripes on top of head and neck; bow-tie-shaped blotches on neck; vertical pupils; rattle at tip of tail

Habitat: wetlands; woodlands; fields

Notes: daytime hunter (small mammals, birds, amphibians, other snake species) but also active at night during hot summers

© Chris Hill/Shutterstock.com

© Patrick K. Campbell/ Shutterstock.com

Common Garter Snake (*Thamnophis sirtalis*)—Family Colubridae

Size: 18-54 in. (46-137 cm)

Key Features: variable morphs; slim dark body with three stripes (yellow, green, or tan), one on back (usually lighter than sides) and on each side (second and third rows of scales above belly); area between stripes often with dark blotches (sometimes red)

Habitat: fields, wetlands, woodlands; parks; often near water

Notes: common snake; mistakenly called garden snake; eats amphibians, worms, slugs, and other small animals

© Steve Bower/ Shutterstock.com

Eastern Ribbon Snake Northern or Common Ribbon Snake **(*Thamnophis sauritus sauritus*)—Family Colubridae**

Size: 15–30 in. (38–76 cm)

Key Features: slim; three bright or light-yellow stripes on dark back (one on top and one on each side at third and fourth rows above belly)

Habitat: wetlands; littoral area of lakes and rivers; bogs

Notes: semiaquatic; basks in sun; not aggressive but releases a musky fluid when frightened; subspecies of ribbon snake from southeastern United States; just like other garter snakes, will produce live young

© Malachi Jacobs/Shutterstock.com

Milk Snake (*Lampropeltis triangulum*)—Family Colubridae

Size: 2.0–3.5 ft. (61–107 cm)

Key Features: variable morphs; mostly light gray to tan alternating with rows of brownish-red saddle-shaped markings outlined in black on back with similar but smaller markings on sides; light "y" or "v" behind head; whitish belly with checkerboard of dark blotches

Habitat: deciduous woods; grasslands; hillsides; rocky areas

Notes: secretive; constrictor found beneath rocks and logs on hillsides (spring and fall) moving to woodlands or fields (summer); coils and strikes when disturbed (tail vibrates); eats small mammals, birds, and bird eggs

© Michiel de Wit/Shutterstock.com

Northern Water Snake (*Nerodia sipedon*)—Family Colubridae

Size: 2.0–3.5 ft. (61–107 cm)

Key Features: stout body; broad head; overall brown to gray with dark blotches or bands on back and sides (often offset from one another); body appears almost one color when skin is dry; belly cream to tan-colored with reddish crescent markings with dark outlines

Habitat: all permanent freshwater; lakes; rivers

Notes: almost always near water; basks in sun; defends itself by releasing a foul smell, flattening its head and neck, strikes; eats fish, crayfish, and amphibians; not venomous

© Eric Isselee/Shutterstock.com

Copperhead (*Agkistrodon contortrix*)—Family Viperidae

Size: 2.0–4.5 ft. (61–117 cm)

Key Features: back copper color or orangish-pink with dark red-brown crossbands narrowing at center of back (hourglass-shape); wedge-shaped head; pit in front of and just below eye; vertical pupils

Habitat: swamp edges; rock outcrops and forest ravines

Notes: should be avoided (venomous); **only in southeast part of region**

© MLHoward/Shutterstock.com

Eastern Fox Snake (*Pantherophis vulpinus*)—Family Colubridae

Size: 2–6 ft. (61–152 cm)

Key Features: tan to yellow with dark spots on back and smaller spots on sides; elongated blotches behind head; tannish-yellow and black checkerboard pattern on belly

Habitat: varied; woodland; field, wetland

Notes: excellent climber, but found mostly on ground; when threatened, shakes tail among dry leaves and sounds like a rattlesnake (mimicry) and releases musk with similar smell to a fox; constrictor of prey (small mammals, frogs, birds, bird eggs); populations threatened due to habitat loss of the eastern Great Lakes region

© Psychotic Nature/Shutterstock.com

Racer or Blue or Eastern Racer **(*Coluber constrictor*)—Family Colubridae**

Size: 2.5–5.0 ft. (76–152 cm)

Key Features: slim body; smooth scales; color varies (slate gray to blue); some with white belly; white chin; large eyes

Habitat: prefers open prairies and fields; edge habitat; woods; rock outcrops

Notes: quick and agile across ground and holds head above surface; good tree climber; raids bird nests; large range of up to 20 acres (8 ha); shakes tail among leaf litter and makes rattling sound (mimicry); contrary to species name, not a constrictor; releases foul substance when disturbed; mid to southern part of region

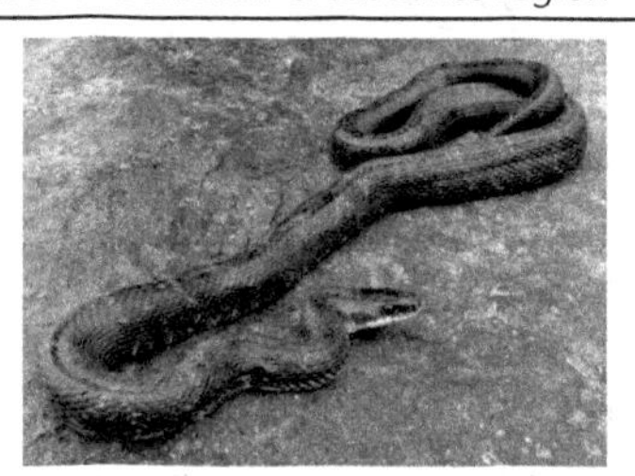

© Matt Jeppson/Shutterstock.com

Black Rat Snake or Eastern Rat Snake **(*Elaphe obsoleta*)—Family Colubridae**

Size: 3.0–5.75 ft. (1.0–1.75 m)

Key Features: stout, large body (square in cross section); nearly all black with perhaps some darkened blotches and flecks (or faint stripes) of white, yellow, or red; white chin and throat; head distinctive from body; gray belly with dark spots or checkerboard pattern

Habitat: wetlands; hardwood forests; rocky outcrops; farms

Notes: large snake; populations declining; climbs trees; controls small rodent populations; will coil and vibrate tail among leaf litter (sounds like a rattler; mimicry) before lunging with an s-shaped curve to its neck (harmless to people)

**Timber Rattlesnake (*Crotalus horridus*)—
Family Viperidae**

© Eric Isselee/
Shutterstock.com

Size: 3–6 ft. (1–2 m)
Key Features: head wider than neck and all one color (usually tan); body color variable but mainly all black to yellow with dark blotches (somewhat zigzagged bands); tail black with rattle
Habitat: wetlands; lowland forest; rocky wooded slopes
Notes: rare; venomous; **not in Michigan, Minnesota, Dakotas, or Canada**

Lizard

A narrow scaly reptile with four limbs and tail

**Five-lined Skink (*Plestiodon fasciatus*)—
Family Scincidae**

© Germain McDaniel/
Shutterstock.com

Size: 5.0–8.0 in. (13–20 cm)
Key Features: brownish-black with thin stripes (fading into adulthood) with "y" on head; males have red-orange jaws; young have light stripes and blue tail
Habitat: mesic woods with leaf litter and rock outcrops; shaded gardens
Notes: digs into soil under rocks and logs; it can detach from tail if threatened while it continues to wiggle for a few minutes; carnivore eating insects, spiders, snails

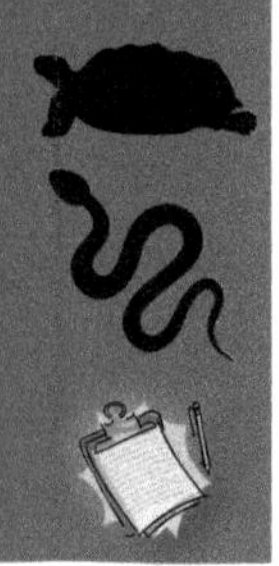

Reptiles—Natural History Journal Pages

Complete the natural history information for a reptile within your Nature Area or region. Consider doing this exercise for a turtle, snake, and lizard.

1. Full Common Name: Blue Racer

   Scientific/Latin Name (*Genus species*): Coluber constrictor

   Family Name: Colubridae

   Order: Sauamata

   Class: Reptillia

   Phylum: Chordata

   Kingdom: Animalia

2. Draw the overall picture of the adult with identifying characteristics diagramed (label: size, colors, and markings).

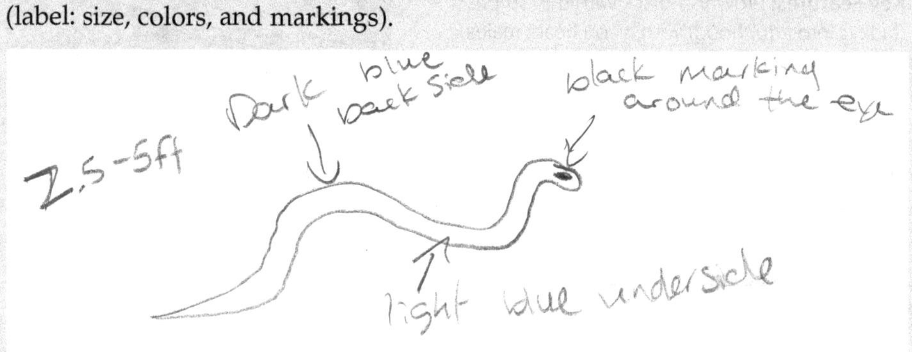

3. Give a description of your reptile.

   Medium sized blue snake with a dark eye patch

4. On the map of North America, shade the range in which your reptile is found.

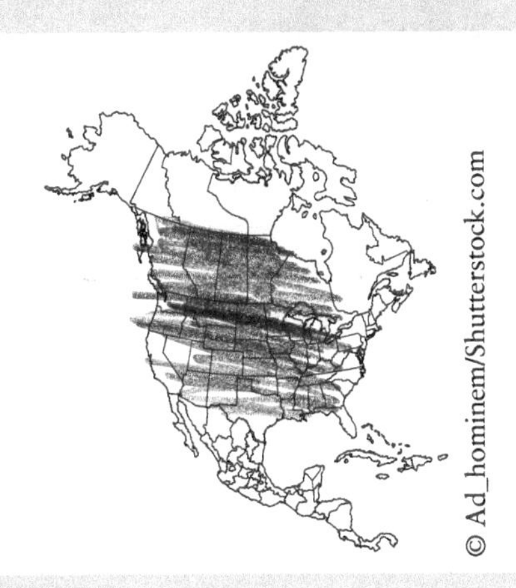

5. a. Where is its preferred habitat?

open Praries and fields, edge habitat
woods, rock outcrops

b. Explain why you think you have a good or poor habitat at your Nature Area for this reptile.

I think I would have a mid habitat
for this reptile because they don't live near
water.

6. What does it eat in its natural habitat?

Small to medium sized rodents
Mice, rabbits, squirrels.

7. Describe an aspect of its behavior that you find interesting—what is it and why do you suspect it does it?

They hunt during the day so they
can hide from predators at night.

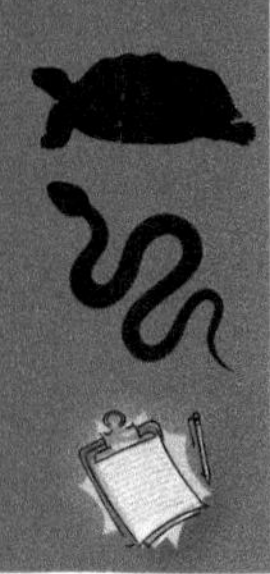

8. Discuss a method of conservation or management for your reptile—to protect/restore/control its population.

9. List all resources used (**books:** author's last name, first initial, year, and book title; **websites:** name of site, date retrieved, and web address):

BBC wildlife magazine

8.4 Birds

There is a growing popularity for birding. It is a great way to get motivated to go outside, focus your attention on what is happening around you, and to start to understand and interpret nature. Having a diverse landscape that offers various habitats will attract a variety of bird species—and birders! **Birders** are people who like to see, observe, identify, and keep a running list of what birds they have encountered. The types of things in a diverse habitat would include vegetation of different types, sizes, ages, and arrangement. To further improve your chances of seeing birds, find a source of water near vegetation. Even a bird bath that you keep cleaned out from time to time will serve as a place for them to cool off and get a drink.

Before we begin thinking formally about birds, let's think about what we already know. Birds are active year-round. Though some do not stay in the same area, you can still find birds regardless of the season. This makes birding a great way to get excited about nature. Having wild animals to focus your attention can pique the interest of people at any age. You will find birds in every habitat.

> ***Nature Journal:*** Read and respond to the prompts to help you think about birds. Recollect the last time you observed what the birds were doing outdoors. Describe that experience. Include the species (if you can) along with a description of the bird, its habitat where you observed it, the behavior you witnessed, and give your thoughts or interpretations of what you saw [or heard].
>
> List the names of other birds that you know. Which birds have you seen in the wild? Identify the birds found in your Nature Area. From your existing knowledge, what are at least five facts about birds (think about what makes a bird a bird)? What do you want to learn about birds?

Overview

There are approximately 9,000 species of birds worldwide, and over 500 species found in this region! Just like the other vertebrates studied so far, birds exist in the taxonomic classifications of Domain Eukarya, Kingdom Animalia, and Phylum Chordata. To help classify the birds further, think about their ability to fly. These aviators are in Class Aves. Most birds have adaptations for flight. This special ability helps account for the success of the species. Flight allows the animals to utilize a large territory to satisfy needs, escape or avoid predation, find food and water, and to occupy niches unreachable by others.

Some of the other unique avian characteristics you may have reflected on in your journal could have included the presence of feathers, wings, flight, beak, "hollow" bones, fewer number of bones, unique respiratory system, bipedal vertebrates, endothermic (warm blooded), amniotic eggs ("land egg"), parental care of offspring, and a well-developed brain. Many naturalists have a deep knowledge and understanding of birds and have spent many hours studying and documenting bird data. A scientist who applies methodological studies to gain knowledge about birds, and is considered an expert in aspects regarding birds is called an **ornithologist.** Do what you can to improve your naturalist intelligence about birds.

Bird Evolution

The evolutionary evidence shows that birds may have come from carnivorous, bipedal (upright, two-footed) reptiles called theropods. Many shared traits between reptiles and birds exist. Scales and feathers are made of the same materials (feathers likely evolved long before flight). Scales on bird legs are reptile-like, and both animal types have an amniotic, or yolked "land egg."

Archaeopteryx

One of the earliest known birds, *Archaeopteryx* is considered the transition animal between reptiles and birds. Fossils have been found in Germany since 1860 that have dated these animals to have lived about 150 million years ago (late Jurassic Period) but are now extinct. It was similar to dinosaurs or reptiles in respect to its teeth, long bony tail, skull, and certain bone structures. However, it also had avian qualities such as feathers (for gliding and weak flapping), wings (claws on tips), and bird-like bone structures. Any reference to the "reptilian character" of *Archaeopteryx* infers modern birds do not have that feature. If referring to its "avian character" indicates modern reptiles do not have that particular feature either.

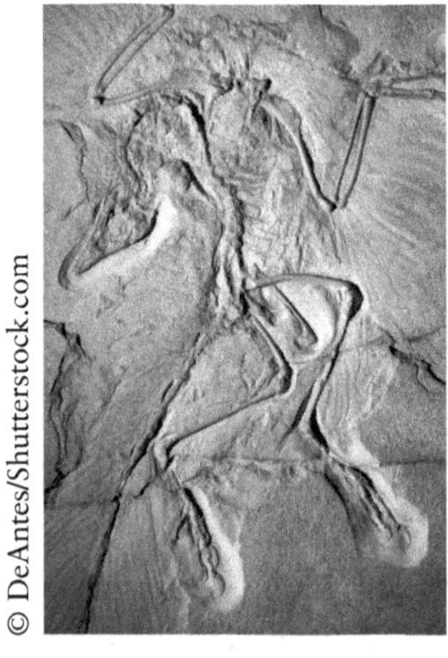

Arachaeopteryx as fossil evidence and an artist's rendition of what it may have looked like.

Form Follows Function

Any animal [or any species] will have adaptations for its preferred habitat. Certain features have obvious connections to a habitat. We say that the "form follows function." The need for a certain feature to survive comes before the presence of that feature. A wetland exists before you see a prevalence of webbed-footed birds.

For birds, feet type and beak shape could arguably provide the greatest insight into their behavior, habitat, and niche. It can also help lead you to an identification of the bird species.

Activity: Now, consider any animal—what is it and where do they live? Describe the conditions of the area. What resources are available for that animal. Consider your chosen animal again. Discuss the features or adaptations of the animal to allow it to survive in its environment. If it did not have those features, would it still live or frequent that habitat?

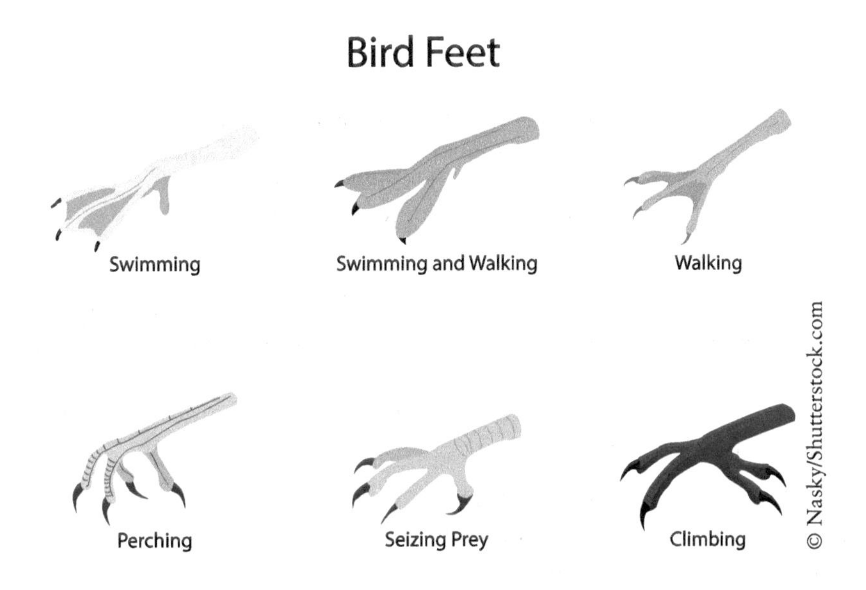

Conditions and resources dictate suitable habitat residents. When considering bird identification, species can be distinguished based upon body shape, beak shape, foot shape, feather coloration, flying style, and behavior. This will be discussed further in the field guide section for birds. Let's first consider shapes of the beak (also referred to as the bill). It has a shape indicative of a species, genus, family, or an entire bird order.

Types of Bird Beaks

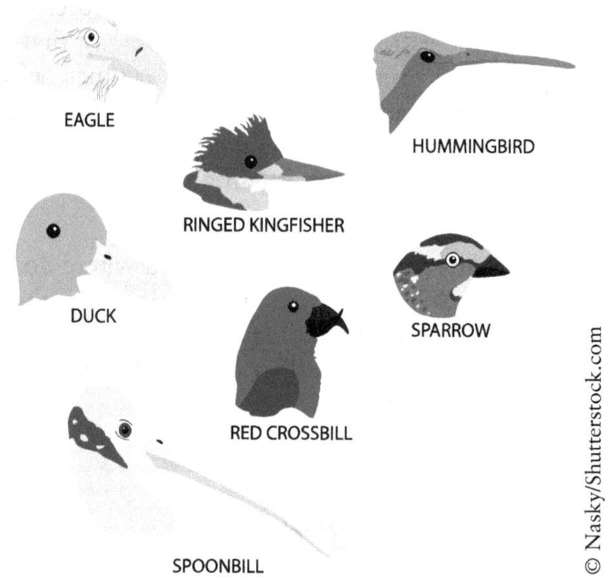

Think of a word to describe each of the beaks pictured here. Compare lengths, tips, and bulkiness. Choose a beak. Hypothesize the type of food it could easily acquire with its beak.

The beak shape is closely tied to its preferred feeding method. Look at the list of beak descriptions, if you used one of these or similar words to describe the bird beaks, then you would be able to discern the diet and niche of that bird. There are many types of beaks, but these are common types found among species (with some variation within groups).

hooked: tear prey; carnivores (hawks, owls)
thick: seed eaters; crush hard husks (finch)
short, slender: insects, small seeds
spoon: strain water; scoop mud for food (duck)
chisel-like: chip away at wood (woodpecker)
long, stout: stab fish (heron, kingfisher)
crossbills: pry apart pine cones
long: probe into mud (sandpiper)
long, slender: siphon flower nectar (hummingbirds)

Learning general bill types can help narrow down the possibilities for identifying the bird.

Anatomy

When thinking about birds, often what comes to mind first is the presence of feathers. Or, perhaps the color of feathers is what captures your attention. The color of feathers results from different sources dependent on species. Light and physiology may play complex roles. The bird's uses of feathers are an important anatomical considerations - especially in flight. The unique skeletal and muscular features of the bird will also be discussed here, and the anatomy of the bird egg and what to know about the young.

Feather Colors

Many feather colors are from the melanophores or pigment cells in the feather. These features produce pigment and color to the developing feather. Sometimes the change in feather color is based on the season. Photoperiodism often triggers internal changes with organisms.

Birds with red, orange, or yellow feather color may result from what food they consume. Food preferences of these brightly colored birds often have bright colors. The color of the bright food (red fruit, colored seeds, insects, etc.) has fat-soluble pigments that enter the feather and changes the **plumage**, or the collection of feathers upon the bird. Typically, bright-colored birds held in captivity may not feast on their preferred diet. The result is a duller plumage than what it would have in the wild. Comparatively, blue feathers are not actually blue. The blue you see in nature results from light reflecting off the feathers due to its structure (**structural coloration**). The feather itself is a brownish color, in other species or circumstances it may have white or iridescent effects. The next time you find a feather, especially ones from a blue jay, hold it up to the light to see its true colors (be sure to wash your hands after handling it).

Sexual Dimorphism

Sometimes, you will find a distinct difference in how the sexes of any animal species looks. This is called **sexual dimorphism**. An obvious difference in the form between males and females of the same species may be observed with the animals' color, size, or presence/absence of body parts (e.g., ornamental feathers, horns, antlers, or tusks). These different features are often utilized in courtship displays, fights, egg/young protection, or predator distraction. For birds, sexual dimorphism occurs mainly with a color difference. Males tend to be more brightly colored than females. This may help attract a mate and can distract potential predators from the nest. Females have more subdued or muted tones. This is useful in staying camouflaged while tending to the nest or the young.

Molting

Other types of plumage characteristics include normal molting (or replacement of feathers), and nuptial, eclipse, and winter plumage displays. Normal molting

Male (left) and female (right) cardinal. The berries they are eating here help keep their plumage vibrant.

happens twice per year and is when birds lose feathers gradually and replaces damaged feathers with new ones. Nuptial, or breeding plumage appears after the first molt in the spring and is when some bird species develop brightly colored breeding feathers, or just simply a healthy replacement of feathers. Some birds will display eclipse plumage, like some waterfowl after breeding season. This is when the male loses all flight feathers at once and his feathers resemble a henlike plumage. He is flightless and vulnerable for several weeks. Winter plumage comes in with the last molt of the summer. The added feathers will help with keeping the animals warm, and for some species to add drab-colored feathers in the winter months. This can help with a more camouflaged appearance among the birds' surroundings. If you observe an unusual display of feathers or lack of plumage on the birds, you may have witnessed the bird going through one of its normal molting phases and will regrow new feathers as the summer season goes on. Of course, doing some research on the bird is always a good idea just to be sure of what you are interpreting.

Feather Types and Characteristics

Feathers are the strongest known material for their size and weight. Learn the different parts of the feather, where it is found on the body, and its function. Use the diagram to help you visualize the parts discussed. The large, stiff feathers you may have found in nature are likely contour feathers of the bird that covers its body. A quill, or the calamus is the hollow shaft found in the follicle (have you ever seen a quill pen for writing?). The web or vane part of the feather consists of a central supporting shaft (rachis) and lateral barbs. The barbs are held in place by a series of hooklets and flanges on proximal and distal barbules (as you will read below, this arrangement serves an important role). You will find afterfeathers attached to the calamus near the web to provide extra insulation in northern birds, and down feathers found under contour feathers to provide insulation with its short shaft and hookless barbs (especially abundant in ducks and geese). Powder down is found under some bird's breast feathers, like herons to clean grease off their feathers. Feathers can also be found at the base of some bills, like those of flycatchers to capture insects – these are called rictal bristles. Lastly, the wing and tail feathers are referred to as remiges and rectrices, respectively.

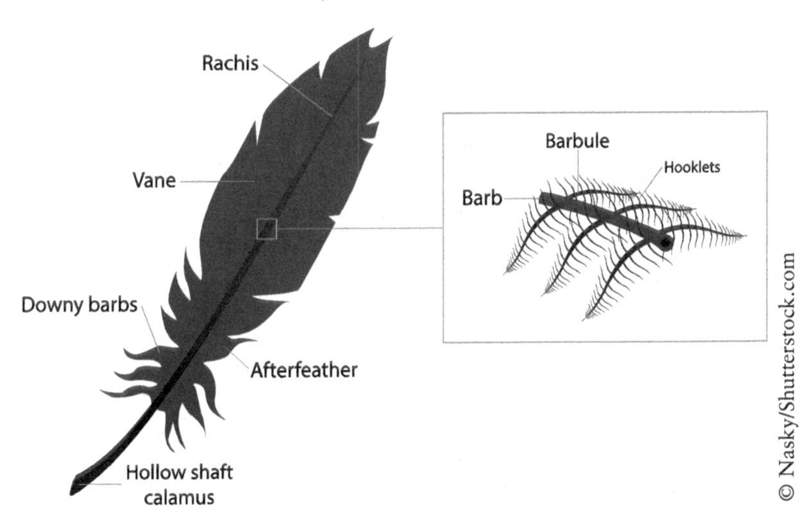

Have you ever seen a bird rub its beak along its feathers? The bird is likely **preening**. Feathers get ruffled or get out of place for various reasons. By preening, the bird smooths the feathers. They can "zip" the barbules together with the hooklets as seen in the diagram. This helps make the bird more streamlined to travel through the air or water more easily. It may also make them wind or water resistant.

Activity: If you have the chance, use your magnifier to examine a bird feather up close. Identify the parts. Can you see the barbules?

Flight

Contour feathers are found on the wing feathers and used for flight. There are slightly different wing feather types in structure and position. Primaries are attached to the bird's wrist bones and the feathers considered the secondaries are attached to the ulna. Small covert feathers cover the quills of these feathers.

Have you also noticed that different bird species have their own flight pattern? Check it out. Describe the observations you make about the patterns. Some birds can fly better than others, some fast, some straight, some erratic, some in groups. There are all kinds. Learn to identify signature flights associated with the birds nearby. Aviation engineers and scientists have studied birds for centuries to discover what avian characteristics create the best lift and soaring capabilities. The wing shape, curved on both the top and bottom with it tapered on the outer edge is called an **airfoil.** Wing shapes in flight can help you identify the birds.

European starlings (*Sturnus vulgaris*) synchronized in their signature murmuration flight pattern.

Skeletal Structures

What do you know about bird bones? The skeleton has unique characteristics to help them achieve a lightweight body frame for flight. Birds have fewer bones in their skeleton compared to other chordates. They are also mostly hollow which contributes to their lightness. Internal bony struts or braces give strength to the skeletal system.

Within the skull, the bones are thin and fused, which reduces the weight in the head. The beak (also referred to as the bill, especially in reference to aquatic birds with their rounded, scooplike bills) also has the bony skeleton making up its jaws but has no teeth. Although, some birds have distinctive ridges along the interior edge to serve as a sieve to strain out vegetation from the

The bony struts within a bird bone (left) and a complete skeleton of a bird (right).

water (i.e., ducks, geese, and swans). A fleshy sheath of keratin (the same material of our fingernails) covers the bony beak. Like our fingernails, it continues to grow so birds constantly need to wear it down. The beak is essentially used like our hands and fingers. Along with the bone in the tongue, the beak is used to help feed, build nests, communicate, defend themselves, and more.

Tall birds will often rest on one leg. Take a look at the picture of the heron's silhouette. When observing the structure of the leg, you may have thought the knee bent backward (compared to human knees). What you are actually looking at is the bird's ankle! The lower part of a bird's leg is the **tarsometatarsus**, a bone formed by the fusing of the **metatarsal** and **tarsal** bones. What we see as a bird's "foot" is really its toes. Think of it as if birds walk around on their tiptoes all the time. Its upper leg corresponds to our shin. Its knee is up next to the body. Within the "arms" or within the wings you will find the wrist and hand bones fused together as the **carpometacarpus** and digits. The reduced number of bones, once again, allows for a lighter structural framework.

The sternum, or breastbone is called the **keel**. This reference pertains to the shape of the bone. Its flat sides allows the chest muscles to attach in a way that enables the bird's body plan to have an aerodynamic or streamlined effect. It somewhat resembles or acts as the keel of a ship that runs along the centerline of the vessel's bottom hull to increase stability without increasing drag. Another skeletal structure important to flight is the **furcula**. This feature is found at the base of their neck to strengthen the pectoral girdle and helps support the crop, their food storage organ. The furcula is what some people call the "wish bone"!

Find the keel and furcula of this bird in flight.

Muscles

Half of a bird's body weight is muscle. Most of the muscle mass exists in the center of the bird's body. This arrangement allows for increased leverage and streamlining for flight. The largest muscle is the **pectoralis**. This powers the downstroke and takes the most muscular activity to pull from a distance – it increases leverage and streamlining. The **supracoracoideus** muscle powers the upstroke.

What are you eating when you eat meat? You are consuming the animal's muscle. Have you ever noticed a difference in the color of bird meat? Or have you been asked if you prefer dark meat or white meat? If you are a meat eater, then you probably know what I am referring to. Consider the birds you typically find at the grocery store or restaurant—chicken and turkey. The "dark meat" will be the "drumsticks" or legs. The "white meat" is the breast muscle.

Dark meat indicates that those muscles were used and depended upon extensively. Dark meat has fat and myoglobin-rich muscle that can transport the necessary oxygen throughout the muscle fibers efficiently. Migratory birds have dark meat throughout their entire body to help them with long, sustained flight. White meat is glycogen-loaded (sugar). These muscles do not get exercised as much yet are used for quick, short flights. Birds rely on these muscles for quick action but they cannot sustain that energy and will fatigue easily.

Bird Eggs and the Young

Another bird characteristic we identify with as having is egg laying. Birds also lay "land eggs" (amniotic eggs) like reptiles, but will have a hardened, calcareous egg shell to protect it. Once the fluid inside the egg is metabolized, the bird starts to peck through the air pocket that had formed. When birds hatch, they first

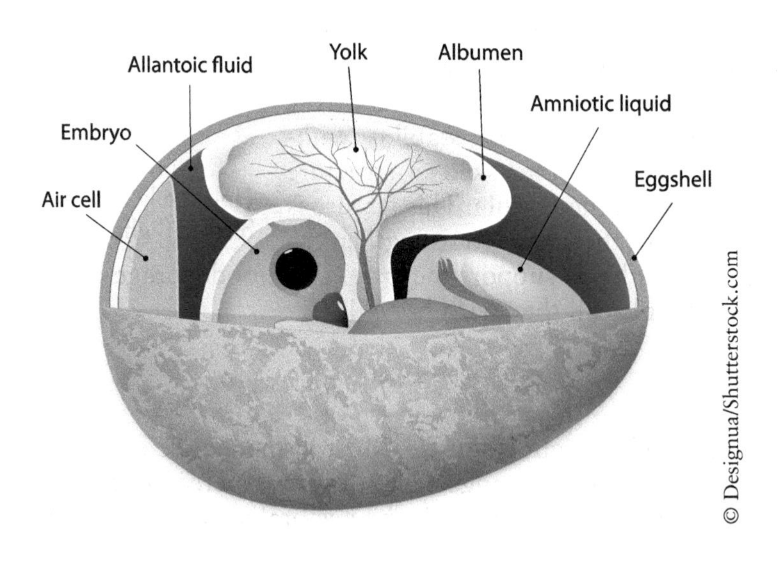

© Designua/Shutterstock.com

poke their bills into the air space, which has formed as water was metabolized. They peck through the shell with their **egg tooth,** which eventually will become absorbed into the beak. Turtles also have this feature. This is another shared characteristic between birds and reptiles.

Eggs often hatch at the same time. Many chick species communicate with clicks to synchronize hatching. One will start clicking and it usually triggers others to begin the hatching process. Why don't we find many bird egg shells laying around? Parents will often remove shells as soon as they can by either consuming them or making them less conspicuous to not attract predators.

Altricial hatchlings waiting to be fed.

Precocial mallards (*Anas platyrhyncos*) with mother.

Regardless of how the baby bird looks when it hatches, the **hatchling** is featherless and helpless (**altricial**) or feathered and raring to go (**precocial**), they will reach adult size within one year. The time of **fledging** varies among species. Fledging is the time between hatching and leaving the nest. **Nestlings** are young who are still in the nest and completely dependent on the parents. Whereas, **fledglings** are those young who have the capability to fly but still depend on parents for food. All songbirds will have altricial young and waterfowl will have precocial young. Most birds tend to their young until they fledge.

Physiology

The circulatory, respiratory, digestive, excretory, and nervous systems are discussed in this section.

Circulatory System

As expected, most animals who expend the energy to care for young will be endothermic, or warm blooded. These are animals able to metabolically control their internal body temperature. How can they control their temperature? In simple terms, they have an effective four-chambered heart with two atria and two ventricles complete with a ventricular septum. This allows for double circulation where the right side of the heart is responsible for pulmonary circulation and carries blood that lacks oxygen to the lungs, and the left side is responsible for systemic circulation and carries blood rich in oxygen to the rest of the body.

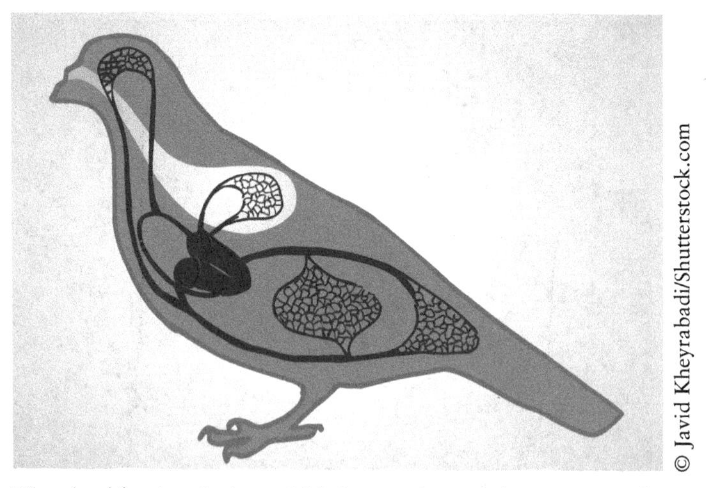

The double circulation of birds consists of deoxygenated (blue) and oxygenated (red) blood flow.

The heart rate is relatively high in birds with most having 220 beats/minute and may be up to 1,000 beats/minute when under stress. Hummingbirds can have a heart rate of up to 1,200 beats/minute. Compare these rates by trying to tap your finger that many times in 1 minute! Humans have an average rate of about 70 beats/minute. Birds' body temperature is about 102–112°F (34–44°C). That's about 10° warmer than human body temperature.

Since birds are warm-blooded, this means they will have a different response to cold temperatures than cold-blooded animals. You will discover that a bird's reaction to cold is very similar to how we would react—like shivering, seeking warmth, and reducing activity. Responses to cold stresses involve shivering, mainly of pectoral muscles and fluffing of feathers to create air pockets for insulation. Sometimes, birds reposition dark feathers toward light to help absorb solar energy for heat. Birds will also select warm microclimates such as roosting in evergreens or tree holes, some burrow into snow, like chickadees and grouse. Many can induce hypothermia (lowering body temperature) or torpor (extreme hypothermia) to avoid losing too much heat (hummingbirds do this at night and can drop their internal temperature to 36–54°F or 2–12°C). Warm-blooded animals also will respond to heat stresses in certain ways. Birds, in particular, will pant to allow for evaporative cooling, seek shade, bathe in available freshwater, and decrease their activity during the heat of the day.

Respiratory System

Birds have the most complex respiratory system of all animals. First, they respire through the nasal openings on top of their beak (and can breathe out of their mouth when necessary). Then, approximately 75% of the air inhaled bypasses

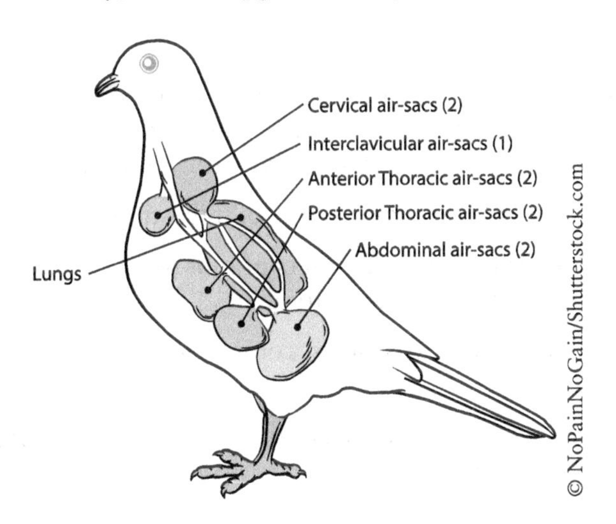

the lungs and flows into posterior air sacs. The air sacs are extensions from the lungs that connects with air spaces in bones and fills them with air. The air sacs also aid in cooling. The other 25% of the inhaled air travels into the lungs. Upon exhalation, air is expelled out of the lungs and the stored air from the air sacs fill into the lungs. The lungs are small and efficient with no dead air spaces.

A constant supply of fresh air is available to the bird during both inhalation and exhalation. This type of respiratory system makes flying at great heights and distances or expending high energy very efficient and almost effortless. The respiration rate of birds is high, 75–300 breaths/minute compared to human's 12-16 breaths/minute. The higher respiration rates are typical of small birds, or birds under stress. Turkeys breathe about 7 breaths/minute, whereas hummingbirds breathe 143 breaths/minute. Lastly, we cannot forget the most beautiful part of the respiratory system—the syrinx. This is an organ found at the junction of the trachea and bronchi, and is used in sound production.

Digestive System

In terms of the digestion system, features and organs exist that may not appear in other animal classes. Familiarize yourself with the parts of this system, know each one's function, and find them on the diagram. You may discover some fun facts along the way.

beak/bill: highly modified by evolution, depending on what animal the eats and its niche

tongue: shape varies with bird's diet; has a bone in it to help manipulate food!

esophagus: an expandable muscular tube to allow food passage

crop: food storage organ

> Notice where the crop is located. Did you remember that the furcula (wishbone) supports the crop? The crop stores the animal's food. The bird would not want this organ jostling around during flight!

glandular stomach: secretes enzymes for food digestion

muscular stomach (gizzard): has three sets of muscle fibers running in different directions; contains grit for grinding food

small intestine: enzymatic digestion is completed here and digested food is absorbed

large intestine: water, minerals, and vitamins absorbed here; compacts/prepares solid waste for elimination

The Digestive System

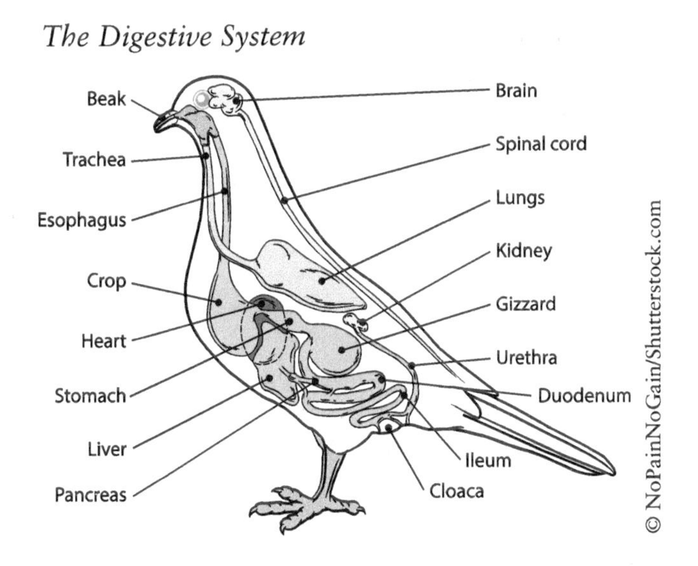

Excretory System

Birds do not have a urinary bladder (like reptiles). Instead, liquid wastes are reabsorbed, wastes are processed to form uric acid paste, and excess water is retained. They have a three-lobed kidney that functions to filter wastes, particularly uric acid from the blood. The cloaca is an orifice for all excrement and reproduction. Find the kidney and cloaca on the diagram. Lastly, the uropygial gland or "oil gland" (on the pygostyle, or short bone in the tail) is an oil gland used by birds in preening their feathers. Their skin, however, contains no sebaceous glands.

Nervous System

For their nervous system, birds have a very well-developed cerebellum, which results in excellent coordination. Have you ever noticed birds on a wire or tree branch? Bird responses to situations are generally stereotypic, mechanical, and instinctive. A bird's aptitude is also species specific. Corvids, especially crows and ravens have the capacity and ability for learning, responding to the environmental conditions regardless of previous knowledge, and communicating interspecifically (between different species). Some studies suggest they have the intelligence level as the average 7-year old human!

Vision plays a prominent role or is considered a dominant sense. Birds have intense vision—their multiple fovea, or vision centers in the retina where the photoreceptors are concentrated allows for incredible focus. Humans have a single fovea in each eye. Predators and potential mates rely on their vision to see color to find prey and partners, respectively. In diurnal predators, birds tend to have a large number of cones within the eye that is responsible for seeing color vision. This is 6–8 times the cone density in the retina than in the human eye. Nocturnal predators have a large number of rods to discern images and shapes. The other senses of touch, taste, or smell do not seem as dominant or highly developed except the sense of hearing in owls. Owls rely on their keen sense of hearing at night to hone in on their prey.

Bird Behavior

Birds have various behaviors worth mentioning that will help you improve your naturalist and interpretation skills. The behaviors discussed here include territoriality, vocalizations, mating, nesting, and migration.

Territoriality

An animal's territory contains cover, shelter, food, and a display area for the male to advertise to a female for mating readiness. In the spring, you will notice birds behaving aggressively. Males of many species defend an area in which they will court or try to attract a female, and potentially breed and raise the young. During these times, birds, typically males will act territorial by flying to the claimed boundaries, making certain vocalizations for communication, and possibly displaying or showing off their puffed-up feathers.

Tom turkeys (males) will display or strut in a great visual and auditory spectacle (left). They puff out their feathers and fan their tail to intimidate rivals and attract potential female mates in the spring. You may also witness similar behavior with male red-winged blackbirds as it displays its red shoulder badges while defending its territory and trying to attract a female (right).

Vocalizations

Why do birds make sounds in the first place? A vocalization indicates communication. It can also indicate the type of bird. Birds "sing," but only within the Order Passeriformes, or songbirds (also referred to as the perching birds), which are nearly half of all bird species. The well-developed syrinx enables them to express a variety of vocalizations with many birds having a melodic song to sing during breeding season to attract a mate. These songbirds and the nonpasserines make other noises, like chip notes or calls for many reasons including to hold a flock together in dense foliage or during nocturnal migrations, to intimidate and drive away enemies or competitors, to convey information about food or predators, and to serve as an identification "password." Try to witness as many vocalizations as you can! What kind of bird made them? Can you imitate the song or call? Learn about other differences between bird orders in the field guide section.

Black-capped chickadees (*Poecile atricapillus*)have a wide repertoire of vocalizations.

Nature Journal: How do you go about learning bird songs and calls? Watch the birds actually vocalize! It can give a person great joy to observe a bird call in nature. It's one thing to hear it, but a whole different experience when you see them sing! Spend time at your Sit Spot. Hear a bird call? Go find it and watch it vocalize! This is a fun and rewarding way to learn your bird calls. Even if you know who is making the call, try to find it. You may need binoculars for this activity, or simply go look for it.

Report your findings. Try to write out what the call sounded like. In which direction did you hear it? What was the habitat? Did you find who was making the call? Where was it calling from? Try to identify the bird. Complete a Natural History Journal Page for your species of bird (found at the end of this section).

Mating Systems

Most birds are monogamous. This means they will form a pair bond with only one individual for a lifetime. The term can also describe the behavior for just one season in some birds. They will be exclusive with that other individual and carry out roles to prepare and maintain a nest and help raise the young before fledging. Sometimes males of certain species are polygamous and will form bonds and mate with multiple females.

Two other bird mating systems also exist—polyandrous and lek mating. Polyandrous is when the female forms bonds and mates with multiple males (<1% of species). Females defend territories, compete and court males, have a brighter color than males, and the males incubate and take care of the young. Contrarily, lek mating is not related to territories, and mating is random with no pair bonding. Males display in "courtship arenas" or in communal areas called leks, and females visit and mate.

Nests and Nesting

It's fun to find nests in nature. Look at where the nest is located. What was used in its construction? What is its shape? All of these nest characteristics will help you identify what bird species made the nest. Regardless of the species, most females construct the nest. The males may help scout, protect, or search for specific items for construction, but the females will put it together and use their body to form the cup-shaped egg space.

Some natural materials you find in nests may include twigs, grass, lichen, leaves, mud, rocks, spider webs, feathers, hair, or snakeskin. Birds will also use whatever is available like twine, ribbon, plastic, or yarn. Some birds dig into the ground for their nest. Many cavity nesters put specific plants in nests to ward off parasites. There are variations of nests based on the individual, circumstances, location, and the materials available.

The height of which a nest is from the ground will indicate the hatchling's level or ability for leaving the nest after it hatches. If the nest is found on the ground, then you know the young are precocial and will have the ability to leave the nest soon after hatching. The nest found up high and secluded indicate altricial young.

Shapes of nests will also indicate who may have constructed it. The main types you will find include ground nests, platform, cup, or pendulous nests. **Ground nests** occur in a shallow depression without nesting material (i.e., killdeer). **Platform nests** will have a splayed arrangement of twigs that are not anchored to anything and have a small depression for the eggs. These nests

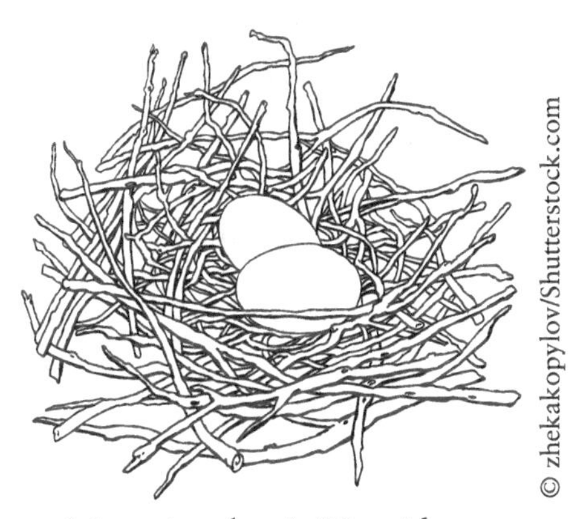

Mourning dove's (*Zenaida macroura*) platform nest.

American robin's (*Turdus migratorius*) cup nest.

Baltimore orioles (*Icterus galbula*) and their pendulous nest.

will be found either on the ground, near or on water, or elevated (i.e., Canada goose, waterfowl, and mourning dove, respectively). Most songbirds will have a **cup nest** supported to a structure like a tree or ledge. Construction begins like a platform, it gets anchored, then a softer lining and depression is made in the center by the female. A **pendulous nest** resembles a pouch hanging from a branch. Woven tightly, it can withstand strong winds and rain. Orioles and kinglets make such nests and access them from a small opening on the top or side. Another type of nest is the **cavity nest** where a tree trunk or branch is excavated so the bird can take refuge. Woodpeckers typically construct these areas and only live in it once (species dependent), but other birds and animals can use it in the future.

Nature Journal: The next time you see a nest, make some interpretations. Try to explain what you see. Sketch it. What is its size and shape? What materials were used? Where was it found? If you do not know who constructed it, try to narrow down who made it by measuring the inner portion, inventory what birds are nearby, make some deductions, and do research on nests of those birds. If you cannot tell if the birds are still in it, spend time watching the nest to make that determination or use a pair of binoculars, a camera, a smartphone, trail camera, or a telescopic mirror to get a view.

Nest predation is a problem. Mammals, snakes, and other birds prey upon bird eggs and young. In response, birds may hide nests and actively protect nest by aggression or distraction displays. Some bird species, about 13% nest in colonies. This seems to reduce predation and increases their foraging capacity. Conversely, other issues may arise with groups like more parasites and disease inflicting upon the population and increased intraspecific competition for nest sites and mates.

Here's an interesting story about the brown-headed cowbird, *Molothrus ater*. Have you seen it before? If you have, I can guarantee you never saw it with its young. That is because it never raises its young. It gets somebody else to do it! These birds are **brood parasites.** The female waits until a bird of another species leaves its nest for food or water, and swoops in and lays her eggs and leaves. The mother bird host returns to the nest and does not realize the additional, foreign eggs. She will continue to warm all the eggs until hatching. The cowbird young typically hatch first, and will be more aggressive and likely outcompete or kill the host offspring. The cowbird parasitizes hundreds of species' nests.

© Agnieszka Bacal/Shutterstock.com

A brown-headed cowbird (*Molothrus ater*) egg in the nest of an indigo bunting (*Passerina cyanea*) and its eggs.

Migration and Navigation

How fast can birds travel? Flight speeds vary from 5–10 mph (8–16 kph) with song birds to >200 mph (322 kph) in the peregrine falcon's dive, but many will travel great distances regardless to escape inhospitable climates. Birds seek areas where they can tolerate the temperature and find food. Realize that not all birds migrate, and some might only travel a couple hundred miles south to meet their needs.

Birds will migrate to escape inhospitable climates like flying south in the winter to escape cold weather and because of a lack of food. They exploit the seasonal feeding opportunities in the south and north when and where food is plentiful. This type of travel is called **complete migration** and will occur at a specific time of year and has a particular pattern of flight. Sometimes a bird species may travel more than 15,000 miles (24,150 km) in a year! However, a bird can still be considered a complete migrator if only traveling from Canada to the southern reaches of our region, or as long as it is a different latitude and occurring at a somewhat predictable time. **Partial migrators** may only move far enough to find enough food to sustain them with variable patterns each year. The American goldfinch (*Spinus tristis*) is an example of this behavior of **seasonal movement**. Some **irruptive migrators** will have mass movements during times of scarce food or adverse living conditions.

In preparation of migration in the early fall and spring, birds will consume great amounts of food to build fat reserves and double its weight (depending on

Canada geese (*Branta canadensis*) flying in formation.

the species' behavior). Some birds will eat throughout migration stopping at wetlands to replenish their energy. The times of day in which they move is also species dependent. Hawks tend to move during the day, and small land birds travel at night to avoid predation. The skies are often calmer at night, and this also makes travel easier.

Birds navigate by many means. The cues will be species and condition dependent. Visual landmarks are used like rivers, coastlines, mountains, and buildings to get bearings. This is very good for use around their home area but not necessarily during migration. Birds will use the sun as a compass by orienting themselves in relation to the sun's position. The stars are also used to orient them in the direction needed to travel. Many planetarium experiments have been conducted to test this hypothesis. Additionally, olfactory cues or the sense of smell can help them find their way back to nests after a very long day of foraging. Studies have also shown geomagnetic fields of the earth are used as a map of horizontal space. Research on the geomagnetic field reversal and wing magnet experiments provide us with insight on the birds' ability to utilize this as a flight reference.

Nature Journal: Start a phenological study of when migrating birds return in the spring. See Chapters 1 and 2 for ideas on making observations and setting up a study. What kinds of behaviors have you noticed taking place between birds? If you have a bird feeding station, you will have greater insight of species present nearby and their tendencies. Field investigations involving birds can be insightful and fun!

Bird Conservation and Management

Birds provide us with a bioindicator of our natural resource health. Having clean water, air, and a diverse native landscape will encourage bird diversity. Birds also contribute to local economies through birding, photography, and hunting opportunities for people. Your state or provincial agencies, and the federal agency, U.S. Fish and Wildlife Service (USFWS) manages migratory birds and their habitats.

What Can We Do to Help the Birds?

Any type of permit you can purchase that relates to our wildlife, land, or water will likely help fund management efforts. You can buy a Duck Stamp through the USFWS to help with funds toward the management of wetlands and other habitats used by waterfowl. The stamp is an annual requirement for waterfowl hunters, but anyone can purchase them.

Want to learn more about birds? Join a birding group. This is a great way to become a better birder. It can also help you stay current with the latest concerns or topics regarding birds. Research the community science opportunities for bird counts. Often, they will take place for a few days in December or February, they may run for months into the spring, or the monitoring may happen during peak migration. They are fun and easy to do for people of all ages. This is an opportunity to learn to identify the birds near you and submit data to help scientists! You can also find many websites, social media pages, and blogs about birds. Discover apps for your devices to learn calls, identify birds, or to record your findings.

Another thing that many of us do is to have bird feeders. Contrary to what some believe, birds do not depend upon bird seed for their only food source. This is a great way to get to know the birds around your home. However, it is important to feed the birds responsibly. This means, keep your feeders clean and feed only the birds.

Choose feeders that are species specific or make it more of a challenge to get the seed out. You do not want a lot of excess seed spilling over. This draws in other types of animals, like mammals (e.g., squirrels, deer, raccoon, bear). These animals may become dependent or comfortable with receiving food from humans, and can attract too many animals to one, small location. These are issues that can lead to them becoming a nuisance, crowding out other animals, spreading diseases, or threatening human safety. You would also lose money by wasting seed or having to replace destroyed feeders. The best practice is to bring in feeders at night. Another thing you can do to help with the preservation of existing bird populations is to keep your pet cat indoors or contained in a yard. Studies have shown that they have a devastating impact on bird populations.

Lastly, consider landscaping for birds so that you do not "need" feeders to attract the birds. Provide diverse vegetation types and sizes, houses, and a water source Use native plants when possible to support healthy habitats. With anything you provide the birds, be sure to also keep it clean on a monthly basis. Research more on what you can do where you live to help the conservation of these animals.

Making Connections

Many ecological connections can be made when studying any group of organisms. Each organism type can be used as the "test subject" or focus of study while enlarging the scope to include other scales of life.

Birds are found everywhere. They have an obvious presence in the habitat with either their physical appearance, sounds, or behavior. Birds can capture the attention of anyone and can be used for learning more about nature. Visit Chapter 2 to help you think about field investigations and how to inventory the birds at your feeder.

Consider the following prompts to help you make connections between birds and the broader ecosystem perspective of "place". How does agriculture and economy play a role in relation to the presence of certain birds? What is the impact of invasive species? Identify the predator-prey relationships in a habitat. What are the food webs present? Are there any endangered species nearby? How are habitats fragmented in the surrounding area? When does migration begin? How does geography affect the birds? And, what are the adaptations of different birds? If you start asking questions about your bird observations, you may soon be wanting to learn and observe more birds, become curious about nature, spend time outdoors, and make choices to improve the environment.

Nature Journal: Wherever you live, discover the birds that frequent your yard. Do what you can to create a diverse habitat. Find out the birds living around your home, and complete Natural History Journal Pages for each one, discover their song in person, and start keeping track of the birds that you find.

Mark the birds you discover in your field guide or in a designated section in your Nature Journal. This can become a souvenir of the places you visit. Make notes on all the things you encounter in nature and record notes about what you observed, when and where you observed it, and make some interpretation of the behavior. Think about field investigation questions and carry out a study (see Chapter 2).

Happy birding!

REVIEW

Review Questions

1. Which came first, flight or feathers? Explain. Feathers are actually highly modified ___ used for __________.
2. *Archaeopteryx* is the earliest known _____. Fossils were found in _____. It lived ______ years ago. Most scientists believe this to be the evolutionary transition animal between ______ and ______. What characteristics were similar to dinosaurs? What features were most like modern birds?
3. Without doing genetic testing, many bird species can easily be distinguished based upon what things?
4. What is an endotherm? How can they tolerate cold and heat stresses?
5. Describe a bird's egg. Describe the hatching process for a bird. What is synchronized hatching? What is the term for egg laying? Differentiate between altricial and precocial. How do their nests differ?
6. Describe each of the feather types by using just *one to two* adjectives. What structures are involved in the "zipper affect" or smoothing/sealing the feathers so that they are air-worthy? What is an airfoil? Diagram a feather.
7. Discuss the different types of plumage that a bird may display (nuptial, eclipse, and winter plumage). What is sexual dimorphism? What structure adds pigment and color to developing feathers? What are biochromes? When a bird appears a different color than it is due to the light's reflection off its plumage, this is called ____.
8. Describe the bones you would find within a bird.
9. What makes up half of a bird's body weight? Where are they mainly located and how are they used? Describe the flight muscles. What is the difference between "dark" and "white" meat?
10. Discuss the three respiratory structures of birds: lungs, air sacs, and syrinx.
11. What is the dominant sense of most birds? Give the function of the digestive system structures: bill, esophagus, crop, glandular and muscular stomach, and small and large intestine. What is the function of the bird's three-lobed kidney? Birds do not have a urinary bladder. What happens to their liquid waste?

12. What is a bird's "territory"? How do male birds defend their territory?
13. Discuss the various mating systems in the bird world (monogamy, polygamy, polyandry, and lek mating).
14. What three things would you examine when trying to identify a bird's nest? What would contribute to variation in what you'd expect for a particular bird's nest? Identify these nest structures: ground nest, platform nest, cup nest, pendulous nest, no nest, and cavity nest. What is nest predation and brood parasitism?
15. What is migration? How do birds prepare for it? Describe five ways birds navigate during migration. Which bird migrates the longest distance per year?
16. From the field guide section, how can you tell the difference between a passerine and nonpasserine? What are some tips and tricks to help with the identification of birds?

Field Guide: Birds

© icon Stocker/
Shutterstock.com

Start birding any time! The times to see the most birds are in the early morning of the spring and summer when birds tend to sing and forage (evening is another good time for viewing). Stay alert and watch and listen for bird activity. Try to make observations of these birds, and definitely try to seek out the ones you hear.

When faced with a bird that is unfamiliar to you, first try to determine what "group" of birds it belongs. The general groups of birds fall into these orders:

ducks, geese, swans
herons
hawks, falcons
owls
woodpeckers
perching birds

Groups are based on individuals in the same taxonomical order or family that tend to have the same or similar shape and posture (silhouette), flight patterns (sweep), and behavior (signs).

Order Passeriformes, the songbirds, or perching birds, comprise approximately half of all bird species. A nonpasserine will not have the singing capabilities and will have other morphological differences. Passerines have the toe structure to grip a perch for extended periods of time and will never have webbed feet. Just because some birds will be in trees, does not indicate a passerine. Hawks, doves, and hummingbirds can be seen on a tree limb and are sometimes mistaken for passerines. Can they sing? No, they make calls but do not have a varied melody. Use the process of elimination to get down to at least the order or family of a bird, **"If it's not a perching bird, then is it a ...?"** (or if it does not fit into the other orders, then it must be a passerine!). Other important features or factors to consider when trying to make a good bird identification are its size, shape, silhouette, field marks, flight pattern, foraging style, and the habitat in which it was found.

Determine the Bird Size

Use familiar birds and objects (your arm works perfectly) to estimate size. Size is closely associated with shape and used to determine the bird group. Once you know which group of birds you are observing, then size can help determine the species. A quick and useful reference for determining size is to use your fist, hand, and forearm to make comparisons to the bird size you are observing!

Once you become familiar with common birds or the birds near you and know their sizes, you can start comparing other birds to those common birds as reference points to help with identification. For example, you could compare the bird in question to the size of your fist or the relative size of the house sparrow (HOSP), or the size of a hand and the American robin (AMRO), and perhaps the size of a forearm to the American crow (AMCR). Is the bird you are looking at the same size as a HOSP? Larger than an AMCR? Is the bird you are looking at larger than an AMRO but smaller than an AMCR?

When you first start identifying the birds near your home or in a nearby park or nature area, you may discover woodpeckers—a common and very recognizable bird type. You will find woodpeckers on the trunk of the tree. If you are looking at a woodpecker, you might see one that looks to be the same size as a HOSP. It may likely be a downy woodpecker. One that is slightly larger than an AMRO might be a northern flicker. One that is the same size as an AMCR might be a pileated woodpecker.

The method of comparing bird sizes to your hand or arm and to other bird species is helpful, and it will take practice to become proficient with your estimates. Lighting and distance can alter one's perception of size, but you will get better with practice. Another approach for sizing birds (and other animals) is to measure it against itself! You do not need another bird for comparison.

For the downy and hairy woodpeckers, you will notice that the colors and markings are quite similar. However, the downy woodpecker is smaller than the hairy. At first, you may have difficulty discerning between them if you cannot see

Female downy (left) and hairy woodpeckers (right)—*Dryobates pubescens* and *Leuconotopicus villosus*, respectively.

them next to each other to compare the sizes. So instead, compare the bill to the size of the bird's head for each of the woodpeckers. The downy's bill is about a third of its head, and the hairy woodpecker's bill is nearly as long as its head. It becomes obvious which bird is which when you have this perspective. Neat trick!

What is the Bird's Shape?

Look at the shape of the bird to determine to which group (or family) the bird belongs. What are the bird's proportions, and do you notice any unusual aspects of the bird? For example, you might ask yourself whether the bird is noticeably chunky or slender. What is the size of the bird's body parts in relation to its other parts? Is the tail longer than the rest of the body, shorter, or the same length as the rest of the body? Is the bill long in relation to the head? Do you consider the bill thick or thin? Consider the European starling (look in the index to find its page number in the field guide quickly). You might agree that the bird is a seemingly chunky, squat bird, and the tail is hardly noticeable at all.

In review, keep these things in mind when studying a bird: What are the proportions of the bird? Describe the size or length of wings, tail, legs, and bill in proportion to the rest of the body. How does the bird perch or stand? Its posture is an important consideration.

Silhouettes

Determine the bird's silhouette or outline while its at rest and in flight. Learn these silhouettes and practice them while birding. Based on a bird's profile from a distance, notice how the bird holds itself. Does the bird hold itself more vertically or more horizontally? In flight, focus on the wing's size and shape. Knowing how to identify a bird (at least to group) based on its silhouette can often allow you to identify birds rather quickly and in difficult lighting situations. With any bird you see, practice describing its size, shape, and silhouette. These are characteristics you can witness almost every time you see a bird. It is only when you can see the bird clearly that you can pick out its key features or field marks.

Field Marks

Field marks are markings or features on a bird that help distinguish them from other birds. Many field guides emphasize or point out the field marks that make a species unique or more noticeable than others of similar species. A key to identifying a bird is to pay attention to details—even take notes about the details before you try to look it up in a field guide or online. Use the following diagram to help you learn how to reference different parts of a bird.

Body Parts of a Bird

Patterns in Flight

Spending the time to observe flight patterns of birds can help you make an identification. This is especially true in the case of birds in which it is difficult to size or see key field marks. Here are a few great examples. The American crows flap slowly and methodically, whereas common ravens take frequent breaks from flapping to soar or glide. Finches and woodpeckers will have a flap and fall pattern to their flight. Hawk groups have distinctive differences in flight. Accipiters will have the pattern of flap–flap–glide–flap–flap, and so on, and buteos will soar–circle–flap–soar, and so on. A group of birds traveling together in general terms is called **flocking.** Flocking is a characteristic or behavior that you will use to narrow down your choices when making identifications. A flock of white birds near the water are probably gulls.

Activity: Learn the flight patterns and behaviors of the common birds around your home.

Foraging Style

Another behavior to take note of when trying to make an identification is the foraging style, or how animals go about acquiring their food. Does the bird spend most

of its time searching for food on the ground in the open, among the base of herbaceous plants or of trees? Or does it forage the heads of plants or tips of branches? On the trunk of a tree? You will often find two similarly colored birds together in our region, the chickadee and nuthatches. However, the chickadee will spend more time at the ends of branches compared to the nuthatch who will spend its time on the tree trunk (this is the only bird who moves head first down the tree!).

Habitat

Pay attention to where birds are typically found, or their preferred habitats. Spend time in different areas and keep a list of the birds you discover in each area. Use this information to apply to other similar habitats. This will help you make quick identifications. Again, diverse habitats will likely offer a plethora of wildlife to observe, especially birds. Have your binoculars at the ready when you visit fields, forests, and wetlands.

Common Birds to Know

The list provided here refers to the birds that tend to frequent bird feeders or may be commonly encountered in this region. This is a great place to start learning your birds. You will likely have some, if not many of these inhabitants near your home. Tailor the list to fit with what you observe. Make it a goal to find all of these in the wild. Treat this as a checklist. From there, move on to the other birds in the field guide.

- ☐ blue jay
- ☐ northern cardinal
- ☐ black-capped chickadee
- ☐ white-breasted nuthatch
- ☐ tufted titmouse
- ☐ American goldfinch
- ☐ house finch
- ☐ house sparrow
- ☐ American tree sparrow
- ☐ song sparrow
- ☐ dark-eyed junco
- ☐ American robin
- ☐ European starling
- ☐ American crow
- ☐ red-winged blackbird
- ☐ rose-breasted grosbeak
- ☐ ring-billed gull
- ☐ mourning dove
- ☐ downy woodpecker
- ☐ hairy woodpecker
- ☐ red-bellied woodpecker
- ☐ red-tailed hawk
- ☐ mallard
- ☐ Canada goose

Bird Species Accounts

Use the following key to find the section where you may find your bird in question. Match the overall shape of your bird with the silhouettes pictured here and go to the corresponding number listed to find the section where you might find your bird.

Bird Identification Key

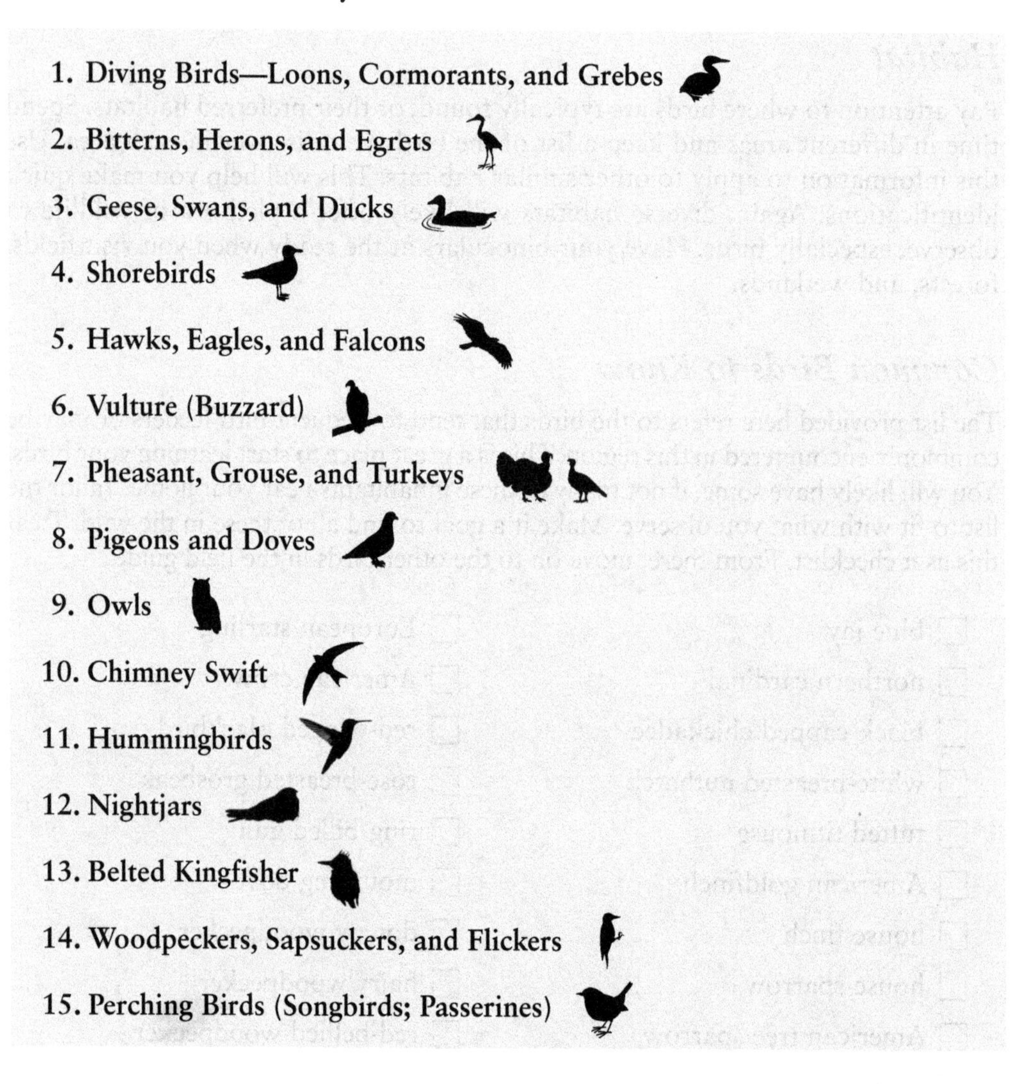

1. **Diving Birds—Loons, Cormorants, and Grebes**
2. **Bitterns, Herons, and Egrets**
3. **Geese, Swans, and Ducks**
4. **Shorebirds**
5. **Hawks, Eagles, and Falcons**
6. **Vulture (Buzzard)**
7. **Pheasant, Grouse, and Turkeys**
8. **Pigeons and Doves**
9. **Owls**
10. **Chimney Swift**
11. **Hummingbirds**
12. **Nightjars**
13. **Belted Kingfisher**
14. **Woodpeckers, Sapsuckers, and Flickers**
15. **Perching Birds (Songbirds; Passerines)**

The songbird order has the following families represented in the field guide: *flycatchers, corvids, swallows and martins, chickadees and titmouse, nuthatches, creepers, wrens, kinglets, thrushes, catbird, starling, waxwing, warblers, sparrows, cardinals and grosbeaks, blackbirds and orioles, finches, and house sparrow*

1. Diving Birds—Loons, Cormorants, and Grebes

Mostly medium-sized aquatic birds. Dives into water from the surface or the air to catch fish and other food.

© Brian Lasenby/Shutterstock.com

Common Loon (*Gavia immer*)—
Order Gaviiformes, Family Gaviidae

Size: 28–36 in. (71–90 cm)
Key Features: heavy body; spear-shaped beak; dark head; distinct neck ring; black and white checkered plumage on its back (mostly gray in winter along coast of continent); white underside; red eyes
Habitat: forested lakes; rivers, coastal areas
Notes: listen for its unique, laughing yodel; swims well and for long distances under water; legs toward rear of body to help dive deeply; almost hopeless on land; flies adequately, slow with take-off; five species worldwide and in North America

© Brian Lasenby/Shutterstock.com

Double-Crested Cormorant
(*Phalacrocorax auritus*)—Order Suliformes,
Family Phalacrocoracidae

Size: 33 in. (84 cm)
Key Features: large, dark body; thin yellowish-orange bill (sometimes gray) with hooked tip; orange throat; long neck
Habitat: coasts; freshwater
Notes: flocks in silent flight with frequent sailing; breeds in colonies along rocky shorelines; dives for fish with extended time below water; makes piglike grunting sounds in colonies; complete migration

© Steve Byland/
Shutterstock.com

Pied-billed Grebe (*Podilymbus podiceps*)—Order Podicipediformes

Size: 13 in. (33 cm)
Key Features: small size; chunky shape; brown body; thick neck; white undertail; short, white bill with black ring near tip (breeding)
Habitat: wetlands with open water; bays
Notes: common in summer; dives for food; sensitive to pollution; awkward on land; voice has a quick, loud whooplike call then slows; complete migration

2. Bitterns, Herons, and Egrets—Order Pelecaniformes, Family Ardeidae

Medium to large-sized water birds found globally; have feet with webbing between all four toes; long legs and found in aquatic habitats. Additionally, these animals are carnivorous and await the opportunity to spear prey while standing motionless or slowly stalking it. Most members are at least partially migratory with complete migration for the examples listed.

© vagabond54/Shutterstock.com

American Bittern (*Botaurus lentiginosus*)

Size: 25–30 in. (64–76 cm)

Key Features: brown and white pattern (excellent camouflage); brown streaking from chin through breast; white throat; black streaks from bill down neck; outer wing areas dark in flight; short tail; yellowish, short, stout bill; yellow legs and feet

Habitat: wetlands

Notes: its motionless "freeze" pose often observed with bill pointing upward; makes a "pumping" sound as it constricts its swollen neck and expels the swallowed gulps of air; more crepuscular or nocturnal of the family

© Kaewta Yong/Shutterstock.com

© Allexxandar/Shutterstock.com

Great Blue Heron (*Ardea herodias*)

Size: 42–52 in. (107–132 cm); wingspan up to 6 ft. (2 m)

Key Features: large, blue-gray body; long bill; long neck (s-shaped especially in flight)

Habitat: freshwater; salt marshes

Notes: common; stalks for fish (or anything of that size) in shallow water; can also float on water (and take off from that position); barks like a dog if startled; makes occasional deep honking sounds when taking off; nests in colonies in treetops

© PACO COMO/Shutterstock.com

Great Egret (*Ardea alba*)
Size: 38 in. (96 cm); wingspan up to 4.5 ft. (137 cm)
Key Features: large white body; long, thin neck (curved in flight); yellowish-orange bill; long, black legs and black feet; during breeding will have white plumes from throat and rump with green skin patch between eyes
Habitat: wetlands
Notes: nests in trees; voices a low, but loud, rapid *cuk-cuk-cuk*

© sandymsj/Shutterstock.com

Green Heron (*Butorides virescens*)
Size: 16–22 in. (40–56 cm)
Key Features: small bird (in relation to other herons); overall brownish-gray with green iridescence; red-brown neck; sharp bill; relatively short, yellowish-green legs (bright orange during breeding); immatures have heavy streaking along neck and underparts
Habitat: lakes; wetlands
Notes: very acrobatic when catching fish and other prey (small land animals); known to use bait (twigs, leaves, etc.) on top of water to catch fish; raises crest when excited; nests singly; generally silent but will vocalize loud and quick in alarm or aggression

3. Geese, Swans, and Ducks—Order Anseriformes, Family Anatidae

All species highly adapted for aquatic existence at water surface; web-footed for efficient swimming (some are more terrestrial); feathers excellent at shedding water (oils); bills flattened (mostly); some species are **dabblers** (upturning their short tail completely as they submerge their head to feed in the water below) whereas others may be diving ducks, or **divers** with shorter legs with large webbed feet to propel them underwater completely; 131 species worldwide, 61 in North America.

© Vladimir Wrangel/Shutterstock.com

Canada Goose (*Branta canadensis*)

Size: 25–43 in. (63–109 cm); wingspan up to 5.5 ft. (168 cm)
Key Features: large, dark gray-tan body with long black neck; white cheek strap; dark tail; white undertail
Habitat: lakes; ponds; bays; parks; grasslands
Notes: common; males often stand at edge of group and may show aggression (bobs head, lunges forward, hisses); eats plants, insects, seeds; tall native grasses helps deter groups congregating; honks in flight; nonmigratory to partial

© Drakuliren/Shutterstock.com

Mute Swan (*Cygnus olor*)

Size: 5 ft. (152 cm); wingspan up to 8 ft. (244 cm)
Key Features: large white body (immature gray); s-curved neck; orange bill with black tip and black knob at base of upper bill
Habitat: lakes; wetlands; sheltered coasts
Notes: nonnative, invasive (introduced in mid 1800s); aggressive toward waterfowl and often displacing natives; threatens common loon and trumpeter swan populations; wings make a loud hum in flight; usually silent unless barking or hissing when disturbed; nonmigratory or seasonal movement

Tundra Swan (*Cygnus columbianus*)

Size: 50–54 in. (127–137 cm); wingspan up to 5.5 ft. (168 cm)
Key Features: all white body (gray immature); neck held straight (not s-shaped); black bill with yellow spot near eye; black legs and feet
Habitat: lakes; wetlands; northern part of region (widespread)
Notes: seen in large family groups of 20 or more; high-pitched, quivering voice constantly repeating *ooo-ooo-whooo* during migration; complete migration

© Agami Photo Agency/Shutterstock.com

Trumpeter Swan (*Cygnus buccinator*)
Size: 6 ft. (183 cm); wingspan up to 7 ft. (213 cm)
Key Features: all white body with large, black bill extending to eyes; black feet; neck kinked at base (standing or swimming) and twice as long as body
Habitat: lakes; wetlands with open water; spotty distribution in region
Notes: once hunted to nearly extinction; loud bugling voice; largest species of waterfowl

© Juan Aunion/Shutterstock.com

Wood Duck (*Aix sponsa*)
Size: 17–20 in. (43–50 cm)
Key Features: colorful green head with patterns of white and black; dark back; white chin and throat; breast purplish-brown; tan sides; black and white shoulder sash; female is mottled gray-brown with white, teardrop-shaped eye patch
Habitat: swamps; forest-edged lakes
Notes: dabblers; females nest in tree cavities or nest boxes; ducklings jump out in response to mother's call; cries weep, weep, weep; complete migration

© Paul Reeves Photography/Shutterstock.com

American Black Duck (*Anas rubripes*)
Size: 20–24 in. (51–61 cm)
Key Features: dark brown body; light brown head and neck; yellow to olive bill (male; female); bright orange feet; violet speculum
Habitat: shallow lakes; wetlands
Notes: once the most common duck, now displaced by mallard (males will mate with female black ducks; offspring less fertile); dabbler; omnivore; vocalizes a croak (male) or quack (female); complete migration

© Zocchi Roberto/Shutterstock.com

Mallard (*Anas platyrhynchos*)
Size: 28 in. (71 cm)
Key Features: head and neck glossy green with white neck ring (male); rusty brown chest; gray and white sides; yellow bill; orange legs and feet
Habitat: shallow ponds; wetlands
Notes: dabbler; hybridizes; voices actual quacks (male softer, while female loud and vocal); body heat from brooding hen generates heat to increase growth rate of surrounding grasses; complete migration or seasonal movement

© Steve Byland/Shutterstock.com

© Agami Photo Agency/Shutterstock.com

Blue-winged Teal (*Anas discors*)

Size: 16 in. (40 cm)

Key Features: blue-gray head with white crescent at base of bill; black spotted breast and sides (female all mottled); black bill; dark tail with small white patch; blue forewing patch usually only noticeable in flight

Habitat: shallow lakes and wetlands; prefers dense, short emergent vegetation

Notes: small and speedy in flight with many twists and turns; feeds at water surface for floating seeds and aquatic plants and invertebrates; soft clicks or quacks; complete migration

© Brian Lasenby/Shutterstock.com

Green-winged Teal (*Anas crecca*)

Size: 12–16 in. (31–40 cm)

Key Features: small; green eye patch and chestnut brown head (male); white shoulder slash; cream-colored breast with black spots; green speculum (not always visible); female mottled brown with light belly

Habitat: shallow water

Notes: speeding flocks of twists and turns will uplift from the water's surface when disturbed and return when threat departs; dabbling omnivore; males have a crisp whistle and females a soft quack; partial migrator staying until water freezes

© Paul Reeves Photography/Shutterstock.com

Bufflehead (*Bucephala albeola*)

Size: 13–15 in. (33–38 cm)

Key Features: small size; dark to iridescent head with broad white patch (bonnetlike); in flight, bold white bands near base of wings; female brown with white patch on cheek

Habitat: lakes; rivers; sheltered seawater (winter)

Notes: common diving duck; nests in old woodpecker tree cavities; female very territorial; male voices a growl and females have loud quack; complete migration

© Gallinago_media/Shutterstock.com

Common Goldeneye (*Bucephala clangula*)
Size: 18.5–20 in. (47–50 cm)
Key Features: puffy dark green head with white patch between bill and each golden eye; black bill; mostly black back and white belly (female with brown head and gray mottled body)
Habitat: rivers; lakes; bays
Notes: loud whistling made by wings in flight; male lays head flat on back then catapults it forward with a forced peent sound (breeding); female often lays eggs in other goldeneye nests and cavity-nesting ducks; nests in forests near fishless lakes and feeds on aquatic invertebrates; diving duck; complete migration to seasonal movement

© Belo/Shutterstock.com

Common Merganser (*Mergus merganser*)
Size: 27 in. (69 cm)
Key Features: long, thin body (ducklike); male head dark green (black in some light); bill thin, reddish, serrated; white breast; female has brown shaggy crest
Habitat: lakes; marshes; forested rivers; open freshwater
Notes: dives deep for fish ("fish duck"); has serrations along inner bill; male voice like a harsh guitar twang; complete migration or seasonal movement

© Eleni Mavrandoni/Shutterstock.com

Hooded Merganser (*Lophodytes cucullatus*)
Size: 16–19 in. (40–48 cm)
Key Features: black head with black-edged, white fanned crest (does not wrap around behind head like bufflehead; often lays flat); body black with rusty brown sides and white breast; thin, dark bill; female has tan crest and gray breast
Habitat: forest-edged lakes; forested riparian areas; marshes
Notes: shy; smallest of mergansers; nests in tree cavities; females known to share egg incubation with wood ducks and goldeneyes; low grunts and croaks; complete migration

© Dennis Jacobsen/Shutterstock.com

Red-breasted Merganser (*Mergus serrator*)
Size: 19–26 in. (48–66 cm)
Key Features: large, elongated body; thin, serrated, orange bill; red eyes; shaggy, swooping crest; males with green head and rusty-spotted breast with white collar, gray sides and black and white shoulders (after breeding, male looks similar to female with gray brown overall)
Habitat: large lakes and rivers; rocky shorelines
Notes: diver; nests on the ground; generally quiet; complete migration

4. Shorebirds—Orders Gruiformes and Charadriiformes

Diverse orders of small to medium-large birds living near water that eats invertebrates or other small animals (i.e., plovers, wading birds, gulls, and terns); about 350 species worldwide.

© Melinda Fawver/Shutterstock.com

American Coot or Mud Hen **(*Fulica americana*)—Order Gruiformes**

Size: 13–16 in. (33–40 cm)

Key Features: dark body; white bill with frontal shield, small red patch between eyes, and dark band near tip; red eyes; green legs and feet

Habitat: lakes; ponds; marshes; fields near water; bays; estuaries (winter)

Notes: not a duck; cannot take off directly from water's surface and will instead patter along until they achieve lift (usually dashes to safety rather than fly away); excellent diver and swimmer (may also dabble); grazes on land; large flocks seen (especially in migration, upward of 1,000 gather); floating nests anchored to vegetation; frequent vocalizer *kuk-kuk-kuk-kuk-kuk*; complete migration

© Melinda Fawver/Shutterstock.com

Sandhill Crane (*Grus canadensis*)—Order Gruiformes

Size: 40–48 in. (102–120 cm); wingspan up to 7 ft. (213 cm)

Key Features: tall gray bird (immature tannish-pink); wings often rusty brown; reddish cap; yellow to red eyes

Habitat: open wetlands surrounded by forests (breeding); fields and shorelines (migration)

Notes: one of the tallest birds and capable of flying high; quick upstroke; spectacular mating dance (often flipping sticks and grass into air); deep, rattling calls (often heard before seen); complete migration

© Ray Hennessy/Shutterstock.com

Piping Plover (*Charadrius melodus*)—
Order Charadriiformes
Size: 7 in. (18 cm)
Key Features: sandy upperparts; white underparts; orange legs; breeding will have black-tipped orange bill (black nonbreeding), black forehead band, black "necklace" (occasionally incomplete)
Habitat: sandy beaches; open lakeshores of Great Lakes; Atlantic Ocean shoreline (winter)
Notes: good camouflage; population numbers threatened from shoreline disturbances; uses a strategy to entice invertebrates to the surface by trembling its foot on the sand near water with wave action; clear, whistling melody of *peep-peep-peep-lo*; complete migration

© Keneva Photography/ Shutterstock.com

Killdeer (*Charadrius vociferus*)—
Order Charadriiformes
Size: 11 in. (28 cm)
Key Features: two black bars across upper chest; white collar; dark brown upperparts; white underparts; white forehead with white spot behind eye; in flight, bright red-orange rump and upper tail and wide white wing stripe
Habitat: open areas; coasts; streambeds; wet meadows
Notes: exhibits a broken-wing act to deter potential predators away from the nest and young; classified as a shorebird but found mostly away from water; males build nest of a depression in the ground and defends it (two broods per year); distinctive *kill-deer* call; complete migration

© Brian Lasenby/Shutterstock.com

Greater Yellowlegs (*Tringa melanoleuca*)—
Order Charadriiformes
Size: 12–15 in. (31–38 cm)
Key Features: long yellow legs; long, slender bill slightly upturned; gray mottled back; in flight, white rump and barred tail
Habitat: wetlands; lakes; tidal flats (winter)
Notes: feeds by rushing forward in water; loud, noisy whistle call (series of three to five notes) when approached by intruders; complete migration

© Al Mueller/ Shutterstock.com

Spotted Sandpiper (*Actitis macularia*)—
Order Charadriiformes
Size: 8 in. (20 cm)
Key Features: olive brown back; underside white with round dark spots (breeding); white line over eyes; long bill and long light-yellow legs; white wing stripe in flight
Habitat: pebbly edges of rivers and lakes; coastal areas during migration, winter
Notes: continually tips forward and backward; when disturbed, flies low with wings at or below horizontal; females mate with multiple males, lays eggs in up to five different nests, and only helps the last male incubate and care for the young; voices a crisp *eat-wheat, eat-wheat, wheat-wheat-wheat-wheat-wheat*; complete migration

© Dalton Rasmussen/Shutterstock.com

American Woodcock (*Scolopax minor*)—Order Charadriiformes

Size: 10–12 in. (25–30 cm)
Key Features: chunky, short-legged; very long, sturdy bill; short neck; short tail; dark mottled plumage with buff underparts; light-colored bars on dark head; wings whistle when suddenly takes flight
Habitat: moist woodlands and thickets with nearby clearings; wet bottomlands
Notes: camouflaged unless breeding season when it vocalizes and launches into air (a few hundred feet while wings make a musical flutter) with a circling flight display then plummets to ground in a zig-zag pattern and chirping; a startled or "flushed" woodcock takes off in a straight line; males have a nasal *peent* voice with high-pitched, whistle twitters during breeding; complete migration

© Erni/Shutterstock.com

Ring-billed Gull (*Larus delawarensis*)—Order Charadriiformes

Size: 19 in. (48 cm); wingspan up to 4 ft. (122 cm)
Key Features: white head; pale gray topwings (**mantle**) with black tips (with white spots); yellow bill with black ring around it; yellow legs; immatures mottled with black band on tail
Habitat: lakes; rivers; coasts; landfills
Notes: very common; eats insect, fish, and scavenges for food; high pitched cackle or low, laughing yooks; complete migration or seasonal movements

© Christian Musat/Shutterstock.com

Herring Gull (*Larus argentatus*)—Order Charadriiformes

Size: 23–26 in. (58–66 cm); wingspan up to 5 ft. (152 cm)
Key Features: large gull; yellow bill with red spot on lower mandible near tip; topside wings (mantle) light gray with black tip wings (and white spots); pinkish legs; immature mottled
Habitat: lakes; rivers; coasts; landfills
Notes: more than a scavenger and will crack mollusks by dropping them on rocks and prey on young birds and eggs; loud buglelike *kleew-kleew* or alarm call *kak-kak-kak*; complete migration or seasonal movement

© Dave Montreuil/Shutterstock.com

Caspian Tern (*Hydroprogne caspia*)—Order Charadriiformes

Size: 19–22 in. (48–56 cm)

Key Features: large size; black cap (streaked with white nonbreeding); heavy red bill with black tip; light gray mantle with outer wings blackish underside; white underparts; black legs; shallowly forked tail

Habitat: lakes; wetlands; estuaries; coasts

Notes: largest tern in North America (almost as big as herring gull with similar soaring flight pattern); low harsh voice; complete migration

© alitellioglu/Shutterstock.com

Common Tern (*Sterna hirundo*)—Order Charadriiformes

Size: 13–16 in. (33–40 cm)

Key Features: black cap (nonbreeding no cap but black nape); tail deeply forked; gray mantle with black on outer wing; white rump; reddish legs; red-orange bill with black tip; in flight, wing tips dark gray

Habitat: wetlands; lakes; coasts

Notes: graceful, buoyant flyer; catches fish by diving headfirst; catches insects in flight; noisy colonies; preyed on by gulls and birds of prey; voices high-pitched *keeeee-are*; complete migration

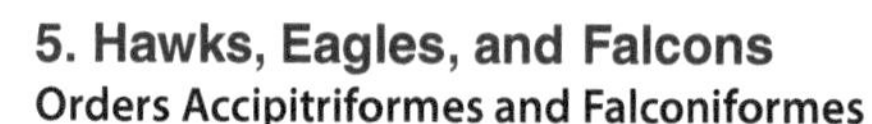

5. Hawks, Eagles, and Falcons

Orders Accipitriformes and Falconiformes

Small to large diurnal birds with strong, hooked beaks; powerful talons and strong legs; feeds on a range of prey; keen eyesight; some are migratory; 233 species worldwide and 28 in North America

Comparing Hawk Types

The different hawks can be subdivided into accipiters (ak-sip'-i-ters), buteos (byoo'-tee-ohs), or falcons. Look at the tail length in relation to the bird's body. In addition, consider wing size. Accipiters and falcons have a long, narrow tail and short wingspan. The accipiters have rounded wings, whereas wings of falcons are streamlined and pointed. Buteos have a short, rounded tail and broad wings. Take note of the taxonomic levels, as you compare and contrast the following bird species.

Osprey or Fish Hawk ***(Pandion haliaetus)*—Order Accipitriformes, Family Pandionidae**

Size: 24 in. (60 cm); wingspan up to 71 in. (180 cm)

Key Features: white underside; dark brown upperparts; white head with dark stripe through eye; long wings, crooked in flight (dark patch underside at bend); dark beak

Habitat: large lakes, rivers, or wetlands

Notes: sights in prey while hovering before plunging feet first into water to pierce fish; carries prey headfirst; nests on dead trees or utility poles; melodious ascending whistles or repeated chips; complete migration

Bald Eagle (*Haliaeetus leucocephalus*)—
Order Accipitriformes, Family Accipitridae

Size: 28–40 in. (70–102 cm); wingspan 6–8 ft. (1.8–2.3 m)

Key Features: large body; white head, neck, and tail; dark plumage; immatures all brown with whitish wing linings

Habitat: open areas; near water; forests

Notes: not bald; eats carrion, waterfowl, and fish; weak squeal or gull-like cackle; partial migrator or season movement

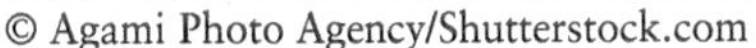

© Agami Photo Agency/Shutterstock.com

© A Zargar/Shutterstock.com

Northern Harrier (*Circus hudsonius*)—Order Accipitriformes, Family Accipitridae

Size: 16–20 in. (41–52 cm); wingspan 38–48 in. (97–122 cm)

Key Features: slim body; dark gray to brown above and white below; long wings and tail; wing tips black; faint narrow bands on tail; white rump; yellow eyes; distinctive owl-like facial disk; immatures with rich brown plumage with streaked underside and dark tail bands

Habitat: open areas; marshes

Notes: glides in zigzags just above ground; wings often held just above horizontal and tilts with wind like a vulture; preens and rests on ground; generally quiet except near nest and courtship with a high-pitched *ke-ke-ke-ke-ke*; complete migration

© Brian E Kushner/Shutterstock.com

Sharp-shinned Hawk (*Accipiter striatus*)—Order Accipitriformes, Family Accipitridae

Size: 9–12 in. (23–30 cm); wingspan 17–23 in. (42–58 cm)

Key Features: small body with gray back and head; rusty red breast; short, rounded wings; long, straight, squared tail with several tail bands; dark bars on pale underwings; red eyes

Habitat: forests; occasionally along rivers and urban areas; nests in bogs or dense conifer stands

Notes: common; swoops at bird feeders; preys almost exclusively on small birds and pursues them at high speeds; in flight, head does not protrude far; silent except in breeding with intense repeated kik-kik-kik-kik; complete migration

© Tom Middleton/Shutterstock.com

Cooper's Hawk (*Accipiter cooperii*)—Order Accipitriformes, Family Accipitridae

Size: 14–20 in. (35–50 cm); wingspan 24–39 in. (62–99 cm)

Key Features: medium-sized body; short wings; long rounded tail with several dark bands; white terminal tail band; rusty breast; dark wing tips; blue gray back; red eyes; immatures brown with streaks on breast and belly

Habitat: mixed forests; wooded riparian areas and urban areas

Notes: known to ambush prey (often birds at feeders) and flies into heavy wooded areas or brush, or runs on the ground in pursuit; has a fast woodpeckerlike *cac-cac-cac-cac*; nonmigrator to partial

© Jesus Giraldo Gutierrez/Shutterstock.com

Northern Goshawk (*Accipiter gentilis*)—Order Accipitriformes, Family Accipitridae

Size: 18–27 in. (46–69 cm); wingspan 35–50 in. (89–127 cm)

Key Features: blue-gray back; fine, gray striping on pale breast and belly; dark crown; dark eye band with white eyebrow; red-orange eyes; rounded wings; long, banded tail with white terminal band; immatures brown with striping on breast and belly and yellow eyes

Habitat: forest edges; open areas; breeds in mature forests; not found as much in southern part of region

Notes: agile and powerful flyer and predator; chases prey in flight or on foot; tends and protects nest and young; silent except during breeding with loud, fast shrill; partial migrator or seasonal movement

© Marina Ninic/Shutterstock.com

Broad-winged Hawk (*Buteo platypterus*)—Order Accipitriformes, Family Accipitridae

Size: 13–17 in. (32–44 cm); wingspan 29–39 in. (74–100 cm)

Key Features: tail with broad, black and white bands; wings broad with pointed tips and mostly whitish below and outlined with dark brown; breast with brownish-red bars; dark brown upperparts; immatures with streaks on underside and finely barred tail

Habitat: deciduous forests (often dense and wet)

Notes: quiet, mostly sedentary; migration is spectacular with spiraling up on a thermal and gliding to next while collecting in large groups (kettles; up to 30,000 in fall migration); high-pitched whistle (quiet in migration); complete migration

© Josef Stemeseder/Shutterstock.com`

© Chris Hill/Shutterstock.com

Red-tailed Hawk (*Buteo jamaicensis*)—Order Accipitriformes

Size: 18–26 in. (45–65 cm); wingspan 4–5 ft. (105–141 cm)

Key Features: bright rusty-red tail, conspicuous in flight; dark upperparts with white throat and underparts usually with dark belly band; immatures highly variable and no red tail

Habitat: fields; forests; prefers open woodlands

Notes: common; hunts from air and exposed perches; eats mostly small mammals; powerful descending scream; year-round

© Ed Schneider/ Shutterstock.com

Roughed-legged Hawk or Rough-legged Buzzard (*Buteo lagopus*)—**Order Accipitriformes**

Size: 18–24 in. (46–60 cm); wingspan 47–60 in. (120–153 cm)

Key Features: mostly brown plumage with much speckling (variable markings); tail white with dark terminal band; feathered feet; wing tips long enough to extend past tail when perched; in flight, dark "wrist" patches

Habitat: winters in region where rodent population is abundant (marshes, prairies; farm fields) and breeds in unforested, open ground of tundra and taiga of North America

Notes: not strongly territorial; hunts opportunistically; hovers in search of prey; vocalizes alarm calls of intruders near nest

© iliuta goean/Shutterstock.com

Golden Eagle (*Aquila chrysaetos*)—Order Accipitriformes

Size: 26–40 in. (66–102 cm); wingspan 6–8 ft. (1.8–2.34 m)

Key Features: large size; dark brown with golden tan crown and nape; dark beak; fully feathered legs; yellow feet; wings held level when soaring; in flight, short neck, long tail, large rectangular wings

Habitat: forests; remote open areas

Notes: very fast fliers and able to take prey on the wing; can attack large mammals like deer; protected under Migratory Bird Act

© Matt Knoth/ Shutterstock.com

American Kestrel (*Falco sparverius*)—Order Falconiformes, Family Falconidae

Size: 9–12 in. (22–31 cm); wingspan 20–24 in. (51–61 cm)

Key Features: small body; rusty brown back and tail; whitish breast with dark spots; two black vertical lines on light face; blue-gray wings; tail with wide black band and white edge at tip of rusty tail

Habitat: open wooded areas; farmlands; suburbs; urban

Notes: regularly seen hunting along fields and in suburbs; hunts while perched upright or hovering with fast-beating wings before dropping; after landing on perches will pump tail; loud, repeated *killy-killy-killy*; nonmigratory to partial

Merlin (*Falco columbarius*)—Order Falconiformes

Size: 9–13 in. (24–33 cm); wingspan 20–29 in. (50–73 cm)

Key Features: blue-gray back (male) and heavily streaked underside; banded tail; one distinct facial stripe; in flight, rapid shallow beats (pigeon-like)

Habitat: open fields and lakeshores; forests with openings (breeding)

Notes: capable of highspeed pursuits of songbirds; complete migration or seasonal movement

© Chris Hill/Shutterstock.com

Peregrine Falcon (*Falco peregrinus*)—Order Falconiformes

Size: 13–23 in. (34–58 cm); wingspan 29–47 in. (74–120 cm)

Key Features: blue-gray topside with dark helmet; light underside with dark flecking and spots

Habitat: open areas; river valleys; urban areas

Notes: dives at incredibly fast speeds to strike its prey that often sends predator and prey tumbling; voices loud, continuous *cack-cack-cack* near nest; migrates north to breed in summer

6. Turkey Vulture or Buzzard (*Cathartes aura*)—Order Cathartiformes

© FotoRequest/Shutterstock.com

© Rainer Lesniewski/Shutterstock.com

Size: 26–32 in. (66–80 cm); wingspan up to 6 ft. (183 cm)

Key Features: large bird; bare red head; long squared tail; in flight, wings appear two-toned with tips fingerlike; ivory bill

Habitat: various land habitats, especially around dead trees; seldom in forests

Notes: scavenger feeding almost exclusively on carrion; flies low enough to detect decay of dead animals; good sense of smell; uses thermal updrafts in the air to soar; nests in caves hollow trees or thickets; raises two chicks; feet suited more for walking not grasping; more closely related to storks than birds of prey; no syrinx but vocalizes with grunts and low hisses; complete migration

7. Pheasant, Grouse, and Turkeys—Order Galliformes, Family Phasianidae

Heavy-bodied ground-feeding domestic or game birds; terrestrial; variable in size—generally plump, with broad relatively short wings; 180 species worldwide, 16 in North America

© Marcin Perkowski/Shutterstock.com

Ring-Necked Pheasant (*Phasianus colchicus*)

Size: male 30–36 in. (76–90 cm) including tail (female smaller)
Key Features: large bronze body; greenish-blue to purple head with naked red face patch and white collar (male); long, pointed and barred tail; female mottled brown
Habitat: grasslands; fertile farm fields; brushy areas
Notes: introduced from Asia in late 1800s as a gamebird for hunters; not adapted for extreme cold (bare legs and feet) but sheltered areas with ample grains available has allowed it to survive; not strong fliers but can run swiftly; males have loud, raspy, roosterlike *ka-squawk*

© Tom Reichner/ Shutterstock.com

Ruffed Grouse (*Bonasa umbellus*)

Size: 16–19 in. (40–48 cm)
Key Features: mottled brownish-gray to reddish above; small crest; blackish ruff on sides of neck (fluffed outward during courtship); long squared tail with wide black band near tip (can fan it out like a turkey)
Habitat: mixed or deciduous forests
Notes: common in deep woods; feeds on aspen buds; grows bristles on feet during winter; dives into snowbanks to roost; populations seem to follow a 10-year cycle; northern goshawks feed on grouse; in springtime, male displays crest, fans tail feathers, and drums wings to produce sound as it stands on logs to attract female

© RelentlessImages/ Shutterstock.com

Wild Turkey (*Meleagris gallopavo*)

Size: 36–48 in. (90–120 cm)
Key Features: very large, plump body; brown to bronze with bare, bluish head and red wattles; copper-colored tail (female duller); males with straight black beard; spurs on legs
Habitat: oak forests; wooded bottomlands
Notes: prefers travel by foot and can run up to 15 mph (24 kph); strong flier for short distances (60 mph; 97 kph); roosts in trees at night; eyesight three times better than human eyesight; hearing excellent; wide array of vocalizations with males (toms) gobbling loudly, alarms are loud *pert*, gathering call is a *cluck*; females are hens and young are poults; eats insects, seeds, berries; nonmigratory

8. Pigeons and Doves—Order Columbiformes, Family Columbidae

Stout-bodied birds with short necks and short slender bills with a fleshy cere; 308 species worldwide, 18 in North America

© Joe Ravi/Shutterstock.com

Rock Dove or Pigeon ***(Columba livia)***

Size: 13 in. (33 cm)

Key Features: colors are highly variable; mostly gray with purplish iridescent neck; white rump; orange feet; in flight, holds wings in deep "v"

Habitat: urban areas; farms; along lakeshores on cliffs ("wild" types)

Notes: introduced in the early 17th century; females provide nutritious liquid to their young produced from glands in her crop; chicks eat the thick, protein-rich fluid by inserting beaks into mother's throat

© IrinaK/Shutterstock.com

Mourning Dove (*Zenaida macroura*)

Size: 12 in. (30 cm)

Key Features: slim body with smooth fawn-colored plumage and pale rosy underside, often with light gray patch on small head; dark patch below ear; long, white-trimmed, tapered tail; black spots on upper wing; dull red legs

Habitat: brushy areas; farmlands; suburbs; woodland edges (avoids heavily forested areas)

Notes: widespread native bird; swift, direct flier with wings that make a whistling sound; parents feed young regurgitated crop-produced liquid first few days of life; mournful cooing *oh-woe-woe-woe* (repeatedly with upward on second syllable and down for last three); seasonal movement to find food

9. Owls—Order Strigiformes, Family Strigidae

Typical owls are small to large solitary nocturnal birds of prey; large forward-facing eyes and ears (excellent sight and hearing); can swivel neck 180° both sides and 90° up and down; a sharp, hooked beak; sharp talons; and a conspicuous circle of feathers around each eye called a facial disk; 195 species worldwide, 21 in North America.

© Mike Mulick/Shutterstock.com

Eastern Screech-Owl (*Otus asio*)

Size: 9 in. (22.5 cm); wingspan up to 20 in. (51 cm)
Key Features: small body; short "ear" tufts'; mottled gray or tan and sometimes reddish; dark streaking on breast; yellow eyes; gray beak
Habitat: mature deciduous forests; orchards; large shade trees with cavities
Notes: nests in tree cavities; male and female likely roosts together; makes an eerie tremulous, descending wail of a call; year-round resident for most of region (seasonal movement in northern reaches)

© Phaeton Place/ Shutterstock.com

Great Horned Owl (*Bubo virginianus*)

Size: 20–25 in. (50–63 cm); wingspan up to 3.5 ft. (107 cm)
Key Features: large body mottled brown-black topside and lighter breast with fine dark barring; widely spaced "ear" tufts; yellow eyes; facial disk rusty orange with black outline and white "chin"
Habitat: fragmented forests; edges; riparian woodlands
Notes: "ears" are just tufts of feathers and nothing to do with ears or hearing; has silent flight (wing feathers have ragged tips); main predator of skunks and porcupines; during breeding season (January or February) will call *hoo-hoo-hoooo hoo-hoo* (or *eat-my-food, I'll-eat-you*); nonmigratory

© RT Images/Shutterstock.com

Snowy Owl (*Nyctea scandiaca*)

Size: 19–25 in. (48–64 cm)
Key Features: large white with some dark flecks; yellow eyes; black beak and talons; feathered to their toes; active in daylight
Habitat: open fields; wetlands; lakeshores; tundra (nesting)
Notes: moves into our range when food is scarce in north; may be seen on a roof building or utility poles in winter

© Don Mammoser/ Shutterstock.com

Barred Owl (*Strix varia*)

Size: 20–24 in. (50–60 cm); wingspan up to 3.5 ft. (107 cm)
Key Features: large round head with big dark brown eyes; chunky body with mottled brown-gray plumage; dark horizontal barring around throat and streaked chest and belly; yellow beak and feet; may hunt during the day
Habitat: wooded swamps; mature forests with sparse understory; near water
Notes: common; although active day or night, they will hunt mostly between midnight and 4 a.m.; young stay with parents four months after fledging; may sound like a barking dog then sounds, *"Who-cooks-for-you? Who-cooks-for-you-all?"*; nonmigratory

10. Chimney Swift (*Chaetura pelagica*)—Order Apodiformes, Family Apodidae

© Paul Reeves Photography/Shutterstock.com

© Daniel Friend/Shutterstock.com

Size: 5 in. (13 cm)
Key Features: small body with dark gray plumage and lighter throat; crescent-shaped wings; short tail; in flight, looks boomerang-shaped and has rapid wingbeats with erratic flight pattern
Habitat: open air above towns, farmlands, woodlands
Notes: they do everything in flight (i.e., eat insects, drink, collect nesting materials, and mate); small weak legs; strong claws that enable it to cling to vertical surfaces; very conspicuous on warm summer evenings and fall migration; some say it looks like a flying cigar; nests in large chimneys or hollow tree; has a rapid, chattering call or series of chipping in flight; complete migration to remote upper region of Amazon

11. Ruby-throated Hummingbird (*Archilochus colubris*)—Order Apodiformes, Family Trochilidae

© Steve Byland/Shutterstock.com

Size: 3.0–3.5 in. (7.5–9.0 cm)
Key Features: tiny, green iridescent body with red metallic throat patch and black chin (male; white if female) and light underside; long, needlelike beak
Habitat: mixed and deciduous forests; gardens; wetlands; orchards
Notes: smallest bird in region; nests made of plant material, spider webs, and lichen on the outside; capable of hovering in mid-air due to rapid flapping of wings; only birds that can fly backward; wings create humming sound and birds will chatter to communicate; 337 species worldwide, 23 North American species; complete migration

12. Common Nighthawk (*Chordeiles minor*)—Order Caprimulgiformes, Family Caprimulgidae

© Steve Byland/Shutterstock.com

© Agami Photo Agency/Shutterstock.com

Size: 9 in. (23 cm)
Key Features: medium-sized nocturnal bird with soft plumage cryptically mottled resembling bark or leaves; barred underparts; white throat; has small feet (little use for walking), and very short, pointed bills; usually nests on ground; in flight, bold white "wrists" on long pointed wings; barred tail shallowly forked; erratic flight pattern
Habitat: forest openings; grasslands; urban
Notes: large mouth rimmed with feathers for capturing insects (at dusk and night); might sleep on a post during the day; gravelly surfaces preferred for laying eggs; calls often with a *beent* sound; 86 nightjar species worldwide, 9 in North America; complete migration

13. Belted Kingfisher (*Ceryle alcyon*)—Order Coraciiformes

© Harry Collins Photography/ Shutterstock.com

Size: 13 in. (33 cm)
Key Features: blue-gray above with blue-gray band on breast (female with additional chestnut band); shaggy crest; heavy, sharp bill
Habitat: shores of rivers and lakes; coasts
Notes: often utters rattling call when leaving perched location; dashes over water while calling; dips or plunges headfirst for prey; burrows in bank near water (tunneling up to 15 ft; 460 cm) by using its beak and feet; complete migration

14. Woodpeckers, Sapsuckers, and Flickers—Order Piciformes, Family Picidae

Small-medium birds with chisel-beaks, short legs, stiff tails (for support) and long barbed tongues used for capturing insects; stores food for future use; feet adapted for tree climbing; some with two toes forward, and two backward (several species only three toes); many tap regular rhythm noisily on tree trunks with beaks; feathered nostrils; reinforced skull with strong neck muscles; 4–5 young in a clutch; 218 species worldwide, 26 in North America

© DMS Foto/Shutterstock.com

Red-headed Woodpecker (*Melanerpes erythrocephalus*)

Size: 9 in. (23 cm)
Key Features: red head and neck; solid black back and tail; white chest, belly, rump; gray legs and beak; in flight, large white patches on wings
Habitat: open woods (especially oaks); parks; edges; swamps
Notes: varied diet; stores grasshoppers, acorns, beechnuts, corn, fruit; mostly pecks dead trees (beak not well adapted for excavating holes); loud *kwrring* notes; complete or partial migration

© Jim Nelson/Shutterstock.com

Red-Bellied Woodpecker (*Melanerpes carolinus*)

Size: 9.25 in. (23 cm)
Key Features: fine black and white barring on back; red nape; red crown (male); white patches on rump and base of primaries; reddish tinge to belly
Habitat: mature deciduous forests; edges; wooded suburbs; orchards; farmland
Notes: excavates holes in search of insects, spiders, centipedes; caches acorns and berries for winter; uses same tree for nesting in the following year but excavates below old hole; call is loud, rolling *churr* and a low *chug-chug-chug*; nonmigratory

© Dennis W Donohue/ Shutterstock.com

Yellow-Bellied Sapsucker (*Sphyrapicus varius*)

Size: 8–9 in. (20–22.5 cm)

Key Features: long white wing stripe on side at rest; black, white, red pattern on head and neck (black "bib"; male with red "chin"; female white "chin"; juvenile brownish overall); yellowish on lower breast and belly

Habitat: deciduous and mixed forests; orchards

Notes: drills neat rows of holes (mainly birches and orchard trees) to remove nutritious inner bark and sap that runs out (plus insects that may come with it); has brushlike tongue and cannot capture insects from boring into trees; other animals benefit from sap released; has an irregular drumming (pecking) similar to Morse code rhythm; mostly quiet bird but calls a catlike *meow*; complete migration

© Fiona M. Donnelly/Shutterstock.com

Northern Flicker (*Colaptes auratus*)

Size: 12 in. (30 cm)

Key Features: brown body with barred body and wings buff belly with black spots; black crescent throat; black "mustache" (male); red spot on nape; long beak; in flight, white rump and yellow underwings and undertail

Habitat: open forests; edges; fields; wetlands

Notes: only woodpecker to forage on ground for insects, especially ants; uses an antacid saliva to neutralize the ants' formic acid (and will use remains to help preen off parasites), also "bathes" in dust depressions to absorb excess oils; nests in trees (takes males 12 days to excavate); loud, rapid laughlike call; nonmigratory to partial migration

© Michael Woodruff/ Shutterstock.com

Hairy Woodpecker (*Picoides villosus*)

Size: 9 in. (23 cm)

Key Features: black-and-white with white belly and white stripe down center of back; black stripe through eye; black tail with unspotted, white outer feathers; long black beak; red mark on back of head (male)

Habitat: deciduous and mixed forests

Notes: very similar to downy woodpecker except size of body and beak; hairy is more aggressive than downy woodpecker (when observed at feeders); loud, sharp call then unbroken trill (drums less than downy); nonmigratory

© Andrea J Smith/ Shutterstock.com

Downy Woodpecker (*Picoides pubescens*)

Size: 6 in. (15 cm)

Key Features: small body; black and white pattern on head with stripe running through eye; white back and belly; black-and-white spotted wings; red patch on back of head (male); short beak; black spots on white outer tail feathers

Habitat: woodlands; parks; suburbs; orchards

Notes: common where trees are present; has soft taps; long, unbroken trill with sharp calls and drumming on logs; nonmigratory and roosts in tree cavities

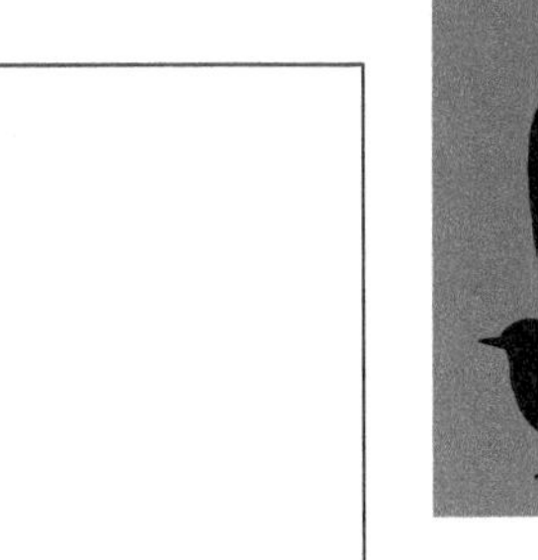

© FotoRequest/Shutterstock.com

Pileated Woodpecker (*Dryocopus pileatus*)

Size: 19 in. (48 cm)

Key Features: large body; red crest; black, white, red pattern on head; body mostly black; long gray beak with red mustache (male); in flight, prominent white wing linings

Habitat: mature forests

Notes: region's largest woodpecker; needs large tracts of woodlands; announces territory by loud drumming; excavates long rectangular to oval holes several feet long (0.5–0.75 m) that many other birds and mammals rely on for nesting; beak becomes shorter as it ages; loud, fast laugh and resonant drumming; nonmigratory

15. Perching Birds (Songbirds)—Order Passeriformes

The Passerines have the most species, with over 5,000 types! The foot arrangement of these perching birds have three toes forward facing and one toe pointing behind. Passerines also are termed "songbirds" because of their diverse vocalization capabilities. The families are pictured in the following sequence and listed from smallest to largest in each family: *flycatchers, corvids, swallows and martins, chickadees and titmouse, nuthatches, creepers, wrens, kinglets, thrushes, catbird, starling, waxwing, warblers, sparrows, cardinals and grosbeaks, blackbirds and orioles, finches, and house sparrow*

Flycatchers—Family Tyrannidae

Largest family of passerines with more than 400 species of varied morphology and behavior; colors are subdued; insectivores

Eastern Phoebe (*Sayornis phoebe*)
Size: 7 in. (18 cm)
Key Features: gray-brown back with darker head; white underside with grayish breast (washed with yellow in fall); no eye ring; no wing bars; sits upright and pumps tail frequently (often seen at the end of a dead branch)
Habitat: woodlands; farmlands; near water
Notes: sits waiting for insects to pass, then flies to catch them and returns to same branch ("hawking"); may reuse same nest for many years (mud nest); helps control insect populations; first bird species to be banded for ornithological study (Audubon 1803 tied silver string to nestlings legs to monitor seasonal visitation); hearty clipped *fee-bee* sang repeatedly and frequently; complete migration

Great Crested Flycatcher (*Myiarchus crinitus*)
Size: 8 in. (20 cm)
Key Features: reddish-brown tail and wing patch; bright yellow belly and undertail coverts; gray head with slight fluffy crest (not always obvious); gray throat and upper breast
Habitat: mixed and deciduous forests; near edges
Notes: nests in a tree cavity and fills with debris if too deep (may leave snakeskin, feathers, string, or plastic hanging out of nesting hole); gleans insects from tree leaves; loud, clear whistled *wheep!* and rolling *prrreeet* (often heard before seen); complete migration

© Michael Dante Salazar/Shutterstock.com

Eastern Kingbird (*Tyrannus tyrannus*)
Size: 8 in. (20 cm)
Key Features: blackish-gray back and white underside; tail black with prominent white band at tip; stiff, shallow wing beats; concealed red crown (seldom seen)
Habitat: fields with occasional tall trees; forest edges
Notes: defends territory aggressively regardless of invader's size; has a fluttering courtship flight of short, quivering wingbeats

Family Vireonidae

Red-eyed Vireo (*Vireo olivaceus*)
Size: 6 in. (15 cm)
Key Features: constant song in spring and summer; light olive green back with white to pale gray underside; pale yellow on sides; dark eye line with white "eyebrow"; grayish-blue crown; no wing bars; red eyes
Habitat: deciduous woodlands with understory; often seen singing at tops of trees
Notes: male sings almost continuously through the day in spring and summer; tends to "hunch over" when hopping (different than most songbirds); hops with body diagonal to direction its traveling; gleans insects from tree leaves, also berries; similar song as robin's; *Look-up, way-up, tree-top, see-me, here-I-am!*; complete migration

© Paul Reeves Photography/Shutterstock.com

Corvids (Jays, Crows, Ravens)—Family Corvidae

Medium to large passerines; intelligent birds with self-awareness, tool-making abilities, and counting skills; total brain-to-body mass ratio equal to primates; young have been known to "play" together; many with strong community groups (capable of different forms of parental care; strong bonds); powerful bill and feet; large wingspan; nostrils covered with bristlelike feathers; found worldwide with over 120 species

© FotoRequest/Shutterstock.com

Blue Jay (*Cyanocitta cristata*)

Size: 12 in. (30 cm)

Key Features: bright blue crest; blue body with white underside and black "necklace"; white bar and flecking on wings; black bars and white corners on tail; black bill

Habitat: woodlands; farmlands; suburbs; parks; common where fruit-bearing trees or shrubs are abundant

Notes: highly intelligent; caches food; does not hesitate to drive away other animals; omnivore; feathers refract light to display blue; known as the alarm of the forest; raucous voice with wide variety of vocalizations (calls, cries, screams, whispers); nonmigratory to partial migrator

© Melinda Fawver/Shutterstock.com

American Crow (*Corvus brachyrhynchos*)
Size: 18 in. (45 cm)
Key Features: large black body with black bill, legs, feet; squared tail
Habitat: forests; open areas; densely forested waterways; farmland; suburbs
Notes: wary and intelligent; complex social structure; studies show capability to count; adapts to a variety of habitats; common; opportunistic; hunts food during day and roosts with its extended family at night; collects shiny objects; reuses nest every year (unless taken over by great horned owl); lives up to 20 years; impressive capability to mimic sound; repetitive *caw-caw-caw*; flocks in the fall and roosts together through winter (100s to 1000s of crows)

© Vladimir Wrangel/Shutterstock.com

Common Raven (*Corvus corax*)
Size: 22–27 in. (56–69 cm)
Key Features: large body, all black with large black bill; shaggy throat (not always obvious); rounded wings; wedge-shaped tail (in flight)
Habitat: mixed and coniferous forests; suburbs; northern part of region
Notes: largest passerine; very intelligent, bold, clever; playful; maintains lifelong pair bond; courtship involves grabbing bills and preening one another while cooing; complex vocalizations; nonmigratory to partial migration

Swallows and Martins—Family Hirundinidae

Passerines adapted for aerial feeding with slender, streamlined body; pointed wings for gliding, and keen eyesight; most have a glossy plumage; swallows typically have long forked tails and martins with more squared tail

© Elliotte Rusty Harold/ Shutterstock.com

© Stubblefield Photography/Shutterstock.com

Tree Swallow (*Tachycineta bicolor*)

Size: 5–6 in. (13–15 cm)

Key Features: glossy blue-black above (or green near fall) and white below; tail slightly forked

Habitat: open areas; wetlands; lakeshores; fence lines; open woodlands

Notes: common; feeds on insects in flight and some berries; nests in tree cavities or nest boxes; uses found feathers to line its nest; male has a chattering twitter song or sounds with metallic, buzzy sound; complete migration but some may stay in some southern parts of region with seasonal movement to find food

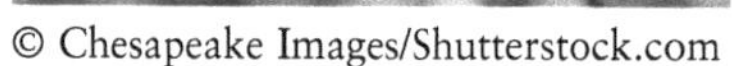

© Chesapeake Images/Shutterstock.com

© Gallinago_media/Shutterstock.com

Barn Swallow (*Hirundo rustica*)

Size: 7 in. (18 cm)

Key Features: in flight, deeply forked tail and pointed wings; glossy dark blue above and buff to light rust belly; dark rust throat and forehead

Habitat: open areas (woodlands, farms, suburbs)

Notes: tends to build mud nests on human-made structures; feeds on insects on the wing; swift, graceful flight pattern; continuous chattering twitter of *zip-zip-zip*; complete migration

© Paul Roedding/Shutterstock.com

Purple Martin (*Progne subis*)

Size: 8.5 in. (22 cm)

Key Features: glossy blue-black body with a purplish head (male; female duller with mottled throat and light belly)—coloration dependent on light; tail slightly forked; pointed wings; small bill

Habitat: open areas; woodlands; suburbs; prefers near water

Notes: largest swallow species; drinks and bathes while flying in rain or skimming water; eats dragonflies and other insects as it spirals around trying to catch them; returns to same nest each year (mostly dependent on martin houses); fluty, robinlike *pew-pew* when flying; complete migration

Chickadees and Titmouse—Family Paridae

Small and active with a variety of calls and songs; social passerines but often territorial during breeding season; highly adaptable and considered very intelligent (some say second to corvids); consumes insects and seeds; often very agile while foraging on twigs; cavity nesters (can excavate hole)

© Paul Roedding/Shutterstock.com

Black-capped Chickadee (*Poecile atricapillus*)

Size: 5 in. (13 cm)

Key Features: black cap and throat; mostly light gray back and wings; white cheek patch; white underside; buff sides and flanks; dark legs; small white edging on wing feathers

Habitat: mixed and deciduous forests; riparian woodlands; urban parks; backyards

Notes: often seen with nuthatches and woodpeckers; eats insects, fruit, seeds; agile when feeding; nest of moss and lined with animal fur; calls *chick-a-dee-dee-dee* or slow song of *sweee-teee*; nonmigratory (feeds every day year-round)

© Al Mueller/Shutterstock.com

Tufted Titmouse (*Parus bicolor*)

Size: 6 in. (15 cm)

Key Features: gray back with white chest and belly; pointed crest; flanks rusty brown; round, dark eyes against white; gray legs

Habitat: deciduous forest; parks

Notes: usually one or two seen at once; agile feeder; strikes acorns and seed with dainty, yet strong bill (also eats fruits); known to pull hair from sleeping squirrels, cats, dogs, or what you may offer it; strong family bonds; calls loud *peter-peter-peter*; nonmigratory

Nuthatches—Family Sittidae

Compact passerines with large heads, stubby tail; short legs and strong toes; long, sturdy bills; advertises with loud song; often seen head pointed toward ground while hopping on tree trunk (if you see a bird oriented this way, then you know it is a nuthatch!)

© Steve Byland/Shutterstock.com

Red-breasted Nuthatch (*Sitta canadensis*)

Size: 4.5 in. (11 cm)
Key Features: gray-backed bird; black-and-white cap (prominent black eye line); rust chest and belly; female gray cap and pale below; body usually oriented horizontal to its surface
Habitat: climbs down tree trunks headfirst; forests (prefers mature conifers)
Notes: similar to white-breasted nuthatch; will crack seeds open by wedging in crevices and pounding it open; active bird; nests in cavities and always smears entrance hole with pitch from conifers (may prevent insects and other animals from entering); nasal call of *yna-yna-yna-yna*; nonmigratory to some seasonal movement

© Doug Lemke/Shutterstock.com

White-breasted Nuthatch (*Sitta carolinensis*)

Size: 5–6 in. (13–15 cm)
Key Features: grayish-blue back with dark cap and nape; white face and belly; chestnut undertail; long, thin bill
Habitat: mixed and deciduous forests; suburbs
Notes: hops headfirst down tree trunk; agile climber; does not use tail to brace itself; has extra long hind tow claw; often seen with chickadees and downy woodpeckers; *whi-whi-whi-whi* spring call; nonmigratory

© Tom Reichner/Shutterstock.com

Brown Creeper (*Certhia americana*)—Family Certhiidae
Size: 5 in. (13 cm)
Key Features: almost completely camouflaged brown bird until you see it creeping around the trunk of the tree (head pointed up); white from chin to belly; white eyebrows and dark eyes; long stiff pointed tail; rusty rump; thin curved bill
Habitat: deciduous forest
Notes: spirals around tree trunk and when startled will flatten on trunk or branch to heighten camouflage capabilities; tends to build nests behind bark of dead trees; faint, high-pitched song *trees-trees see the trees* (often sounds like a wood warbler); nonmigratory to partial (only creeper species found in North America)

© Tom Reichner/Shutterstock.com

House Wren (*Troglodytes aedon*)—Family Troglodytidae
Size: 5 in. (13 cm)
Key Features: small body with small tail often held erect; all brownish-gray with lighter brown underside with markings on tail and wings; thin brown bill slightly curved downward
Habitat: open woodlands; forest edges; parks
Notes: energetic; males leave small twigs in several possible nesting cavities and female chooses site to finish building (female fills it with twigs and creates a cup for grass and pine needles at back of the cavity); will nest in almost anything; bullies other birds; two broods per season; male changes partners; bubbly, smooth singer nonstop all day in mating season *tsi-tsi-tsi-tsi oodel-oodle-oodle-oodle*; complete to impartial migration

© Mircea Costina/Shutterstock.com

Golden-crowned Kinglet (*Regulus satrapa*)—Family Regulidae
Size: 3–4 in. (8–11 cm)
Key Features: olive-gray back and white underparts; dark "cheek"; two white wing bars; black stripe through eyes; orange stripe set within yellow crown with black border (male; female with yellow crown); thin bills; short tails
Habitat: all forest types (nonbreeding); mixed forest but prefers coniferous dominated by spruce (breeding)
Notes: habit of flicking wings; drinks tree sap and feeds on insects from trees; swings from tree branches; unusual hanging nest of spider webs, moss, lichen, lined with feathers and bark; numerous eggs often in two layers in nest; faint, high-pitched increasing *tsee-tsee-tsee-tsee, why do you shilly-shally*; nonbreeding range in southern part of region and breeds in the northern range

Thrushes—Family Turdidae

Small to medium-sized plump passerines with soft-plumage found; mostly seen on the ground feeding on invertebrates and fruit; builds cup-shaped nests sometimes lined with mud with both parents raising young; may have two or more clutches per year

© Rabbitti/Shutterstock.com

Eastern Bluebird (*Sialia sialis*)

Size: 7 in. (18 cm)
Key Features: bright blue back, head, and tail; rusty red breast and white belly
Habitat: open areas; fenceposts
Notes: feeds on grasshoppers and other insects, fruit; competes with house sparrow and European starlings (both introduced species) for natural nesting sites; *chur-lee chur chur-lee* song; complete migration or partial in southern region

© steve52/Shutterstock.com

American Robin (*Turdus migratorius*)

Size: 9–11 in. (22.5–28 cm)
Key Features: dark gray back with rusty red breast; white broken eye ring; yellow beak; head and tail dark gray; white-tipped tail; some have noticeable dark gray-and-white striped chin; female duller gray head and duller breast color
Habitat: open forests; suburbs; parks; farmlands; fruit trees in sheltered areas (winter)
Notes: widespread and abundant; looks for movements in soil for worms and insects; sings all night in spring; very territorial; evenly spaced warble of *cheerily cheer-up cheerio* (calls rapid *tut-tut-tut*); complete migration to partial in southern region and wintering in low swampy areas (only food source available, i.e., berries and insect eggs)

© Brian Lasenby/Shutterstock.com

Gray Catbird (*Dumetella carolinensis*)—Family Mimidae

Size: 9 in. (23 cm)
Key Features: dark gray bird with black crown; chestnut patch under tail; long, thin black bill; often lifts tail up
Habitat: woodland understory; bushy areas; suburbs; parks
Notes: nests in thick foliage; quickly recognizes brown-headed cowbird eggs in its nest and destroys them; call sounds like mewing cat, mimics other birds; complete migration

© Eric Isselee/Shutterstock.com

© xpixel/Shutterstock.com

European Starling (*Sturnus vulgaris*)—Family Sturnidae

Size: 7.5 in. (19 cm)

Key Features: iridescent purplish-black green body (spring, summer); gray to black with white speckles (fall, winter); long, pointed yellow bill (spring; gray in fall); tail short

Habitat: open woodlands; towns; cities; farmlands

Notes: 100 birds introduced to New York's Central Park in late 1800s and now over 200 million in North America; pries open crevices to find insects; gathers in huge roosts and may displace other hole-nesting species (Eastern bluebird, tree swallow, red-headed woodpecker); has variety of whistles, squeaks, and mimics other bird sounds (killdeer, red-tailed hawk); nonmigratory

© Double Brow Imagery/Shutterstock.com

Cedar Waxwing (*Bombycilla cedrorum*)—Family Bombycillidae

Size: 7.5 in. (19 cm)

Key Features: sleek looking, grayish-brown bird with pointed crest; black mask (banditlike); light yellow belly; bright yellow band across tail tip; wings with waxy-looking red tips

Habitat: coniferous forests; tops of tall trees; wooded swamps; suburbs

Notes: mostly in flocks looking for berries (also feeds on insects); shares food with others in flock (sometimes passes fruit beak-to-beak); courtship ritual of male offering female berry as it hops toward her and she hops away after accepting it; constant high-pitched trilled *sreee* whistle; nonmigratory to seasonal movement to find food

Warblers—Family Parulidae

Small, often colorful tropical passerines many of which breed in our region; males are usually brighter than females and immatures; mainly arboreal (found in trees); diverse and numerous species; tends to lay large clutch of eggs (typically six)

© Frode Jacobsen/Shutterstock.com

Black-and-White Warbler (*Mniotilta varia*)
Size: 5 in. (13 cm)
Key Features: black-and-white stripes above and white below; striped cap; black chin and cheek
Habitat: forests
Notes: constantly moving searching for insects on bark; brisk, high-pitched song *weetsee weetsee weetsee weetsee*

© Mircea Costina/Shutterstock.com

Northern Parula (*Parula americana*)
Size: 4.5 in. (11 cm)
Key Features: blue above with greenish-yellow patch on back; white wingbars; yellow breast and throat with dark band (male)
Habitat: usually near water; forests (especially mature woods with lichen present)
Notes: very active as it forages for insects; nests in lichen; young spend a few weeks enclosed in a delicate, socklike nest hanging from branch; sings a buzzy trill increasing in pitch then snaps off, *zzzzzzzz-zup*

© Melinda Fawver/Shutterstock.com

American Redstart (*Setophaga ruticilla*)
Size: 5 in. (13 cm)
Key Features: black with orange patches on wings and tail; black chest; white belly
Habitat: second-growth forests of unbroken tracts; parks
Notes: common; moves quickly; contrasting colors conspicuous; movement and coloration mistaken for butterflies at a glance as it catches insects in flight; variable song often ending on a high note *zee-zee-zee-zee-zee-zeeo*

© Paul Reeves Photography/Shutterstock.com

Ovenbird (*Seiurus aurocapilla*)
Size: 6 in. (15 cm)
Key Features: olive brown above with orange crown bordered in black; white underside with heavy, dark streaks; pinkish legs; white eye ring
Habitat: deciduous woodlands and riparian shrubbery during migration; mature forests (breeding)
Notes: walks on ground; has covered nest on forest floor; voice very obvious in woods and often heard rather than seen, *teacher, teacher, teacher, teacher!*

© Mircea Costina/Shutterstock.com

Common Yellowthroat (*Geothlypis trichas*)
Size: 5 in. (13 cm)
Key Features: black mask edged with white above; greenish above and bright yellow below; long, thin pointed black bill
Habitat: wet, shrubby area
Notes: easy to recognize and common; bounces in and out of tall grasses in courtship display; sometimes nests in colonies; rhythmic *witchery, witchery, witchery*

© Hanjo Hellmann/Shutterstock.com

Yellow-rumped Warbler (*Setophaga coronata*)
Size: 5–6 in. (13–15 cm)
Key Features: gray with black streaks and yellow crown, rump, shoulder patch; dark "cheek"; white throat and belly; thin white eyebrow; black bib; two faint white wingbars; white tail patches in flight
Habitat: mixed and coniferous forests; thickets
Notes: abundant; bright loud *chip* call, sometimes difficult to recognize trilling song

© Steve Byland/Shutterstock.com

Yellow Warbler (*Setophaga petechia*, formerly *Dendroica petechia*)
Size: 5 in. (13 cm)
Key Features: mostly yellow with slight greenish above; male has brownish streaking on breast; long, pointed dark bill
Habitat: wooded wetlands; suburbs; parks
Notes: common; prolific insect eater; nests in willows and shrubs in wetlands at low forks with neat cup of woven plant fibers; two persistent songs from male, *pip-pip-pip-sissewa-is-sweet* and *wee-see-wee-see-wiss-wiss-u*

© Frode Jacobsen/Shutterstock.com

Blackburnian Warbler (*Setophaga fusca* formerly *Dendroica fusca*)
Size: 4.5–5.5 in. (11–14 cm)
Key Features: bright, fiery orange throat and breast; yellowish-orange head with two black stripes; angular black "mask"; striped black back and sides, flanks; pale to yellow underside; broad white wingbars
Habitat: deep woods, especially conifers
Notes: hidden most of summer at coniferous treetops (orange throat usually will catch your eye); thin, buzzy song ending in a sliding high note

© FotoRequest/Shutterstock.com

Chestnut-sided Warbler (*Setophaga pensylvanica*, formerly *Dendroica pensylvanica*)
Size: 5 in. (13 cm)
Key Features: male with yellow cap, black "mask," and chestnut sides; gray wings with two yellow wingbars; white below
Habitat: open woodlands; farmlands; brushy fields
Notes: hops high on branches while searching for insects; fast mover; loud, distinctive song of *I-wish-to-see-Miss-Beecher*

© Philip Rathner/Shutterstock.com

Pine Warbler (*Setophaga pinus*)
Size: 5 in. (13 cm)
Key Features: olive green above and yellowish and streaks below; white wingbars
Habitat: deciduous woodlands during migration; breeds in pine forests
Notes: often difficult to see while it forages and nests high in pine trees; often observed with smears of pine resin; loose trill of a song

© Jeff Rzepka/Shutterstock.com

Kirtland's Warbler (*Setophaga kirtlandii*)
Size: 6 in. (15 cm)
Key Features: blueish-gray upper with black streaks; yellow underside; black streaks on sides and flanks; whitish undertail; bold, broken white eye ring; black patch in front of eyes
Habitat: young jack pine stands (Michigan)
Notes: rare warbler nests almost exclusively in jack pine forests (3–25 ft; 1–8 m) of northern Michigan; population declines due to fire suppression (less jack pine) and brown-headed cowbird nest parasitism; loud, low *chip-chip-che-way-o*

Sparrows [American]—Family Passerellidae

Small brown or gray passerines, seed and insect eaters with conical bills and many with distinctive head patterns; found often feeding on ground and displaying a "double-scratch" of both feet hopping backward at same time in search of food

© Charles Brutlag/Shutterstock.com

Eastern Towhee (*Pipilo erythrophthalmus*)
Size: 7–8 in. (18–20 cm)
Key Features: mostly black with rusty brown sides and white belly; long black tail with white tip; in flight, white wing patches
Habitat: open forests; thickets; parks
Notes: loud buzzy *shreee!* or calls *drink-your-tea!*; migratory

© Paul Reeves Photography/Shutterstock.com

American Tree Sparrow (*Spizelloides arborea*)
Size: 6 in. (15 cm)
Key Features: mottled brown body with rusty crown and tannish-gray breast (unstreaked); black spot on center of chest; lower bill yellow; gray eyebrows with rusty stripe behind eye; two white wingbars
Habitat: brushy thickets; semiopen fields; farmlands
Notes: nests in the north on ground in dense thickets; high pitched bubbly bright whistle of *tseet-tseet* then slurred whistle series; year-round with some seasonal movement

© Nancy Bauer/Shutterstock.com

Chipping Sparrow (*Spizella passerina*)
Size: 5 in. (13 cm)
Key Features: reddish cap; black eye streak; white eyebrow; light gray below
Habitat: open woodlands; forest edges; farmlands; conifers favored for singing and nesting
Notes: common; nest lined with animal hair in varied places (conifers preferred); male call has slow "chip" and song with a trill of varied musical chips

© Steve Byland/ Shutterstock.com

White-throated Sparrow (*Zonotrichia albicollis*)
Size: 5.5–6.5 in. (14–17 cm)
Key Features: black-and-white striped crown; often with yellow patch in front of eye; white throat; mottled brown upperparts and gray below; gray "cheek"; grayish bill
Habitat: woodlands with understory; brushy areas; forest edges
Notes: similar appearance to white-crowned sparrow (*Zonotrichia leucophrys*) that typically migrates through region to northern nesting area (this species lacks white chin and yellow eye patch); two broods; song sung loud and slow like *oh, sweet Canada, Canada, Canada*; partial migrator especially in northern region

© vagabond54/Shutterstock.com

Dark-eyed Junco (*Junco hyemalis*)
Size: 5.5 in. (14 cm)
Key Features: all dark gray with white belly; pale bill (pinkish to light yellow); tail forms white "v" in flight
Habitat: mixed and coniferous forests; forest edges
Notes: spends most of the time on the ground; dominant birds chase less dominant; song is long, musical trill; common winter resident

Cardinals, Grosbeaks, Buntings, and Tanagers—Family Cardinalidae

Passerines with typically robust, seed-eating bills associated with open woodlands; sexes often have distinctive morphology

© Bonnie Taylor Barry/ Shutterstock.com

Northern Cardinal (*Cardinalis cardinalis*)
Size: 8–9 in. (20–22.5 cm)
Key Features: all red body with black mask extending into chin and throat; large crest; large, orange-red bill
Habitat: open woods; forest edges; thickets; suburbs; parks
Notes: male feeds female during courtship; two broods; male territorial; female known to call while on nest communicating to male; both male and female sing and has a bubbly song of *What cheer! What cheer! Birdie-birdie-birdie what cheer!*; year-round resident

© Jayne Gulbrand/ Shutterstock.com

Rose-breasted Grosbeak (*Pheucticus ludovicianus*)
Size: 7–8 in. (18–20 cm)
Key Features: black-and-white chunky bird with large, triangular rose patch in center of chest; large ivory bill; wing linings rosy
Habitat: deciduous and mixed forests; suburbs
Notes: touch bills during courtship and after absences; nests low in shrubs; both sexes sing a rich, melodious robinlike song; complete migration with males arriving first in late spring (seen in small groups until females arrive)

© Bonnie Taylor Barry/Shutterstock.com

Indigo Bunting (*Passerina cyanea*)

Size: 5.5 in. (14 cm)
Key Features: bright blue; dark markings on tail and wings
Habitat: woodland edges; orchards
Notes: sometimes appears black when up high in treetops (refracts blue light like other blue birds); feeds on insects; returns to last year's nest site (favors raspberry thickets); sings from tree-tops, *fire-fire, where-where, here-here, see-it-see-it*; complete migration

© Debra Anderson/Shutterstock.com

Scarlet Tanager (*Piranga olivacea*)

Size: 7 in. (18 cm)
Key Features: bright red with black wings and tail; dark eyes; ivory bill
Habitat: mature, unbroken woodlands
Notes: hunts insects atop trees or in understory; although different beaks than northern cardinal, this bird has a similar DNA structure, song, and plumage; male sheds bright plumage in fall; similar to a robin's song but with a distinctive burr in it, *chip-burrr*; complete migration

Blackbirds, Grackles, Cowbirds, and Orioles—Family Icteridae

Small to medium-sized birds mostly black with often bright colorations; species vary in morphology and behavior; displays sexual dimorphism with males brighter and different colors than females

© Jim Nelson/Shutterstock.com

© SunflowerMomma/Shutterstock.com

Red-winged Blackbird (*Agelaius phoeniceus*)

Size: 8.5 in. (22 cm)
Key Features: all black with red and yellow shoulder patches; pointed black bill
Habitat: wetlands and adjacent open areas; farmlands
Notes: numerous; insect and seed eater; two to three broods per year; weaves nest among cattails; males song is a raspy, loud *konk-a-ree* from cattail tops and females have a loud *che-che-che chee chee chee*; complete migration (males return first; sometimes observed congregating until females arrive)

© FotoRequest/Shutterstock.com

Common Grackle (*Quiscalus quiscula*)

Size: 11–13 in. (28–33 cm)
Key Features: large glossy blackish-purple-bluish-green plumage; bright yellow eyes; long black keeled tail (wedge-shaped in flight); long, pointed bill
Habitat: wetlands; edges; farmlands; parks
Notes: mail holds tail in a vertical, keeled position during flight (twisted); at night will commonly roost with other members of Family Icteridae; two broods per year; loud raspy song from male of *gri-de-leeek*; partial migration

© Steve Byland/Shutterstock.com

© Steve Byland/Shutterstock.com

Brown-headed Cowbird (*Molothrus ater*)

Size: 7.5 in. (19 cm)
Key Features: black, glossy body with dark brown head; sharp pointed gray bill
Habitat: farmlands; forest edges; near river woodlands
Notes: brood parasite of over 140 species (lays eggs in a host bird's nest so the other bird can incubate and raise its young); once followed bison across prairies to feed on the insects on and around them (now present around livestock); high liquid gurgle of a song, *bubbloozeee*; migratory

© Al Mueller/Shutterstock.com

Baltimore Oriole (*Icterus galbula*)

Size: 7–8 in. (18–20 cm)

Key Features: bright orange plumage with all black head; black wings with white and orange wing-bars; tail orange with black streaks; gray bill

Habitat: open woodlands; forest edges; shade trees

Notes: song often heard before seen; feeds on caterpillars atop trees; female (lighter in coloration) builds socklike nest at outermost part of branches; returns to same nesting area; sings slow, clear whistle of *peter peter peter here peter*; complete migration

Finches—Family Fringillidae

Medium-sized passerines with conical, strong, stubby bills adapted to eat seeds; colorful plumage with males brighter and more colorful; flight typically has an undulating roller-coasterlike pattern

© FotoRequest/Shutterstock.com

Purple Finch (*Haemorhous purpureus*)

Size: 6 in. (15 cm)

Key Features: rosy red head, back, and chest; white belly; rosy-colored rump; brownish wings and tail

Habitat: open woodland; edges

Notes: similar to house finch appearance but purple finch is more reddish than the orange-red of the house finch; loud, rich song; irruptive migration

© Kristi Blokhin/Shutterstock.com

© Harold Stiver/Shutterstock.com

House Finch (*Haemorhous mexicanus*)

Size: 5 in. (13 cm)

Key Features: red-orange face, breast, rump (male; female light brown in those areas); brown cap; brown across eyes; brown wings with white streaks; whitish belly with brown streaks

Habitat: open forests; farmlands; suburbs

Notes: introduced in New York in 1940s, now across North America; messy nest of sticks; often will nest in hanging flower pots; can suffer from a fatal eye disease; warbling song; common year-round resident for most of range

© Brian Lasenby/Shutterstock.com

© karen burgess/Shutterstock.com

American Goldfinch (*Spinus tristis*)

Size: 5 in. (13 cm)

Key Features: yellow body with black cap; wings black with white wing bars (male loses bright color and cap in winter, resembles female)

Habitat: open fields; edges; woodlands

Notes: lines its late season nest with thistle down; flight pattern erratic up and down; twitters in flight; travels in small flocks; song consists of long, varied series of warbles, trills, and twitters while the call can sound like *po-ta-to-chip*; nonmigratory to seasonal movement

© FotoRequest/Shutterstock.com

Evening Grosbeak (*Coccothraustes vespertinus*)

Size: 8 in. (20 cm)
Key Features: stocky bird; yellowish head with bright yellow eyebrows; large ivory or greenish bill (spring); yellow belly and rump; black-and-white wings and tail
Habitat: forests; parks; seen on gravel roads (gets minerals and grit); breeds in forests
Notes: eats seeds, insects, fruit; wandering warbling song; irruptive migrator

© Pyshnyy Maxim Vjacheslavovich/ Shutterstock.com

House Sparrow (*Passer domesticus*)—Family Passeridae

Size: 6 in. (15 cm)
Key Features: brown back with gray crown and black throat extending to chest; gray belly; single white wing bars
Habitat: urban; suburban; farmlands; not typically found in undeveloped or wooded areas
Notes: very abundant; introduced in New York in 1850, now throughout North America; not really a sparrow (in Finch Family); mostly a herbivore; large, oversized nests made with just about anything it can find; aggressive bird; sings *cheep-cheep-cheep-cheep*; nonmigratory

Birds—Natural History Journal Page

Complete a natural history for birds found near your Nature Area.

1. Full Common Name: Bold Eagle

   Scientific/Latin Name (*Genus species*): Haliacetus leukocephalus

   Family Name: Accipitridae

   Order: Accipitriformes

   Class: Aves

   Phylum: Chordata

   Kingdom: Animalia

2. **Sketches**

   Body: draw the bird's typical posture, beak detail, colors, markings, size, and tail length

   Flight: Sketch the bird's silhouette in add a line to show its flight and flight pattern or describe it

   Track: Show what the bird's feet look like; include the size

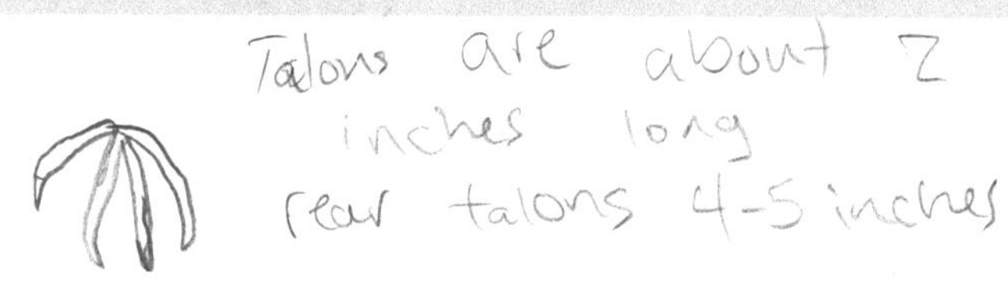

Morphology: Describe the differences in the bird's coloration or looks.

Male: white head and tail, dark large body

Female: Similar to the male, but larger.

Juvenile: full brown body, yellow beak, slightly smaller than adult

Winter: when young eagles are independant of their parents

Breeding: Pressing cloaca toghether, male releases sperm from his to hers

Similar species/look-alikes: Golden eagle

3. What is your bird's size? (in inches or centimeters; and compare to your finger size, fist, hand, forearm size—this will help you quickly communicate/remember)

   28-40 (in.)/cm   Compared to Elbow to finger tips

   What are your bird's identifying characteristics or key features (i.e. colors, the presence and description of an eye circle, crown, chin, breast, bars, striations, tail, feet, etc.)?

   white head and tail. large body.

4. **North America Range Maps** – Shade the regions for where your bird species is found during breeding season, winter, and year-round. Use different colors or patterns to help create a legend.

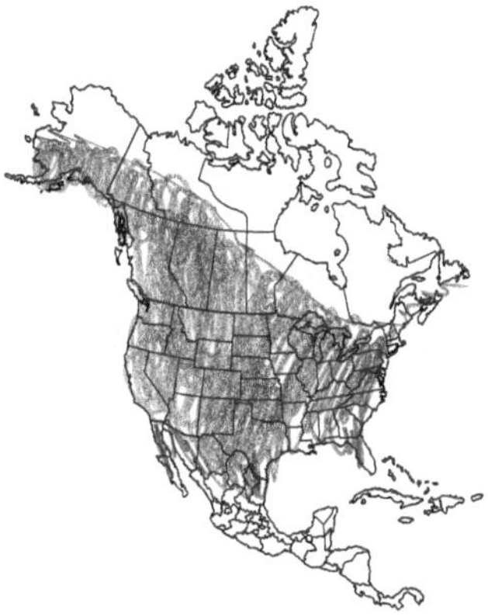

5. Describe the bird's preferred or required habitats.

Within two and a half miles of the coast, bays, rivers, lakes, or other bodies of water

6. Discuss your bird's typical habits and behaviors (may include seasonal changes in behavior, courtship, and feeding behavior).

build large nests, enjoys to eat fish from lakes, rivers. will also eat other waterfowl, Turtles, Rabbits and Snakes

7. What is the bird's preferred diet.

fish.

8. Discuss the predator types and threats to this species.

Humans, Great horned owls, racoons, crows Mainly prey on bald eagle eggs and young.

9. What is the ecological status of the species' population (stable, special concern, threatened, or endangered): Stable What can be done to protect, manage, *or* restore this population?

recovery efforts have been successful as bald eagles are no longer listed as endangered.

10. Discuss your bird's reproductive cycle (When does it breed?) and nesting habits (Where is it built? Who builds and tends it?):

Breeds: late January to march

Nesting habits: builds large nests, protected by both but mostly the female

Nest and egg details (sketch and label nest composition, size, egg color and size, etc.)

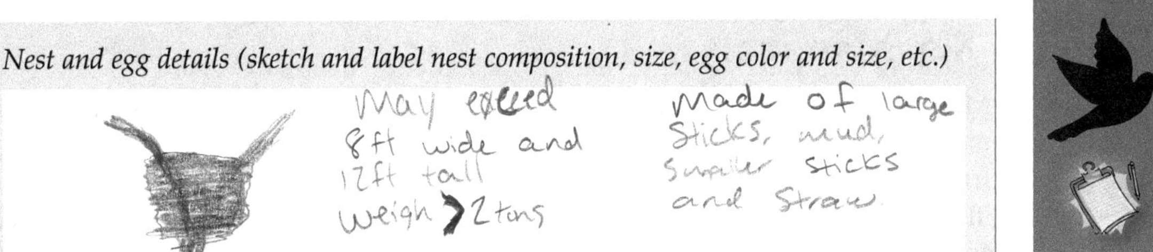

11. Write a description of the sound for the bird's song or call.

High pitched screetch, short and often.

12. Provide other notes or fun facts you discovered but have not yet discussed.

The bald eagles talons are made of the same materials as human and dog nails (very similar to dog nails)

13. List all resources used (**books:** author's last name, first initial, year, and book title; **websites:** name of site, date retrieved, and web address):

Youtube- Tristain Findley
WHPBS.org
Fitzwilliams-Heck, C.J., A practical Guide to nature Study

8.5 Mammals

Mammals live on every continent and in all habitats. Some have specific environmental requirements, whereas others can fit into a variety of places. They range from the size of your finger (a shrew) to the size of a supermarket (a whale—not in our region though!). Some are familiar, some you may have never noticed, or some you do not know.

The smaller mammals are likely the most common and may prove to have very important niches (see Chapter 4 for discussions on ecological relationships and wildlife management). Primary consumers, like deer, also have important roles in the ecosystem but not if their population exceeds its carrying capacity, which can lead to starvation or disease. Wildlife management helps keep herds down in areas of no natural predators.

In this last section of the animal chapter, we delve into Class Mammalia. How are these animals classified otherwise? What is their domain, kingdom, and phylum? At this point, that much of their taxonomy should come naturally as Eukarya, Animalia, and Chordata.

> ***Nature Journal:*** To get in the mindset for learning about mammals, determine what you already know. Reflect on your last experience observing mammals (other than humans). What do you think separates mammals from all other vertebrate animals (in other words, what makes a mammal a mammal)? List as many examples of mammals as you can in 1 minute. How many are found in your Nature Area? What would you like to know about nearby mammals?

Overview

What makes a mammal a mammal? The main classification distinction from other chordates lies with the presence of skin glands, and one in particular. The word "mammal" refers to the female's ability to provide milk for the young by way of her mammary glands. The young are also nourished inside the mother and she gives birth to the young fully developed. Another unique characteristic to mammals is the presence of hair. Mammals are also warm-blooded with an efficient circulatory system including a four-chambered heart. Additionally, mammals have a complex brain capable of learned behavior and a specialized area for seeing and hearing. What's more, they have a variety of adaptations fit for diverse settings.

Mammal Evolution

The appearance of mammals in the fossil record occurs last of any of the vertebrate animals. Or, in other words, the most recent animal evolution are found with the mammals. The first mammals were likely no larger than a modern mouse or shrew, and probably insectivorous. Some present-day mammals likely descended from reptilelike animals, called synapsids from the Triassic Period about 200 million years ago. These animals were small, hairy, nocturnal insect eaters, and laid eggs. They remained very small and had little diversification until the dinosaurs became extinct. The first true mammals arose in the Jurassic, about 150 mya. About 65 mya the mammals expanded to fill new niches. Continental shifting and the appearance of new plant life and conditions caused further speciation and the Cenozoic era became known as the Age of Mammals. By 50 mya, three main groups of mammals were present.

Nature Journal: Research the mammals of the Cenozoic and focus on the last Ice Age (late Pleistocene) for the Great Lakes basin and Upper Midwest region. Were there saber-toothed cats, giant beaver, mammoths, mastadons, giant bison, and others? Which ones still exist, and are there examples for direct ancestors of modern mammals? Discuss possible reasoning for your answers.

Groups of Mammals

Modern mammals have three reproductive strategies and are grouped accordingly from most primitive to most advanced—**monotremes, marsupials**, and **eutherians (placental)** groups. The monotremes are egg layers and have very low populations. Marsupials have young born incompletely developed with further development happening in their mother's ventral pouch. Eutherians, or true mammals, have a placenta to protect the embryo and umbilical cord to nourish it while developing inside the mother.

Monotremes

There are three species of monotremes that currently exist on the planet. This group is considered the oldest and most primitive of the three groups. You will find monotremes in Australia and New Guinea, and nowhere else. These mammals still lay eggs with a leathery shell and have a cloaca. Their milk production secretes from glands found in the stomach region. As the mother sits back, the milk will remain on her stomach while the two newborns suckle her hair and lap up the milk. Monotremes also do not have teeth. Examples include the platypus and a few species of spiny anteaters.

Marsupials

Marsupials, mammals with ventral pouches where the young develop soon after gestation, include opossums, kangaroos, koalas, and more. Slightly less than 300 species exist, most of which live in Australia and New Zealand, such as the kangaroo and koala. We have one type of marsupial in our region, the Virginia opossum or North

American opossum (*Didelphis virginiana*). When the fetus is about half the size of your pinky-finger and has developed its forearms, the mother births the baby and the young then uses its arms to navigate toward the smell of its mother's milk within her front pouch. The young will suckle as it continues to grow and develop.

Placentals

Eutherians, sometimes called "true mammals" or placentals is when the young fully develop within the placenta of the mother. There is a direct, constant connection between the growing fetus and the mother via the umbilical cord used as the young's source of nourishment. In terms of the reproductive parts of the male, they can either exist outside of the abdominal cavity (as in humans) or housed within the body until mating as in many other mammals. Approximately 4,500 species of eutherians exist globally. Unless you encounter an opossum, you know the mammal group you observed in our region is a placental.

Skin Glands

The presence of glands within Class Mammalia separates these animals from other vertebrates. The general groups of skin glands that exist are mammary, sweat, oil, and scent.

Mammary Glands

The most unique gland, and the basis for this animal group's name, is the presence of **mammary glands.** Mammary glands provide milk as nourishment for the young to help them develop and grow after birth. Late in pregnancy, milk production occurs due to the signaling from estrogen levels. Milk storage occurs throughout the breast tissue and travels through pathways or ducts before getting expressed or suckled from the mother's nipple. Once the milk leaves the mammary glands, it triggers an estrogen response to produce more milk. The continued emptying of the milk will stimulate the production of more milk to ensure a constant supply. The milk will cease production if the ducts within the mother's breasts do not get drained.

Not all mammary glands are the same across the three mammal groups. Some mammals lack nipples (monotremes), marsupials may have multiple nipples within the female's pouch, and others have muscles to eject milk with force (whales, porpoises—not found within our region). The number of nipples varies from 2 to 29, but all are arranged along a primitive milk line that runs along both sides of the thorax (or chest) of any mammal type.

Milk has basic ingredients of fat, proteins, lactose, vitamins, minerals, and water but different proportions among species. Breastmilk has highly bioavailable nutrients for the young. For this reason, it is critical that milk given to newborns is from its own species. Breastfed babies get hungry sooner than babies who are formula-fed (synthetic breast milk) because species-specific milk proteins are digested very efficiently. It does not take as much energy to digest its species' milk as it does to digest formulated milk. Frequent feedings also ensure that newborns get attention from their mothers that may be important for development. Additionally, formulated milk for a species has all the necessary vitamins, minerals, and macromolecules needed for healthy development. Formula-fed newborn humans have the potential advantage of bonding with more than one person having the opportunity to feed them. Feeding formulated milk to newborn farm animals or pets may sometimes be necessary, but people should never attempt to provide or feed milk to wild animals.

Sweat Glands

Another unique type of gland within Class Mammalia are the sweat glands. Some of these glands promote evaporative cooling and help eliminate waste. **Apocrine** sweat glands are mostly found in the armpit and perianal areas of humans, and in nonprimate mammals, cover the greater part of their body. In that case, like in canine and felines, the apocrine glands exist at the base of each hair follicle. Modified apocrine glands include mammary glands, **ceruminous** glands that produces ear wax, and **ciliary** glands in the eyelids. The secretion of an oily and odorless substance will coat the hair. Parts of dead skin cells and pheromones may also be released. This substance accumulates bacterial growth, which often causes a unique odor to the individual. These glands do not have a significant effect on cooling humans, but for other mammals it may be their only effective means for sweating and evaporative cooling.

The **eccrine** glands are the most numerous type of skin glands and is scattered over the entire surface of the human body in varying densities. The greatest density is found on palms of hands and soles of feet, then the head, and less on the body trunk and extremities. Sweating through these glands is the primary form of cooling for humans. Nonprimate mammals will only have eccrine glands on their palms and soles.

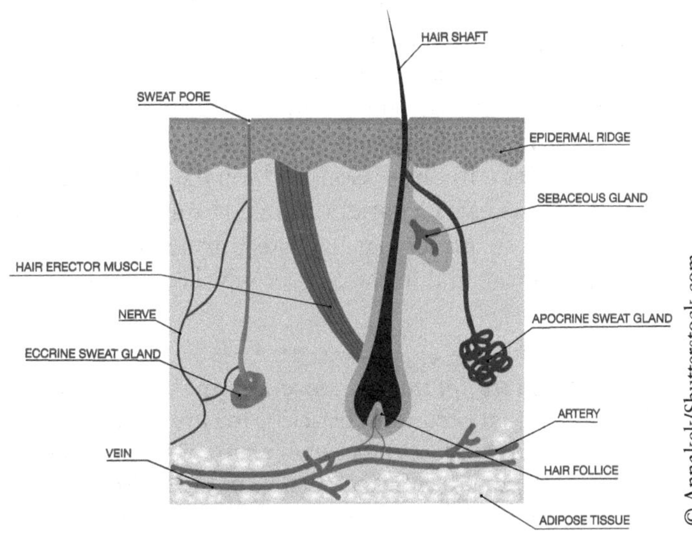

© Annakek/Shutterstock.com

Other varieties of these glands exist that serve a wide array of purposes. Many of which helps mammals survive in their environment. Some mammals use their glands to mark territory, communicate, or defend themselves.

Hair Structures and Functions

Glands are closely tied to hair. The oils produced from the sebaceous glands will help nourish each hair follicle. The hair has a tough tissue made of a protein called keratin. Unlike other parts of a mammal's body, hair consists of dead cells. The living cells exist at the base of the hair at its root. Cell division occurs here and allows for continual growth of the hair.

Each hair consists of an outer layer, or cuticle of cells arranged in a scalelike pattern. The pattern is unique to each species. Have you ever watched or read about a crime-scene investigation? If hair is found as evidence, then it can tell a story. Each mammal species has hair with a unique, scalelike pattern. If you are lucky enough to find hair pulled from the root, then you can have the DNA evidence to help support where the hair originated.

On a friendlier note, have you ever petted a dog or cat? Did you pet it in one direction or two? Probably just one direction. That collection of hair on mammals is called the **pelage** or **pelt**, but the latter term mainly references the fur when it is removed and processed for human use. **Fur** refers to the collection of hair that covers the skin of mammals.

You will notice on most mammals, longer hair extending upward from the body growing toward the back. These are the **guard hairs** and keep moisture from reaching the skin. They also offer protection by allowing the animal to sense the proximity of things nearby and can provide a difference in color or pattern in the coat to enhance camouflage in certain habitats. Beneath the guard hairs, you will find the **underfur** in many mammals. The underfur provides insulation for the animal to keep it warm with its dark, dense hair. In aquatic fur-bearing mammals, the underfur is especially thick for increased warmth. Most mammals will shed or molt in spring and fall, some change color with the season, and some will have sexually dimorphic features. Overall, fur offers insulation; variations in color and patterns to provide potential camouflage capabilities or attract mates; protection against abrasions, sunburn, cold temperatures, and predation; and in the case of whiskers it can help with sensory perception.

> ***Nature Journal:*** Choose any mammal. Think about the type of pelage the animal possesses. Describe the appearance of the hair—color, length, and pattern. Does the appearance ever change? Discuss the suspected insulating properties for the animal throughout the year. How else can the pelage protect this animal?

Mammal Physiology

This section covers the basic functions of internal processes of mammals, including fat and energy storage, circulation, respiration, digestion and excretion, and the nervous system.

Fat and Energy Storage

Although many humans try ridding their body of fat, mammals in the wild rely on fat to serve very important functions for survival. Fat is not unique to mammals but serves similar vital functions across animal groups. Fat is needed for normal physiological functioning. This essential part of the body is found in the skeleton, muscles, cells, the nervous system, and other internal organs. Fat is a source of heat insulation, and a water source for the functioning of the animal's metabolism. This is especially important during cold temperatures and times of dormancy. It is used for energy storage when there are food shortages or during rest and exercise. Fat can also provide padding against trauma.

Circulatory System

All mammals are endothermic (warm-blooded), capable of maintaining a constant body temperature. They have a very efficient, four-chambered heart to help with circulation. There are two types of circulation that exist in mammals, pulmonary and systemic circulation. In **pulmonary circulation**, the blood enters the right atrium, goes to the right ventricle, pumps blood through the pulmonary artery to the lungs. This blood lacks oxygen (deoxygenated) but is rich in carbon dioxide. With **systemic circulation**, the

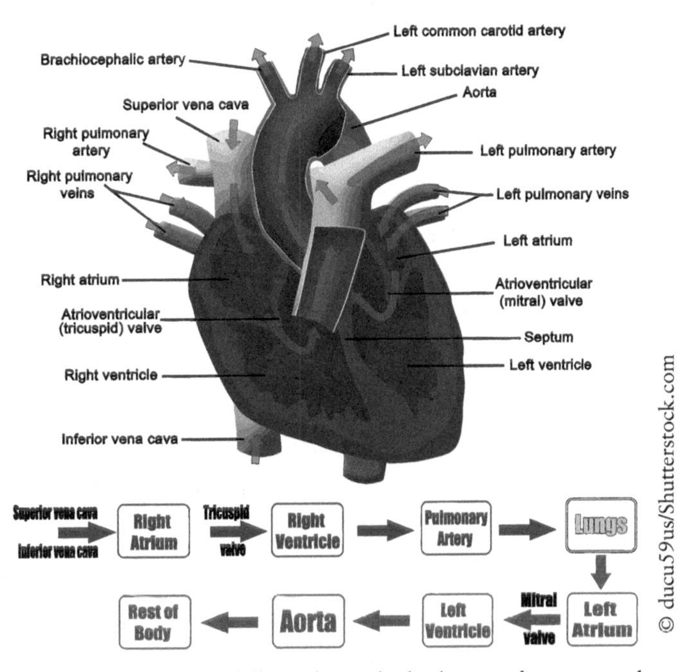

Follow the path of blood flow through the heart of a mammal.

oxygenated blood exits the lungs, and enters the left atrium then travels to the left ventricle to be pumped through the systemic aorta to the rest of the body. The blood is oxygen rich, but lacking carbon dioxide. The red blood cells in mammals are shaped like a biconcave disk and do not have a nucleus. Hemoglobin is a protein found inside the cell and is an iron-based compound with a high affinity for oxygen. Without hemoglobin, the transfer of oxygen through the body to organs and tissues would not occur.

Respiratory System

Something unique to the mammal respiratory system is the presence of the **diaphragm.** This is a muscular layer separating the thoracic (chest) cavity from the abdominal cavity. It helps the lungs breathe with the inhalation and exhalation of air. Breathe in—The air is forced into your lungs by the diaphragm's muscular action (see the diagram for reference). At this time, your ribs raise and the diaphragm contracts straight across the base of your lungs allowing air to go into them. Breathe out—Your ribs lower and the diaphragm relaxes to assume a domelike shape, which forces air out of the lungs.

The respiratory system in mammals facilitates gas exchange. The process begins in the nasal cavity where air is humidified and warmed. The hairs and mucus in the nose help trap particulate matter before entering the system. From the inhalation of air into the nose, the air travels down the pharynx, trachea, and into the lungs. Within the lungs, air goes through the network of **bronchi** and **bronchioles**

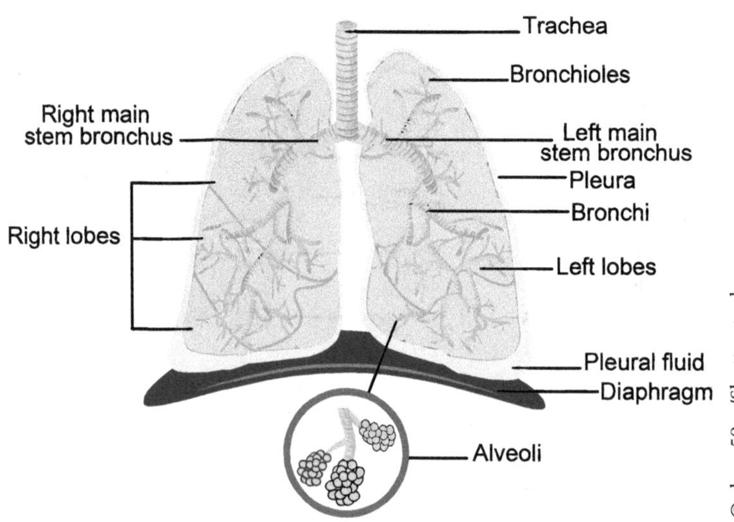

The respiratory system functions similarly for most mammals.

where gas exchange occurs. Those bronchioles open into the **alveoli**. Due to an abundance of alveoli in the lungs, the surface area for gas exchange is great.

Digestive System

Many organs referenced in previous animal sections of the book are also found within mammals. Differences of digestive system physiology among mammal species is mainly dependent on their diet. Although some organ types or functions may differ, the purpose is the same. The digestive system works to digest food and liquids, and to absorb nutrients. Digestion begins in the **mouth** with saliva that starts to break down carbohydrates of the food. It then travels down the **esophagus** to the **stomach** where acids digest proteins. The **liver** is an accessory organ where bile is produced for the **small intestine** to help breakdown fat (its other roles include regulating glycogen, decomposition of red blood cells, and producing hormones). The **pancreas** is another accessory organ in which digestive enzymatic juices are produced for the small intestine to break down macromolecules and regulates blood sugar. The **small intestine** is responsible for further digestion and absorption, and the **large intestine** reabsorbs water from any undigested food

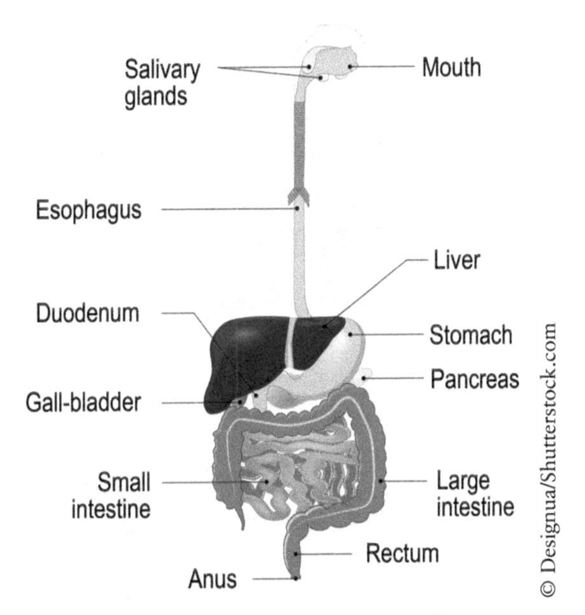

The digestive system in many mammal species. The diagram depicts a human system, and it should be noted that not all mammals have a gallbladder (deer and rodents are common examples).

and stores the waste before elimination. The **cecum**, **colon**, and **rectum** are all parts of the large intestine and are the sequential stages in which the undigested food passes.

Nervous System and the Senses

In this highly complex part of the animal, sensory information is transmitted throughout the body and creates a coordinating response. The nervous system detects environmental stimuli impacting the body and works with other parts of the body to respond to the events. It can be divided into sensory and motor functions.

The brain and spinal cord make up the central nervous system, and the peripheral nervous system are the nerves found everywhere else in the body. In mammals, the brain is typically large in relation to its head size and compared to other chordates. Its covering, called the **neocortex**, has an important role in many functions of the brain. Mammal brains have many folds and ridges that increase the surface area to allow for more neurons. Experiential learning is demonstrated in mammals where actions and behaviors are based on previous experiences, and they have the ability to learn new things.

All tissue of the nervous system has cells called **neurons** that carry electrical signals. This is information being conducted throughout the body up through the central nervous system to the brain from which a response is made. Neurons consist of and receive signals from the **dendrites**, to the **cell body**, then to its **axon**. The brain responds to the signals throughout the stimulated muscles and reacts accordingly through its motor functions.

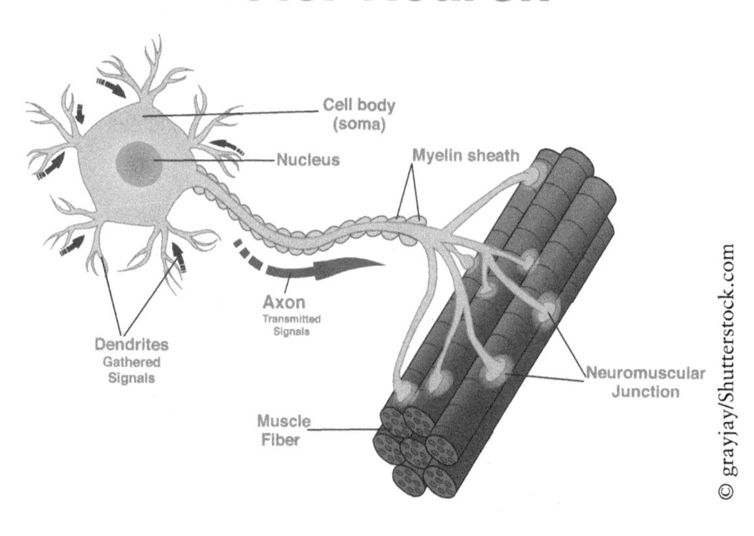

In terms of sensory perception, take a look at any mammal. Which sense organ is the most pronounced or obvious? This is usually an indication of what the animal relies on the most for navigating and sensing its environment. For wild dogs (i.e., wolf, coyote, and fox), the snout or nose is very pronounced. The sense of smell for dogs is 1 million times more acute than it is for humans. The mammal brain evolved to accommodate advanced olfactory capabilities, and for most mammals it is their dominant sense. However, this is not true for humans and other primates. Vision and tactile discrimination, or sense of touch is most important with these mammals. Hearing is also essential to many mammals as well. The overall sensory perception of mammals is relatively quite advanced.

Many mammals do not have color vision, and seeing color has less importance for the nocturnal animals. The eye is like other chordates but has special adaptations. A unique, flexible lens within the eye can adjust to view near and far. It also has a reflective layer, called the **tapetum lucidum** found in nocturnal mammals that shines light back onto the retina. This increases the amount of light gathered to aid in night vision. Have you ever seen an animal's eyes glow or shine at night when exposed to bright light? The eyeshine of animals at night is the result of excess light bouncing off that tapetum lucidum. Eyeshine will glow a color indicative of the species. It may be sometimes hard to discern and can have variation due to your angle of perception. For the most part, felines will shine green, deer will have a golden-greenish white with a bluish periphery, and canines will have a whitish eyeshine with a blue periphery.

The sense of hearing is also highly developed and serves as a substitute for vision in 20% of mammals. To help intensify sound, they have three tiny middle ear bones that detect the sound vibrations. Mammals have an outer ear, the **pinna** to capture the sound waves effectively. Bats have an exceptionally developed ear anatomy used to accurately collect sound waves (read more about bats in the field guide). Yet another unique sensory perception is through vibrissae or whiskers. When present, these also may act as organs of touch to let animals know when things are close to their face, especially in the dark.

Skeletal System

Another form of protection is the mammal's complex skeleton. It protects the animal's vital organs and provides structural strength. By definition, all vertebrate animals, or chordates, have an internal bony support structure that muscles and ligaments are attached. The basic plan consists of a bony skull that protects the brain and supports facial features. The single lower jaw, another feature of mammals, is separate from the skull but directly hinged to it instead of attached to a separate bone as in the other chordates. The skull is at one end of a vertebral

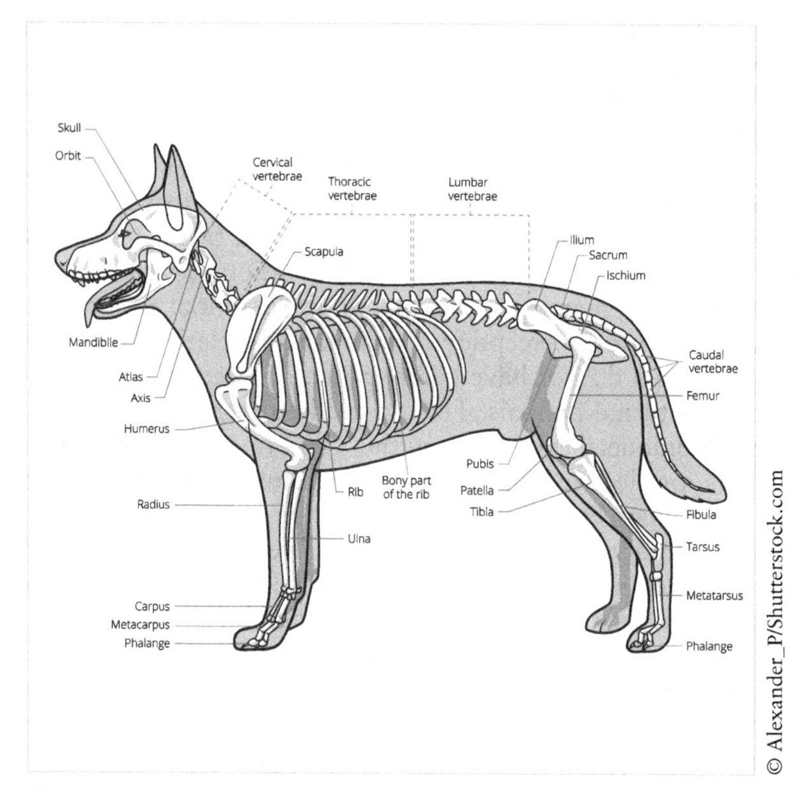

The skeletal anatomy of a canine provides an example of the bone types in mammals.

column of which has the distinctive regions that follow the cervical, thoracic, lumbar, sacral, and caudal vertebrae. Did you know that most mammals have seven neck vertebrae regardless of neck length?

The Teeth

Teeth are also considered part of the skeletal system. You can determine an animal's niche (job or role in its environment) just by looking at its teeth. There are different types of teeth used for different types of food. Each mammal order will have certain teeth indicative of their diet.

Mammals' teeth have been modified through evolutionary time according to food availability. Two important teeth types exist for carnivores. Canines are long pointed teeth used for killing and piercing flesh. Carnassial teeth are sharp, nonoccluding cheek teeth for cutting meat. Herbivores do not have canines!

Carnivores have a limited number of molars. The skulls of herbivores have strong, flat molars for grinding plant matter. Some herbivores move their jaws sideways when chewing to help grind their food further. Omnivores will have a combination of sharp teeth in the front and molars in the back for grinding.

Most mammals do not have the same teeth they are born with in their adulthood (this is not true for the moles in our region)! Mammals exhibit **diphyodonty**, or simply the process of tooth replacement. The weak milk teeth (baby teeth) are replaced by stronger, permanent teeth by adulthood.

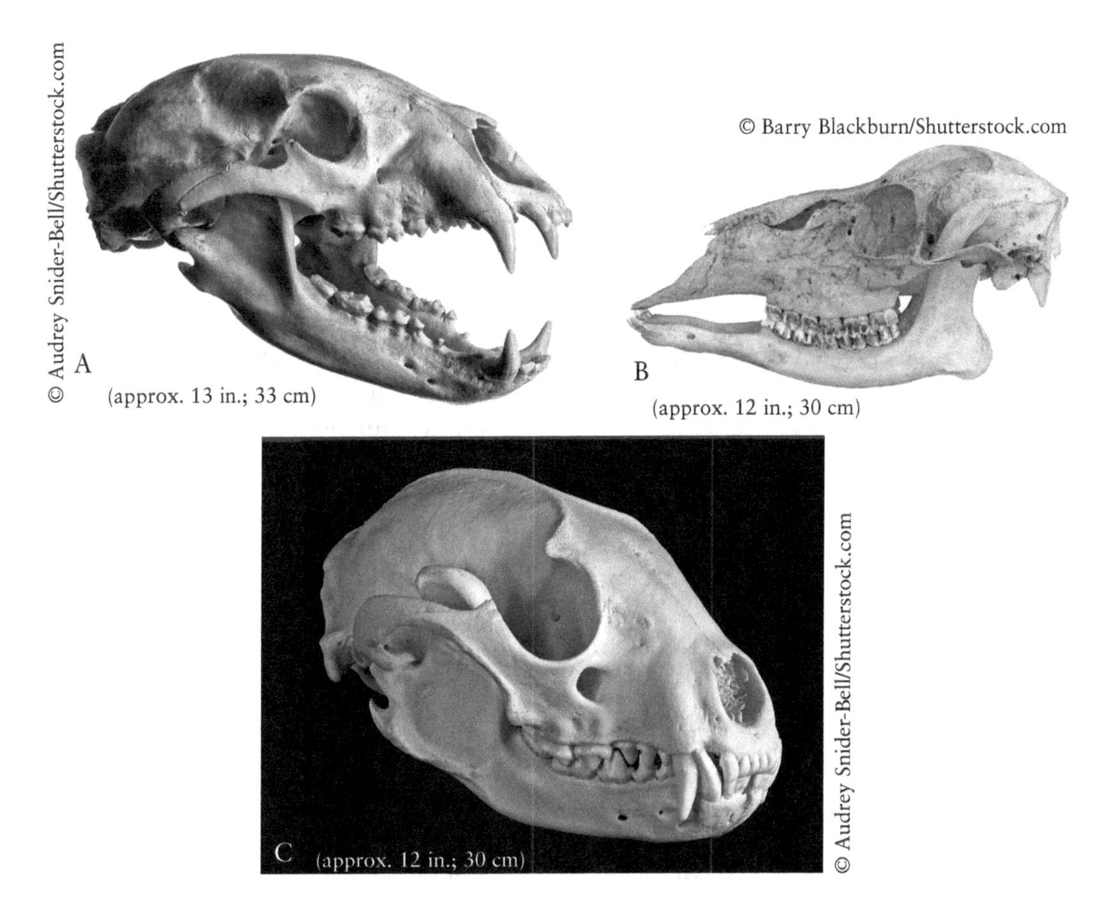

Which of the skulls are from an herbivore? Carnivore? Omnivore? In A, all the teeth are pointed, whereas in C there are some more squared molars for chewing plants. What regional animals do you think these skulls were from? Notice that B does not have canines, it has squared and flat molars to chew plants, and incisors (only on the bottom of this species) to strip off vegetation.
Answers: (A) black bear (carnivore), (B) white-tailed deer (herbivore), and (C) raccoon (omnivore)

Skull Comparison

Have you ever found a skull while on a nature walk? The flesh, membranes, and soft tissue will be eaten by its predator, scavengers, and detritivores and decomposers. The bone is all that remains. Usually the skull is left behind by predators or scavengers because not much meat exists on the head and it is also difficult to carry off because of its shape. Other bones can get crushed and carried off to feast upon. Eventually, the bones will break down by the weather and decomposers and get covered by the soil and leaf litter.

How will you know what type of skull you found? The types of teeth present will help you identify the animal group it belongs. Rodents have enlarged incisors, and canines indicate a carnivore. Other things to notice to help you learn more about what animal it was include its size measured from front to back, the length of the snout (nose) in relation to the overall size of the head, and the size and position of the eye sockets. Considering the animal's dominant sense when it was alive can help you determine the skull's identification. Also, if the eye sockets are set in front of the skull, then it is likely a predator ("eyes in front, likes to hunt"); a prey animal will have its eye sockets positioned on the side ("eyes on the side, likes to hide"). Other intricacies may also help with identification to a particular species (ridgelines present on top of the skull – a single is likely an opossum, whereas connecting ridgelines could be a canine; fenestra or filigree structure on the sides is found on rabbits and hares, etc.). You can look at your field guide to help you narrow down your potential animals and then look on the internet for its skull.

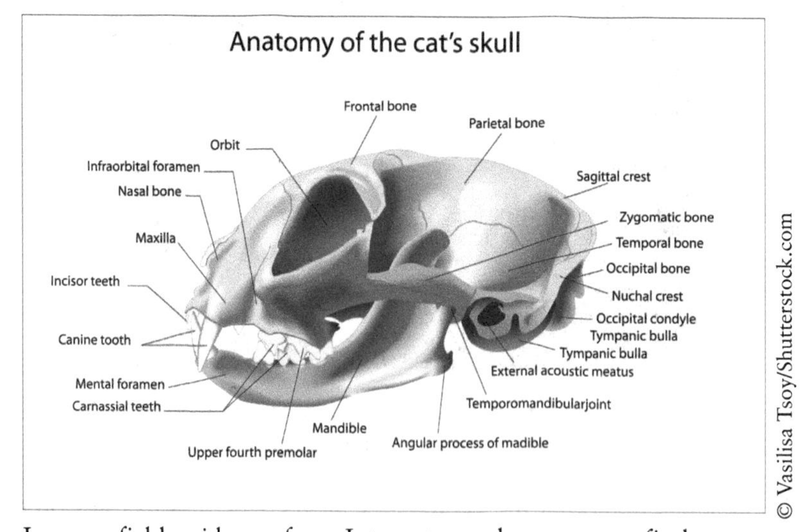

In some field guides or from Internet searches, you can find reliable and detailed references for identifying skulls. You can use this diagram to help learn the features of the skull.

What Can You Do?

As you gain a greater insight on mammal anatomy, physiology, morphology, and its orders; let's think about how you can make a positive difference in the conservation of these animals.

Many people appreciate nature and its inhabitants even if they do not ever witness it firsthand. Just knowing that things exist can bring a peace of mind. This cannot happen without conservation and management of our natural resources. Investigate the types of organizations near you like watershed councils, land conservancies, and government fish and wildlife agencies.

Get to know the mammals near where you live. Spend time outdoors and find sit spots wherever you go. Keep a list of what you observe along with photographs or drawings, and do the research on its natural history and conservation to learn more. You can also upload your nature observations (of any plant or animal) on free apps for community science programs or for your own purposes. On many sites, you can simply upload a photograph and have the option to provide its species name or ask the community to help with its identification. Some sites can also track where your observations occurred. This is a great way to document your inventories.

Furthermore, research your local wildlife conservation efforts, species, sightings, nuisance species, disease, and management practices.

Nature Journal: Use the template in this section to complete Natural History Journal Pages for a common mammal in or near your Nature Area. Consider conducting a field investigation about mammals. A descriptive study could involve discovering as many mammals (or signs of mammals) around your site during a certain time (morning or evening is best). See Chapters 1 and 2 for more ideas.

Spend time outside. Consider trail cameras on your property. Try visiting different locations (sides of home, habitats, and places), times of day, weather, and seasons. National Parks and state/provincial refuges and forests offer land for you to explore. Refer to Chapter 4 for discussions on ecological relationships and conservation approaches.

Learn about the different mammal orders and familiarize yourself with some common species in the field guide section that follows.

Review Questions

1. Discuss the basic evolutionary history of mammals.
2. How are the three groups of mammals classified? Explain the differences between these three groups of mammals. Where are each type found? What animals are represented in each group?

3. What characteristics do mammals have that make them a mammal?
4. What is the function of mammary glands? Discuss the significance of species-specific milk. When and how is milk production stimulated?
5. Give two reasons why sweat glands are important. Describe the difference between eccrine and apocrine glands. Do all mammals have the same number of sweat glands and where are they located? Discuss the following gland types: sebaceous, scent/musk, and ceruminous.
6. What is the general composition of hair as discussed in the text? What are the names and functions of the two layers of a mammal's pelage? When does molting typically occur? What are three types of camouflage that may exist on a mammal's pelage? How would each help protect the animal in their habitat?
7. Fat and energy storage is not unique to mammals, but fat servers three important functions. Explain each function.
8. Differentiate between pulmonary and systemic circulation. How does heartbeat rate differ between sizes of species? What is the importance of hemoglobin and where is it found? What is the structure and function of alveoli? Where is the diaphragm located? How does the muscular action of the diaphragm function in inspiration and expiration?
9. Describe the structure of a mammal's brain. How is a mammal's actions and behaviors different than the other animal types?
10. Discuss mammalian senses. The mammalian eye has two main modifications—what are they and how are they used? How are vibrissae used?
11. Draw and label the types of teeth you would typically find in carnivores, herbivores, and omnivores (use the terms canine, incisor, premolar, and molar). What is each type of tooth used for specifically? Discuss the composition and characteristics of rodent incisors (see the field guide for more details on rodents).
12. From the field guide, discuss general characteristics for each group of mammals the order, family, and genus.

Field Guide: Mammals

Finding mammals in the wild requires patience, keen observational skills, and staying alert. Unlike birds or flowers, mammals are inconspicuous. They may stay out of sight by burrowing or having a nocturnal niche. Mammals have keen senses (and many humans have noisy behaviors) and are alerted to you approaching. They will likely leave an area before you observe them. Moreover, many have concealing qualities like cryptic coloration.

Signs of Mammals

Look for the clues in nature of what may have passed through the habitat before you did. Have you ever noticed these signs: bones, fur, tracks (footprints), scat (feces), vegetation that has been browsed, or twigs or tree trunks chewed? You

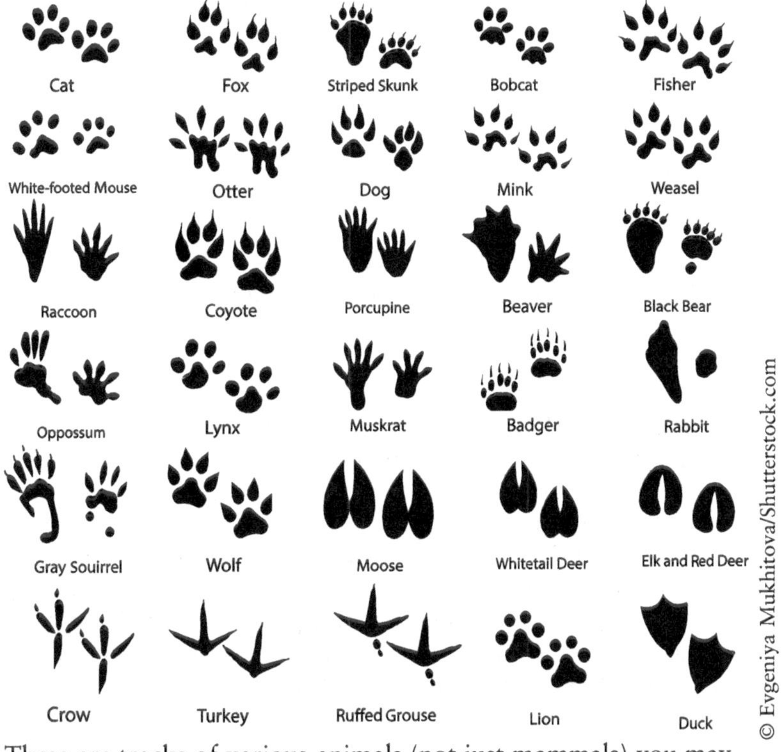

These are tracks of various animals (not just mammals) you may or may not find in the region. Please realize it is not to scale and not to scale between species (i.e., the white-footed mouse footprints are not bigger than a bobcat!). The chart will provide a reference for the general shape of the track for each species displayed. Which tracks pictured are not mammals? Compare closely related species. Which are not found where you live?

can determine a great deal by making close observations of these discoveries, in particular, skulls, as mentioned previously, and mostly tracks and scat. These things can be used for an absolute identification of a species! By learning the basics about the mammals in our region, you can start to make an educated deduction of who may have been in the habitat, the color of their fur, their size, relative shape of their track, and what they may eat.

Try to find and identify tracks. The best time to find them is after a fresh snow, or any time the ground is soft and at least a little damp to hold shape of the imprint.

Nature Journal: During times of snow or mud, venture outside to look for tracks. You can also designate a small area with about a half-meter square of sand near a potential food source, smooth it out, and dampen it before dark. If you have a trail camera, then set it up to view if any animals walk through the sand to get to the food source. When you find prints, try to figure out which direction it was traveling, how long ago do you think it was in the area, was it alone, what do you think it was doing, and try to determine the possible species who made it. Take measurements of tracks and the animal's gait. Sketch your findings. Within your Nature Journal, keep all records of tracks in your Nature Area.

Mammal Species Accounts

The presence of hair or fur will provide the biggest clue that an animal belongs to Class Mammalia. Size and behavior will be further indicators of the type of mammal.

Next, consider the mammal orders listed here to discover a more specific identification. Note the key features presented in the field guide for each order to get an idea of what to notice when you encounter those mammals in nature. In general, get an impression of the animal's shape, color, ear size, and tail length. From there, the options become narrow as you apply range, habitat, characteristics and behavior.

Realize that all mammals are protected and regulated in most regions except for the Norway rat and house mouse. Familiarize yourself with laws in your area.

Mammal Identification Key

Use the key below to help you quickly find the mammal order in which your animal in question may belong. The animals are mainly organized by relative size between and within the orders and family. In some instances, animals are arranged within the family based on similarities. The sizes provided include the head and body length with the tail length separately (except bats).

1. Bats 
2. Shrews and Moles 
3. Rodents—mice, voles, chipmunk, squirrels, muskrat, porcupine, woodchuck, and beaver 
4. Rabbit and Hare 
5. Opossum 
6. Carnivores—raccoon, skunk, weasels, foxes, coyote, wolf, bobcat, lynx, cougar, and black bear 
7. Hoofed Animals—white-tailed deer, elk, moose, wild boar, and bison  

1. Bats—Order Chiroptera, Family Vespertilionidae

Chiroptera includes the only flying mammal—bats, who are far more maneuverable than birds. Bats have a hairy body with membranous, naked wings and tail. Bats are nocturnal and eat mostly insects (e.g., mosquitoes). You will find a greater density of bats near freshwater sources. Bats roost in colonies to conserve heat with mostly females and young; they are usually separate from males. You will often find piles of dark brown scat under these roosting sites. They go into a torpor state when temperatures are near and below freezing (in prolonged periods of cold, they will enter hibernation to conserve energy). To help with the bat population and lower mosquito densities, consider putting up a bat house. Bats are prey for some owl species and raccoons have been observed raiding colonies during the time of the bats' roosting.

We have many different species of bats. In the region, approximately 10 bat species exist. They typically roost in sheltered warm and dark places during the day like hollowed trees, buildings, or rock outcrops. During the winter months, females will migrate long distances to reach a **hibernacula**. This is where many bats colonize to hibernate, and the ambient temperature remains constant with a humidity greater than 90% and above freezing.

Due to the **white-nosed syndrome (WNS)**, many bats have died. WNS causes a fungal growth to grow on the bats' muzzle and wings while they are hibernating in the winter. This causes damage to their wings, tail, and ears, which affects their balance and perception of their surroundings. For more information on WNS, research online for whether it is in your state or province.

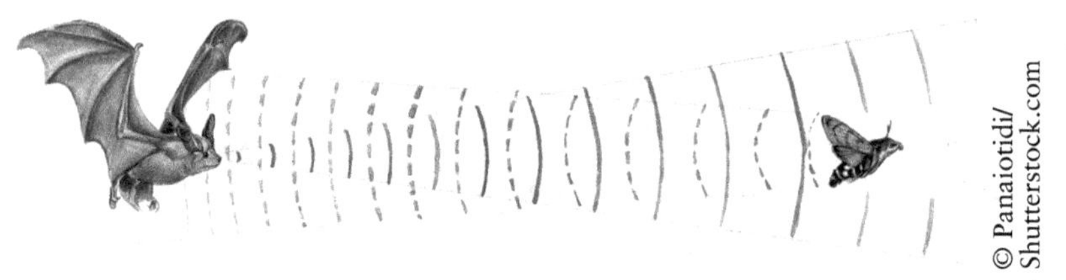

Bats use **echolocation** to detect their prey. They will send out a high-frequency sound that pulses through the air. The sound waves will bounce off nearby objects and if the frequency matches a potential prey, the bat will increase the number of sounds emitted so it can hone in on the prey.

Watch a bat fly at night and notice its flight pattern and other behaviors. Recognizing a bat in flight can become easy with practice, but identifying bat species in flight can be very difficult. Although bats seem to have an erratic flight pattern compared to birds, bats have more maneuverability and efficiency due to their highly articulated arm structure. Bats also tend to hunt by following a detailed internal map of their habitat.

Research the common bats that live in your region to help determine the possible types you might encounter. Roosting bats make it easier to determine species, but never handle bats—especially with bare hands. Seek medical attention if ever bitten by a bat (or by any animal). This is the only way rabies can be transmitted (less than 1% of all bats have the disease).

© Liz Weber/Shutterstock.com

Little Brown Bat or Little Brown Myotis (***Myotis lucifugus***)
Size: 3.1–3.7 in. (8.0–9.5 cm); wingspan of 8.7–10.6 in. (22.2–26.9 cm)
Key Features: small; low zigzagging flight; dark wings; variable glossy (backside) fur color from pale tan, reddish, to dark brown; belly lighter than back; relatively short snout; gently sloped forehead
Habitat: near woodlands and water; roosts in tree cavities, buildings, bat houses
Notes: most abundant; common around cities; winters in mines and caves (migrates)

© Jay Ondreicka/Shutterstock.com

Big Brown Bat (*Eptesicus fuscus*)
Size: 4–5 in. (11–13 cm); wingspan 13–14 in. (33–35 cm)
Key Features: large size; strong, steady flight; dark wings; dark body with lighter belly; pointed snout
Habitat: meadows; forests; suburbs; city traffic; near water; active at twilight; erratic flight
Notes: common; feeds on many crop and forest pests; prefers to roost alone in winter or in small groups; pictured is a wildlife biologist checking wings for signs of WNS; tends to winter in buildings

© Elliotte Rusty Harold/Shutterstock.com

Eastern Red Bat (*Lasiurus borealis*)

Size: 4 in. (11 cm); wingspan 13–14 in. (33–36 cm)
Key Features: very fast flier (only out when completely dark); distinctive, dense, red fur on back (sometimes frosted with white); white underside and shoulders sometimes appear white
Habitat: deciduous forests; edges; suburbs; cities; farmlands
Notes: abundant tree bat where trees are present; roosts alone on south side of trees or cliffs during the day; of all the regional bats, has the most young per year (1–5; average 3 pups); migrates south in flocks

Silver-Haired Bat (*Lasionycteris noctivagans*)

Size: 4.5 in. (10 cm); wingspan 13–14 in. (33–36 cm)
Key Features: medium-sized; slow-flying around treetops; mostly black with white-tipped hair (frosted appearance); flat skull; broad rostrum
Habitat: deciduous forests (old growth; may require at least 40 dead standing trees per acre); suburbs; cities; farmlands
Notes: females and juveniles spend summer in our region, whereas the males tend to live in western North America; both overwinter in southern states

2. Shrews and Moles—Order Eulipotyphla

This order includes shrews and moles. These animals have a high metabolic rate and a voracious appetite. They move almost all the time, mainly underground and eat often. While digging tunnels, some species use echolocation to find invertebrates. Shrews and moles have great senses of touch, hearing, and smell. They have ears and eyes, but you will notice they are barely visible.

If you were to run your hand down their pelt and then up again, you would discover on some species that the hair is bidirectional or smooth in both directions (unlike your dog or cat's fur!). This helps these animals move forward and backward in the tunnels they dig without much resistance. Finding networks of ridges near the surface is an indicator of these animals burrowing underground. They benefit the environment by aerating the soil and allowing moisture to penetrate deeply.

Shrews—Family Soricidae

Looks like a long-nosed mouse but it is not a rodent. Mainly terrestrial omnivores with some species that are arboreal or hunt in water. They do not hibernate but will enter states of torpor (long sleep periods during harsh conditions). Shrews do not lose their milk teeth, and some will have dark red teeth because of iron in the enamel. The iron reinforces the teeth, which is helpful with the nearly nonstop eating. A venom is conducted into prey (insects, worms) through chewing it with its unique grooves in their teeth. Recent studies show that the venom has promising uses in the treatment of human conditions and diseases like high blood pressure, migraines, and ovarian cancer. Shrews are territorial, have up to 10 litters a year, and live 1–2.5 years.

Masked Shrew (*Sorex cinereus*)
Size: 1 ¾–2 ¼ in. (4.5–5.5 cm); tail 1–2 in. (2.5–5 cm)
Key Features: grayish-brown (lighter underside); long tail; pointed snout
Habitat: farmlands; moist soil of fields; brushy areas; wetlands; mesic forests
Notes: energetic, secretive and solitary; active day and night; does not burrow as much as others but will use their tunnels; one of smallest mammals of region and most widespread in North America

© Melinda Fawver/Shutterstock.com

Northern Short-Tailed Shrew (*Blarina brevicauda*)
Size: 3–4 in. (7.5–10 cm); tail ¾–1 in. (2–2.5 cm)
Key Features: dark gray; large size; pointed snout; relatively short to medium tail
Habitat: wide variety—forests; fields; yards
Notes: abundant; very carnivorous eating insects, spiders, earthworms, mice, toads

Moles—Family Talpidae

Similar to shrews, but will be a larger size with big front feet with soles facing outward for digging; tails are stubby with or without hair

© Agnieszka Bacal/Shutterstock.com

Star-Nosed Mole (*Condylura cristata*)

Size: 3–7 in. (7.5–18 cm); tail ¾–1 ¼ in. (2–3 cm)
Key Features: dark gray fur; large, fleshy pink nose with many projections; long tail with sparse hairs; large, naked front feet
Habitat: moist soil of woodlands; fields; wetlands; lakes; rivers
Notes: nose able to detect electrical fields given off by prey underground, and may help manipulate food; tail fattens during warmer seasons as an energy reserve; eats mostly earthworms and insects; often in colonies; most parts of region (mainly northern); the picture here shows animal standing on hind feet

© Melinda Fawver/Shutterstock.com

Eastern Mole or Common Mole
(*Scalopus aquaticus*)

Size: 4–7 in. (10–18 cm); tail ¾–1 ¼ in. (2–3 cm)
Key Features: soft fur gray (north) to golden or dark brown (south); long, pointed, pink snout; large, hairless front feet; tail hairless
Habitat: loamy soil of woodlands and fields
Notes: spends most of time underground; eats mostly earthworms; most parts of region (mainly southern); contrary to species name it is not aquatic (it was presumably found in a well when first identified)

3. Rodents—Order Rodentia

mice, vole, chipmunk, squirrels, porcupine, woodchuck, muskrat, and beaver

Order Rodentia, or the rodents, is the largest mammal order found all over the world and in every habitat. You can recognize rodents by their enlarged incisors that appear orange in color.

Most rodents are omnivores, and dependent on species, will continually gnaw on objects or food sources. The rodents' incisors have a front, hard enamel, and a softer **dentine** on the backside. The dentine wears faster than the enamel resulting in always being sharp. If rodents did not constantly chew on objects, the incisors would continually grow making them unable to eat. Many rodents hibernate where body temperature approaches freezing temperatures and heart rate slows to 4 beats/minute. They may awaken to eat during the winter, depending on species and the conditions—otherwise, species will be active unless extreme weather conditions forces them to the nest and may enter torpor. Young are born hairless and helpless in spring or midsummer of a litter between 2–8, weaned at 4–10 weeks; mature at end of first year; and live 5–10 years in the wild (each dependent on species).

© Landshark1/Shutterstock.com

Meadow Jumping Mouse (*Zapus hudsonius*)—Family Dipodidae
Size: 2–3 in. (5–7.5 cm); tail 4–6 in. (10–15 cm)
Key Features: reddish-brown to yellowish-brown back with lighter sides; white belly; large hind feet; very long tail (dark above and white below); large round ears; long snout
Habitat: moist open field; clearings
Notes: widely distributed; can leap up to 3 ft. (1 m) to escape predation but mostly walks on all fours or with a series of small jumps (will remain motionless after jumping); uses tail to help jump; feeds on an underground fungus (*Endogone* sp.) and does not store food in winter; true hibernator; rarely enters buildings

© Szasz-Fabian Jozsef/Shutterstock.com

House Mouse (*Mus musculus*)—Family Muridae
Size: 2.5–4 in. (6–10 cm); tail 2–4 in. (5–10 cm)
Key Features: small size; grayish brown on back; light gray belly; pointed snout; large rounded ears; long hairless tail gray above and about same length as body
Habitat: buildings; farmlands (sometimes)
Notes: abundant; will inhabit human dwellings; wild populations less common

© Karel Bock/Shutterstock.com

Deer Mouse or North American Deermouse (***Peromyscus maniculatus***)**—Family Cricetidae**

Size: 3–4.5 in. (7.5–11 cm); tail 2–4 in. (5–10 cm)
Key Features: variable colors gray to reddish-brown; white below; white feet; tail dark above and white below and same length as body; bulging eyes; prominent ears
Habitat: all habitats; buildings
Notes: widespread across continent except southeast states; vector for diseases like Lyme disease and hantavirus; not aggressive; climbs trees and shrubs; caches food for winter; usually solitary except in winter

© Rudmer Zwerver/Shutterstock.com

Meadow Vole (*Microtus pennsylvanicus*)—Family Cricetidae

Size: 4–5 in. (10–13 cm); tail 1.5–2.5 in. (4–6 cm)
Key Features: dark brownish-gray back and sides with reddish highlights; gray below; short snout; small ears; medium-sized tail (half size of body)
Habitat: moist grasslands, farmlands, and woodland edges
Notes: active year-round, usually nocturnal; digs burrows; aggressive toward one another; damages gardens and crops

Squirrel—Family Sciuridae

Small to medium-sized rodents with slender bodies, bushy tails, and large eyes; their coat color can be variable between and within species; hind limbs longer than forelimbs; paws have a poorly developed thumb but strong claws for grasping and climbing; they can descend a tree headfirst; live in most habitats; mainly herbivores; excellent vision and sense of touch; mate once or twice per year; altricial young; **diurnal** or **crepuscular**

© Steve Byland/Shutterstock.com

Eastern Chipmunk (*Tamias striatus*)

Size: 6–8 in. (15–20 cm); tail 3–4 in. (7.5–10 cm)

Key Features: reddish brown with one white stripe bordered by two dark stripes running length of back to the rump; underside pale white to gray; reddish-brown rump and hairy dark tail; loud vocalizations

Habitat: forests; forest edges; gardens; rock piles; near human dwellings

Notes: ground-dwelling squirrels but can also climb trees; burrows extensive up to 12 ft. long (4 m) that has multiple rooms for various purposes and concealed entrances; mainly solitary; piles found of cracked seeds and acorns on logs or clearings; transports large quantities of food in its cheek pouches; caches food in its underground chambers; very tolerant of people; partially hibernates in winter

© Svetlana Foote/Shutterstock.com

Thirteen-lined Ground Squirrel or Federation Squirrel **(*Ictidomys tridecemlineatus*)**

Size: 6–8 in. (15–20 cm); tail 2–5 in. (5–13 cm)

Key Features: narrow brownish body with three alternating brown and whiteish lines (or spotted lines) on back and sides; short round ears; short legs; thin hairy tail

Habitat: fields; woodlands; brushy areas

Notes: ground squirrels; larger than chipmunks with more stripes; does not fear humans; semisocial with other ground squirrels; live in large colonies in separate burrow; fast runner with zigzagging patterns; population may be lacking in southeast part of region

American Red Squirrel or Red Squirrel **(*Tamiasciurus hudsonicus*)**

Size: 7–9 in. (18–22.5 cm); tail 4–7 in. (10–18 cm)

Key Features: rusty reddish-brown above with brighter fur on sides and bright white belly; distinctive white ring around eyes; fluffy tail with black tip; tufted ears in winter

Habitat: coniferous and deciduous forests

Notes: smallest of the three tree-dwelling (arboreal) squirrels; defends a year-round territory; wheezy barks or raspy chatters; nests in **drey** of mostly bark and twigs with some leaves close to the tree trunk or in tree cavity or takes over leaf nest of Eastern gray squirrel; eats pine cone seeds, nuts, corn, fungus (can consume poisonous *Amanita* sp.), bird eggs, baby birds, and carrion; creates a **midden** of discarded pine cone parts and other nut fragments in a pile (acorns will have a ragged opening with nutmeat missing); diurnal and year-round (stays in nest during extreme weather)

© Paul Reeves Photography/Shutterstock.com

© My Generations Art/Shutterstock.com

© Paul Reeves Photography/Shutterstock.com

Eastern Gray Squirrel (*Sciurus carolinensis*)

Size: 9–12 in. (23–30 cm); tail 7.5–10 in. (19–25 cm)

Key Features: mostly gray fur but can have brownish color and a black morphology (higher cold tolerance); bushy tail (sometimes reddish); white fur underside

Habitat: deciduous forests; suburbs; parks

Notes: creates numerous caches of food (often retrieved by very accurate spatial memory using landmarks and smell); an important forest regenerator; known to appear like making a cache; prolific and adaptable (invasive in Britain!); communicates through varied vocalizations and posturing; diurnal and year-round (stays in nest during extreme weather); diurnal (mostly midday) and year-round

© VasekM/Shutterstock.com

Eastern Fox Squirrel (*Sciurus niger*)

Size: 10–15 in. (25–38 cm); tail 8–13 in. (20–33 cm)

Key Features: mostly dark gray fur with rusty orange highlights on side; rusty orange chin and underside; large, fluffy tail dark gray top with rusty orange sides and highlights; some variations in morph can occur

Habitat: suburbs; parks; woodlands

Notes: drey (leaf nest) in summer (up to six nests) in fork of tree trunk; up to 2 ft. (61 cm) wide lined with plant material; omnivore; compared to other tree squirrels spends more time away from trees; large home range upward of 50 acres (20 ha)

© Agnieszka Bacal/Shutterstock.com

Northern/Southern Flying Squirrel (*Glaucomys sabrinus* and *Glaucomys volans*)

Size: Northern 7–9 in. (18–22 cm); tail 4–7 in. (10–18 cm); Southern 5–7 in. (13–18 cm); tail 3–5 in. (7.5–13 cm)

Key Features: grayish-brown on back; white below; wide flat tail (gray above, white below); folds of skin between front and hind legs; large eyes

Habitat: mixed and deciduous forests; parks

Notes: nocturnal tree-dwelling squirrel; does not fly but glides with its arms and legs outstretched from tree to tree at regular high launching points (20–50 ft.; 6–15 m); lands at a lower point on another tree then scurries to opposite side; most carnivorous of tree squirrels (kills mice, baby birds and feeds on eggs, carrion); year-round (torpor when extremely cold)

© Stuart Monk/Shutterstock.com

Woodchuck or Groundhog or Whistle Pig (***Marmota monax***)

Size: 18–28 in. (45–71 cm); tail 3–6 in. (7.5–15 cm)
Key Features: chunky body; various shades of brown, gray, black; large head; short legs; small bushy tail; small round ears; dark feet; no distinctive markings
Habitat: fields; open forests; edges; roadsides; around homes
Notes: largest member of squirrel family; a type of marmot; does not emerge on February 2 to look for its shadow (or anything else!) and continues to hibernate until late winter; extensive burrow system with tunnels, chambers, and entrances; feeds mostly on grass; gives a sharp whistlelike call and has a stout stature when alarmed; diurnal, mainly crepuscular

© Geoffrey Kuchera/Shutterstock.com

North American Porcupine (*Erethizon dorsatum*)—Family Erethizontidae

Size: 20–26 in. (50–66 cm); tail 6–12 in. (15–30 cm)
Key Features: chunky body; dark brown or black with white-tipped guard hairs; arching back; stiff quills; slow moving; short legs with small feet and long claws; tiny eyes
Habitat: forests (mostly seen in trees); brushy areas (occasionally)
Notes: does not shoot quills; raises its 30,000 quills when threatened, turns its back and strikes with tail, often the loosely attached barbed quills get embedded into the predator's flesh; dens in trees; nocturnal in trees where it gnaws bark, twigs, buds; one offspring per year born with quills, eyes open, teeth present; active year-round

© Mircea Costina/Shutterstock.com

© Tony Moran/Shutterstock.com

Muskrat (*Ondatra zibethicus*)—Family Cricetidae

Size: 8–12 in. (20–30 cm); tail 7–12 in. (18–30 cm)
Key Features: glossy dark brown to reddish; lighter on sides and belly; tail hairless, black, vertically flattened, scaly; small round ears; tiny eyes
Habitat: wetlands; ponds; lakes; slow streams
Notes: aquatic mammal; varied diet; sometimes observed hundreds of feet from water to harvest vegetation; several litters per year each 6–11 young weaned at three weeks; builds lodges smaller than beaver's made of plant matter rather than sticks and mud

© Jody Ann/Shutterstock.com

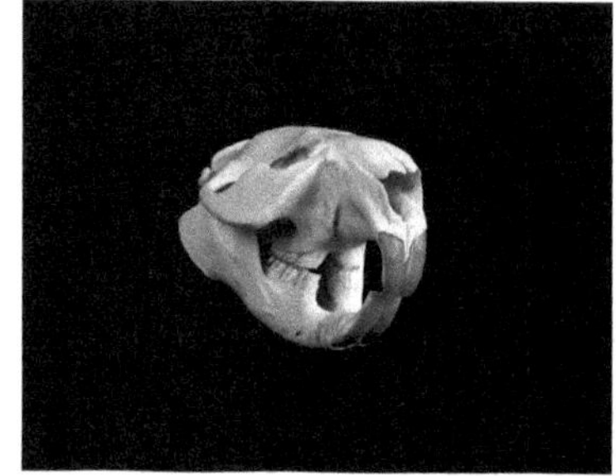

© Audrey Snider-Bell/Shutterstock.com

© Procy/Shutterstock.com

© Sgeneralov/Shutterstock.com

© Adwo/Shutterstock.com

American Beaver (*Castor canadensis*)—Family Castoridae

Size: 3–4 ft. (1–1.2 m); tail 7–14 in. (18–36 cm)

Key Features: large body; reddish-brown fur with lighter head; tail large, flat and paddelike, naked, scaly; large, orange incisors (exposed); tiny eyes; webbed hind toes

Habitat: lakes and rivers bordered with poplar, birch, or other food tree

Notes: largest member of rodent family; body adapted for swimming (has ear and nose valves to close out water, clear membrane over eyes for seeing underwater, special lip sealant so incisors remain exposed while carrying sticks in water); can stay submerged for 15 minutes; slap water with tail when alarmed then dives underwater; loud chewing and gnawing sounds when felling or feeding on trees; incredible dam-building ability with underwater foundation of stone and mud; leaves cone-shaped stumps and drags or floats cut trees to dam site; secretes pungent castor oil from glands near tail and marks territory and boundaries (castor mounds); uses its specialized claw on hind feet for grooming; lodge of sticks and mud is built by a beaver pair behind the dam in the formed pond; has inside platform above water and underwater entrances; makes repairs on dams based on sound of moving or falling water; dams alter ecosystem in significant ways (keystone species); kits born furred with eyes open and able to swim in one week; monogamous and mates for life; nocturnal, mainly crepuscular; active year-round (in lodge in winter and active under ice)

4. Rabbit and Hare—Order Lagomorpha, Family Leporidae

Order Lagomorpha has two species in our region, the Eastern Cottontail and Snowshoe Hare. The ears, legs, and feet of the hare are larger than rabbits. Hares can run faster. They live in open habitats, so the long legs and swiftness is a necessity for survival. The large feet help maintain position on top of deep snow. The long ears help with hearing, but the snowshoe's ear size is smaller than other hares due to the colder weather where it lives.

Although it may be difficult to see in the pictures or in the wild, rabbits differ from hares in very distinctive ways. First, consider their ranges. The cottontail is found all over most of the region except mainly in the northern reaches. Snowshoe hares will be found in the northern ranges of the region and also in areas of higher elevations. The hare adaptations you will read make sense when you think about where they live. Both species molt or shed their coat, but the hare will have a more dramatic seasonal color change than rabbits. The hare will go from a brown to a bright white to provide camouflage in the winter. The rabbit will have a slight color change.

Baby rabbits (kits or kittens) are born hairless, blind, and helpless needing attention of the mother for eight weeks. At birth, hares (the leverets) are fully furred with eyes open and ready to hop. They can live on their own hours after birth and weaned by three weeks.

Rabbits prefer grasses and vegetables with leafy tops, whereas hares eat heartier substances like plant shoots, twigs, and bark. However, both eat woody twigs in winter when vegetation is scarce. These herbivores are also **coprophagic**, meaning that they will eat their own feces until no further nutrients can be acquired. Their homes also differ. Eastern cottontails burrow in dens made by other species like woodchucks. Hares live above ground and will den in hollowed logs or will make a simple nest by trampling down vegetation. Lastly, hares are generally solitary except during breeding. This is unlike most rabbits, except for the eastern cottontail who also often prefers the solitary life.

© JamesChen/Shutterstock.com

© Chinch/ Shutterstock.com

Hind tracks register above front feet in the rabbit's hopping gait.

Eastern Cottontail or Cottontail Rabbit ***(Sylvilagus floridanus)***

Size: 14–18 in. (36–45 cm); tail 1–2 in. (2.5–5 cm)

Key Features: grayish-brown body; black-tipped ears; large pointed ears; distinctive reddish nape; brown tail with conspicuous white cottonlike underside when running

Habitat: open grassy areas or old fields with shrubs; wooded thickets; wetlands with adequate cover; seldom in deep woods

Notes: most common species of rabbit in North America; important prey species; lives in small area of about two acres; flattens ears when threatened then runs in zigzagged patterns circling back to starting point (bounds up to 12–15 ft.; 3.7–4.5 m); not territorial; up to five litters per year with three to seven hairless born blind and hairless in a small excavated burrow made by the mother after mating; young may breed that same season; signs of rabbit include low woody twigs with clean, angular cuts from their incisors (unlike deer browse with ragged twig tips); scat is light brown, woody, pea-sized; nocturnal, mostly crepuscular

© Studiotouch/Shutterstock.com

© Jim Cumming/Shutterstock.com

Snowshoe Hare or Varying Hare ***(Lepus americanus)***

Size: 15–20 in. (38–50 cm); tail 1–2 in. (2.5–5 cm)

Key Features: dark brown body with black-tipped hair giving it a shaggy look (white in winter with brown nose and black-tipped ears); belly light gray or white; long pointed ears (black along edges); tail brown above and white or gray below; large hind feet; feet often have a yellowish color

Habitat: boreal forests; wetlands; fields; thickets

Notes: runs in circles when chased (up to 30 mph; 48 kph) and bounding up to 14 ft. (4.3 m); wallows in a dust bath in summer; takes up to three months to transform coat color creating many unique color patterns; dense fur develops on pads of feet in autumn

5. Opossum—Order Didelphimorphia, Family Didelphidae

In this next order, Didelphimorphia, we only have one representative—the Virginia opossum, the only marsupial north of Mexico (learn about marsupials within the chapter)! The Virginia opossum prefers wet habitats but will be found almost anywhere. Considered nocturnal and omnivorous, they will frequently feed on dead animals along roadways at night and get hit by cars. They are considered survivors because they seem to have an incredible healing ability, and have a resistance to venomous snake bites, rabies, and plague. Although not fast runners, they will show aggression when threatened by showing their teeth (the most teeth than any mammal) and hissing. They also often play dead when feeling defenseless against a potential threat. Have you ever heard the phrase "playin' possum"? It refers to when somebody is motionless, intentionally unresponsive, or pretending to sleep. This is what the opossum does—they will roll over, close their eyes, open their mouth, and let the tongue hang out. They can do this for a relatively long time. The Virginia opossum can also swim and climb to escape potential danger or to take advantage of resources. Their tail is semiprehensile, meaning they can use it to grab ahold of things like branches while climbing. They cannot, however, hang by their tail.

© Rosa Jay/Shutterstock.com

Virginia Opossum or North American Opossum or Opossum (***Didelphis virginiana***)
Size: 25–30 in. (63–76 cm); tail 10–20 in. (25–50 cm)
Key Features: white face with long, pointed pinkish nose; grayish-brown to black body; white underside; wide mouth with many teeth; black, hairless, oval ears; tail long, scaly pinkish; pink toes
Habitat: usually near water; farmlands; forests; suburbs
Notes: after young leave pouch (2–13) where they nurse for many months, they will ride on the mother's back for many weeks; with their 50 teeth (more than any mammal), they have a diverse diet and also prey on ticks; nocturnal, year-round; oftentimes, their naked tail and ears get frostbite, turn black, and fall off

6. Carnivores—Order Carnivora

raccoon, skunk, weasels, foxes, coyote, wolf, bobcat, lynx, cougar, and black bear

Carnivora includes the animals who eat predominantly meat. Found globally, you will also observe sexually dimorphic animals in size with males larger than females. Members of this order have keen senses for hunting and often have means of communicating through chemical signaling. These predators help control prey population numbers in what is often called, "top-down" control.

As you learn about the major families of this order, take note of the animals' eye position. You will find them in the front and not the sides of the head—"eyes in front, likes to hunt." These are the hunters. Eyes in the front allows the animal to hone in on and triangulate depth perception to help with targeting prey. When observing the skull of a member of Order Carnivora, the presence of carnassial teeth indicate a meat-eater with the adaptation of efficient sheering and cutting of flesh.

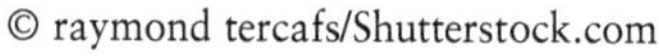

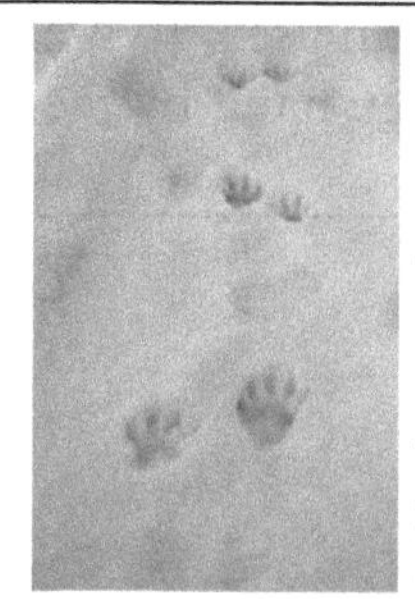

Northern Raccoon (*Procyon lotor*)—Family Procyonidae

Size: 24–25 in. (60–63 cm); tail 7–16 in. (18–40 cm)

Key Features: bushy body; grayish-brown; dark eye mask; dark-ringed tail

Habitat: all habitats wetlands; fields; forests; edges; suburbs; urban

Notes: common; highly adaptable, opportunist (taking advantage of easy food sources rather than hunting); with highly developed tactile sense, fingerlike paws help capture and eat prey and to climb; handles food prior to eating it (often manipulating it in water to help get most edible parts); excellent hearing and sense of sight; signs include distinctive tracks, piles of half-digested berries in conspicuous places, and cylindrical scat (2 in., 5 cm long; ¾ in., 2 cm wide); three to six offspring born with eyes closed; nocturnal; do not hibernate but may sleep for about a month from January–February

Striped Skunk (*Mephitis mephitis*)—Family Mephitidae

Size: 20–24 in. (50–60 cm); tail 7–14 in. (18–36 cm)

Key Features: black with white "V" stripes on back; white face stripe; large, mottled bushy tail

Habitat: open forests; farmlands; suburbs; urban; usually near water

Notes: common; omnivore; defense mechanism of releasing a potent smelling spray at potential predators; gives clues it will spray by raising its tail and hairs, stomping, presenting its backside, and sprays toward eyes of its predator (can aim with relatively good accuracy up to 15 ft.; 4.5 m); four to seven offspring born hairless with black and white skin; nocturnal; mostly active spring-fall (dens in winter during coldest parts of season; up to 15 individuals)

Weasels—Family Mustelidae

The Family Mustelidae have the most carnivorous members of all the carnivores. They often will have a surplus of food nearby to satisfy their insatiable appetite and high metabolism. Found in almost all habitats, these proficient hunters are recognized by their elongated, sometimes broad agile body and short legs and snout. The mustelids have well developed anal scent glands to mark territory and for defense.

After partaking in copulation that can take up to several hours to induce ovulation, the female weasel has the ability for delayed implantation of the embryo up to 10 months. Examples of mustelids, or weasels in the region include **ermine, long-tailed weasel, mink, marten, fisher, American badger, and northern river otter.** Contradicting evidence exists that wolverines ever inhabited or had a reproducing population in the area.

© FotoRequest/ Shutterstock.com

© Thierry de Villeroche/ Shutterstock.com

Ermine or Stoat or Short-Tailed Weasel (***Mustela erminea***)

Size: 7–10 in. (18–25 cm); tail 2–4 in. (5–10 cm)

Key Features: long, slim body; light to dark brown body; tail black-tipped; white chin and underside; white feet; all white in winter except for black tail tip

Habitat: open areas within forests, with brushy edges; wetlands; farmlands

Notes: ferocious; energetic; climbs well; preys on rodents and known to chase squirrels (sometimes killing them); bites base of skull severing spinal cord; laps blood before eating it; dramatic color change; loud chatters and shrills; nocturnal except in winter; known as ermine during winter and stoat during summer

© Dubsma93/Shutterstock.com

Long-Tailed Weasel or Big Stoat or Bridled Weasel (***Mustela frenata***)

Size: 8–16 in. (20–40 cm); tail 3–6 in. (7.5–15 cm)

Key Features: long, slim body; brown body with white or pale yellow below; tail brown with black tip; brownish feet; body turns white with black tail tip in winter (does not change color in southern part of range)

Habitat: open areas; forests; usually near water

Notes: larger and stronger than ermine; preys on similar animals as ermine plus rabbits, ground nesting birds, and chickens

© shauttra/Shutterstock.com

Mink or American Mink (***Neovison vison***)

Size: 14–20 in. (36–50 cm); tail 6–8 in. (15–20 cm)

Key Features: long, slender body; dark brown or black (sometimes reddish-brown or tan); often shiny fur; chin white or pale color; some white spots on underside

Habitat: wetlands; rivers; lakes; farmlands and forests near water

Notes: always near water; excellent swimmers; prey on aquatic and terrestrial animals (i.e., fish, frogs, snakes, chipmunks, rabbit, and prefers muskrat); 40 acre territory (16 ha); fierce fighter; screams, hisses, spits, and emits pungent musk (as all weasels); 2–10 kits; diurnal and nocturnal (hunts then rests for many hours)

© Green Mountain Exposure/Shutterstock.com

Northern River Otter or North American River Otter or Common Otter ***(Lontra canadensis)***
Size: 2.5–3.5 ft. (76–107 cm); tail 11–20 in. (28–50 cm)
Key Features: dark brownish-black fur with lighter underside; elongated body with thick, heavy tail tapered at tip; short snout and white whiskers; small eyes and ears
Habitat: rivers; large lakes; and neighboring aquatic areas
Notes: semiaquatic; playful and social; treads water; slides down riverbanks; rolls in vegetation leaving its musky odor; variety of vocalizations; not afraid of humans; feeds on slow-moving fish; sensitive to water pollution; diurnal and nocturnal; active year-round

© Matt Knoth/Shutterstock.com

American Badger (*Taxidea taxus*)
Size: 20–30 in. (50–76 cm); tail 3–6 in. (7.5–15 cm)
Key Features: flattened body; shaggy fur of brownish-black and white; distinctive black-and-white pattern on its triangular face; white cheeks and ears; small gray tail; huge foreclaws
Habitat: fields; farmlands; woodland edges; dry, treeless areas
Notes: secretive; very vocal when threatened; persistent fighter; hunts cooperatively with coyotes; digs fast, large burrows used while waiting for prey to be chased into it and to escape danger; nocturnal; torpor during winter

Foxes, Coyote, Wolf—Family Canidae

Within Order Carnivora, we have canines or dogs as part of Family Canidae. The canines found wild in nature include the red fox, gray fox, coyote, and gray wolf. In appearance, they have the general features of a deep-chested body and a long muzzle (snout). Although these animals are within the carnivore order, they have more omnivorous tendencies. If you ever observe a canine in pursuit of its prey, you will witness an aptitude for endurance over speed. They can last long distances at a steady pace with the intent of outlasting the prey's endurance. Have you ever watched a pet dog play with a small stuffed animal or toy? What does it do when it gets it in its mouth? Shakes his head back and forth, right? This is what wild canines do once they have their prey. The target is the prey's nape, or base of its neck. It will then tackle the prey and violently shake it with the goal to dislocate the neck (like the toy!). In instances where the prey is much larger than the canine, like the moose and the gray wolf, the wolf will attack the soft underbelly of the moose or will target the young or weak. Canines also have an acute sense of smell and hearing, which prove critical to their existence and success. Dogs mark their territory by repeated urination throughout the area in which they claim. They have anal glands on the underside of the tail that will emit pheromones to communicate to other canines an element of defense, breeding readiness, or something relatively insignificant. All are elusive and avoid humans. They are curious and will take advantage of easy food sources like small pets, livestock, and garbage. Red fox has the most distinctive coat whereas the other canines' coats can vary. Gray foxes are the only tree climbers! The coyote is the most ubiquitous, found most anywhere. The gray wolf has made a tremendous comeback in its population and is found in the northern parts of MN, WI, MI and in Ontario with potential of further expanding. Foxes are not pack animals, meaning they will travel alone except during breeding season while living with and hunting for its family. The wolf and coyote typically form groups or packs to help with hunting and raising young.

Have you ever wondered if that rectangular-shaped track you observed in the mud or snow came from somebody's large pet or a coyote? Something to look for is if a "center lobe" exists at the base of the track. If so, you may have a coyote in the neighborhood. What if it was a small track? First, look for the presence of claw marks. If it has claws and is shaped like a dog track you might be looking at a fox print. Look at the position of the two outer toes. If they are behind and away from the inner toes, then you have a fox nearby! Another feature to consider with canines is the inability to retract their claws. Contrarily, cats can retract their claws so you would typically not find any claw marks associated with cat tracks and their prints would have a rounder shape than canines.

© Tory Kallman/ Shutterstock.com

Red Fox (*Vulpes vulpes*)

Size: 22–24 in. (56–60 cm); tail 13–17 in. (33–43 cm); height 15–16 in. (38–40 cm)

Key Features: reddish to rust color with dark highlights; white underparts; bushy tail tipped white; black legs and feet

Habitat: farmlands; open forests; suburbs; cities

Notes: common; intelligent; catlike pouncing behavior; sleeps at base of trees or large rocks; opportunist eaters, preying mainly on small mammals; barks, yelps, high-pitched screams or cries; scat tapered at end; straight line of tracks with direct register; dens in logs, old woodchuck den, or digs in stream bank or hillside with tall mound in front of entrance; 1–10 kits per years; nocturnal, mostly crepuscular; active year-round

© Holly Kuchera/Shutterstock.com

© Holly Kuchera/Shutterstock.com

Gray Fox or Treefox **(*Urocyon cinereoargenteus*)**

Size: 22–24 in. (56–60 cm); tail 10–17 in. (25–43 cm); height 14–15 in. (36–38 cm)

Key Features: coarse coat of salt-and-pepper highlighted with orange at nape, shoulders, sides, and chest; cheeks, chin, and underside white; large pointed ears with white inner; large bushy black-tipped tail with black stripe

Habitat: open woodlands; rock outcrops; river valleys

Notes: tree climber that can jump from branch to branch (backs down or runs headfirst out of tree); burrows in ground to escape or give birth; dens similar to red fox; mates for life; nocturnal and crepuscular, but diurnal in winter

© Jim Cumming/Shutterstock.com

Coyote (*Canis latrans*)

Size: 3–3.5 ft. (1–1.1 m); tail 12–15 in. (30–38 cm); height 2 ft. (61 cm)

Key Features: grayish-tan fur with orangish-red on flanks; tan legs, feet, and ears (white inner); long snout with white upper lip; long bushy tail tipped black and held between legs when running

Habitat: open forests; fields; farmlands; brush; suburban, urban

Notes: very common (except where wolves live); opportunists; barks and calls to others with howling and yipping; can cooperatively kill a deer; been known to mate with domestic dogs; monogamous; four to six pups a year; seen all times of day and night; active year-round

© Volodymyr Burdiak/Shutterstock.com

Gray Wolf (*Canis lupus*)

Size: 4–5 ft. (1.2–1.5 m); tail 14–20 in. (36–50 cm); height 26–38 in. (66–96 cm)

Key Features: gray with dark and silver highlights (variations of white to black); large bushy tail tipped black; ears short, pointed; long legs

Habitat: open forests; prairie

Notes: complex social structure; travels great distances with territory of up to 300 square miles (780 sq. km); can go weeks without food; 1–10 pups per year; nocturnal but more diurnal in winter

Bobcat, Lynx, Cougar—Family Felidae

Felines, or cats have a specialized method of hunting and prefer to kill their own prey rather than scavenge on another's kill. Cats will stalk prey until its quick, final rush or pounce. Aiming for the prey's neck, they will sever the cervical vertebrae with their canine teeth and will kill or paralyze the prey.

Members of Family Felidae will have acute senses of sight, smell, and hearing. They also depend on their whiskers to sense their surrounding environment. Cats do not travel in packs and live most of the year solitary. Their tracks typically do not show claws (retractable) and have a rounded shape with direct register. They have five toes in front with four toes in back. Wild felines found in the Great Lakes and Upper Midwest region include the bobcat, Canada lynx, and cougar or mountain lion. The bobcat is found throughout the area. Canada lynx has a population growing smaller as the climate warms and habitats change. They may still be found in northern MN, WI, perhaps in the northwestern part of the Upper Peninsula of MI, and in Ontario. The cougar populations have a wider range than the lynx.

© Sam Carrera/Shutterstock.com

Bobcat (*Lynx rufus*)

Size: 2.5–3.5 ft. (69–107 cm); tail 3–7 in. (7.5–18 cm); height 2 ft. (61 cm)

Key Features: fur varies from tawny brown (summer) to light gray (winter) and from dark (forests) to light (open areas); spots on belly; stripes and spots on legs (especially in winter); tapered hair hanging down from jowls; triangular ears tipped with black tufts; white spot on back of ears; short stubby tail with black tip and white sides and underneath

Habitat: mixed forests; wetlands; farmlands; fields

Notes: common; rabbit favorite food; can go several weeks without eating; swims well; dens in logs or under branch piles; one to seven kittens per year; signs of scratch posts 3–4 ft. (1–1.2 m) above ground; active day and night

© Erni/Shutterstock.com

Canada Lynx (*Lynx canadensis*)

Size: 2.5–4 ft. (76–122 cm); tail 2–5 in. (5–13 cm); height 2–2.5 ft. (61–69 cm)

Key Features: fur color varies, usually grayish-tan; black tufts on triangular ears; long stiff fur from hanging down from jowls; black-tipped tail; scattered spots (mostly summer; less than bobcat); long legs and large feet

Habitat: boreal forests; swamps; rock outcrops

Notes: elusive; agile climber; good swimmer; swift traveler; populations fluctuate in 10-year cycles based on prey; snowshoe hare preferred food; defines boundaries with scent markings; nocturnal, mostly crepuscular

Cougar or Mountain Lion or Puma ***(Puma concolor)***

Size: 5–6 ft. (1.5–1.8 m); tail 2–3 ft. (61–91 cm); height 2.5–3 ft. (76–91 cm)

Key Features: tawny brown to gray (spotted when young); small head; long tail tipped black; white upper lip and chin; dark spot at base of white whiskers; pink nose; long legs; large feet

Habitat: woodlands; wetlands; river valleys; rock outcrops

Notes: rare; only hunts about once per week; caches prey under leaf pile or branches; excellent climber; leaps up to 20 ft. (6.1 m); home range of male up to 115 square miles (300 sq. km) and female half the range; usually three cubs every two years; mainly nocturnal (rests in trees during day usually near recent kill); active all year

Bears—Family Ursidae

Black Bear or American Black Bear ***(Ursus americanus)*—Family Ursidae**

Size: 4.5–6 ft. (1.4–1.8 m); tail 3–7 in. (7.5–18 cm); height 3–3.5 ft. (1–1.1 m)

Key Features: mostly black (sometimes tan or cinnamon); brown snout; may have white patch on chest; short, round ears; short tail

Habitat: forests; swamps; farmlands; prefers inaccessible terrain with thick understory vegetation and abundant food (trees/shrubs, berries, dead trees for insects); extirpated in south part of region (IA, IL, IN, OH), but some reports indicate populations expanding

Notes: only bear species in region; large, very intelligent, curious, and exploratory; hardly eats meat, especially from any prey they have taken (carrion or from another's kill or roadkill constitutes most of its meat); prefers vegetation, berries, and frequents dead trees for insects; feeds heavily in summer to prepare for hibernation when it does not eat, drink, urinate, or pass feces; bears can be roused during winter; cubs born during hibernation in January or February; female (sow) will have one to five young (typically two to three) and will nurse through the winter within the den; nocturnal but often feeds during day, considered crepuscular; if you suspect you are in bear territory, make noise (whistle, talk loudly) so that the animals know you are coming and do not feel surprised; bears can run fast, climb, and swim.

7. Hoofed Animals—Order Artiodactyla

deer, elk, moose, wild boar, and bison

The hoofed, wild animals in our region belong to Order Artiodactyla, meaning even-toed ungulates that include deer, elk, moose, wild boar, and bison (not truly wild anymore). The two hooves, or toes of these animals bears the weight of the animal. Worldwide, the order includes over 200 species such as deer, antelope, sheep, goats, cattle, pigs, hippopotamuses, giraffes, camels, llamas, and alpacas. Many of these animals have a dietary, economic, and cultural importance to humans. The artiodactyls live mainly in open habitats. Some members of this order have a simple stomach for digestion, like pigs and hippos. Most other members can **ruminate**—or regurgitates food and chews it for greater digestive efficiency.

Deer—Family Cervidae

Our main focus for Order Artiodactyla (even-toed ungulates) include the wild, native species in Family Cervidae. Deer are found throughout the region, elk are found in the northeast part of Michigan's lower peninsula, and moose inhabit the northern part of the region. Cervids are **ruminants** with a four-chambered stomach and bacteria to digest plant matter. Ruminants can consume a great amount of food and digest it later.

In the field guide photographs, notice the size of the animals' ears and snout. This indicates the significance or dependency of these senses for its survival. The eyes are relatively small in comparison. They can see fairly well but will hear or smell potential danger much sooner than seeing it. When observing deer from the side, you notice the presence of an eye. Their eyes are on the side of the head—"eyes on the side, likes to hide." This is an indicator the animal is a prey species. As identified in the section about mammal skulls, the deer's teeth arrangement clearly indicates an herbivorous lifestyle. Having no canines, and no incisors on the top make it perfect for stripping vegetation and the square molars are great for chewing the plants.

Antlers will be found on males within Family Cervidae. The antlers are shed annually and made of bone that typically has a branching pattern. Notice the difference between the antlers seen for this family to that of a bison's permanent horns. As the antlers grow each spring, they will have a soft, velvety covering that helps nourish the antler as it grows from a cartilage into bone. The velvet is eventually rubbed off after the bone matures. At this time, the male is ready to "show off" his new set off antlers to nearby females and defend against other males. Mating occurs in fall with about six to seven months gestation and one to two fawns (calves for elk and moose). The Cervids are active year-round.

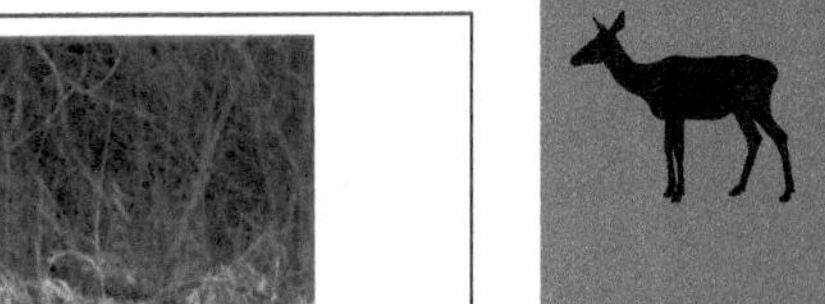

Single deer track oriented in the direction it was headed

Scat

White-Tailed Deer (*Odocoileus virginianus*)

Size: 4–7 ft. (1.2–2.1 m); tail 6–12 in. (15–30 cm); height 3–4 ft. (1–1.2 m)

Key Features: fur reddish (summer) and grayish-brown (winter); tail white on underside (raised when alarmed); large ears rimmed black with white inners; white eye ring, nose band, chin, throat, underside; antlers (male; buck) have main beam with several tines (spread of 12–36 in.; 30–90 cm); females (doe) with thinner neck

Habitat: forests; wetlands; farmlands; fields

Notes: very common with widespread range; herds will gather near food supply; eats 5–9 lb per day (2–4 kg); eats acorns in fall, grasses and grains in spring and summer, twigs and buds in winter; tracks, scat, tree rubs, beds are common signs; nocturnal, mostly crepuscular

© Ghost Bear/Shutterstock.com

Elk or Wapiti ***(Cervus canadensis)***

Size: 7–9.5 ft. (2.1–2.9 m); tail 3–8 in. (7.5–20 cm); height 4.5–5 ft. (1.4–1.5 m)

Key Features: large size; reddish-brown fur; pale rump; short tail; male (bull) with huge antlers (spread of 4–5 ft.; 1.2–1.5 m); female (cow) with thinner neck and lighter mane than male

Habitat: mixed forest; fields; farmlands; northeast Lower Peninsula of MI

Notes: gregarious; highly territorial; wallows in dust in summer to protect against insects; fast animal (35 mph; 56 kph); strong swimmer; most polygamous animal; bugles during rut that can be heard miles away; nocturnal, mostly crepuscular and seen during day; active year-round

© Jack Bell Photography/Shutterstock.com

Moose (*Alces alces*)

Size: 7–9 ft. (2.1–2.7 m); tail 4–7 in. (10–18 cm); height 6.5–7.5 ft. (2–2.3 m)

Key Features: large size; dark brown to reddish fur (lighter in winter); dark bulbous muzzle; obvious humped shoulders; dark belly and legs; large light brown ears; small brown tail; bulls (male) have dewlap (flap of skin under chin); bulls with flattened antlers (palmate shaped) with many points (spread up to 5 ft.; 1.5 m)

Habitat: mixed forests; boreal forest; wetlands; rivers; lakes; found in northern MN, WI, MI's Upper Peninsula and Isle Royale, upstate NY, and Ontario

Notes: bulls can weigh up to a 1,000 lb (454 kg); feeds on many types of vegetation and aquatic plants, nuts, buds, twigs (Algonquian name translates to "twig eater"); cold-adapted animals (seeks shade and water in summer); good swimmer; nocturnal, mostly crepuscular

Pigs—Family Suidae

© Eric Isselee/Shutterstock.com

Wild Boar or Wild Pig or Feral Swine **(*Sus scrofa*)**

Size: males 59 in. (150 cm); height 31 in. (80 cm); females somewhat smaller

Key Features: large head; dark to variable fur color; light underfur; long, straight, narrow snout; erect ears; small and deep-set eyes; straight tail; males weigh up to 220 lb (100 kg); grunting noises, alarm calls

Habitat: forests; farmlands; fields

Notes: nonnative, invasive species intentionally released or escaped captivity; wreaks havoc on farm fields and forests, and competes with native species; uses long snout and sharp tusks to dig up food and wallow in dirt; causes erosion, degrades water quality, destroys crops and native plants; aggressive toward people and can carry disease; runs up to 25 mph (40 kph); same family as domestic pig; opportunistic diet including crops, eggs, insects, young deer and livestock, vegetation, seeds, and nuts; male mates with many sows year-round with two litters a year and 4–12 piglets each; *research whether wild boars live near you!*

Bovines—Family Bovidae

In North America, a few native species of Family Bovidae exist. These examples include the American bison, mountain goats, bighorn sheep, Dall's sheep, and muskoxen. Of these, only the American bison were once found wild in the Great Lakes basin and Upper Midwest region. Today, bison are found in the west in National Parks, kept on private or government property free-ranging throughout the country, and sometimes harvested for their meat. Bovids eat grass and other vegetation. They use bacteria in their guts to ferment the plant matter. They are ruminants with a four-chambered stomach and microorganisms to decompose cellulose (plant matter) into digestible parts. They will regurgitate what does not initially get digested and rechew it for further processing. Males of all bovid species have horns and often females have horns as well. Horns differ from antlers, which are found in the deer family (Cervidae). Horns never shed from the animal's head. They have a bony core covered by a sheath of keratin (the same material of your fingernails). Horns do not branch as you would observe of the cervid's antlers.

American Bison or Bison or Buffalo ***(Bison bison)***

Size: 8–12 ft. (2.4–3.6 m); tail 12–19 in. (30–48 cm); height 5–6 ft. (1.5–1.8 m)

Key Features: massive animal; brown fur with lighter-colored body; large humped shoulders; long shaggy mane; long tufted tail; both sexes with horns (not shed like antlers)

Habitat: tall grass prairie; farmlands; open forests

Notes: extirpated from region, but some managed herds exist (research your state; never found in eastern Canadian provinces); herbivore; often quiet; highly mobile and can run up to 45 mph (72 kph); one calf birthed every one to two years; crepuscular; active year-round; *research where you can visit to view bison in the wild in the Midwest*

Mammals—Natural History Journal Pages

Select a mammal that is likely to live near your Nature Area and complete a natural history report for it. If necessary, use other reputable sources for your research. Strive to complete a report for all the mammals you have in the area.

1. Full Common Name: White-Tailed Deer

   Scientific/Latin Name (*Genus species*): Odocoileus Virginianus

   Family Name: Cervidae

   Order: Artiodactyla

   Class: Mammalia

   Phylum: Chordata

   Kingdom: Animalia

2. **Sketches**

   Draw the full-size animal in its typical position. Label size, colors, and identifying characteristics

4-7ft

Darker brown in fall/winter
Tan in Summer

White neck patch

3-4ft

Skull (include sizes and identifying features)

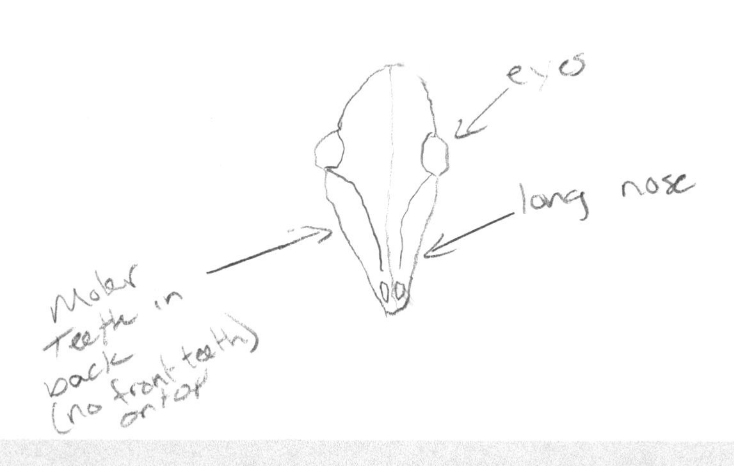

Tracks: front and rear paws (include sizes); include typical gait pattern (way of walking)

Scat (include size, content, and identifying features)

Small round brown pellets

3. Provide a detailed description of the mammal (size, weight, shape, and identifying characteristics).

3-4 ft tall, brownish color with a white belly and underside of tail, males have antlers in the fall/winter females do not

What are, if any, similar species or look-alike animals?

Mule deer

4. On the map of North America, shade the range of where your mammal is found.

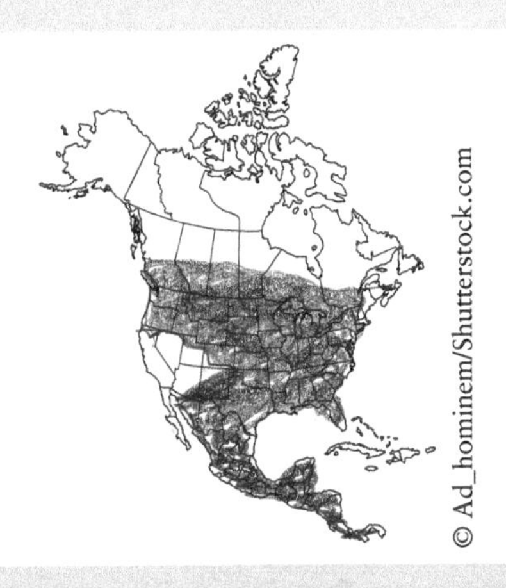

5. Discuss the dominant senses used by your mammal (e.g. relatively large-sized eyes, ears, and nose can be indicators; defined "fingers" can indicate frequent tactile uses; pronounced whiskers can also provide sensory perception; other behaviors or special features could also apply).

large ears so hearing is good but their most dominant sense is their smell. A long nose helps with that.

6. Describe the preferred or required habitats of the mammal.

forests, farmlands, wetlands, feilds.
They like thick and heavy cover

7. Explain the typical habits and behaviors of your mammal (this may include seasonal changes in behavior, courtship, and feeding behavior).

During mating season the males get teritorial and will fight other males. Will move at night in high pressured areas (Hunters)

8. What is the preferred diet of the mammal?

corn, acorns, grasses, apples.

9. Discuss the potential predators or threats to this species.

Coyotes, bears, wolves

10. What is the ecological status of this species' population (stable, special concern, threatened, or endangered): Stable What can be done to protect, manage, *or* restore this population?

Right now in michigan, there is an overpopulation happening so a way to fix that could be harvesting more females.

11. Provide information about the mammal's reproductive cycle (breeding age, breeding season, and typical number of young):

Breeding age: M: 1.5 F: .5 *Breeding season:* Mid-late fall

Typical number of young 1-2 *Breeds* 1 *time(s) per year*

Describe the type of parental care given to offspring: Will stay with young until next breeding season.

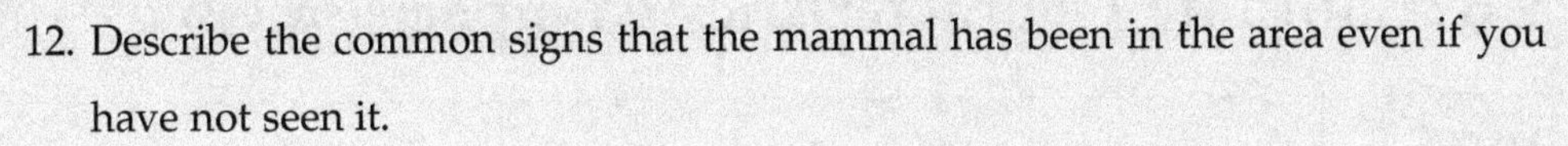

12. Describe the common signs that the mammal has been in the area even if you have not seen it.

Tracks, scat, scrapes on trees.

13. What are other notes or fun facts you discovered but have not yet discussed?

Newborn deer have spots on them and will lose them after about 6 months.

14. List all resources used (**books:** author's last name, first initial, year, and book title; **websites:** name of site, date retrieved, and web address):

Animal Diversity web www.animaldiversity.org

Fitzwilliams-Heck, C., 2020/2021, A practical guide to nature study

8.6 Chordate Characteristics Comparison Chart

Nature Journal: *Choose local vertebrate animals potentially found near your Nature Area.* Complete the chart on the next page for each animal type you choose or observe-a fish, amphibian, reptile, bird, and mammal. You will describe what it looks like and how it moves. In the book, research its anatomy and physiology. Be sure to identify it by the animal's full common name and all its taxonomic levels. Provide concise, useful answers so that you can easily identify the similarities, differences, and unique characteristics among the animals, and reflect on what you learned about. Use information found in this book, in other field guides or reputable Internet sources to supplement your answers.

Kingdom Animalia	Taxonomy	Morphology	Skeletal Structure and Locomotion	Dominant Senses -Nervous System	Circulatory System	Digestive System	Respiration	Reproduction	Excretory System	Behavior or Fun Facts
select a local species for the following animal types below (provide full common names)	provide phylum, class, and order	size, general body shape, color, and external characteristics	bony and unique skeletal characteristics; means for movement and other related adaptations	sense used mostly and unique characteristics of nervous system	info. about heart, blood flow, cold/warm blooded	teeth type; diet; organs/ unique characteristics	mode of breathing and various adaptations	means for, structures, mating behavior, raising young	organs/ unique characteristics	what you want to remember; unique features or actions
Fish:										
Amphibian:										
Reptile:										
Bird:										
Mammal:										

Conclusion: Wrapping the Bundle

Reflecting on your Nature Study. . .

After you have spent time going through the book and have started your nature study, reflect on what you have learned. Package everything you covered into a reflective entry in your Nature Journal ... wrap the bundle. This is a way to remember the concepts covered as you move forward in your own personalized nature study. Flip through the book and your notes to help you bring everything together.

Depending on how deeply you studied and how you applied what you learned, will determine how well you can follow through with the nature study skills listed below. As you read through the skills, give yourself a personal rating on how well you developed that skill—a 5 would indicate that you feel completely confident with the skill and a 1 is very little confidence in the topic. The skills you were introduced to included:

1. Making good nature **observations**
2. Implementing the **scientific process**
3. Conducting **field investigations**
4. Understanding **taxonomic levels** and connections
5. **Identifying** fungus, plants, trees, and animals (fish, amphibians, reptiles, birds, and mammals)
6. **Interpreting** the landscape in terms of potential ecological relationships

How did you do? Revisit Chapters 1 and 2 to see what you still need to develop as a naturalist. Retake the Nature Appreciation Survey. Reflect on differences between your results from when you first started your nature study and now. Can you answer the review questions in all the chapters? Have you completed all the Natural History Journal Pages provided in the book for species you encountered?

How about completing all the activities and the Nature Journal prompts? There is enough material here to last you a very long time.

Throughout the book, *A Practical Guide to Nature Study*, the ecological concepts that pertain to environmental protection and wildlife conservation have also been discussed.

Environmental Protection

In your nature study, you learned about the four spheres of nature—atmosphere, hydrosphere, geosphere, and the biosphere. You discovered the roles of the natural resources found within these spheres in various habitats, connections between them, how humans have negatively impacted them, and how we can also protect these gifts of nature required for modern society.

Read through the reminders of things we discussed that can be applied to your life to help protect the environment.

1. **AIR Quality**: Reduce emissions (on personal and industrial levels), minimize burning, plant native trees, drive hybrid or electric vehicles, carpool, walk, bike
2. **WATER**: Prevent erosion and runoff by planting native plants, eliminate invasives, monitor quality, dispose medicines/toxins properly, conduct regular runoff assessments where you live and work—do what you can to minimize or eliminate the runoff of water
3. **SOIL**: Prevent erosion with native plants, minimize/eliminate pesticide/fertilizer use, minimize urban sprawl
4. **EDUCATE** others about how to protect our natural resources (i.e., forests, water, fertile land, plants, animals)
5. **VOTE** in favor of environmental protection
6. Contact **environmental agencies** with concerns; permits
7. **BUY** annual park permits, and fishing and hunting licenses to help protect our natural resources

Now, go back and read the environmental protection reminders again. This time, reflect on what you already do and what you could do to address each effort.

Wildlife Conservation

Throughout your nature study, we also discussed wildlife conservation. Not only are wildlife protected by preserving or restoring habitats, but the quality of an ecosystem can be improved for all. The list below provides some reminders for wildlife conservation we covered to apply as you continue your nature study.

1. Create/protect **diverse habitats with native plants** and trees
2. Eliminate and prevent the spread of **invasive species**
3. **EDUCATE** others about our natural resources
4. **VOTE** in favor of environmental protection
5. **PARTICIPATE** in citizen advisory meetings and surveys to learn about current and local issues, and to voice your opinions
6. **DO community science**: water quality, monarch watch, bud burst, creel survey, frog and toad survey, backyard bird count
7. Contact **natural resource agencies** with concerns and subscribe to email alerts from these organizations to learn more about topics of interest
8. **BUYING** annual park passes and hunting and fishing licenses will help manage entire habitats

Think about ways you can incorporate these ideas into your life and how to share them with others.

On a personal level. . .

Try these five things listed below to help make a positive difference in nature wherever you live.

What can we do *personally* to make a positive difference in nature?

1. Pay attention—**be observant!**
2. **Get outside**—walk, destress, and journal
3. Be aware of **environmental/wildlife issues** (read/watch reputable sources)
4. **Do what you can** to protect nature and conserve our natural resources
5. Vote

Thank you for reading this practical guide to help build your naturalist skills. I hope your personal nature study journey brings you joy that lasts a lifetime.

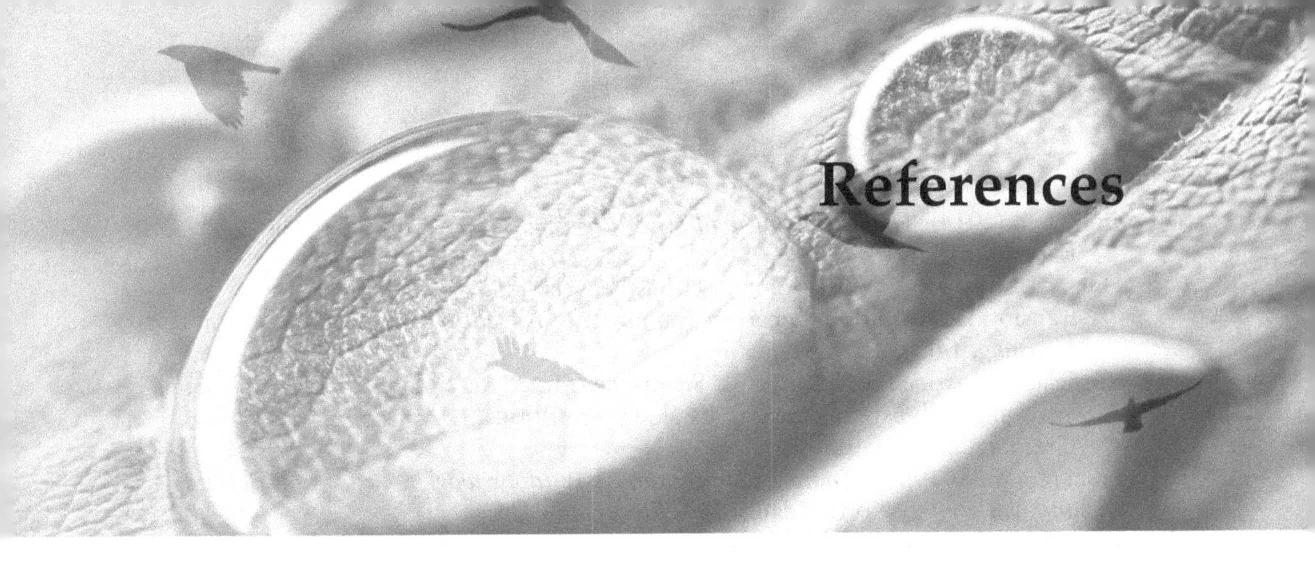

References

The author has used the resources listed below over the years to help build what was incorporated into each chapter. Using these references could prove invaluable for going deeper into your nature study.

Introduction

Lorsbach, A., and J. Jinks. (2013). What early 20th century nature study can teach us. *Journal of Natural History Education and Experience*, 7, 7–15.

Robbins, J. (2020, January). Ecopsychology: How immersion in nature benefits your health. *Yale Environment 360.*

Robertson, D., & Saad, D. (2011). Nutrient inputs to the Laurentian Great Lakes by source and watershed estimated using SPARROW Watershed Models 1. *Journal of the American Water Resources Association,* 47, 1011–1033. doi:10.1111/j.1752-1688.2011.00574.x

Chapter 1—Building Your Naturalist Skills

Bailey, L.H. (1905). *The outlook of nature.* Norwood, MA: Norwood Press.

Barbiero, G., & Berto, R. (2018). From biophilia to naturalist intelligence passing through perceived restorativeness and connection to Nature. *Annals of Reviews and Research, 3*(1), 555604.

Bishop, P.J. (2020). *Wisdom of the animals: A nature-led journey into the heart of transformational leadership.* Indy Pub.

Brown, T., Jr. (1986). *Nature observation and tracking.* New York, NY: Berkley Publishing.

Callenbach, E. (1998). *Ecology: A pocket guide.* Berkeley, CA: University of California Press.

Comstock, A. B. (1967). *Handbook of nature study.* Ithaca, NY: Comstock Publishing Associates. (Original work published 1911)

Dudderar, G. (1991). *Nature from your back door.* E. Lansing, MI: Cooperative Extension Services, MSU.

Gooley, T. (2014). *The lost art of reading nature's signs*. New York, NY: The Experiment, LLC.

Hill, A., & Brown, M. (2014). Intersections between place, sustainability and transformative outdoor experiences. *Journal of Adventure Education & Outdoor Learning, 14*(3), 217–232.

Korpela, K., Savonen, E. M., Anttila, S., Pasanen, T., & Ratcliffe, E. (2017). Enhancing wellbeing with psychological tasks along forest trails. *Urban Forestry & Urban Greening, 26*, 25–30.

Leslie, C. W. (2010). *The nature connection: An outdoor workbook for kids, families, and classrooms*. North Adams, MA: Storey Publishing.

Louv, R. (2006). *Last child in the woods: Saving our children from nature-deficit disorder*. Chapel Hill, NC: Algonquin Books of Chapel Hill.

Mertins, B. (2019). *How nature-based education can inspire brilliant minds*. Retrieved from https://nature-mentor.com/nature-based-education/

Meyer, M. (1998). Learning and teaching through the naturalist intelligence. *Clearing, 102*, 7–11.

Monkman, J., & Rodenburg, J. (2016). *The big book of nature activities: A year-round guide to outdoor learning*. Gabriola Island, BC: New Society Publishers.

Nicol, R. (2014). Entering the fray: The role of outdoor education in providing nature-based experiences that matter. *Educational Philosophy and Theory, 46*(5), 449–461.

Otto, S., & Pensini, P. (2017). Nature-based environmental education of children: Environmental knowledge and connectedness to nature, together, are related to ecological behaviour. *Global Environmental Change, 47*, 88–94.

Rinehart, K. (2006). *Naturalist's guide to observing nature*. Mechanicsburg, PA: Stackpole Books.

Seng, P. T. (Ed.) (2008). *Stewardship education best practices education guide*. Washington, DC: Association of Fish and Wildlife Agencies.

Silver, D. M. (1993). *One small square: Backyard*. New York, NY: Learning Triangle Press.

Tekiela, S., & Shanberg, K. (1995). *Nature smart: A family guide to nature*. Cambridge, MN: Adventure Publications.

Wood, C. J., & Smyth, N. (2020). The health impact of nature exposure and green exercise across the life course: A pilot study. *International Journal of Environmental Health Research, 30*, 226–235.

Young, J. (2007). *Kamana one: Exploring natural mystery*. U.S.A.: Owlink Media.

Chapter 2—Nature as Your "Place"

Benyus, J. (1989). *Wildlife habitats of the Eastern United States*. New York, NY: Simon & Schuster.

Blanchan, N. (1897). *The bird book: Bird neighbors and birds that hunt and are hunted*. Garden City, NY: Doubleday, Doran & Company, Inc.

Commoner, B. (1971). *The closing circle: Nature, man, and technology*. New York, NY: A. A. Knopf.

Crews, J., Braude, S., Stephenson, C., & Clardy, T. (2002). *The ethogram and animal behavior research: Science curriculum for grades 5-8*. St. Louis, MO: Washington University Science Outreach.

Daniel, G., & Sullivan, J. (1981). *A Sierra Club naturalist's guide: The North Woods*. Sierra Club.

Otto, P., Robbins, K., Sotak, B., & Gabler, C. (2015). *Field investigations: Using outdoor environments to foster student learning of scientific practices*. Olympia, Washington: Pacific Education Institute.

Schlee, J. (2000). *Our changing continent*. United States Geological Survey. Retrieved from https://pubs.usgs.gov/gip/continents/

Sobel, D. (2005). *Place-based education: Connecting classrooms & communities*. Great Barrington, MA: Orion Society.

Worth, K. (2010). Science in early childhood classrooms: Content and process. *Early Childhood Research & Practice (ECRP)*, *12*(2).

Chapter 3—Parts of Nature: Terrestrial Habitats and Freshwater Systems

Benyus, J. (1989). *Northwoods wildlife: A watcher's guide to habitats*. St. Paul, MN: NorthWord Press, Inc.

Grady, W. (2007). *The Great Lakes: The natural history of a changing region*. Vancouver, BC: Greystone Books.

Kricher, J. C., Morrison, G., & Peterson, R. T. (Eds.). (1998). *A Peterson field guide to Eastern forests: North America*. New York, NY: Houghton Mifflin Company.

Lynch, D. R., & Lynch, B. (2010). *Michigan rocks and minerals: A field guide to the Great Lake State*. Cambridge, MN: Adventure Publications, Inc.

Reid, G. K., Latimer, J. P., Nolting, K. S., & Brooks, J. L. (2001). *Pond life*. New York, NY: St. Martin's Press.

Sargent, M. S., & Carter, K. S. (Eds.). (1999). *Managing Michigan wildlife: A landowner's guide*. East Lansing, MI: Michigan United Conservation Club.

Tarbuck, E. J., & Lutgens, F. K. (1993). *The earth: An introduction to physical geology* (4th ed.). New York, NY: Macmillan Publishing Company.

Travis, T., & Brown, S. *Pocketguide to eastern streams*. Mechanicsburg, PA: Stackpole Books.

Wetzel, R. G. (2001). *Limnology: Lake and river ecosystems*. San Diego, CA: Academic Press.

Chapter 4—Ecology Essentials

Leopold, A. (1949). *A Sand County almanac*. New York, NY: Oxford University Press.

Pollock, S. (2005). *Eyewitness ecology*. New York, NY: DK Publishing, Inc.

Smith, T. M., & Smith, R. L. (2015). *Elements of ecology* (9th ed.). San Francisco, CA: Pearson Education, Inc.

Tallamy, D. W. (2007). *Bringing nature home: How you can sustain wildlife with native plants*. Portland, OR: Timber Press.

Wright, R. T., & Boorse, D. F. (2017). *Environmental science: Toward a sustainable future* (13th ed.). San Francisco, CA: Pearson Education, Inc.

Chapter 5—Fungus

Arora, D. (1986). *Mushrooms demystified* (2nd ed.). Berkeley, CA: Ten Speed Press.

Lincoff, G. (1981). *National Audubon Society field guide to North American mushrooms*. New York, NY: A. A. Knopf.

Marrone, T., & Yerich, K. (2014). *Mushrooms of the Upper Midwest: A simple guide to common mushrooms*. Cambridge, MN: Adventure Publications.

Ostry, M. E., Anderson, N. A., & O'Brien, J. G. (2011). *Field guide to common macrofungi in Eastern Forests and their ecosystem functions*. Newtown Square, PA: U.S. Forest Service.

Smith, A. H., & Weber, N. S. (1980). *The mushroom hunter's field guide*. Ann Arbor, MI: The University of Michigan Press.

Weber, L., & Mollen, C. (2006). *Fascinating fungi of the North Woods*. Duluth, MN: Kollath-Stensaas Publishing.

Welewski, J. (2007). *Lichens of the North Woods*. Duluth, MN: Kollath-Stensaas Publishing.

Chapter 6—Plants

Alexander, T. R., Burnett, R. W., & Zim, H. S. (1970). *Botany*. New York, NY: Western Publishing Co., Inc.

Barnard, E. S., & Yates, S. F. (Eds.). (1998). *North American wildlife: Trees and nonflowering plants*. Pleasantville, NY: Reader's Digest Association, Inc.

Chadde, S. W. (2013). *Midwest ferns: A field guide to the ferns and fern relatives of the North Central United States*. Createspace Independent Publishing Platform.

Cox, D. D. (2005). *A naturalist's guide to field plants: An ecology for Eastern North America*. Syracuse, NY: Syracuse University Press.

Elpel, T. J. (2008). *Botany in a day: The patterns method of plant identification* (5th ed.). Pony, Montana: HOPS Press, LLC.

Kershaw, L. (2006). *Trees of Michigan*. Auburn, WA: Lone Pine Publishing International.

Lund, H. C. (1988). *Michigan wildflowers*. Traverse City, MI: Village Press.

Mikolas, M. (2017). *A beginner's guide to recognizing trees of the Northeast*. New York, NY: The Countryman Press.

Newcomb, L. (1977). *Newcomb's wildflower guide*. New York, NY: Little, Brown and Company.

Petrides, G. A. (1986). *Peterson field guides: Trees and shrubs*. Boston, MA: Houghton Mifflin Company. (Original work published 1958)

Wernert, S. J. (Ed.). (2012). *North American wildlife: An illustrated guide to 2,000 plants and animals*. Pleasantville, NY: Reader's Digest Association.

Williams, M. D. (2007). *Identifying trees: An all-season guide to Eastern North America*. Mechanicsburg, PA: Stackpole Books.

Chapter 7—Invertebrate Animals

Borror, D. J., & White, R. E. (1998). *Insects* (2nd ed.). Boston, MA: Houghton Mifflin Company.

Hahn, J. (2009). *Insects of the North Woods.* Duluth, MN: Kollath+Stensaas Publishing.

Kaufman, K., Sayre, J., & Kaufman, K. (2015). *Field guide to nature of the Midwest.* New York, NY: Houghton Mifflin Harcourt.

Weber, L. (2003). *Spiders of the North Woods.* Duluth, MN: Kollath+Stensaas Publishing.

Chapter 8—Vertebrate Animals

Barnard, E. S., & Yates, S. F. (Eds.). (1998). *North American wildlife: Mammals, reptiles, and amphibians.* Pleasantville, NY: Reader's Digest Association, Inc.

Bennet, D., & Tiner, T. (2003). *The wild woods guide: From Minnesota to Maine, the nature and lore of Great North Woods.* New York, NY: HarperCollings Publishers Inc.

Black, T., & Kennedy, G. (2003). *Birds of Michigan.* Auburn, WA: Lone Pine Publishing International Inc.

Bosanko, D. (2007). *Fish of Michigan field guide.* Cambridge, MN: Adventure Publications.

Harding, J. H., & Mifsud, D. A. (2017). *Amphibians and reptiles of the Great Lakes region* (rev. ed.). Ann Arbor, MI: The University of Michigan Press.

Hubbs, C. L., & Lagler, K. F. (1958). *Fishes of the Great Lakes region.* Ann Arbor, MI: The University of Michigan Press.

Larson, A. (2013). *Integrated principles of zoology* (16th ed.). New York, NY: McGraw-Hill Education.

McCormac, J. S., & Kagume, K. (2009). *Great Lakes nature guide.* Auburn, WA: Lone Pine Publishing International, Inc.

Proctor, N.S., & Lynch, P.J. (1998). *Manual of ornithology: Avian structure and function.* Yale University Press.

Sibley, D.A. (2002). *Sibley's birding basics.* New York, NY: Knopf Doubleday Publishing Group.

Sibley, D. A. (2016). *The Sibley field guide to birds of Eastern North American* (2nd ed.). New York, NY: Knopf Doubleday Publishing Group.

Tekiela, S. (1999). *Birds of Michigan.* Cambridge, MN: Adventure Publications.

Tekiela, S. (2005). *Mammals of Michigan.* Cambridge, MN: Adventure Publications.

Tekiela, S. (2014). *Reptiles & amphibians field guide.* Cambridge, MN: Adventure Publications.

Wernert, S. J. (Ed.). (2012). *North American wildlife: An illustrated guide to 2,000 plants and animals.* Pleasantville, NY: Reader's Digest Association.

Whitaker, J. O., Jr. (1996). *National Audubon Society field guide to North American mammals.* New York, NY: Chanticleer Press, Inc.

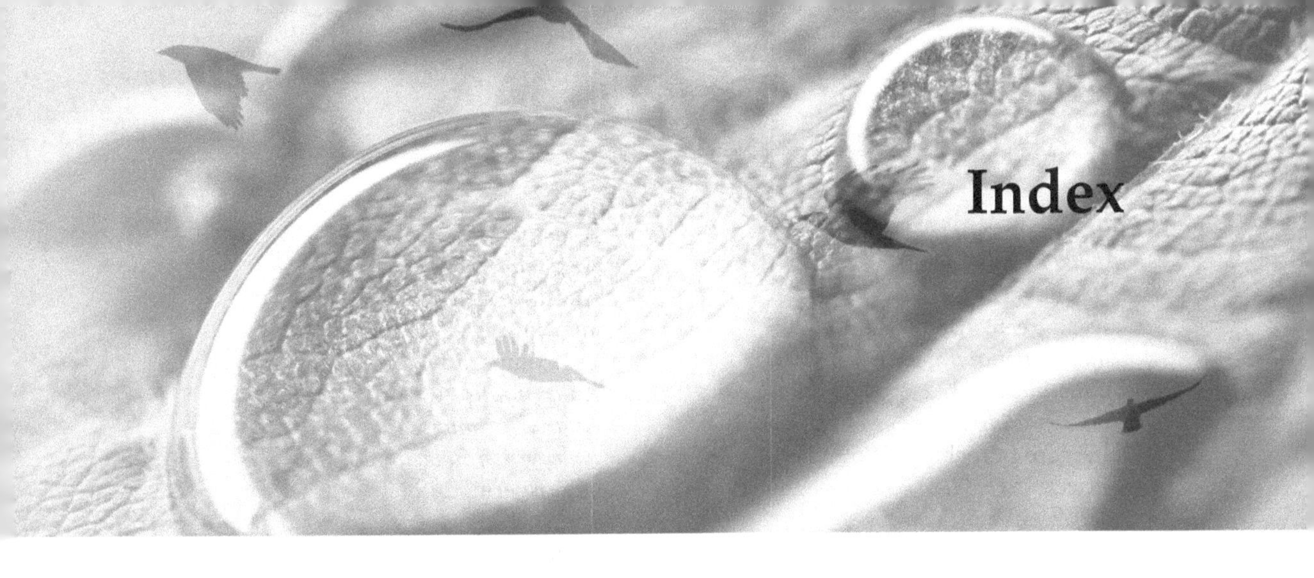

Index

For the Nature Study Instructor

A Practical Guide to Nature Study is an all-in-one reference book dedicated to methods for discovering and interpreting nature, the fundamental sciences of nature study, and natural histories and field guides of species. It can serve as a solid foundation for anyone with an interest in studying nature. For those who find themselves in the role of teacher of nature, either by choice or chance, this is the perfect book to start your journey and for building your repertoire of content to teach audiences of any age.

As an instructor, you may be a formal educator who works for an academic institution and meets with the same class over an extended period, a nonformal educator who engages with an audience one or a few times usually covering just one or two topics, or you might be an informal educator who shares your knowledge and skills with others when the right time and place presents itself. Regardless of who you teach, the content in this book will help you meet the nature-based education mission and objectives of your school, organization, facility, or agency.

Alignment. For those educators needing to align with the Next Generation Science Standards (**NGSS**), you will find flexible, cross-cutting relationships across disciplines and content to meet NGSS requirements at any grade level throughout the book. If **STEM** – *science*, *technology*, *engineering*, and *math* are your focus, you will find the material in the book ties to these topics. You will find inspiration from the book to apply STEM in nature right outside your door. The outdoors is the perfect classroom for piquing and retaining the interest of your students. To take it further to help reach all students and objectives, add the *arts* to meet STEAM goals, get the students moving with outdoor *recreation* in a STREAM focus, and connect what is being learned about the environment to the people and *social science* in STREAMS. Ask yourself what needs and aims you have and use the book to help you to reach those expectations.

Preparation. To teach nature study effectively, spend time outside and allow students to explore and discover their interests. Before heading out, help ensure your class is prepared for the outdoor conditions and that any of their personal needs are met. When working with a group, always set boundaries of where they can roam and for how long and discuss all safety concerns and instructions. Once these things are taken care of, you and your students are ready to commence. Very few materials are needed to carry out a meaningful nature study.

Implementation. Use *A Practical Guide to Nature Study* in ways that make sense for your audience. Adopting it as a textbook will likely be welcomed by your students as a fresh change of pace. The content is filled with scaffolded activities and content-related journaling activities for building their naturalist skills while also presenting the science and natural histories in a readable, concise format. This approach helps build a person's confidence in getting outside and making connections in and to nature. There are also many colorful images and photographs that will help make the text appealing and more understandable. For additional sources of information relevant to each chapter, refer to the References section of the book. The book's content and format will encourage self-directed learning as students pore over topics that interest them.

As you use the book, you will recognize the most relevant content for your audience and how to package what you present. The best approach is to get the students comfortable with the outdoors and starting to develop their naturalist skills. Keep it simple in the beginning. Start with the *Nature Appreciation Survey* and Chapter 1.1. The survey helps a person assess their naturalist abilities and Chapter 1.1 helps increase the students' nature awareness. Ideally, the self-assessment is taken before spending too much time in the book and at the end of a set length of time with their nature study. The survey can also be modified to meet your needs and interests. Chapter 1.1 can be broken into multiple parts depending on your group's abilities and your time frame. From there, combining sections of the book will likely make the most sense. For example, connect Chapter 1.2's naturalist activities along with 2.1's ecology introduction, 2.2's landscape shapers, and 3.1 and 3.2 for an overview on the regional landscape and looking more closely at habitats. These sections will help provide a broad perspective on nature to help interpret and make deeper connections on their naturalist's journey. Assigning the Animal Observation Chart from 1.2 can bring together some of the concepts learned so far while also getting students into the field guides. From there, move through the book and try combining parts of Chapters 1.1, 1.2, and 2.4 to create a tailored nature study. If you are learning about populations and communities in nature (Chapter 4.1), you might want to include field investigations, an animal behavior, or a species diversity study as described in Chapter 2.4. Yet another approach to the book could have students focus on one habitat at a time as described in Chapter 3. For any given lesson, the students could apply their naturalist skills from Part I, ecological concepts learned in Part II, and use the natural histories and field guides from Part III to learn and apply what they discover outside. To deepen the students' understanding, choose the Nature Journal activities in the chapters that fit and exemplify your objectives.

The Nature Journal. Application and reflection are effective methods in teaching and learning. It is strongly encouraged to have students create a Nature Journal to demonstrate their nature study knowledge and skills. Students will take pride in their product. The possibilities of what to have them include are endless. Starting with simple things, from what is presented in Chapter 1, will help build their confidence in getting outside and in the subject of science. Consider a weekly Nature Journal assignment that aligns with the topic being discussed. Another great consideration is to have students complete Natural History Journal Pages. These are found at the end of Chapter 4 and following each field guide section in Part III. The activity is essentially a template for the nature study student to follow as they research a local species of interest. The students' compilation of nature-related activities could be submitted as a physical copy of their Nature Journal or parts of it could be uploaded into a digital portfolio template (a slideshow of uploaded images can easily be used). It is helpful for everyone if you provide a scoring rubric to your students ahead of time and follow it for grading their project. There are enough Nature Journal activities in *A Practical Guide to Nature Study* to last a lifetime – and beyond. This book is a timeless reference as the content will remain relevant for years to come.

Nature Study Professionals. Lastly, to help further your teaching skills and network with like-minded nature-based educators, consider joining a professional organization. Here are a few highly recommended groups to check out that are reputable, resourceful, welcoming, and inspiring: the North American Association for Environmental Education (NAAEE), National Association for Interpretation (NAI), Association for Experiential Education (AEE), National Science Teaching Association (NSTA), and seek out similar organizations on a local level for your state or province. Becoming part of at least one of these communities can lead to teaching nature-based education more effectively and keeping your mind and soul refreshed.